The publisher gratefully acknowledges
the generous contribution to this book
provided by the Ahmanson Foundation
Humanities Fund and the General
Endowment Fund of the University
of California Press Associates.

MAPS OF TIME

THE CALIFORNIA WORLD HISTORY LIBRARY

Edited by Edmund Burke III, Kenneth Pomeranz, and Patricia Seed

DAVID CHRISTIAN

MAPS OF TIME
AN INTRODUCTION TO BIG HISTORY

UNIVERSITY OF CALIFORNIA PRESS
BERKELEY   LOS ANGELES   LONDON

Epigraph: the final verse of the Diamond Sutra, as translated by
Kenneth Saunders, in *The Wisdom of Buddhism,* ed. Christmas
Humphreys (London: Curzon Press, 1987), p. 122.

University of California Press
Berkeley and Los Angeles, California

University of California Press, Ltd.
London, England

Library of Congress Cataloging-in-Publication Data

Christian, David, 1946–.
   Maps of time : an introduction to big history / David Christian.
      p.   cm. — (The California world history library ; 2)
   Includes bibliographical references and index.
   ISBN 0-520-23500-2 (alk. paper)
      1. Civilization—Philosophy.   2. Human evolution.   3. World
history.   I. Title.   II. Series.
CB19 .C477   2003
901—dc21   2003012764

Manufactured in the United States of America

13   12   11   10   09   08   07   06   05   04
10 9 8 7 6 5 4 3 2 1

The paper used in this publication meets the minimum requirements
of ANSI/NISO Z39.48–1992 (R 1997) (*Permanence of Paper*).♾

Thus shall ye think of all this fleeting world:
A star at dawn, a bubble in a stream,
A flash of lightning in a summer cloud,
A flickering lamp, a phantom, and a dream.

—*Diamond Sutra, ca. fourth century* CE

CONTENTS

ILLUSTRATIONS

TABLES

FOREWORD

Maps of Time unites natural history and human history in a single, grand, and intelligible narrative. This is a great achievement, analogous to the way in which Isaac Newton in the seventeenth century united the heavens and the earth under uniform laws of motion; it is even more closely comparable to Darwin's nineteenth-century achievement of uniting the human species and other forms of life within a single evolutionary process.

The natural history that David Christian deals with in the first chapters of this book is itself radically extended and transformed from the natural history of earlier ages. It starts with the big bang some 13 billion years ago, when, according to twentieth-century cosmologists, the universe we inhabit began to expand and transform itself. Processes thereby inaugurated are still in course, as time and space (perhaps) began, allowing matter and energy to separate from one another and distribute themselves throughout space in different densities and with different rates of energy flows in response to a variety of strong and weak forces. Matter, gathering into local clots under the influence of gravity, became radiant stars, clustered into galaxies. New complexities, new flows of energy arose around such structures. Then, some 4.6 million years ago, around one star, our sun, planet Earth formed and soon became the seat of still more complicated processes, including life in all its forms. Humankind added yet another level of behavior a mere 250,000 years ago, when our use of language and other symbols began to introduce a new capacity for what Christian calls "collective learning." This in turn made human societies uniquely capable of concerting common effort so as to alter and sporadically expand widely varying niches in the ecosystem around each of them and, by now, surround us all in the single, global system.

The human history that Christian thus fits into the recently elaborated natural history of the universe is also an intellectual creation of the twentieth century. For while the efforts of physicists, cosmologists, geologists, and biologists were making the natural sciences historical, anthropologists, archaeologists, historians, and sociologists were busy enlarging knowledge about the human career on earth. They extended it back in time and expanded it pretty well across the face of the earth to embrace foragers, early farmers, and other peoples who left no written records and had therefore been excluded from document-based "scientific" history in the nineteenth century.

Most historians, of course, paid no attention to "prehistory," or to the lives of illiterate peoples, busy as they were with their own professional debates. Across the twentieth century, those debates, and the study of abundant Eurasian and a few African and Amerindian texts, added substantially to the sum of historical information and to the scope of our ideas about the accomplishments of the urbanized, literate, and civilized peoples of the earth. A few world historians, like myself, tried to weave those researches together into a more adequate portrait of humanity's career as a whole; and some also explored the ecological impact of human activity. I even wrote a programmatic essay, "History and the Scientific Worldview" (*History and Theory* 37, no. 1 [1998]: 1–13), describing what had happened to the natural sciences and challenging historians to generalize boldly enough to connect their discipline with the historicization of the natural sciences that had taken place behind our backs. Several scholars are, in fact, working toward that end, but only when I began to correspond with David Christian did I discover a historian who was already writing such a work.

The truly astounding dimension of Christian's accomplishment is that he finds similar patterns of transformation at every level. Here, for example, is what he says about stars and cities:

> In the early universe, gravity took hold of atoms and sculpted them into stars and galaxies. In the era described in this chapter, we will see how, by a sort of social gravity, cities and states were sculpted from scattered communities of farmers. As farming populations gathered in larger and denser communities, interactions between different groups increased and the social pressure rose until, in a striking parallel with star formation, new structures suddenly appeared, together with a new level of complexity. Like stars, cities and states reorganize and energize the smaller objects within their gravitational field. (p. 245)

Or weigh the words with which he closes this extraordinary book:

> Being complex creatures ourselves, we know from personal experience how hard it is to climb the down escalator, to work against the universal

slide into disorder, so we are inevitably fascinated by other entities that appear to do the same thing. Thus this theme—the achievement of order despite, or perhaps with the aid of, the second law of thermodynamics—is woven through all parts of the story told here. The endless waltz of chaos and complexity provides one of this book's unifying ideas. (p. 511)

I venture to say that Christian's discovery of order amid "the endless waltz of chaos and complexity" is not just one among other unifying themes, but the supreme achievement of this work.

Here, then, is a historical and intellectual masterpiece: clear, coherent, erudite, elegant, venturesome, and concise. It offers his readers a magnificent synthesis of what scholars and scientists have learned about the world around us in the past hundred years, showing how strangely, yet profoundly, human societies remain a part of nature, properly at home in the universe despite our extraordinary powers, unique self-consciousness, and inexhaustible capacity for collective learning.

Perhaps I should conclude this introduction with a few words about who David Christian is. First of all, he has an international identity, being the son of an English father and an American mother who met and married in Izmir, Turkey. His mother, however, returned to Brooklyn, New York, for the birth of her son in 1946, while her husband, after discharge from his wartime duties in the British army, joined the colonial service and became a district officer in Nigeria. His wife quickly joined him there, so David's childhood was spend up-country in Nigeria until, at age 7, he went away to boarding school in England. Then, in due course, he went up to Oxford, getting a B.A. in modern history in 1968. (At Oxford this means mastering isolated segments from the history of England since Roman times along with a scattering of other fields in European history and even a few decades sliced from the American past: the very antithesis of "big history.") For the next two years, he took a job as a tutor at the University of Western Ontario in Canada, and earned an M.A. degree there. By then he had decided to specialize in Russian history and returned to Oxford, where a thesis on administrative reforms under Tsar Alexander I won him a D.Phil. in 1974. Like his father, he married an American wife; they have two children.

Between 1975 and 2000 he taught Russian history at Macquarie University in Sydney, Australia, along with other courses in Russian literature and European history. Influenced by the Annales school in France, his interests shifted to everyday aspects of Russian lives. Two books resulted, both dealing with what Russians put into their mouths: *Bread and Salt: A Social and Economic History of Food and Drink in Russia* (1985, coauthored with R. E. F. Smith) and *Living Water: Vodka and Russian Society on the Eve of Emancipation* (1990). These books soon attracted invitations to write more general works: first *Power and Privilege: Russia and the Soviet Union in the Nineteenth and Twentieth Centuries* (1986), then *A History of Russia, Central Asia, and Mongolia*, volume 1, *Inner Eurasia from Prehistory to the Mongol Empire* (1998).

The broad geographical and temporal sweep of the last of these books already reflected a teaching venture he launched in 1989 when, in the course of a discussion about what sort of introduction to history the department at Macquarie ought to provide for its students, David Christian blurted out something like "Why not start at the beginning?" and promptly found himself invited to show his colleagues what that might mean. Unlike every other historian who ever tried to teach human history on a world scale, Christian decided to begin with the universe itself; and with help from colleagues in other departments of the university, who lectured on their own scientific specialties, he staggered through the first year of what he jestingly chose to call "big history."

From the start, big history attracted a large and what soon became an enthusiastic student following. But his most responsive professional audience first arose in the Netherlands and in the United States, where news of what David Christian was doing persuaded a handful of venturesome teachers to launch parallel courses. The World History Association as well as the American Historical Association took note by devoting a session to big history at their annual meetings in 1998. Three years later David Christian decided to accept an invitation to come to San Diego State University and bring big history with him.

Other professional interests remain active. A second volume of his *History of Russia, Central Asia, and Mongolia* is in the works; so is an account of the Russian campaign to ban alcohol that peaked in the early 1920s. In his spare time David Christian has also written several important articles on scale in the study of history and a variety of other subjects. He is, in short, a historian of altogether unusual energy, daring, and accomplishment.

You, who are about to peruse this book, have a great experience before you. Read on, wonder, and admire.

William H. McNeill
22 October 2002

ACKNOWLEDGMENTS

A project like this turns a person into a magpie. You collect ideas and information voraciously; and after a bit you can start to forget each particular act of intellectual larceny. Fortunately for me, most scholars are (despite popular reputation) astonishingly generous with their time and ideas. I have benefited from this generosity particularly at the two institutions at which I have spent most of my career: Macquarie University in Sydney and San Diego State University in California. I will try to acknowledge as many debts as I can, but there are many more that I cannot acknowledge because I cannot remember them. Suggestions, approaches, book references have been tucked away in my mind so securely that I cannot remember where I got them; sometimes I may even be tempted to think of them as my own discoveries. Where such memory lapses have happened (and I am sure they often have), I can only apologize and express a generalized thanks to those friends and colleagues who have had the patience to discuss with me the large-scale historical problems that have fascinated me now for more than a decade.

I would particularly like to thank Chardi, who is a professional storyteller and a Jungian. She persuaded me that I was really teaching a creation myth. I also want to thank Terry (Edmund) Burke, who teaches a course on "big history" at Santa Cruz, in California. He persuaded me that the time had come to try to write a textbook on big history, in the hope that it might encourage others to teach similar courses. He has also given me invaluable (if sometimes painful) criticism on earlier drafts of this manuscript. And he has been a constant source of encouragement.

I am extremely grateful to all those who lectured or tutored in the big history course I taught at Macquarie University between 1989 and 1999. I list them in alphabetical order: David Allen, Michael Archer, Ian Bedford, Craig Benjamin, Jerry Bentley, David Briscoe, David Cahill, Geoff Cowling, Bill Edmonds, Brian Fegan, Dick Flood, Leighton Frappell, Annette Hamilton, Mervyn Hartwig, Ann Henderson-Sellers, Edwin Judge, Max Kelly, Bernard Knapp, John Koenig, Jim Kohen, Sam Lieu, David Malin, John Merson, Rod Miller, Nick Modjeska, Marc Norman, Bob Norton, Ron Paton, David Phillips, Chris Powell, Caroline Ralston, George Raudzens, Stephen Shortus, Alan Thorne, Terry Widders, and Michael Williams. I would also like to thank Macquarie University for the period of academic leave during which the first draft of this book was written.

Several people have been particularly supportive of the idea of big history, and some have taught other courses on the big history scale. John Mears started teaching such a class at about the same time as I did, and has always been an enthusiastic supporter of the idea. Tom Griffiths and colleagues taught a big history course at Monash University in the early 1990s. Johan Goudsblom began teaching one at the University of Amsterdam, and has been an enthusiastic supporter of the project. His colleague, Fred Spier, wrote the first book on big history, an ambitious and brilliantly argued case for the construction of a "grand unified theory" embracing the social sciences as well as the natural sciences (*The Structure of Big History: From the Big Bang until Today*). Others who have expressed interest and support for such an approach or have taught similar courses include George Brooks, Edmund Burke III, Marc Cioc, Ann Curthoys, Graeme Davidson, Ross Dunn, Arturo Giráldez, Bill Leadbetter, and Heidi Roupp. At the American Historical Association conference in Seattle in January 1998, Arnold Schrier chaired a panel on big history that included papers from myself, John Mears, and Fred Spier, as well as a perceptive and supportive commentary from Patricia O'Neal. Gale Stokes invited me to discuss big history on a panel on "the play of scales" at the American Historical Association conference in San Francisco in January 2002.

A number of other people have read or commented on drafts of parts of this book. In addition to some of those already listed, they include Elizabeth Cobbs Hoffman, Ross Dunn, Patricia Fara, Ernie Grieshaber, Chris Lloyd, Winton Higgins, Peter Menzies, and Louis Schwartzkopf. Professor I. D. Koval'chenko invited me to give a talk on big history at Moscow University, and Valerii Nikolayev invited me to talk at the Institute of Oriental Studies in Moscow, both in 1990. Stephen Mennell asked me to speak about big history at a conference he convened almost ten years ago, and Eric Jones gave me invaluable feedback on that paper. Ken Pomeranz sent me a draft chapter from his then-unpublished book on the "great divergence" and invited me to speak on big history at the University of California, Irvine. Over the years, I have given talks on big history at many other universities, including Macquarie and Monash Universities and the Universities of Sydney, Melbourne, Newcastle, Wollongong, and Western Australia in Australia; at the University of California, Santa Cruz; at Minnesota State University, Mankato; and at Indiana University, Bloomington, in the United States; at Victoria University in Canada; and at Newcastle and Manchester Universities in the United Kingdom. I worked with John Anderson for almost two years on a theoretical article on power- and wealth-maximizing societies. The article never saw the light of day, but collaborating with John gave me many new insights into the transition to modernity.

Since sending out copies of an earlier draft of this text in September 1999, I have received generous criticisms and comments from several other colleagues. They include (in alphabetical order): Alfred Crosby, Arturo Giráldez, Johan Goudsblom, Marnie Hughes-Warrington, William H. McNeill, John Mears, Fred Spier, and Mark Welter. I am also grateful to at least two anonymous reviewers recruited by the University of California Press. Marnie Hughes-Warrington taught with me in my big history course in 2000 and offered many invaluable suggestions. As a historiographer, she was able to alert me to historiographical implications of the subject that I had missed. William McNeill engaged in a long and generous correspondence with me about an earlier draft of this manuscript. His comments were both encouraging and critical, and they have shaped my own ideas substantially. In particular, he persuaded me to take more seriously the role of networks of exchange in world history.

I would also like to thank the many students whom I have taught at Macquarie in HIST112: An Introduction to World History, and at San Diego State University in HIST411: World History for Teachers and HIST100: World History. Their questions kept me focused on what is important. I am particularly thankful to those students who provided me with information or told me about new discoveries they found in books I didn't know about or on the Internet, and also to those who, by enjoying the course, made me feel it was worthwhile.

I feel I owe particular thanks to the staff of the University of California Press, including Lynne Withey, Suzanne Knott, and many others. Alice Falk copyedited my manuscript with terrifying thoroughness. Their professionalism, courtesy, and good humor greatly eased the complex and sometimes difficult passage from manuscript to book.

In a book on this scale, it goes without saying that no one I have thanked for their help or support is in any way to blame for errors in the text; nor can it be assumed that any of them necessarily agree with the book's argument. I remained stubborn enough to resist many of the kindly criticisms that have been offered of its earlier drafts, so I must remain wholly responsible for all remaining errors of fact, interpretation, or balance.

I hope Chardi, Joshua, and Emily will think of this as a gift from me, a small return for the gift that they have been to me over many years.

David Christian
January 2003

INTRODUCTION
A MODERN CREATION MYTH?

"BIG HISTORY": LOOKING AT THE PAST ON ALL TIMESCALES

[T]he way to study history is to view it as a long duration, as what I have called the *longue durée*. It is not the only way, but it is one which by itself can pose all the great problems of social structures, past and present. It is the only language binding history to the present, creating one indivisible whole.

Universal history comprehends the past life of mankind, not in its particular relations and trends, but in its fullness and totality.

> A Moment's Halt—a momentary taste
> Of BEING from the Well amid the Waste—
> And Lo!—the phantom Caravan has reached
> The NOTHING it set out from—Oh, make haste!

Like merchants in a huge desert caravan, we need to know where we are going, where we have come from, and in whose company we are traveling. Modern science tells us that the caravan is vast and varied, and our fellow travelers include numerous exotic creatures, from quarks to galaxies. We also know a lot about where the journey started and where it is headed. In these ways, modern science can help us answer some of the deepest questions we can ask concerning our own existence, and that of the universe through which we travel. It can help us draw the line we all must draw between the personal and the universal.

"Who am I? Where do I belong? What is the totality of which I am a part?" In some form, all human communities have asked these questions. And in most human societies, educational systems, formal and informal, have tried to answer them. Often, the answers have been embedded in cycles of creation myths. By offering memorable and authoritative accounts of how everything began—from our own communities, to the animals, plants, and landscapes around us, to the earth, the Moon and skies, and even the universe itself—creation myths provide universal coordinates within which people can imagine their own existence and find a role in the larger scheme of things. Creation myths are powerful because they speak to our deep spiritual, psychic, and social need for a sense of place and a sense of belonging. Because they provide so fundamental a sense of orientation, they are often integrated into religious thinking at the deepest levels, as the Genesis story is within the Judeo-Christian-Islamic tradition. It is one of the many odd features of modern society that despite having access to more hard information than any earlier society, those in modern educational systems do not normally teach such a story. Instead, from schools to universities to research institutes, we teach about origins in disconnected fragments. We seem incapable of offering a unified account of how things came to be the way they are.

I have written this book in the belief that such intellectual modesty is unnecessary and harmful. It is unnecessary because the elements of a modern creation myth are all around us. It is harmful because it contributes to the subtle but pervasive quality of disorientation in modern life that the pioneering French sociologist Émile Durkheim referred to as "anomie": the sense of not fitting in, which is an inescapable condition of those who have no conception of what it is they are supposed to fit into.

Maps of Time attempts to assemble a coherent and accessible account of origins, a modern creation myth. It began as a series of lectures in an experimental history course taught at Macquarie University in Sydney. The idea of that course was to see if it was possible, even in the modern world, to tell a coherent story about the past on many different scales, beginning, literally, with the origins of the universe and ending in the present day. Each scale, I hoped, would add something new to the total picture and make it easier to understand all the other scales. Given the conventions of the modern history profession, this was an extremely presumptuous idea. But it turned out to be surprisingly doable, and even more interesting than I had originally supposed. Part of the task of my introduction will be to justify this distinctive way of thinking and teaching about the past.

I began teaching "big history" in 1989; two years later I published an essay in which I attempted a formal defense of this approach.[1] Though aware of the oddity of the project, those of us trying to teach big history were soon convinced that these large questions made for interesting classes and encouraged fruitful thinking about the nature of history. Teaching this large story persuaded us that beneath the awesome diversity and complexity of modern knowledge, there is an underlying unity and coherence, ensuring that different timescales really do have something to say to each other. Taken together, these stories have all the power and richness of a traditional cycle of creation myths. They constitute what indigenous Australians might call a modern "Dreaming"—a coherent account of how we were created and how we fit into the scheme of things.

We found something else that most premodern societies have known: there is an astonishing power to any story that attempts to grasp reality whole. This power is quite independent of the success or failure of any particular attempt; the project itself is powerful, and fulfills deep needs. Trying to look at the whole of the past is, it seems to me, like using a map of the world. No geographer would try to teach exclusively from street maps. Yet most historians teach about the past of particular nations, or even of agrarian civilizations, without ever asking what the *whole* of the past looks like. So what is the temporal equivalent of the world map? Is there a map of time that embraces the past at all scales?

This is a good moment to raise such questions, because there is a growing sense, across many scholarly disciplines, that we need to move beyond the fragmented account of reality that has dominated scholarship (and served it well) for a century. Scientists have moved fastest in this direction. The success of Stephen Hawking's *A Brief History of Time* (1988) also shows the great popular interest in trying to understand reality whole. In Hawking's own field, cosmology, the idea of a "grand unified theory" once seemed ridiculously overambitious. Now it is taken for granted. Biology and geology have also moved toward more unified accounts of their subject matter, with the consolidation, since the 1960s, of modern paradigms of evolution and plate tectonics.[2]

Scholars at the Santa Fe Institute in the United States have been exploring such interconnections for many years. An associate of the institute, the Nobel Prize–winning physicist Murray Gell-Mann, has eloquently stated the arguments for a more unified account of reality as they appear to a physicist.

> We live in an age of increasing specialization, and for good reason. Humanity keeps learning more about each field of study; and as every specialty grows, it tends to split into subspecialties. That process happens over and over again, and it is necessary and desirable. However, there is also a growing need for specialization to be supplemented by integration. The reason is that no complex, nonlinear system can be adequately described by dividing it up into subsystems or into various

aspects, defined beforehand. If those subsystems or those aspects, all in strong interaction with one another, are studied separately, even with great care, the results, when put together, do not give a useful picture of the whole. In that sense, there is profound truth in the old adage, "The whole is more than the sum of its parts."

People must therefore get away from the idea that serious work is restricted to beating to death a well-defined problem in a narrow discipline, while broadly integrative thinking is relegated to cocktail parties. In academic life, in bureaucracies, and elsewhere, the task of integration is insufficiently respected.

At the Santa Fe Institute, he adds, "People are found who have the courage to take a *crude look at the whole* in addition to studying the behavior of parts of a system in the traditional way."[3]

Should historians look for a similar unifying structure, perhaps a "grand unified story" that can summarize the best modern knowledge about origins from a historian's perspective? The rise of the new subdiscipline of world history is a sign that many historians also feel the need for a more coherent vision of their subject. Big history is a response to this need. In the late 1980s, John Mears, at Southern Methodist University (in Dallas, Texas), began teaching a history course on the largest possible scales at about the same time as I did. And since then, a number of other universities have offered similar courses—in Melbourne, Canberra, and Perth in Australia; in Amsterdam; and also in Santa Cruz in the United States. Fred Spier, from the University of Amsterdam, has gone one step further and written the first book on big history. In it, he offers an ambitious defense of the project of constructing a unified account of the past at all scales.[4]

Meanwhile, there is a growing sense among scholars in many fields that we may be close to a grand unification of knowledge. The biologist E. O. Wilson has argued that we need to start exploring the links between different domains of knowledge, from cosmology to ethics.[5] The world historian William McNeill has written:

> Human beings, it appears, do indeed belong in the universe and share its unstable, evolving character. . . . [W]hat happens among human beings and what happens among the stars looks to be part of a grand, evolving story featuring spontaneous emergence of complexity that generates new sorts of behavior at every level of organization from the minutest quarks and leptons to the galaxies, from long carbon chains to living organisms and the biosphere, and from the biosphere to the symbolic universes of meaning within which human beings live and labor, singly and in concert, trying always to get more of what we want and need from the world around us.[6]

I intend this book to contribute to the larger project of constructing a more unified vision of history and of knowledge in general. I am well aware of the difficulties of that project. But I am sure that it is both doable and important, so it is worth attempting in the hope that others may eventually do it better. I am also convinced that a modern creation myth will turn out to be as rich and as beautiful as the creation myths of all earlier communities; it is a story that deserves telling even if the telling is imperfect.

STRUCTURE AND ORGANIZATION

> utterly impossible as are all these events they are probably as like those which may have taken place as any others which never took person at all are ever likely to be   E A R T H
>
> If the Eiffel Tower were now representing the world's age, the skin of paint on the pinnacle-knob at its summit would represent man's share of that age; and anybody would perceive that that skin was what the tower was built for. I reckon they would, I dunno.

Erwin Schrödinger, one of the pioneers of quantum physics, described the difficulties of constructing a more unified vision of knowledge in the preface to a book he wrote on a biological topic—the origins of life. His preface also offers the best justification I know for presuming to undertake such a project.

> We have inherited from our forefathers the keen longing for unified, all-embracing knowledge. The very name given to the highest institutions of learning reminds us, that from antiquity and throughout many centuries the *universal* aspect has been the only one to be given full credit. But the spread, both in width and depth, of the multifarious branches of knowledge during the last hundred odd years has confronted us with a queer dilemma. We feel clearly that we are only now beginning to acquire reliable material for welding together the sum total of all that is known into a whole; but, on the other hand, it has become next to impossible for a single mind fully to command more than a small specialized portion of it.
>
> I can see no other escape from this dilemma (lest our true aim be lost forever) than that some of us should venture to embark on a synthesis of facts and theories, albeit with second-hand and incomplete knowledge of some of them—and at the risking of making fools of ourselves.
>
> So much for my apology.[7]

Some of the most daunting problems posed by big history are organizational. What shape will a modern creation myth take? From what standpoint should it be written? What objects will take center stage? What time-scales will dominate?

A modern creation myth will not and cannot hope to be "neutral." Modern knowledge offers no omniscient "knower," no neutral observation point from which all objects, from quarks to humans to galaxies, have equal significance. We cannot be everywhere at once. So the very idea of knowledge from no particular point of view is senseless. (Technically, this statement reflects a philosophical position, associated with Nietzsche, known as *perspectivism*.) In any case, what use could such knowledge have? All knowledge arises from a relationship between a knower and an object of knowledge. And knowers expect to put knowledge to some use.

Creation stories, too, arise from a relationship between particular human communities and the universe as these communities imagine it. They offer answers to universal questions at many different scales, which is why they sometimes appear to have a nested structure similar to a Russian *matryoshka* doll—or to the Ptolemaic vision of the universe, with its many concentric shells. At the center are those trying to understand. At the outer edge is a totality of some kind: a universe or a deity. In between are entities that exist at different chronological, spatial, and mythic scales. It is thus the questions we ask that dictate the general shape of all creation myths. And because we are humans, humans are guaranteed to occupy more space in a creation myth than they do in the universe as a whole. A creation myth always belongs to *someone;* and the story recounted in this book is the creation myth of modern human beings, educated in the scientific traditions of the modern world. (Curiously, this means that the *narrative* structure of the modern creation myth, like all creation myths, may appear pre-Copernican, despite its definitely *post*-Copernican content.)

Though its scope is vast, *Maps of Time* aims at not overwhelming the reader with detail. I have tried (without complete success) to stop the book from growing too large, in the hope that the details will not obscure the larger picture. Those with a particular interest in any one part of this story will have no difficulty finding out more, and the brief guides to further reading at the end of each chapter provide some starting points.

The exact balance of topics and themes in this book reflects the fact that this is an attempt at big history from a historian's perspective, not that of an astronomer, a geologist, or a biologist. (Some alternative approaches to big history are listed at the end of this introduction.) This means that human societies loom larger than they do in, for example, Stephen Hawking's books, or in Preston Cloud's *Cosmos, Earth, and Man* (1978). Nevertheless, the first five chapters cover topics that normally fall within the sciences of cosmology, geology, and biology. They discuss the origins and evolution of the universe, of galaxies and stars, of the solar system and the earth, and of life on earth. The rest of the book surveys the history of our own species and its relationship to the earth and to other species. Chapters 6 and 7 discuss the origins of human beings and the nature of the earliest human societies. They attempt to identify what is distinctive about human history, and what distinguishes humans from other organisms inhabiting this earth. Chapter 8 examines the earliest agrarian societies, which existed without cities or states. With the emergence of agriculture, about 10,000 years ago, humans began for the first time to live in dense communities, in which exchanges of information and goods became more intensive than ever before. Chapters 9 and 10 describe the emergence and evolution of cities, of states, and of agrarian civilizations. Chapters 11 to 14 try to construct a coherent interpretation of the modern world and its origins. Finally, chapter 15 looks to the future. Big history is inevitably concerned with large trends, and these do not stop suddenly in the present moment. So a large view of the past inevitably raises questions about the future, and at least some answers are available, both for the near future (say, the next 100 years), and the remote future (the next few billion years). Raising such questions should be a vital part of modern education, for our assessments of the future will affect decisions taken today; these, in turn, may shape the world inhabited by our own children and grandchildren. They will not thank us if we take such tasks lightly.

A second organizational difficulty is thematic. It may seem there can be little coherence in a narrative that spans so many different scholarly disciplines. But there are phenomena that cross all scales. Above all, it turns out that the main actors are similar. At every level, we will be interested in ordered entities, from molecules to microbes to human societies to large chains of galaxies. Explaining how such things can exist, how they are born, how they evolve, and how, eventually, they perish is the stuff of history at all scales. Of course, each scale also has its own rules—chemical in the case of molecules, biological in the case of microbes—but the surprise is that some underlying principles of change may be universal. This is why Fred Spier has argued that at a fundamental level, big history is about "regimes." It is about the fragile ordered patterns that appear at all scales, and the ways in which they change.[8] So a central theme of big history is how the rules of change vary at different scales, despite some fundamental similarities in the nature of all change. Human history *is* different from cosmological history; but it is not *totally* different. I discuss some of the general principles of change in appendix 2, but the book as a whole will explore some of the different rules of change that appear at different scales.

FOR AND AGAINST BIG HISTORY

Specialists in many fields, from geology to archaeology and prehistory, will find it quite natural to look at the past on very large scales. But not everyone will be persuaded that big history is worth doing. Particularly to professional historians, the idea of exploring the past on such huge timescales can seem overambitious and perhaps simply impossible, a diversion from the real tasks of historical scholarship. In the last part of this introduction, I will respond to four main reservations that I have encountered.

The first is common, particularly among professional historians. It is that on large scales, history must thin out. It must lose detail, texture, particularity, and substance. Eventually, it must become vacuous. To be sure, on large scales, themes and problems familiar to professional historians may vanish, just as the details of a familiar landscape may disappear as one looks down from an airplane as it climbs. In a big history course, the French Revolution may get no more than a passing mention. But there are compensations. As the frame through which we view the past widens, features of the historical landscape that were once too large to fit in can be seen whole. We can begin to see the continents and oceans of the past, as well as the villages and roadways of national and regional histories. Frames of any kind exclude more than they reveal. And this is particularly true of the conventional time frames of modern historiography, which normally extend from a few years to a few centuries. Perhaps the most astonishing thing the conventional frames hide is humanity itself. Even on time frames of several thousand years, it is difficult to ask questions about the broader significance of human history within an evolving biosphere. Yet in a world with nuclear weapons and ecological problems that cross all national borders, we desperately need to see humanity as a whole. Accounts of the past that focus primarily on the divisions between nations, religions, and cultures are beginning to look parochial and anachronistic—even dangerous. So, it is not true that history becomes vacuous at large scales. Familiar objects may vanish, but new and important objects and problems come into view. And their presence can only enrich the discipline.

A second possible objection is that to write big history, historians will have to move beyond the boundaries of the discipline. Of course, this is true. Synoptic studies like this book are risky because the author depends on secondary sources and on *other* synoptic studies. As a result, there will inevitably be blunders and misunderstandings: error is built into the project. Indeed, it is part of the process of learning. To understand your own country, you must travel beyond its borders at least once in your life. You will not understand everything you see; but you may begin to see your own country in a new light. The same is true of history. To understand what is distinctive about human history, we must have some idea of how a biologist or a geologist might approach the subject. We cannot *become* biologists or geologists, and our understanding of these fields will have its limits; but we do have to use as skillfully as we can the expertise of specialists in other fields. And we have much to learn from their different perspectives on the past. Excessive respect for disciplinary boundaries has hidden many possibilities for intellectual synergy between disciplines. I will argue, for example, that we need the vision of a biologist to see what is truly distinctive about our type of animal, *Homo sapiens*.

Third, it may be objected that big history proposes to create a new "grand narrative" just when we have learned the futility, even the danger, of grand narratives. Will not a big history metanarrative crowd out alternative histories—of minorities, of regions, of particular nations or ethnic groups?[9] Perhaps a fragmented vision of the past (a "jeweler's-eye" view, in the phrase used by the anthropologists George Marcus and Michael Fischer) is the only one that can do real justice to the richness of human experience.[10] Natalie Zemon Davis makes the point well:

> The question remains whether a single master narrative is an adequate goal for global history. I think not. Master narratives are especially vulnerable to be taken over by patterns characteristic of the historian's time and place, however useful they may be for accounting for some of the historical evidence. If a new decentred global history is discovering important alternative historical paths and trajectories, then it might also do well to let its big stories be alternate or multiple. The challenge for global history is to place these narratives creatively within an interactive frame.[11]

Once again, the charge is at least partly true. Narratives of some kind seem unavoidable when looking at the past on large scales, and they will certainly be shaped by contemporary concerns. Nevertheless, it is a mistake for historians to shun these large narratives, however grand they may seem. Like it or not, people will look for, and find, large stories, because they can provide a sense of meaning. As William Cronon has written of environmental history: "When we describe human activities within an ecosystem, we seem always to tell *stories* about them. Like all historians, we configure the events of the past into causal sequences—stories—that order and simplify those events to give them new meanings. We do so because narrative is the chief literary form that tries to find meaning in an overwhelmingly crowded and disordered chronological reality."[12] If paid intellectuals are too finicky to shape these stories, they will flourish all the same; but the intellectuals will be ignored and will eventually disenfranchise themselves. This is an abdication of responsibility, particularly as intellectuals have played such a crucial role in creating many of today's metanarratives. Metanarratives exist, they are powerful, and they are potent. We may be able to domesticate them; but we will never eradicate them. Besides, while grand narratives are powerful, subliminal grand narratives can be even more powerful. Yet a "modern creation myth" already exists just below the surface of modern knowledge. It exists in the dangerous form of poorly articulated and poorly understood fragments of modern knowledge that have undermined traditional accounts of reality without being integrated into a new vision of reality. Only when a modern creation myth has been teased out into a coherent story will it really be possible to take the next step: of criticizing it, deconstructing it, and perhaps improving it. In history as in building, construction must precede deconstruction. We must see the modern creation myth before we can criticize it. And we must articulate it before we can see it. Ernest Gellner made this point well in the introduction to his attempt at a synoptic view of history, *Plough, Sword, and Book* (1991):

> The aim of the present volume is simple. It is to spell out, in the sharpest and perhaps exaggerated outline, a vision of human history which has been assuming shape of late, but which has not yet been properly codified. The attempt to bring it to the surface is not made because the author has any illusions about *knowing* it to be true: he does not. Definitive and final truth is not granted to theories in general. In particular, it is unlikely to attach to theories covering an infinite diversity of extremely complex facts, well beyond the reach of any one scholar. The vision is formulated in the hope that its clear and forceful statement will make possible its critical examination.[13]

Besides, a "grand narrative" of the kind offered in this book may prove surprisingly capacious. In the global "truth" market of the twenty-first century, all narratives face stiff competition. The many detailed stories of the past already taught in our schools and universities ensure that a modern creation myth will emerge not as a single monolithic story but rather as a large and ramshackle cycle of stories, each of which can be told in many ways and with many variants. Indeed, it may turn out that the very large narratives create more space for alternative accounts of the past that struggle to survive within existing (and less ample) history syllabi. As Patrick O'Brien has written, "Hopefully as more historians risk writing on a global scale, the field will achieve a reputation and produce competing metanarratives to which the overwhelming flow of parish, regional and national histories could be reconnected."[14]

The fourth objection is closely related to the third: is not a narrative on this huge scale bound to make exaggerated truth claims? I have found in teaching big history that students struggle to find a balance between two extreme positions. On the one hand, they are tempted to suppose that a modern, "scientific" account of origins is true, while all earlier accounts were more or less false. On the other hand, faced with some of the uncertainties of modern accounts of the past, they may be tempted to think that this is "just one more story."

Thinking of a big history narrative as a modern creation myth is a good way of helping students to find the epistemological point of balance between these extremes. For it is a reminder, first, that *all* accounts of reality are provisional. Many of the stories we tell today will seem quaint and childish in a few centuries, just as many elements of traditional creation myths seem naive today. But by acknowledging this, we do not commit ourselves to a nihilistic relativism. All knowledge systems, from modern science to those embedded in the most ancient of creation myths, can be thought of as maps of reality. They are never just true or false. Perfect descriptions of reality are unattainable, unnecessary, and too costly for learning organisms, including humans. But workable descriptions are indispensable. So knowledge systems, like maps, are a complex blend of realism, flexibility, usefulness, and inspiration. They must offer a description of reality that conforms in some degree to commonsense experience. But that description must also be useful. It must help solve the problems that need to be solved by each community, whether these be spiritual, psychological, political, or mechanical.[15]

In their day, all creation myths offered workable maps of reality, and that is why they were believed. They made sense of what people knew. They contained much good, empirical knowledge; and their large structures helped people place themselves within a wider reality. But each map had to build on the knowledge and fulfill the needs of a particular society. And that is why they don't necessarily count as "true" outside their home environments. A modern creation myth need not apologize for being equally parochial. It must start with modern knowledge and modern questions, because it is designed for people who live in the modern world. We need to try to understand our universe even if we can be certain that our attempts can never fully succeed. So, the strongest claim we can make about the truth of a modern creation myth is that it offers a unified account of origins *from the perspective of the early twenty-first century.*

FURTHER READING ON BIG HISTORY

Listed below are a number of works in English that explore the past on scales larger than those of world history, or try to see human history in its wider context, or provide methodological frameworks for such attempts. This is a wide definition of "big history," and there are doubtless many other works that could be included under it. The authors come from many different fields, and the books vary greatly in approach and quality, so there is plenty of room for argument as to which do and which do not really count as big history books. This preliminary bibliography is based on a list first compiled by Fred Spier. It excludes books so technical that they cannot possibly be of use to historians or general readers. It also excludes a vast number of books that operate at large scales, and have much to offer historians, but do not try to move across multiple timescales.

Asimov, Isaac. *Beginnings: The Story of Origins—of Mankind, Life, the Earth, the Universe.* New York: Walker, 1987.

Blank, Paul W., and Fred Spier, eds. *Defining the Pacific: Constraints and Opportunities.* Aldershot, Hants.: Ashgate, 2002.

Calder, Nigel. *Timescale: An Atlas of the Fourth Dimension.* London: Chatto and Windus, 1983.

Chaisson, Eric J. *Cosmic Evolution: The Rise of Complexity in Nature.* Cambridge, Mass.: Harvard University Press, 2001.

———. *The Life Era: Cosmic Selection and Conscious Evolution.* New York: W. W. Norton, 1987.

———. *Universe: An Evolutionary Approach to Astronomy.* Englewood Cliffs, N.J.: Prentice-Hall, 1988.

Christian, David. "Adopting a Global Perspective." In *The Humanities and a Creative Nation: Jubilee Essays,* edited by D. M. Schreuder, pp. 249–62. Canberra: Australian Academy of the Humanities, 1995.

———. "The Case for 'Big History.'" *Journal of World History* 2, no. 2 (fall 1991): 223–38. Reprinted in *The New World History: A Teacher's Companion,* edited by Ross E. Dunn (Boston: Bedford/St. Martin, 2000), pp. 575–87.

———. "The Longest Durée: A History of the Last 15 Billion Years." *Australian Historical Association Bulletin,* nos. 59–60 (August–November 1989): 27–36.

Cloud, Preston. *Cosmos, Earth, and Man: A Short History of the Universe.* New Haven: Yale University Press, 1978.

———. *Oasis in Space: Earth History from the Beginning.* New York: W. W. Norton, 1988.

Crosby, Alfred W. *The Columbian Exchange: Biological and Cultural Consequences of 1492.* Westport, Conn.: Greenwood Press, 1972.

———. *Ecological Imperialism: The Biological Expansion of Europe, 900–1900.* Cambridge: Cambridge University Press, 1986.

Delsemme, Armand. *Our Cosmic Origins: From the Big Bang to the Emergence of Life and Intelligence*. Cambridge: Cambridge University Press, 1998.

Diamond, Jared. *Guns, Germs, and Steel: The Fates of Human Societies*. London: Vintage, 1998.

———. *The Rise and Fall of the Third Chimpanzee*. London: Vintage, 1991.

Emiliani, Cesare. *Planet Earth: Cosmology, Geology, and the Evolution of Life and Environment*. Cambridge: Cambridge University Press, 1992.

Flannery, Tim. *The Eternal Frontier: An Ecological History of North America and Its Peoples*. New York: Atlantic Monthly Press, 2001.

———. *The Future Eaters: An Ecological History of the Australasian Lands and People*. Chatswood, N.S.W.: Reed, 1995.

Gould, Stephen Jay. *Life's Grandeur: The Spread of Excellence from Plato to Darwin*. London: Jonathan Cape, 1996. [The U.S. edition is titled *Full House*.]

———. *Wonderful Life: The Burgess Shale and the Nature of History*. London: Hutchinson, 1989.

Gribbin, John. *Genesis: The Origins of Man and the Universe*. New York: Delta, 1981.

Hawking, Stephen. *A Brief History of Time: From the Big Bang to Black Holes*. New York: Bantam, 1988.

Hughes-Warrington, Marnie. "Big History." *Historically Speaking*, November 2002, pp. 16–20.

Jantsch, Erich. *The Self-Organizing Universe: Scientific and Human Implications of the Emerging Paradigm of Evolution*. Oxford: Pergamon Press, 1980.

Kutter, G. Siegfried. *The Universe and Life: Origins and Evolution*. Boston: Jones and Bartlett, 1987.

Liebes, Sidney, Elisabet Sahtouris, and Brian Swimme. *A Walk through Time: From Stardust to Us: The Evolution of Life on Earth*. New York: John Wiley, 1998.

Lovelock, James C. *The Ages of Gaia*. Oxford: Oxford University Press, 1988.

———. *Gaia: A New Look at Life on Earth*. Oxford: Oxford University Press, 1979.

———. *Gaia: The Practical Science of Planetary Medicine*. London: Unwin, 1991.

Lunine, Jonathan I. *Earth: Evolution of a Habitable World*. Cambridge: Cambridge University Press, 1999.

Macdougall, J. D. *A Short History of Planet Earth: Mountains, Mammals, Fire, and Ice*. New York: John Wiley, 1995.

Margulis, Lynn, and Dorion Sagan. *Microcosmos: Four Billion Years of Microbial Evolution*. London: Allen and Unwin, 1987.

———. *What Is Life?* Berkeley: University of California Press, 1995.

Maynard Smith, John, and Eörs Szathmáry. *The Origins of Life: From the Birth of Life to the Origins of Language*. Oxford: Oxford University Press, 1999.

McNeill, J. R., and William H. McNeill. *The Human Web: A Bird's-Eye View of World History*. New York: W. W. Norton, 2003.

McNeill, W. H. "History and the Scientific Worldview." *History and Theory* 37, no. 1 (1998): 1–13.

———. *Plagues and People.* Oxford: Blackwell, 1977.

McSween, Harry Y., Jr. *Fanfare for Earth: The Origin of Our Planet and Life.* New York: St. Martin's, 1997.

Morrison, Philip, and Phylis Morrison. *Powers of Ten: A Book about the Relative Size of Things in the Universe and the Effect of Adding Another Zero.* Redding, Conn.: Scientific American Library; San Francisco: dist. by W. H. Freeman, 1982.

Nisbet, E. G. *Living Earth—A Short History of Life and Its Home.* London: HarperCollins Academic Press, 1991.

Packard, Edward. *Imagining the Universe: A Visual Journey.* New York: Perigee, 1994.

Ponting, Clive. *A Green History of the World.* Harmondsworth: Penguin, 1992.

Prantzos, Nikos. *Our Cosmic Future: Humanity's Fate in the Universe.* Cambridge: Cambridge University Press, 2000.

Priem, H. N. A. *Aarde en leven: Het leven in relatie tot zijn planetaire omgeving / Earth and Life: Life in Relation to Its Planetary Environment.* Dordrecht: Kluwer, 1993.

Rees, Martin. *Just Six Numbers: The Deep Forces That Shape the Universe.* New York: Basic Books, 2000.

Reeves, Hubert, Joël de Rosnay, Yves Coppens, and Dominique Simonnet. *Origins: Cosmos, Earth, and Mankind.* New York: Arcade Publishing, 1998.

Rindos, David. *Origins of Agriculture: An Evolutionary Perspective.* New York: Academic Press, 1984.

Roberts, Neil. *The Holocene: An Environmental History.* 2nd ed. Oxford: Blackwell, 1998.

Simmons, I. G. *Changing the Face of the Earth: Culture, Environment, History.* 2nd ed. Oxford: Blackwell, 1996.

Smil, Vaclav. *Energy in World History.* Boulder, Colo.: Westview Press, 1994.

Snooks, G. D. *The Dynamic Society: Exploring the Sources of Global Change.* London: Routledge, 1996.

———. *The Ephemeral Civilization: Exploding the Myth of Social Evolution.* London: Routledge, 1997.

Spier, Fred. *The Structure of Big History: From the Big Bang until Today.* Amsterdam: Amsterdam University Press, 1996.

Stokes, Gale. "The Fates of Human Societies: A Review of Recent Macrohistories." *American Historical Review* 106, no. 2 (April 2001): 508–25.

Swimme, Brian, and Thomas Berry. *The Universe Story: From the Primordial Flaring Forth to the Ecozoic Era: A Celebration of the Unfolding of the Cosmos.* San Francisco: HarperSanFrancisco, 1992.

Wells, H. G. *The Outline of History: Being a Plain History of Life and Mankind.* 2 vols. London: George Newnes, 1920.

———. *A Short History of the World.* London: Cassell, 1922.

Wright, Robert. *Nonzero: The Logic of Human Destiny.* New York: Random House, 2000.

PART I

THE INANIMATE UNIVERSE

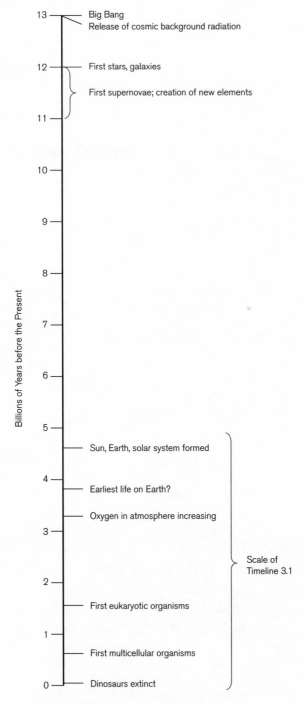

Timeline 1.1.   The scale of the cosmos: 13 billion years.

1

THE FIRST 300,000 YEARS
ORIGINS OF THE UNIVERSE, TIME, AND SPACE

Viola: What country, friends, is this?
Captain: This is Illyria, lady.

THE PROBLEM OF BEGINNINGS

How did everything begin? This is the first question faced by any creation myth and, despite the achievements of modern cosmology, answering it remains tricky.

At the very beginning, all explanations face the same problem: how can something come out of nothing? The problem is general, for beginnings are inexplicable. At the smallest scales, subatomic particles sometimes emerge instantaneously from nothingness. One moment there is nothing; the next moment there is something. There is no in-between state. Quantum physics can analyze these odd jumps into and out of existence with great precision, but it cannot explain them in ways that make sense at the human level. These paradoxes are captured beautifully in a modern Australian Aboriginal saying: "Nothing is nothing."[1]

Awareness of the difficulty of explaining origins is as old as myth. The following passage poses these questions with great sophistication and a surprisingly modern skepticism. It comes from one of the ancient Indian hymns known as the Rig-Veda, and was probably composed ca. 1200 BCE. It describes a pre-creation realm that was not really present, but was not entirely absent either.

> There was neither non-existence nor existence then; there was neither
> the realm of space nor the sky which is beyond. What stirred?
> Where? In whose protection? Was there water, bottomlessly deep?
> There was neither death nor immortality then. There was no distin-
> guishing sign of night nor of day. That one breathed, windless,
> by its own impulse. Other than that there was nothing beyond. . . .

> Was there below? Was there above? There were seed-placers; there were
> powers. There was impulse beneath; there was giving-forth above.
> Who really knows? Who will here proclaim it? Whence was it produced?
> Whence is this creation? The gods came afterwards, with the creation
> of this universe. Who then knows whence it has arisen?
> Whence this creation has arisen—perhaps it formed itself, or perhaps it
> did not—the one who looks down on it, in the highest heaven, only
> he knows—or perhaps he does not know.[2]

Here we have a hint that there was, first, a sort of potent nothingness—waiting, like clay in a potter's yard, to be formed into something. This is very much how modern nuclear physics views the idea of a vacuum: it is empty but can nevertheless have shape and structure, and (as has been proved in experiments with particle accelerators) "things" and "energies" can pop out of the emptiness.

Perhaps there was a potter (or potters) waiting to shape the vacuum. And perhaps the potter and the clay were somehow identical. According to the *Popol Vuh*, or "Council Book," a sixteenth-century Mayan manuscript, "Whatever might be is simply not there: only murmurs, ripples, in the dark, in the night. Only the Maker, Modeler alone, Sovereign Plumed Serpent, the Bearers, Begetters are in the water, a glittering light. They are there, they are enclosed in quetzal feathers, in blue-green."[3] But where did the Maker come from? Each beginning seems to presuppose an earlier beginning. In monotheistic religions, such as Christianity or Islam, the problem arises as soon as you ask, How was God created? Instead of meeting a single starting point, we encounter an infinity of them, each of which poses the same problem.

There are no entirely satisfactory solutions to this dilemma. What we have to find is not a solution but some way of dealing with the mystery, some way of "pointing at the moon," in the Zen metaphor. And we have to do so using words. Yet the words we reach for, from *God* to *gravity*, are inadequate to the task. So we have to use language poetically or symbolically; and such language, whether used by a scientist, a poet, or a shaman, can easily be misunderstood. A French anthropologist, Marcel Griaule, once questioned a Dogon wise man, Ogotemmeli, about a mythic detail according to which many animals were crowded together onto a single, small step (like the animals in Noah's ark). Ogotemmeli replied, with some irritation: "All of this has to be said in words, but everything on the step is a symbol. . . . Any number of symbols could find room on a one-cubit step." The word translated here as "symbol" could also be translated as "word of this lower world."[4] At the very beginning of things, language itself threatens to break down.

One of the trickiest problems concerns time. Was there a "time" when there was no time? Is time a product of our imagination?[5] In some systems of thought, time does not really exist. Places become the source of everything significant, and the paradoxes of creation take different forms.[6] But for communities that see time as central, there is no way of avoiding the paradox of origins. The following is an Islamic summary of a Zoroastrian attempt to deal with these riddles. In it, the creator is an unchanging entity called Time, who creates a universe of change. It is dominated by two opposite principles, those of the gods Ohrmazd and Ahriman.

> Except Time all other things are created. Time is the creator; and Time has no limit, neither top nor bottom. It has always been and shall be for evermore. No sensible person will say whence Time has come. In spite of all the grandeur that surrounded it, there was no one to call it creator; for it had not brought forth creation. Then it created fire and water; and when it had brought them together, Ohrmazd came into existence, and simultaneously Time became Creator and Lord with regard to the creation it had brought forth. Ohrmazd was bright, pure, sweet-smelling, and beneficent, and had power over all good things. Then, he looked down, he saw Ahriman ninety-six thousand parasangs away, black, foul, stinking, and maleficent; and it appeared fearful to Ohrmazd, for he was a frightful enemy. And when Ohrmazd saw this enemy, he thought thus: "I must utterly destroy this enemy," and he considered with what and how many instruments he could destroy him. Then did Ohrmazd begin the work of creation. Whatever Ohrmazd did, he did with the aid of Time; for all the excellence that Ohrmazd needed, had (already) been created.[7]

Time, like pattern, means difference, if no more than the difference between then and now. So this story, like most creation stories, is really about the emergence of difference from an original sameness. In this version, as in many creation myths, difference begins with a fundamental clash of opposites.

One of the more poetic solutions to these paradoxes is to think of creation as a sort of awakening. A story from the Karraru people of southern Australia describes how, originally, the earth was still, silent, and dark. However, "Inside a deep cave below the Nullarbor Plain slept a beautiful woman, the Sun. The Great Father Spirit gently woke her and told her to emerge from her cave and stir the universe into life. The Sun Mother opened her eyes and darkness disappeared as her rays spread over the land; she took a breath and the atmosphere changed, the air gently vibrated as a small breeze blew." The Sun Mother then goes on a long journey during which her rays awaken all the various creatures and plants that have been sleeping.[8] Such a story suggests that creation is not a single event but has to be constantly repeated; and, as we will see, this is a truth we all experience. The paradoxes of creation are repeated each time we observe something new, from galaxies to stars to solar systems and life. And many of us also experience our own personal origins, the moments of our earliest memories, as a sort of awakening from nothingness.

Modern science has approached the problem of origins in many different ways, some more satisfying than others. In *A Brief History of Time* (1988), Stephen Hawking suggests that the question of origins is just badly posed. If we think of time as a line, it is natural to ask about its beginning. But what if the universe has a different shape? Perhaps time is more like a circle. There is no sense in asking if a circle has a beginning or an end, just as there is no point in asking what is to the north of the North Pole. There is no beyond, no boundary, and everything about the universe is perfectly self-contained. As Hawking puts it: "The boundary condition of the universe is that it has no boundary."[9] Many creation myths adopt a similar approach, perhaps because they arise in societies that do not think of time as a straight line. As we look back in time, the past seems to fade away into what modern Aboriginal myths call a "Dreamtime." It is as if the past turned a corner beyond which we cannot see it anymore, however hard we try. The same is true if we look forward, so it seems as if in some sense the future and the past may meet.[10] Mircea Eliade describes similar visions of time in a difficult but fascinating work, *The Myth of the Eternal Return* (1954).[11]

In modern societies, which usually envisage time as a line rather than a curve, such solutions may seem artificial. Perhaps, instead, the universe is eternal. We can look back along the line of time as long as we like, but we will always find a universe, so the problem of origins does not really arise. Religions of the Indian subcontinent, in particular, have tended to adopt this strategy. So has the steady state theory, the most serious modern alternative to big bang cosmology. And so does a recent theory, proposed by Lee Smolin, that suggests the existence of universes that breed other universes whenever they create black holes, in a repetitive or "algorithmic" process analogous to Darwinian evolution, which ensures that they "evolve" in ways that increase the possibility of creating complex entities such as ourselves (see chapter 2).[12] Similar arguments are common in modern cosmology, and what they imply is that the universe we see may be merely one tiny atom in a much larger "multiverse." But such approaches are also unsatisfying, because they still leave the nagging question, How did such eternal processes themselves begin? How was an eternal universe created?

Or we can return to the idea of a creator. Within Christianity, it was generally agreed that the Creator made the universe a few thousand years ago.

In one famous calculation, a Dr. Lightfoot from Cambridge "proved" that God had created humans at exactly 9:00 AM on 23 October 4004 BCE.[13] Many other creation myths also introduce deities who created the world, working like potters, or builders, or clockmakers. This approach solves much of the problem, but leaves open the basic question of how the gods themselves were created. Once again, we seem forced back to an infinite regress.

A final position is skepticism. This entails a frank admission that at a certain point, we must run out of knowledge. Human knowledge, by its nature, has limits, so some questions must remain mysteries. Some religions treat such mysteries as secrets that the gods choose to hide from humans; others, such as Buddhism, treat them as ultimate riddles that are not worth pursuing. We will see that modern cosmology also opts for skepticism at the beginning of its story, though it offers a very confident account of how our universe evolved once it was created.

EARLY SCIENTIFIC ACCOUNTS OF THE UNIVERSE

Modern science tries to answer questions about origins using carefully tested data and rigorous logic. Though many pioneering scientists, like Newton, were Christians who believed deeply in the existence of a deity, they also felt the Deity was rational, so their task was to tease out the underlying laws by which the Deity had created the world. This meant trying to explain the world *as if there were no deity.* Modern science, unlike most other traditions of knowledge, tries to explain the universe as if it were inanimate, as if things happened without intention or purpose.

The Christian view of the universe owed much to the ideas of the Greek philosopher Aristotle. Though some Greeks had argued that the earth orbited the Sun, Aristotle placed the earth at the center of the universe and surrounded it with a series of transparent spheres, each revolving at a different speed. The spheres held the planets, the Sun, and the stars. This model sounds quaint today, but it was given a rigorous mathematical basis by Ptolemy in the second century CE, and in this form it proved good at predicting planetary motions. Christianity added the further idea that this universe had been created perhaps 6,000 years ago by God, in the course of five days. In sixteenth- and seventeenth-century Europe, the Ptolemaic story began to break down. Copernicus gave some powerful reasons for thinking that the earth revolved around the Sun, and the heretical monk Giordano Bruno argued that stars were suns and that the universe was probably infinite in extent. In the seventeenth century, scientists such as Newton and Galileo explored many of the implications of these ideas, while retaining as much as they could of the biblical creation story.

During the eighteenth century, the Ptolemaic view of the universe finally collapsed. In its place, there emerged a new picture of a universe operating according to strict, rational, and impersonal laws that could, in principle, be discovered by science. God may have created it, perhaps in time; perhaps, in some sense, out of time. But then he left it to run almost entirely according to its own logic and rules. Newton assumed that both time and space were absolutes, providing the ultimate frames of reference for the universe. It was widely accepted that both might be infinite, and thus the universe had neither a definable edge nor a time of origin. In this way, God was moved further and further away from the story of origins.

But there were problems. One arose from the theory of thermodynamics, which suggested that the amount of usable energy in the universe was constantly diminishing (or that entropy was constantly increasing; see appendix 2). In an infinitely old universe the consequence would be that no usable energy was left to create anything—yet clearly that was not true. Perhaps, this might have suggested, the universe was not infinitely old. The night sky posed another problem. As early as 1610, the astronomer Johannes Kepler pointed out that if there were an infinite number of stars, the night sky should be infinitely bright. The problem is now known as *Olber's paradox,* after a nineteenth-century German astronomer who publicized the problem more widely. One possible solution was to suppose that the universe was not infinitely large. That would solve Olber's paradox—but would create another; for as Newton had pointed out, if the universe were *not* infinitely large, then gravity ought to draw all the matter into the center of the universe, like oil in a sump. And that, fortunately, was not what astronomers observed when they studied the night sky.

Of course, all scientific theories contain problems. But as long as the theories can answer most of the questions put to them, such difficulties can be ignored. And the problems faced by the Newtonian theory were largely ignored in the nineteenth century.

THE BIG BANG: FROM PRIMORDIAL CHAOS TO THE FIRST SIGNS OF ORDER

In the first half of the twentieth century, evidence began to accumulate for an alternative theory that we now know as *big bang cosmology.* It solved the problem of entropy by suggesting the universe was not infinitely old; it solved Olber's paradox by describing a universe that was finite in both time and space; and it solved the paradox of gravity by showing that the universe was expanding too fast for gravity to gather everything into a single lump (yet!). Big bang cosmology described a universe with a beginning and a history, so it turned cosmology into a historical science, an account of change and evolution.

According to this view, the universe began as an infinitesimally small entity, which expanded rapidly and continues to expand today. In form, at least, this account is similar to the traditional creation myths known as *emergence myths*. In such accounts, the universe develops, like an egg or an embryo, through distinct stages from a remote and perhaps undefinable point of origin, and under the control of internal laws of development. In 1927, one of the pioneers of big bang cosmology, Georges Lemaître, referred to the early universe as the "primordial atom." Like all emergence myths, the modern account implies that the universe was created at a particular time, that it has a life story of its own, and that it may die in the distant future. The new theory could explain many of the difficulties encountered by previous theories. For example, it could explain Olber's paradox by showing that the universe had not existed forever; and because light has a finite speed (as Einstein had shown), light from the most distant galaxies might not reach us during the entire life of the universe. The theory was also consistent with the torrent of new information and data about stars, matter, and energy that was generated in the early twentieth century. But at its very beginning, it too has to fall back on a sense of inexplicable mystery.

The modern story of origins goes something like this.[14] The universe was created about 13 billion (13,000,000,000) years ago.[15] (How long ago is that? If each human being were to live exactly the biblical span of 70 years, it would take about 200 million human life spans laid end to end to reach back this far in time. For more on these huge timescales, see appendix 1.) About the beginning, we can say nothing with any certainty except that something appeared. We do not know why or how it appeared. We cannot say whether anything existed before. We cannot even say that there was a "before" or a "space" for anything to exist in, for (in an argument anticipated by St. Augustine in the fifth century CE) time and space may have been created at the same time as matter and energy. So, we can say nothing definite about the moment of the big bang, or about any earlier period.

However, beginning a tiny fraction of a second after the big bang, modern science can offer a rigorous and coherent story, based on abundant evidence. Many of the most interesting "events" occurred within a fraction of a second. Indeed, it may be helpful to think of time itself as stretched out during these early moments, so that a billionth of a billionth of a second then was as significant, in its way, as many billions of years in the later history of the universe.[16]

In the beginning, the universe was tiny, perhaps smaller than an atom. (How small is that? The physicist Richard Feynman illustrated the size of an atom by saying that if you blew up an apple until it was the size of Earth, each of the atoms it was made from would now be the size of the original apple.)[17] The temperature of this atom-sized universe was many trillions of degrees. At this temperature, matter and energy are interchangeable—as Einstein showed, matter is really little more than a congealed form of energy. Here, in this fantastically dense flux of energy/matter, we come close to the primordial chaos of so many traditional creation myths. But in the modern account, this tiny universe was expanding at a staggering speed, and it was this expansion that gave rise to the first differences and the first patterns.[18] The theory of inflation asserts that for a fraction of a second, between ca. 10^{-34} and 10^{-32} seconds after the big bang, the universe expanded faster than the speed of light (which is about 300,000 kilometers per second), driven apart by some form of "antigravity." The magnitudes involved in such processes are inconceivable: before inflation, the entire universe may have been smaller than an atom; after inflation (a fraction of an instant later), it may have been larger than a galaxy. Inflation seems to ensure that most of the universe is beyond our observation, as light from most of the universe will be too distant ever to reach us. The parts of the universe we can see may be only a tiny part of the real universe. As Timothy Ferris puts it: "If the entirety of an inflationary universe were the surface of the earth, the observable part would be smaller than a proton."[19]

As the universe expanded, it became less homogenous. Its original symmetry was broken, distinct patterns appeared, and matter and energy began to assume forms that we can recognize today. Modern nuclear physics can tell at what temperatures particular types of energy or matter appear, just as most of us can tell at what temperature water will turn into ice. So, if we can estimate how fast the universe cooled, then we can estimate when different forces and particles emerged from the flux of the early universe. Within the first second, quarks appeared, and from these were constructed protons and neutrons, the main constituents of atomic nuclei. Quarks and atomic nuclei are held together by the strong nuclear force, one of the four fundamental forces that rule our universe.

At this point in the modern creation story (still less than $\frac{1}{1000}$ of a second after the big bang), there occurs a display of extravagance that is remarkable even by the extravagant standards of most creation myths. Particles appeared in two forms, to make up almost equal amounts of matter and antimatter. Particles of antimatter are identical to particles of matter except for having the opposite electrical charge. Unfortunately, when the two meet, they annihilate each other and 100 percent of their mass is transformed into energy. So, during the first second after the big bang there played out a perverse subatomic game of musical chairs, in which quarks were the players, antiquarks were the chairs, and the winner was the one quark in a billion that couldn't find an antiparticle chair. The matter left to construct our universe was made from the one in a billion particles that didn't find an antimatter partner. The particles that *did* find a partner were transformed into pure energy, and that energy pervades the universe today, in the form of cosmic background radiation.[20] And this process may explain why there are about a billion photons of energy for every particle of matter in the universe today.

Now the pace slows. Some seconds after the big bang, electrons appeared. Electrons carry a negative electrical charge, while protons (which are made up of quarks) carry a positive charge. Relations between electrons and protons were controlled by a second fundamental force, the electromagnetic force, which also appeared within the first second of the universe's history. In the hot early universe, the photons of energy that carry the electromagnetic force were entangled with charged particles of matter. The universe was rather like the interior of the Sun today: a white-hot sea of particles and photons in constant interaction. The entire universe would have been crackling with the energy generated by constant interactions between positive protons and negative electrons and light. In this "era of radiation," as Eric Chaisson explains, matter existed as no more than "a relatively thin microscopic precipitate suspended in a macroscopic, glowing 'fog' of dense, brilliant radiation."[21]

After perhaps 300,000 years, the average temperature of the universe fell to ca. 4,000°C above absolute zero, and this cooling made possible one of the most fundamental of all transitions in the history of the universe.[22] Moments of transition are as mysterious as beginnings, and they will occur throughout our story. One of the most familiar examples in daily life is the transition that takes place when water turns into steam. Water is heated, and for a time all that seems to happen is that it gets warmer. Change occurs gradually, and we can watch it happening. Then, abruptly, a threshold is crossed; something new is created and the whole system enters a new phase. What had been liquid becomes gas. Why should a threshold occur at this particular point, in this case at 100°C (at sea level)? Sometimes we can explain transitions from one state to another, and the answer generally turns on a changing balance between different forces—between gravity, pressure, heat, electromagnetic forces, and so on. Sometimes we simply do not know why a threshold is crossed at a particular point.

The ending of the radiation era is a transition that physicists can more or less explain as a result of a balance between the falling energy of light photons as the universe expanded and the electromagnetic forces acting at the subatomic level. As the universe expanded, it cooled, and the energy of the light flowing through it fell sufficiently to enable positive protons to capture negative electrons and create stable, and thus electrically neutral, atoms. Because of that neutrality, atoms no longer interacted strongly with photons (though subtle interactions could still occur). As a result, photons of light could now flow freely through the universe. For most purposes, matter and energy ceased to interact. They became separate realms, like matter and spirit in the cosmologies of the Judeo-Christian-Islamic world. The era after this decoupling can be described as the "era of matter."[23]

The first atoms were extremely simple. Most were hydrogen atoms, consisting of one proton and one electron. But there also appeared about one-third as many helium atoms, each with two protons and two electrons, as well as a trace of even larger atoms. All atoms are tiny, with diameters of roughly one-ten-millionth of a centimeter. But they consist mostly of empty space. The protons and neutrons huddle together in the nucleus, while the electrons orbit far away from them. As Richard Feynman puts it: "If we had an atom and wished to see the nucleus we would have to magnify it until the whole atom was the size of a large room, and then the nucleus would be a bare speck which you could just about make out with the eye, but almost *all the weight* of the atom is in that infinitesimal *nucleus.*"[24] Three hundred thousand years after its creation, the universe was still simple. It consisted mostly of empty space, within which there drifted huge clouds of hydrogen and helium, and through which there poured an immense amount of energy.

Table 1.1 is a brief chronology of the early history of the universe. About 300,000 years after the big bang, all the ingredients of creation were present: time, space, energy, and the basic particles of the material universe, including protons, electrons, and neutrons, now mostly organized into atoms of hydrogen and helium. Since that time, nothing has really changed. The same energy and the same matter have continued to exist. All that has happened is that for the next 13 billion years these same ingredients have arranged themselves in different patterns, which constantly form and dissipate. From one perspective, the rest of the modern creation myth is merely the story of these different patterns.

But for us the patterns are all-important because we are pattern-detecting organisms. The patterns that emerged include the galaxies and stars, the chemical elements, the solar system, our earth, and all the living organisms

TABLE 1.1. A CHRONOLOGY OF THE EARLY UNIVERSE

Time since Big Bang	Significant Events
10^{-43} seconds	"Planck time"; the universe is smaller than the "Planck length," the smallest length that has any physical meaning; we can say nothing about what happened before this point, but gravity appears already as a distinct fundamental force.
10^{-35} seconds	"Strong" and "electromagnetic" forces begin to appear as distinct fundamental forces.
10^{-33}–10^{-32} seconds	"Inflation": the universe expands faster than the speed of light and cools to near absolute zero.
ca. 10^{-10}–10^{-6} seconds	As fundamental forces separate, the universe heats up again; quarks and antiquarks are created and annihilate each other; surviving quarks are confined in protons and neutrons (their total mass representing about one-billionth of the previous mass of quarks and antiquarks).
1–10 seconds	Electron–positron pairs form and annihilate (leaving a residue equivalent to perhaps one-billionth of the previous mass of electrons and positrons).
3 minutes	Nuclei of hydrogen and helium form from protons and neutrons.
300,000 years	Atoms form as negative electrons are captured by positive protons; the universe becomes electrically neutral, and radiation and matter separate; radiation is released in a huge "flash" now detectable in background microwave radiation.

SOURCES: Cesare Emiliani, *The Scientific Companion: Exploring the Physical World with Facts, Figures, and Formulas,* 2nd ed. (New York: John Wiley, 1995), p. 82; and see the similar chronology in Stephen Hawking, *The Universe in a Nutshell* (New York: Bantam, 2001), p. 78.

that inhabit our earth. Finally, of course, they include ourselves. As an anonymous wit is supposed to have put it: "Hydrogen is a light, odorless gas which, given enough time, changes into people."[25] From this perspective, the modern creation myth is as paradoxical as any other early creation myth. Nothing changes; but everything changes. Though things seem to exist independently of each other, and to have particular and distinctive characteristics, it is also true that everything is really the same. The idea that form and matter are different expressions of the same underlying essence was proposed by the Italian Giordano Bruno as early as 1584, in a book called *Concerning the Cause, Principle, and One.* But the same idea occurs in much deep religious and philosophical thought. According to one of the holiest of Buddhist texts, the Heart Sutra, "Form is emptiness; emptiness also is form. Emptiness is no other than form; form is no other than emptiness."[26] How patterns were created out of the apparent chaos of the early universe will be one of the central themes of the next chapter.

EVIDENCE FOR BIG BANG COSMOLOGY

From these metaphysical speculations, we must return to the prosaic but crucial issue of evidence. Why do modern astronomers accept what seems, at first sight, such a bizarre creation story? Why should we take this story seriously? The short answer is that for all its oddity, the modern story of creation is based on a colossal amount of hard evidence.

Hubble and the Redshift

The first crucial piece of evidence emerged from studies of the size and shape of the universe. Mapping the universe meant trying to determine the distance between the stars, the way in which stars were arranged, and how they moved relative to each other. Modern attempts to map the universe scientifically date to the late nineteenth century.

Finding the distance to stars is extremely difficult. With nearby stars, it is possible to estimate distances using elementary trigonometry and exact measurements of a star's parallax. The largest baseline available to Earthbound astronomers is Earth's orbit around the Sun, so astronomers look for stars whose positions appear to shift when observed at six-month intervals. But even this approach requires measurements that were too precise for any astronomers before the nineteenth century (see figure 1.1).

For more distant stars, we have to rely on methods that are even less precise. In the first decade of the twentieth century, an American astronomer, Henrietta Leavitt, studied variable stars—that is, stars whose brightness varies in a regular cycle. She discovered that in a particular type of variable star, the so-called Cepheid variables, the cycle reflected the size and the brightness of the stars. What made the Cepheids seem to grow by turns brighter and then darker was their expansion and contraction. Leavitt showed that the larger (and therefore brighter) Cepheids expand and contract more slowly. So, by measuring the length of the cycle, astronomers could estimate the size and therefore the *real* (or "intrinsic") brightness of each Cepheid variable. Then, by measuring the brightness it *appeared* to have to an observer on the earth, they could estimate how much light had been lost in the journey to our earth, and therefore how far away the star really was.

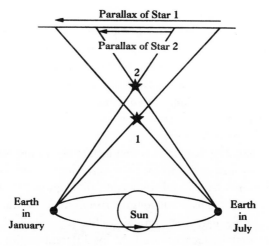

Figure 1.1.   Parallax: measuring the distance of
stars using elementary trigonometry. In the course
of six months, the earth changes its position in the
sky as it orbits the Sun. As a result, the positions of
nearby stars seem to shift slightly during the year;
and the closer stars are, the greater their apparent
change in position. (A shift in the apparent position
of an object caused by movements of the observer
is known as *parallax*.) By measuring these shifts
carefully, you can use elementary trigonometry
to determine a star's real distance from the earth.
This was the first way of determining the real scale
of the cosmos. With more distant stars, the angles
are too tiny for this method to work, so other
methods have to be used. From Ken Croswell,
The Alchemy of the Heavens (Oxford: Oxford
University Press, 1996), p. 16. Used by permission
of Doubleday, a division of Random House, Inc.

In the 1920s, another American astronomer, Edwin Hubble, using the
Mount Wilson telescope outside Los Angeles, relied on Cepheid variables
as he tried to map large areas of the universe. He found, first, that many
Cepheids apparently existed outside our galaxy, the Milky Way. This meant
that the universe consisted not just of one galaxy but of many, thereby proving an idea that the German philosopher Immanuel Kant had proposed almost two centuries before. (Specifically, Kant suggested quite correctly that
the objects astronomers call *nebulae* often consisted of separate galaxies, well
beyond our own.) This idea, which Hubble announced in 1924, already

marked a revolution in modern astronomy. Within a few years, Hubble's work led him to an insight that was even more revolutionary, and much more profound. In the late 1920s, he found that most distant galaxies seemed to be moving away from us. Indeed, the farther away they were, the faster they seemed to be moving away from our galaxy. We now know that the most distant observable galaxies are moving away from us at more than 90 percent of the speed of light. How could Hubble know this? And what did this strange observation mean?

Oddly, it is easier to measure whether distant objects are moving toward or away from us than to determine their exact distance from us. The techniques involved are elegant, and not too difficult to grasp. If we take the light from a distant star and pass it through a spectrometer, we can analyze the various parts of the light spectrum. This is like watching sunlight through a prism. Different frequencies are bent by different angles as they pass through a prism; thus, as they leave the prism, they are displayed in bands of different colors like a rainbow. Each band, or color, represents light of a certain energy or frequency; and once they are split up in this way, each energy level can be studied separately. In star spectra, including that of our sun, narrow dark lines appear at particular frequencies. Studies in laboratories have shown that these lines occur because as light travels toward us, it passes through materials that absorb energy at particular frequencies, ensuring that those frequencies reach us in a weakened form. These darker lines are known as *absorption lines*. Each absorption line corresponds to a particular element, which absorbs light energy at specific frequencies. Remarkably, this means that by studying the absorption lines in starlight, we can estimate what elements are present in stars and in what quantities. Indeed, modern knowledge of how stars work (see chapter 2) is based largely on such studies.

Even more remarkably, star spectra can tell us whether a star is moving toward or away from us, and at what speed. The principle here is that of the Doppler effect—the phenomenon that makes an ambulance siren seem to drop in pitch as it passes by us. If a moving object (such as an ambulance) emits energy in waves (such as sound waves), those waves appear to be squashed up if the object is moving toward us, and stretched out if it is moving away from us. On a beach, if you walk into the surf, the wave crests will seem to strike your legs more frequently than if you stand still. But if you walk *toward* the beach, the crests will strike your legs less frequently. The same principle applies to light spectra. In the light from stars, absorption lines often seem to be shifted slightly from the position you would expect in a laboratory. Thus, the absorption line that represents hydrogen might be shifted to a higher frequency, making its light waves appear to be squashed up (or closer to the blue end of the spectrum). Or it might be shifted to a lower frequency (closer to the red end of the spectrum), in which case its light waves would appear to be stretched out. Hubble found both types of shift. But as he worked on the remotest objects, he realized that these were all shifted toward the red end of the spectrum. In other words, they appeared to be stretched out as if they were moving away from us. And the farther away they were, the greater the extent of the redshift.

The implications of Hubble's discovery are spectacular but simple to comprehend. The farther a galaxy is from the earth, the faster it is moving away from us, although stars in our own galaxy and some neighboring galaxies are held together by gravity. We have no reason to think that we live in an abnormal part of the universe. Indeed, modern maps of the distribution of galaxies suggest that the universe really is pretty homogenous on the largest scales. So we have to assume that other observers in any other part of the universe would also observe that other parts of the universe seemed to be moving away from them. And this must mean that the universe as a whole is expanding. If the universe is expanding, then in the past it must have been much smaller than it is now. If we follow this logic back in time, we will soon see that at some point in the distant past, the universe must have been infinitesimally small. This argument leads directly to the basic conclusion of modern big bang cosmology: the universe was once infinitesimally small, but it then expanded, and it continues expanding to the present day. Hubble's work provided the first and still the most basic evidence for big bang cosmology.

Hubble also showed that by measuring the rate of expansion, scientists should be able to estimate how long the universe has been expanding. This was an astonishing conclusion, for it seemed to imply something totally unexpected. Hubble had found a way of measuring the age of the universe! Originally, he calculated that the rate of expansion (or the *Hubble constant*) was ca. 500 kilometers per second for every megaparsec of distance between two objects. (A megaparsec is the distance traveled by light in 3.26 million years, which is ca. 30.9×10^{18} km, or ca. 30 billion billion km.) This figure meant that the universe could only be about two billion years old. We now know that this is an impossible date, as the earth itself is at least twice this age. Modern estimates of the Hubble constant are lower, and imply an older universe. But determining exactly how old remains tricky, mainly because of the difficulty of calculating the real distance to remote galaxies. Modern attempts, which use several other types of distance markers in addition to Cepheid variables, suggest that the Hubble constant lies between 55 and 75 km per second per megaparsec. These figures imply that the universe is between 10 and 16 billion years old, and the most recent estimates seem to be converging on a figure of about 13 billion (13,000,000,000) years.[27] For simplicity's sake, that is the date used throughout this book.

Relativity and Nuclear Physics

In the early twentieth century, most astronomers still assumed that the universe was infinite, homogenous, and stable. Hubble's conclusions would have seemed very odd if it had not been for some other developments that were undermining the traditional picture. One was the publication of Einstein's theory of relativity. The details of his theory are not important here, but one implication is that at the largest scales, the universe was probably unstable. Einstein's equations suggested that the universe, like a pin standing on its end, had to fall to one side or the other. It had to be either expanding or contracting; a perfectly balanced universe was very unlikely. Einstein himself resisted this conclusion. Indeed, in what he later described as the greatest error of his life, he altered his theory by proposing the existence of a force he called the "cosmological constant," in order to preserve the idea of a stable universe. This force he imagined as a sort of antigravity, which could counterbalance gravity and thus prevent the universe from collapsing in on itself. However, in 1922 a Russian, Alexander Friedmann, showed that the universe really might be either expanding or contracting. Eventually, even Einstein accepted the idea of an unstable and evolving universe.

But it took time to work out the ramifications of these discoveries. In the 1940s, the idea of an expanding universe still seemed odd to most astronomers. Then, between the 1940s and the 1960s, new evidence accumulated in support of the idea until, by the late 1960s, the big bang theory had become the standard account of the origins of the universe. In the late 1940s, using some of the knowledge gained from work on the atomic bomb, a number of physicists in the United States—including the Russian American physicist George Gamow—began to work their way through the implications of this new view of the universe. What would a tiny universe look like? It was clear that it would have been extremely hot: just as a bicycle tire becomes hotter when more air is pumped in, so the universe must have been extremely hot when all its matter and energy was squashed into a tiny space. The details of how matter would behave under such conditions do not concern us here. What matters is that scientists such as Gamow and later Fred Hoyle (who was to become a fierce critic of big bang cosmology) soon realized that it was possible, using existing ideas about how energy and matter worked at different temperatures, to start doing some calculations about the behavior of the early universe. And the answers made sense. They found they could construct a surprisingly plausible picture of how the early universe was constructed under the assumptions of the big bang theory. In particular, it was possible, roughly, to work out what forms of energy and matter would have existed in the early universe, and determine how that universe would have changed as it expanded and cooled. It soon became apparent that the idea of an early, dense, and hot universe was perfectly consistent with all that was known in the emerging field of particle physics.

Cosmic Background Radiation

What finally persuaded most astronomers to accept the big bang theory was the discovery of cosmic background radiation, or CBR. Early theories of how a big bang might have worked suggested that as temperatures fell during the early history of the universe, distinct particles and forces would acquire a stable existence as soon as temperatures were low enough for them to survive. As we have seen, for several hundred thousand years the early universe was too energetic and too hot for atoms to form. But eventually temperatures fell low enough for protons (with their positive electrical charges) to capture electrons (which have a negative charge). At this point, matter became electrically neutral, and energy and light could flow freely through the universe. Some of the earliest theorists of big bang cosmology predicted that there ought at that moment to have been a huge release of energy, whose remnants might be detectable today.

It is a sign of the caution with which scientists still approached the idea of a big bang that no one actually looked for this background energy. It was found accidentally, in 1964, by Arno Penzias and Robert Wilson, two scientists working for Bell Laboratories in New Jersey. They were trying to build extremely sensitive radio antennae, but found it was impossible to eliminate all the background "noise" they picked up. Eventually, they realized that wherever they pointed their antennae, there was always a faint hum of weak energy. What could possibly be emitting energy from all directions of the sky at the same time? Energy coming from a particular star or galaxy made sense, but energy coming from *everywhere*—and *so much* energy—seemed to make no sense at all. Though the signal was weak, the total when all the energy it represented was added up was colossal. They mentioned their discovery to a radio astronomer who had heard a talk by a cosmologist, P. J. E. Peebles, predicting the existence of remnant radiation at an energy level equivalent to a temperature of ca. 3°C above absolute zero.

This was remarkably close to the temperature of the radiation found by Penzias and Wilson. They had found the flash of energy predicted by early theorists of the big bang.

Their discovery was decisive because no other theory could explain such a universal and powerful source of energy, while big bang cosmology could explain it naturally and easily. Since 1965, few astronomers have doubted that the big bang theory is the best current explanation for the origins of the universe. It is now the central idea of modern astronomy, the paradigm that unifies the theories and ideas of modern astronomy. And the cosmic background radiation is central to modern cosmology: attempts to map tiny variations in it should provide us in the near future with the best information available on the nature of the early universe. (One cosmologist, Dr. Max Tegmark, has even suggested that "the Cosmic microwave background is to cosmology what DNA is to biology.")[28] A new satellite, the Wilkinson Microwave Anisotropy Probe, or WMAP, which was launched in June 2001, is designed to describe these tiny variations more precisely than ever before.[29]

Other Forms of Evidence

More evidence for the big bang has accumulated since the discovery of the CBR. For example, the big bang theory predicts that the early universe will consist mainly of simple elements, above all hydrogen (ca. 76 percent) and smaller amounts of helium (ca. 24 percent). These are about the ratios we observe in the universe today (though the amount of hydrogen has fallen to ca. 71 percent as reactions within stars have converted hydrogen into helium, which now accounts for ca. 28 percent of all matter). The chemical dominance of hydrogen and helium is not immediately obvious to us, because we live in a corner of the universe that happens to have high concentrations of other elements (see chapters 2 and 3), but the evidence is all around us nonetheless. Hydrogen is by far the most common element, even in our own bodies. As Lynn Margulis and Dorion Sagan write: "Our bodies of hydrogen mirror a universe of hydrogen."[30] Especially precise measurements have also been made of the tiny amounts of lithium created in the big bang. These, too, are remarkably close to the figure predicted by theories of element formation during the big bang.

Then there is the fact that neither astronomical observations nor radiometric dating techniques (see appendix 1) can identify any objects that are much more than 12 billion years old. If the universe had in fact existed for much longer than this (perhaps for several hundred billion years), the absence of any objects older than a cutoff date of 12 billion years would be extremely surprising.

Finally, the big bang theory—unlike its main rival, the steady state theory—implies that the universe has changed over time. This means that the most distant parts of the universe ought to seem different from those closer to us; for in looking at objects, say, 10 billion light-years away, we are in effect looking at the universe as it was 10 billion years ago. And, as we will see, distant objects *are* different from the modern universe in important ways. For example, the early universe contained many more quasars (see chapter 2) than does the modern universe.

How Trustworthy Is Big Bang Cosmology?

Is big bang cosmology true? No scientific theory can claim absolutely certainty. And there remain problems with the theory, some of which are highly technical. But at present, none of these problems seems insurmountable.

For a time in the early 1990s, it appeared that some stars were older than the apparent age of the universe—evidence, according to some astronomers, that cast serious doubt on the entire theory. Observations using the Hubble telescope have since shown that this is not true. The oldest stars now seem to be about one billion years younger than the date of the universe as determined by the latest estimates of the Hubble constant. This is good news for big bang cosmology! But there was less welcome news when, in the late 1990s, evidence began to accumulate from studies of distant Ia type supernovae (see chapter 2) that the rate of expansion of the universe, rather than decreasing under the influence of gravity, is in fact increasing. If these observations are correct, they are startling, for they seem to imply that there exists some hitherto unknown force that has operated constantly since the big bang to maintain and accelerate the rate of expansion, but that is too weak to have been detected before. One possibility is that this force consists of "vacuum energy," a force predicted by quantum mechanics that would act in a way opposite to gravity, driving matter and energy apart rather than drawing them together. If so, its effects may be almost identical to those of Einstein's speculative cosmological constant.[31] This evidence may throw a largish wrench into the machinery of big bang cosmology. On the other hand, it may provide an unexpected solution to the problem of dark matter (see chapter 2), because vacuum energy, like all energy, has mass, which may account for a substantial amount of the matter that astronomers have been looking for. There is also the tricky problem of beginnings. At the beginning of the big bang, all our scientific knowledge seems to go haywire. The density of the universe seems to move toward infinity, as does its temperature, and modern science has no good way of dealing with such phenomena, though it has many promising ideas.

What encourages us to take the theory seriously despite these difficulties is its consistency with most of the empirical and theoretical knowledge assembled by modern astronomy and modern particle physics. And no other theory of origins can explain so much. That scientists have constructed a logical theory consistent with so much evidence, and one that seems to tell us what happened during the first few minutes of our universe's history, is itself an astonishing achievement. It is no less remarkable when we realize that future research is likely to modify the current theory, perhaps in quite significant ways.

NOTE ON EXPONENTIAL NOTATION

Modern science often deals with large quantities and large numbers. Writing out, say, a billion billion billion would take a lot of space (to see how much, look at the second paragraph of this note), so scientists use what they call *exponential notation*; a number of figures in this chapter use this convenient mathematical shorthand. Here is how it works.[32] One hundred is 10 multiplied by 10, or two 10s multiplied together. In exponential notation, 100 can be written as 10^2. One thousand is three 10s multiplied together, or 10^3, and so on. To convert a number in exponential notation to one in normal notation, write down a 1, then add the number of zeros that appear in the exponent. One thousand (10^3), therefore, is 1 with three zeros after it; one billion is 10^9, or 1 with 9 zeros after it—that is, 1,000,000,000. We can use the same notation for small numbers, too. One hundredth ($\frac{1}{100}$ or 1 percent) is written as 10^{-2}; and one thousandth ($\frac{1}{1000}$) is written as 10^{-3}. The system also works well for numbers that are not an exact multiple of ten. Thus 13 billion years can be thought of as 13 times a billion years. In exponential notation this becomes 13×10^9 years.

The crucial thing to note is that increasing the exponent by one *multiplies* the size of the previous number ten times. So 10^3 is not just slightly bigger than 10^2; it is, in fact *ten times as large*. In the same way, 10^{18} (or a billion billion) is not double the size of 10^9; it is one billion times (10^9 times) as large; and it is ten times as large as 10^{17}. Exponential notation provides a deceptively simple way of describing colossal numbers, which can easily lull us into forgetting how large these numbers really are. The mass of a hydrogen atom can be written in exponential notation as 1.7×10^{-27} kilograms. In ordinary script, this is a simple, but lengthy, fraction: 1.7/1,000,000,000,000,000,000,000,000,000 kilograms, or 1.7 times one billionth of a billionth of a billionth of a kilogram. To understand what this really means is trickier. Try to imagine something so small that it weighs just one-billionth of a kilogram. (We cannot do it, of course—our minds are not designed to deal with such calculations; but we can make the effort.) Then try to imagine something that weighs one billionth of this; then repeat the experiment a third time, and you are imagining the size of a hydrogen atom. To weigh the Sun, you multiply instead of dividing. The Sun has a mass of about 2×10^{27} tons, or 2,000,000,000,000,000,000,000,000,000 tons, which is two times a billion billion billion tons. It contains about 1.2×10^{57} atoms. The universe contains about 10^{22} stars. To roughly estimate the number of atoms in the universe, we can multiply these two numbers together, which means *adding* the two exponents, to get 1.2×10^{79} atoms. This may not seem so impressive until we start writing the number out in ordinary notation, and even then, most of us cannot really understand what we are writing down. In the final chapter of this book, we will come across numbers much, much larger than even these huge figures.

SUMMARY

Before about 13 billion years ago, we can say nothing with any confidence about the universe. We do not even know if space and time existed. At some point, energy and matter exploded out of the emptiness, creating both time and space. The early universe was fantastically hot and extremely dense, and it expanded extraordinarily fast in a sort of cosmic explosion. As it expanded, it cooled. Matter and antimatter annihilated each other, leaving a tiny residue of matter. Out of the violent flux of the early universe, there appeared distinct entities—protons, neutrons, photons of light, electrons— and distinct forces, including the strong force, the weak force, and the forces of gravity and electromagnetism. After a few hundred thousand years, the universe was cool enough for protons and electrons to form stable atoms, and the matter in the universe became electrically neutral. As a result, matter and energy ceased to interact constantly, and radiation began to flow freely through the universe. As the universe expanded, the temperature of the radiation fell; it is now detectable as the cosmic background radiation.

This story, as strange as it may seem, is based on a colossal amount of scientific research, and it is compatible with most of what we know today about astronomy and particle physics. Big bang cosmology is now the central idea of modern cosmology. It is the paradigm that unites modern ideas on the nature and history of the universe, and it dominates the first chapter of the modern creation myth.

FURTHER READING

Barbara Sproul's *Primal Myths* (1991) is a collection of creation myths from many different cultures, accompanied by an introductory essay. There are now many popular accounts of big bang cosmology, some by authors who helped construct the modern story of the origins of the universe. The following are some of the books I have found most helpful. Stephen Hawking's *A Brief History of Time* (1988) is one of the best known, and has recently been followed by his *The Universe in a Nutshell* (2001); even more technical is Steven Weinberg, *The First Three Minutes* (2nd ed., 1993). John Gribbin's *Genesis* (1981) is a superb introduction for the general reader (and one of the inspirations for this book), though it's beginning to show its age. More up-to-date, though equally readable, are Timothy Ferris, *The Whole Shebang* (1997); John Barrow, *The Origin of the Universe* (1994); Peter Coles, *Cosmology* (2001); and Armand Delsemme, *Our Cosmic Origins* (1998). Delsemme's book is relevant for much of the first half of this book. Cesare Emiliani's *Scientific Companion* (1995) is a useful handbook for those who want more precise information about the ideas and terminology of modern astronomy, chemistry, and physics. Eric Chaisson's *Cosmic Evolution* (2001) is an attempt to think through the meaning of order and entropy at many different scales, from stars to microbes, and Martin Rees's *Just Six Numbers* (2000) is also about the fundamental structures of the universe. Lee Smolin's *Life of the Cosmos* (1998) is a readable book that consists of grand speculations about the possibility that our universe is one of a vast population of universes that change according to some form of cosmic evolution. Charles Lineweaver's short essay "Our Place in the Universe" (2002) is a marvelous introduction to the challenge of thinking about scale and orientation within the universe. Nigel Calder's *Timescale* (1983) is a remarkable chronology for the whole of time, though it is now old enough to be slightly dated.

2

ORIGINS OF THE GALAXIES AND STARS
THE BEGINNINGS OF COMPLEXITY

If one had to summarize, in just one sentence, "What's been happening since the Big Bang?," the best answer might be to take a deep breath and say: "Ever since the beginning, gravity has been moulding cosmic structures and enhancing temperature contrasts, a prerequisite for the emergence of the complexity that lies around us ten billion years later, and of which we are part."

Look at the sky on a clear night, and it seems obvious that stars are the most important inhabitants of our universe. But stars, like humans, do not exist in isolation. They gather into the huge cosmic societies we call *galaxies*, each of which may contain 100 billion stars. Our home galaxy is the Milky Way. Unlike other galaxies, which appear to us as faint stars or blurs, the Milky Way looks like a pale river of light flowing across the night sky, because we see it from inside. What is less obvious to the naked eye, and was not apparent even to most astronomers until a decade or two ago, is that galaxies gather into even larger communities. These include *groups* (usually a few million light-years in diameter, containing perhaps twenty galaxies) and *clusters* (up to 20 million light-years broad and holding hundreds, even thousands, of galaxies). Groups and clusters of galaxies are held together by gravitational forces. But there exist even larger structures, structures so large that they are stretched out by the expansion of the universe. These include *superclusters* (up to 100 million light-years across, with perhaps 10,000 galaxies) and the huge chains of superclusters enclosing vast bubbles of empty space that were first detected by astronomers in the 1980s. At even larger scales, the universe appears to be remarkably homogenous. This homogeneity shows up in the uniformity of the cosmic background radiation. So the complex pat-

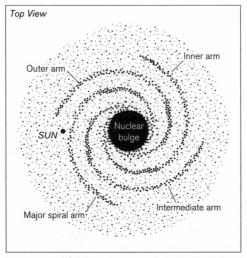

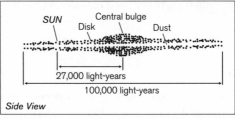

Figure 2.1.   The position of the Sun within
the Milky Way. The Sun lies within an arm
of the Milky Way, about 27,000 light-years from
its center. Clouds of dust obscure our view of the
central parts of the galaxy. Adapted from Nikos
Prantzos, *Our Cosmic Future: Humanity's Fate
in the Universe* (Cambridge: Cambridge University Press, 2000), p. 97.

terns that will interest complex observers such as ourselves seem to appear
only at scales smaller than chains of superclusters.

At present, these seem to be the largest ordered structures in the observable universe. Their discovery pushes us even farther from the center
of the universe than Copernicus's discovery that the earth revolved around
the Sun. Our sun, it seems, is situated in an undistinguished suburb in a
second-rank galaxy (the Andromeda Galaxy is the largest in our local group),
in a group of galaxies that lies toward the edge of the Virgo Supercluster,
which contains many thousands of other galaxies (see figure 2.1).[1]

More recently, it has become clear that even superclusters may be mere bit players in the history of the universe. It seems that most of the mass of the universe (90 percent or more) is *not* visible, and the exact nature of this mass (known appropriately as *dark matter*) remains a mystery. In other words, we are in the embarrassing position of not knowing what most of the universe is made of.[2] This chapter will touch on theories about the nature of dark matter, but it will focus mainly on those parts of the universe that we know most about—those parts that are visible.

We take up the history of the early universe where we left it in the previous chapter: about 300,000 years after its creation, as energy and matter went their separate ways.

THE EARLY UNIVERSE AND THE FIRST GALAXIES

In the first minutes of its existence, the universe cooled so rapidly that it was impossible to manufacture elements heavier or more complex than hydrogen, helium, and (in minute amounts) lithium: elements 1, 2, and 3 in the periodic table. In the heat and chaos of the early universe, nothing more complex could survive. From a chemical point of view, the early universe was very simple, far too simple to create complex objects such as our earth or the living organisms that inhabit it. The first stars and galaxies were constructed from little more than hydrogen and helium. But they were a sign of our universe's astonishing capacity to build complex objects from simple building blocks. Once created, stars laid the foundations for even more complex entities, including living organisms, because in their fiery cores they practiced an alchemy that turned hydrogen and helium into all the other elements of the periodic table.

So far, the story of the universe has been dominated by the expansionary force of the big bang. Now we must introduce a second large-scale force: that of gravity. Gravity is the force that Newton described so successfully in the seventeenth century and that Einstein described even more precisely early in the twentieth century. While the force of the big bang drives energy and matter apart, gravity pulls things together. Newton argued that all forms of matter exert a tug on all other forms of matter. Einstein maintained that the effects of gravity arise because of the way that large masses can warp the geometry of space-time. Einstein also showed that gravity acts on energy as well as on matter. This conclusion was not entirely surprising, for Einstein had already demonstrated that matter is really a sort of congealed energy. But he went further, offering an ingenious proof that gravity can warp energy as well as matter. The Sun is the largest object in our solar system and has the greatest mass. He argued that its huge mass ought to bend the space-time around it enough to alter the trajectory of light rays passing close to the Sun's edge. The best opportunity for detecting this effect was during a solar eclipse, the only time when it was possible to see stars close to the Sun. If stars at the edge of the Sun were photographed just before a solar eclipse, he predicted that their movement would appear to slow down just before they passed behind the Sun. As they appeared on the other side, they would also seem to hover momentarily at the Sun's edge before moving away from it. This effect would result from beams of starlight being bent by the Sun's mass, just as a stick seems to bend when placed in water. In 1919, Einstein's prediction was tested during a solar eclipse and found to be astonishingly accurate.

By pulling on both matter and energy, gravity can give the universe shape and structure. It may be easiest to see how it does so if we stick with Newton's intuitively simpler notion of gravity as a "force." Newton showed that gravity can work at very large scales but is most powerful close-up. To be precise, the gravitational attraction between two objects is proportional to the (square of the) mass of the two objects, and is inversely proportional to the (square of the) distance between them. This means that gravity can pack closely packed masses even more closely together, but has less effect on objects separated by large distances. Gravity has even less impact on light, fast-moving objects such as the particles that carry energy, and thus it shapes matter more effectively than energy. Because its effects vary in these ways, gravity has managed to create many complex structures at a number of different scales. This is a remarkable conclusion, for it suggests that in some sense, and at some scales, gravity can temporarily counter the second law of thermodynamics, the fundamental law that seems to guarantee that over time, the universe will become less ordered and less complex (see appendix 2). Instead, as gravitational energy is released (as gravity clumps matter together), the universe appears to become *more* ordered. Gravity is thus one of the major sources of order and pattern in our universe. In the rest of this chapter we will see how gravity created many of the complex objects studied by astronomers.

Much of the history of the early universe, and of the galaxies and stars, can be thought of as a product of competition between the force of the big bang, which drives the universe apart, and the force of gravity, which tends to draw the universe back together again. There is an unstable and shifting balance between these two forces, with expansion winning at the largest scales and gravity winning on smaller scales (up to the level of clusters of galaxies). But gravity needs some initial differences to work with. If the early universe had been perfectly smooth—if, say, hydrogen and helium had been distributed with absolute uniformity throughout the universe—gravity could have done little more than to slow down the rate at which the universe expanded. The universe would have remained homogenous; and complex, lumpy objects such as stars, planets, and . . . human beings could never have formed.

So it is important to know how homogenous the early universe was. Astronomers try to measure the "smoothness" of the early universe by looking for tiny differences in the temperature of the cosmic background radiation. Any "bumpiness" ought to show up as slight temperature differences in the cosmic background radiation. The COBE (Cosmic Background Explorer) satellite, launched early in the 1990s, was designed to look for such differences, and the WMAP (Wilkinson Microwave Anisotropy Probe) satellite, launched in June 2001, is mapping these variations with even greater precision. COBE has shown that although the cosmic background radiation is extremely uniform, there *are* tiny variations in its temperature. Apparently, some areas of the early universe were slightly hotter and denser than others. These "wrinkles" gave gravity some differences to work with, and it did so by magnifying them, making dense regions even denser. Within a billion years after the big bang, gravity had created huge clouds of hydrogen and helium. These may have been as large as several clusters of galaxies, and locally, their gravitational pull would have been sufficient to counteract the expansionary drive of the universe. At larger scales, the expansionary force of the big bang remained dominant, so that over time the gaps between these massive clouds of matter increased.

Under the pull of their own gravity, the clouds of hydrogen and helium began to collapse in on themselves, as atoms of hydrogen and helium were packed ever more closely together. As the gas clouds shrank, some regions became denser than others and began to collapse more rapidly; in this way, the original clouds broke into smaller and smaller clumps at many different scales, from that of whole galaxies to single stars. As gravity packed each cloud into ever smaller spaces, pressure built up in the center. Increasing pressure means increasing temperatures, and so, as they shrank, each gas cloud began to heat up. Within the smaller clumps, which contained a mass equivalent to several thousand stars, there appeared regions of enormously high density and extreme heat; it was in pockets within these cosmic nurseries that the first stars were born.[3]

As the core regions heated up, the atoms within them moved faster and faster, and collided more and more violently. Eventually, the collisions were violent enough to overcome the electric repulsion between the positively charged nuclei of hydrogen atoms. (These repulsive forces depend partly on the number of protons, or positive charges, in the nuclei, so this reaction occurs most easily in hydrogen atoms and becomes progressively more difficult to achieve with larger atoms.) Wherever temperatures reached 10 million degrees C, pairs of hydrogen atoms fused to form helium atoms, each of which has two protons in its nucleus. This nuclear reaction, known as *fusion*, is what happens at the center of a hydrogen bomb. As hydrogen atoms fuse into helium, a tiny amount of matter is transformed into a huge amount of energy according to Einstein's formula, $E = mc^2$: the energy released is equal to the mass that is transformed multiplied by the speed of light *squared*. Because the speed of light is an enormous figure, Einstein's equation tells us that an enormous amount of energy is released by the transformation of even tiny amounts of matter. To be precise, when hydrogen atoms fuse to form an atom of helium, they lose about 0.7 percent of their mass; we know this because a helium atom weighs less than the hydrogen atoms used to construct it. The lost mass has been converted into energy.[4] Stars are like massive hydrogen bombs with so much fuel that their "explosions" can continue for millions or even billions of years. And this is how the first stars lit up the billion-year-long night of the early universe.

The colossal heat and energy generated by fusion reactions resist the force of gravity, so as they light up, young stars stop collapsing. And it is this balance between the expansionary force of the nuclear explosions at their center and the attractive force of gravity that tames the violent energies at the heart of all stars. Stars form durable structures because they are the result of a negotiated compromise between gravity, which crushes matter together, and the explosive force of fusion reactions, which forces matter apart. The negotiations are continuous; if the center heats up, the star expands and thus cools down—so it contracts again, in a negative feedback cycle analogous to that in an air-conditioning system. (If the air gets too hot, the system switches on and cools the air down again.) We can watch these negotiations in the pulsations of variable stars. But normally, the underlying truce endures for millions or billions of years, as long as the star exists.

The lighting up of the first stars was a momentous turning point in the history of the universe, for it marked the appearance of a new level of complexity, of new entities operating according to new rules. What had been billions and billions of atoms, drawn together by the force of gravity, suddenly became a new organized structure—one that could last for millions or billions of years. The moment of transition occurred when a slight increase in temperature ignited fusion reactions throughout the core of the proto-star, thereby transforming gravitational energy into heat energy and creating a new and more stable system of energy flows. Stars organize the atoms they contain into new, durable configurations, which can handle huge energy flows without disintegrating. This, we will see, is the characteristic pattern of all such thresholds. New configurations emerge quite suddenly as once independent entities are drawn into new and more ordered patterns, held together by an increasing throughput of free energy (see chapter 4). But, as is true of all these structures, they are held together only with difficulty, so none is eternal. New levels of complexity are characterized, therefore, by a certain fragility and by the certainty of eventual collapse. The second law of thermodynamics ensures that all complex entities will eventually die; but the simpler the structure, the better its survival chances, which is why stars live so much longer than humans (see appendix 2).

Many of the first stars are still around today, 13 billion years later. Most can be found in the centers of galaxies, or in the huge balls of stars known as *globular clusters*, which orbit most galaxies in large spherical tracks. The earliest stars probably formed during the chaotic and rapid collapse of relatively formless clouds of gas. They can be detected today by their erratic orbits and by the absence of elements heavier than hydrogen and helium, because those were the only elements available when they were formed. In the crowded early universe, embryonic galaxies often blended into each other, and these mergers help explain the erratic orbits of many of the oldest stars.

As galaxies formed and merged in the early universe, gravity went to work on them, sculpting many into a shape that is surprisingly common in the universe. As the ragged galaxies of the early universe were pulled together by gravity, different parts were dragged toward the center in huge arcs; and minor variations in the movements of these arcs ensured that each cloud began to spin, like water going down a drain. As each cloud contracted, the rotation accelerated, as happens when skaters fold in their arms. Like a spinning ball of dough, the areas spinning fastest were flung out by centrifugal force, and the entire cloud began to flatten into a sort of cosmic pizza. These simple processes, all dominated by the force of gravity, explain why so many of the largest clouds of matter in the universe, even at the scale of galaxy clusters, take the form of spinning disks, which the Soviet theorist Yakov Zel'dovich has called "crepes." We will see that the same rules also operate at smaller scales, which is why our solar system would also look like a huge, flat disk if we could see it from a distance.

By the time a second generation of stars began to form, these processes had transformed some of the larger galaxies, such as the Milky Way, into huge and more or less regular disks. This change is reflected in the more orderly orbits of younger stars, such as our own sun, which, traveling at the stately speed of 800,000 kilometers an hour, takes about 225 million years to process once around the center of the Milky Way. Similar mechanisms shaped other galaxies, creating a universe populated by galaxies of stars, constructed in different ways but often forming regular, rotating disks. Star formation continues to the present day. In the Milky Way, about ten new stars are formed every year.

A COSMOLOGICAL MENAGERIE: BLACK HOLES, QUASARS, AND DARK MATTER

The early universe contained stranger objects than stars. At the center of most galaxies, densities were so great that huge clouds of matter and energy kept collapsing even at temperatures high enough to start fusion reactions. Here, gravity acquired such momentum that it crushed matter and energy out of existence, thereby forming the bodies called *black holes*. Black holes are regions of space so dense that no matter and no energy can escape their gravitational pull, not even light. This means we can never directly observe what goes on inside a black hole, except by entering it—and then, of course, we could never return to report our findings. Black holes are so dense that to form one from our earth, we would have to crush it into a ball with a diameter of about 0.7 inches.[5]

There has been much fascinating speculation about the true significance of black holes. Recently, for example, it has been suggested that black holes may be what new universes look like from the outside. Each may represent a separate universe, beginning with its own big bang. Lee Smolin has argued that if this is true, we may have an explanation for some other oddities of our universe. In particular, we may be able to explain why so many crucial parameters—such as the relative strength of the fundamental physical forces, or the relative size of fundamental nuclear particles—seem precisely tuned to create a universe capable of producing stars, elements, and complex entities such as ourselves. On Smolin's assumptions, only universes that can produce black holes can have "offspring." If we add a further assumption, that new universes differ only slightly from their "parent" universe, we see that a process akin to Darwinian selection may be at work.[6] After many generations, the hyperspace in which these many universes exist is likely to be dominated by those universes that have the precise qualities needed to produce black holes, however statistically improbable these universes may be, because all other universes will be sterile. But if a universe can produce black holes, it can probably produce other large objects as well, such as stars, and many other kinds of complex structures besides. Such ideas suggest that there may be new levels to our modern creation myth above the level of the universe, and that a "hyperuniverse" could exist that is far older than 13 billion years, and much bigger than our universe. But at present, we have no way of proving or disproving these grandiose ideas.

So we can safely return from these speculations to the universe we know. Black holes can tell us some important things about our own universe and the galaxies that populate it. They are so dense that the gravitational forces they exert can generate energies much larger than those produced within stars. It is likely that a black hole lurks at the center of the Milky Way, 27,000 light-years away in the direction of the constellation Sagittarius. It may be identified with a powerful source of radio waves known as Sagittarius A, and it probably has a mass about 2.5 million times that of our own sun.

The existence of black holes at the center of many galaxies may help explain another strange object, the quasar, or "quasistellar radio source." The first quasars, the brightest objects known to modern astronomy, were detected by Australian astronomers in 1962. They shine more brightly than even the largest galaxies, though they are no larger than our solar system. They are also extremely remote. Most are more than 10 billion light-years away, and none is closer than 2 billion light-years from us. So when we look at quasars, we are seeing objects that existed early in the universe's life. Currently, it seems likely that their energy comes from huge black holes that suck in large amounts of matter from the galactic material surrounding them. Quasars thus consist of black holes plus star food. Quasars were particularly numerous early in the life of the universe, because at that time galaxies were crowded more closely together, and black holes were better fed. Since then, the universe has expanded, galaxy clusters have moved farther apart, and the pickings have become leaner for galactic black holes. So, though most galaxies may still have black holes at their centers, few of these beasts now consume enough to create quasars. And as most quasars do not live more than a few million years because of their prodigious appetite for star dust, they are rare in the modern universe. Quasars are the astronomical equivalent of dinosaurs, though the black holes that powered them still survive at the centers of most galaxies, waiting for unwary stars to fall into their clutches.

Galaxies and stars make up most of the visible universe. But observations of the movements of galaxies and galaxy clusters have led to the embarrassing conclusion that we are seeing only a tiny part of what is actually out there. Indeed, what we can *see* may constitute no more than 10 percent, and perhaps as little as 1 percent, of the matter in the universe. Using the basic laws of gravity, astronomers can calculate roughly how much matter is in a group of galaxies by studying the way they rotate, and such studies show that galaxies contain perhaps ten times as much matter as we can see. Astronomers refer to this matter as *dark matter*, which is really a way of expressing their puzzlement.

Finding out what makes up this huge amount of material is one of the central projects of modern astronomy. At present, there are two main types of candidate. First, it may consist of tiny particles, each much smaller than electrons but collectively more massive than all other forms of matter. These are known as WIMPs, or "weakly interacting massive [in the sense of *having mass*] particles." The current best bet for such a particle is the *neutrino*, a particle that may or may not have any mass. If it does have mass, it is no more than 1/500,000th the mass of an electron. However, there are about one billion neutrinos for every other particle, so that even if their individual mass is tiny, they may make up most of the material in the universe. If we could see neutrinos, the universe would seem like a huge neutrino fog, contaminated by tiny specks of matter. Alternatively, there may be many large objects that we cannot see because they do not emit light or other forms of radiation. These may consist of the corpses of stars, or planetlike objects. These are referred to as MACHOs, or "massive compact halo objects." Recently, a third possibility has emerged, which may offer an elegant solution to the problem of dark matter: dark matter may really be dark energy. As we have seen, energy also exerts a gravitational pull. So perhaps the so-called vacuum energy discovered in the late 1990s, which seems to be accelerating the rate at which the universe is expanding, constitutes as much as 70 percent of the mass/energy of the universe. If so, it may account for most of the extra gravitational pull observed by astronomers. In this scenario, dark matter may account for no more than 25 percent of all the stuff in the universe, while the visible universe makes up merely 5 percent.[7]

THE LIFE AND DEATH OF STARS

Stars, like people, have biographies. They are born, they live, they change, and they die. And today we know a great deal about the typical life cycles of stars. This knowledge is derived largely from studying the spectra of starlight. As we saw in the previous chapter, careful analysis of the absorption bands (the frequencies at which energy has been absorbed as it travels through stars) can tell us much about the materials within stars. It can also tell us how hot they are. As astronomers have studied the spectra of more and more stars over the last century, they have built up a picture of the different stages of a star's life, and of the different types of stars that can exist.

The most important single feature of stars is their size, or rather the size of the initial cloud of material from which they form. This determines many features of a star, including its brightness, temperature, color, and life span. If the initial cloud is smaller than about 8 percent of our own sun's size, its center will never get dense or hot enough to fuse hydrogen, and no star will form. At best, there may appear a *brown dwarf:* an object with a faint glow, a bit like the planet Jupiter. Brown dwarves are half planet and half star, though recent observations of orbiting material around them suggest they are formed in much the same way as stars even if they are never quite large enough to ignite.[8] On the other hand, if the initial cloud is more than 60 to 100 times the size of our sun, it is likely to split into two or more regions of star formation as it collapses, which explains the large number of double or multiple star systems observed by astronomers. Between these extremes, it is helpful to think of two main sizes: the majority of stars, which range from much smaller than the Sun to about 8 times its size, and those that are between 8 and 60 times the size of our sun.

The amount of material in the embryonic star cloud determines the gravitational pull of the cloud, the speed with which it shrinks, and the density and heat at its center. The heat at the center of a new star determines the speed at which it burns up its available fuel. Thus large stars are much hotter than small stars; though they contain more material, they also burn their fuel more rapidly, live more dangerously, and die sooner. Stars 10 times larger than our sun may have life spans of a mere 30 million years, while even larger stars may live for only a few hundred thousand years. Smaller stars, from twice the size of our sun down to $\frac{1}{10}$ its size, have less dense and therefore cooler cores. So they burn their fuel more prudently. The smallest stars may have life spans of hundreds of billions of years, many times the present age of the universe.

Most stars, like our sun, burn their fuel more slowly than the giants. But eventually, they use up their hydrogen, and their cores fill with helium. At that point, the fusion reactions using hydrogen, which have sustained the star during most of its life, can continue no longer. The central region starts to cool and to collapse in on itself. But the collapse increases the internal pressure, which causes the star's interior to heat up again, so that the star seems to rebound, swelling to many times its previous size. If the star is large enough, the initial collapse may raise the temperatures at its core to 100 million degrees C. At this temperature, new fusion reactions begin that use helium as their initial fuel. But these reactions convert much less mass to energy than hydrogen fusion, so they don't last as long. Stars run out of helium quite quickly; and when they do, the center collapses again and the outer layers swell even further, sometimes being thrown out into space. A series of reactions of this kind, each requiring higher temperatures, produces many new elements, the most abundant of which are carbon, oxygen, and nitrogen. Our sun, for example, will maintain this sequence until it starts producing carbon, but will go no further, while slightly larger bodies may be able to maintain the sequence up to oxygen. In this way, aging stars create many of the elements in the early parts of the periodic table; the largest stars of all manage, in their final stages, to produce iron (atomic number 26), whose creation requires temperatures of between 4,000 and 6,000 million degrees. This sequence of reactions ends with iron. When old stars die, their ashes, containing all these new elements, are scattered around them, creating stellar graveyards that are more complex, chemically, than any region of the early universe.

In their dying phases, many stars swell into red supergiants; an example is Betelgeuse in the constellation of Orion. When our sun reaches this phase, in about 5 billion years, it will expand so much that both Earth and Mars will be within its outer layers. (Betelgeuse is so large that if it were placed where our sun is, the earth would be buried halfway between its center and its surface.) Eventually, when they run out of fuel, small- to medium-sized stars begin to cool, turning eventually into star cinders known as *white dwarfs*. They are extremely dense, and closer in size to earthlike planets than to stars. Over billions of years, most of them cool down, and their lives as stars ends.

Giant stars, larger than about 8 times the size of our sun, have even more dramatic life histories. Because these stars are large, the pressure and temperature at their cores are much higher, so they can manufacture elements up to silicon and, as noted above, even iron. In their final stages, they make different elements, layer by layer, in a frenetic attempt to keep pumping out energy and to avoid gravitational collapse. But when they finally run out of fuel, their end is much more spectacular than that of medium-sized stars. When they have no more energy to support themselves, gravity takes over and folds them up in a sudden and catastrophic collapse that may last no more than a second. This creates the phenomenon known as a *supernova*. So much energy is generated in a supernova explosion that, for a few weeks, it may shine with the energy of 100 billion stars, or an entire galaxy. If the original star is less than 30 times the size of our sun, its collapse may leave behind a neutron star. These are objects whose atoms have been crushed so tightly together that their electrons fuse with protons and turn them into neutrons. In neutron stars, an object with the mass of our sun may be concentrated into an area the size of a large modern city. Neutron stars can spin at rates as fast as 600 times a second. They were first detected on Earth (in

1967) as *pulsars*, because, as they spin (if that happens to be at an angle convenient for Earth-bound astronomers), the energy they emit seems to strike Earth in short pulses. The Crab Nebula contains a neutron star that spins 30 times each second, and is the remnant of a supernova explosion detected by Chinese astronomers in 1054 CE.

Stars more than about 30 times the size of our sun collapse even more violently, and their core regions may be crushed into a black hole. Just outside the core, protons and electrons are combined to form neutrons, and a huge stream of neutrons and neutrinos flees the dying star. This vast pulse creates a cauldron with temperatures of several thousand million degrees. Momentarily, in the extraordinarily high temperatures of a supernova, a new threshold of some kind is crossed, for in this furnace it is possible to bake elements much heavier than iron. Indeed, in a few moments, supernova explosions can manufacture all the elements of the periodic table, up to uranium. These are then blasted deep into space. The elements that dominate this galactic alchemy are oxygen and lesser amounts of neon, magnesium, and silicon, which, as a result, are some of the commonest heavier elements in interstellar space. The most recent supernova of this type was detected on Earth in February 1987; it was the brightest supernova to be observed from Earth since 1604, when one exploded within the Milky Way. The supernova whose light reached us in 1987 was in the Large Magellanic Cloud, a neighboring galaxy that can be seen in the southern sky. It marked the death throes of a star previously known as Sanduleak -69 202; in its final, red giant phase, it had a diameter about 40 times that of our sun. This supernova blew up about 160,000 light-years away from us, which means that it actually occurred 160,000 years ago. Many "new stars" recorded earlier in human history may also have been supernovae, including the star reported at the birth of Jesus. Because large stars have short lives, supernovae have been fertilizing interstellar space with new chemicals since the formation of the earliest galaxies. The gold or silver ring you may be wearing was made in a supernova. Without supernovae, we could not exist.[9]

There is a second kind of supernova, known as a *Ia supernova*, which is created by the explosion of white dwarf stars fed with new material from neighboring stars. These explosions are even brighter than the supernovae created by the death of large stars, and they eject particularly large amounts of iron, along with other heavy elements.

The story of star deaths is a vital part of the story of life on earth, for stars created both the raw materials from which our world is created and the energy that fuels the biosphere. The heavier elements scattered throughout the galaxy were first formed in stars and in supernovae. As the universe aged, the proportion of new elements (other than hydrogen and helium) has steadily increased. Without the chemically rich environment created by stars and supernovae, our earth could not have been born, and life could not have evolved. So, the chemicals we are made from were created in three distinct stages: while most hydrogen and helium was created during the big bang, most elements from carbon (atomic number 6) to iron (atomic number 26) were formed inside medium and large stars, and most other elements were formed in supernovae. The first generation of stars, formed in the early universe, could not possibly have sustained life. But later generations of stars, such as our own sun, could.

The energy that drives the biosphere is also derived largely from stars. Direct sunlight is one of the most important of all sources of energy on Earth. But for humans in the past two centuries, the sunlight stored long ago in coal and oil has been almost as important. In addition, many important processes on the earth are driven by the earth's internal heat engine, whose heat was generated partly during the formation of our sun and comes partly from radioactive elements created in supernovae. In all these ways, the life histories of stars are a vital component of the story of life on Earth.

CREATION OF OUR SUN

Like all stars, our sun was born in the gravitational collapse of a cloud of matter. A nearby supernova probably triggered the collapse. The shock waves from this huge explosion rippled through gas clouds that had formed in a region of our galaxy's spiral arm, about 27,000 light-years from its center, or 40 percent of the way to its edge. As they did so, the material in these clouds was rearranged, like sand on the vibrating surface of a drum. In this way, a whole stellar tribe was born, with hundreds of new stars.

All of them count as second- or third-generation stars, for they formed from material that contained many other elements in addition to hydrogen and helium. The primordial gases accounted for 98 percent of the material in the cloud from which our sun formed (hydrogen made up ca. 72 percent; helium, ca. 27 percent). But many other elements were also present, including carbon, nitrogen, and oxygen (which now account for 1.4 percent of all matter in the universe), and also iron, magnesium, silicon, sulfur, and neon (which account for another 0.5 percent). These ten elements, all created either during the big bang or within large stars, represent all but 0.03 percent of the mass of atomic matter in our part of the galaxy, while the remaining elements were created in supernovae.[10] The presence of elements heavier than hydrogen and helium, as well as many simple chemicals formed from these elements, explains why, unlike the first generation of stars, our sun (and perhaps many of its sibling stars) was born with an attendant group of satellites. These are the planets of the solar system (see chapter 3).

As with all stars, many features of the Sun are determined by its size. It is a yellow star (spectral type G2), which means that it falls in the middle of the range of brightness of stars. However, most stars (about 95 percent) are smaller and cooler than the Sun.[11] It is extremely big compared to the earth. Its diameter of almost 1.4 million kilometers is more than 4 times the distance from the earth to the Moon. Nevertheless, our sun is not large enough to collapse into a supernova when it dies. But it is not small enough to live to a very old age. It was born about 4.6 billion years ago, and will live for another 4 to 5 billion years. It has existed for about a third of the life of the universe and is halfway through its own life cycle. Like all stars, it is powered by a huge, stable, nuclear explosion in its core, whose temperature is about 15 million degrees. Here, atoms of hydrogen fuse into helium, releasing large amounts of radiant energy. It can take a million years for the photons of energy produced in these reactions to fight their way through the dense core of the Sun and appear at the surface. There, the temperature is a modest 6,000°C. From the surface, energy is radiated throughout the solar system and deep into space. Once photons reach the surface of the Sun they begin to move with the speed of light. After struggling through a traffic jam of subatomic particles for 1 million years, it takes photons of light just over 8 minutes to reach Earth, 150 million kilometers away.

Without the Sun, our earth could not have existed and life could not have evolved. All the planets of our solar system were made from the Sun's debris and constructed within its gravitational field. And the Sun provides most of the light and heat energy that sustains life on Earth. It is the battery that drives the complex geological, atmospheric, and biological processes on the surface of our planet.

THE SCALE OF THE UNIVERSE

The universe started out unimaginably small, and is now unimaginably large. Somehow or other, to make sense of the story of its creation, we have to try to grasp the spatial as well as the temporal scales on which this story has been told. We can never fully comprehend these scales, but it is worth making the effort.

If the universe is 13 billion years old, this means we cannot see anything farther away from us than 13 billion light-years, for nothing can travel faster than light, and that is as far as light could have traveled since the origins of the universe. But the universe may, in fact, be bigger than this, because the notion of inflation suggests that in the first second of its existence, the space-time in which the universe is embedded expanded much faster than the speed of light. If so, the real universe may be billions of billions of times larger than the observable universe. Indeed, if different parts expanded in different ways, there may be billions of different universes, each with slightly differently physical laws.

In practice, of course, even the size of the visible universe is impossible to grasp. To get from the size of the smallest subatomic particle to the largest known cluster of galaxies, we must multiply by ten 36 times. The largest galaxy cluster is 10^{36} times the size of the smallest known particle.[12] Such statements mean little to us; to even begin thinking about these scales, we have to make a special imaginative effort. It may help to use a thought experiment that can shock us into some appreciation of very large scales.

Large galaxies such as the Milky Way contain something like 100 billion stars each. Larger galaxies may contain up to 1,000 billion, while the more numerous dwarf galaxies may contain as few as 10 million, so 100 billion may adequately approximate the average galaxy size. As far as we know, there are also something like 100 billion galaxies in the observable universe. How large is 100 billion? Imagine a pile of 100 billion rice grains: it would be large enough to fill a building the size of the Sydney Opera House.[13] That suggests how many stars there are in just our own galaxy. To represent the number of stars in the entire visible universe, you would have to build a hundred billion opera houses, each filled with rice grains. (This total number of rice grains may be roughly equivalent to the number of grains of sand on all the deserts and beaches of the earth.)[14] But let us concentrate on the one opera house, imagining that it represents our own galaxy, the Milky Way. If we now use the rice grains to make a scale model of the Milky Way, what would be the distance from our star, sitting in the center of the Sydney Opera House, to the closest rice grain? The nearest star to us is Proxima Centauri, which is part of the triple star system of Alpha Centauri, the third-brightest starlike object in the night sky. If our sun were the size of a rice grain in the Sydney Opera house, Proxima Centauri would be near the Australian city of Newcastle, about 100 kilometers away, which represents a mere 4.35 light-years (more than 40,000 billion km or ca. 25,000 billion miles). Altogether, about twenty-six stars lie within 12 light-years of Earth. (One is Sirius, which appears as the brightest star in our sky partly because it is so close—only 8.6 light-years away—and partly because it is more than twice as massive as our sun and twenty-three times as bright.) To begin grasping the size of just our own galaxy, we must imagine all the rice grains in the Opera House spaced out on this scale.

Here's another way of trying to grasp these scales. A jumbo jet takes about four or five hours to cross a large continent like Australia or the continental United States. How long would it take the same jet to reach the Sun? (How many airline dinners would we have to eat before we got there?) In a Boeing 747 cruising at ca. 900 kilometers (550 miles) per hour, it would take us almost twenty years to reach the Sun, which is ca. 150 million kilometers (ca. 95 million miles) away. To reach our closest neighbor, Proxima Centauri, would take the same jumbo jet over 5 million years! This is the distance between next-door neighbors in a galactic city of 100 billion stars. To get a feeling for the size of the entire galaxy, the Milky Way, remember that it takes light only 8 minutes to reach Earth from the Sun, but would take about 4 years and 4 months to reach Proxima Centauri. The same light beam would have to travel for another 30,000 years, or 10,000 times the distance to Proxima Centauri, before it would reach the center of our galaxy.

Though rough-and-ready, these thought experiments may help us begin to imagine how large the universe is. They also suggest how small, in absolute terms, are the scales that normally concern us as human beings. On the scale of the universe, our sun and earth are infinitely small specks of matter.

These calculations suggest something else that is important for understanding human history. The placing of our earth within the universe is by no means random. We can exist only because we are in an atypical region. Most of space is empty and cold. Indeed, our thought experiments were concerned with a galaxy, a region of space that contains unusual amounts of matter. Beyond galaxies, matter is much less dense. Our earth exists within a region unusually rich in matter, in a large galaxy, in which supernovae have generated a broad variety of elements. Within that galaxy, we live in a region of star formation, close to a mature star. Even in the densest part of the galaxy, the disk, regions of empty space normally contain only about one atom in each cubic molecule. But in the earth's atmosphere, there may be 25 billion billion molecules in the same space.[15] And pouring through this matter is the energy emitted every second by the Sun. In other words, human history has taken place in a pocket of the universe that is dense in matter and packed with energy. It is the extraordinary richness and complexity of this environment that made life possible.

SUMMARY

After about 300,000 years, the early universe consisted mainly of huge clouds of hydrogen and helium. These contained the raw materials from which future stars and galaxies were to be created. About a billion years after the universe formed, the first stars appeared in regions where hydrogen and helium were more concentrated. Gravity pulled these dense clouds of gas into flat, rotating disks at many different scales. At the smallest scales were clouds of matter roughly the size of our solar system. As they collapsed, their centers heated up until the hydrogen at the center began to fuse into helium. These nuclear reactions released energy that prevented the center from collapsing further, and created the stable cores of stars. Stars burn hydrogen. Once that is used up, large stars can burn helium and even more complex elements up to iron, at which point further fusion demands more energy than is available. The largest stars burn their fuel fast and eventually collapse in the huge explosions known as supernovae. It is within supernovae that most of the more complex chemical elements are created. Smaller stars burn more slowly and at lower temperatures, they live longer, and they eventually cool more gradually, like cinders, when they have run out of fuel.

We live in a universe that has been made much more complex chemically by the life and death of stars. Indeed, the complex objects that dominate our earth and our history could not have existed in the much simpler environment of the early universe.

FURTHER READING

Ken Croswell's *Alchemy of the Heavens* (1996) is a good introduction to the life of stars, while Timothy Ferris's *Coming of Age in the Milky Way* (1988) is a superb history of modern astronomy. Armand Delsemme, *Our Cosmic Origins* (1998), is a good introduction, while Cesare Emiliani, *The Scientific Companion* (2nd ed., 1995), gives some technical details in a highly accessible form. The works of Isaac Asimov are very readable, but becoming dated. John Gribbin's *Genesis* (1981) is a superb popular history of the universe and our place within it, but cosmology is changing so rapidly that it, too, is getting dated. Martin Rees, *Just Six Numbers* (2000), and Lee Smolin, *Life of the Cosmos* (1998), give a sense of some of the more speculative ideas around in modern astronomy. In *Cosmic Evolution* (2001), Eric Chaisson offers a fascinating attempt to define the level of complexity we find in stars. Charles Lineweaver's short essay "Our Place in the Universe" (2002) provides a sense of the universe's "geography" and spatial scales.

3

ORIGINS AND HISTORY OF THE EARTH

The previous two chapters surveyed regions so vast that light can take billions of years to cross them and the stars they contain may be as numerous as grains of sand on a beach. Toward the end of chapter 2, we zoomed in on one region of a single galaxy, the Milky Way. In this chapter, we shift to a more intimate scale, that of a single star and one of its planets. On this tiny scale, we think of our local star as "the Sun"—and it seems to dominate our universe. So it is not surprising that many earthly religions regard the Sun as the supreme god. But the earth is where we live, and in many religions the earth is the mother and nurturer. The Greeks called her "Gaia."

Our earth, like the other planets and moons of our solar system, is a by-product of the creation of our sun. Gravity, though not the only active agent, dominates this story as it does the story of star formation in general. Our understanding of how the solar system was created has been revolutionized since the 1960s, through the use of satellites that have let us travel, at second-hand, through much of the solar system.

THE SOLAR SYSTEM

The planets of our solar system, including our own earth, were created at the same time as our sun, about 4.56 billion years ago. They are all about one-third of the age of the universe. Studies of the composition and motion of the Sun, the planets, and the moons and meteorites within the solar system, together with recent observations of planet formation around nearby stars, give us considerable confidence in modern explanations of how our solar system formed. But there remains some uncertainty about the details.

The Sun contains perhaps 99.9 percent of the matter in the solar system. What interests us now is the remaining 0.1 percent, for it is from that tiny

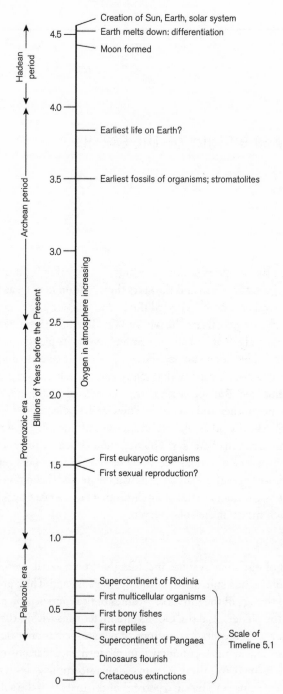

Timeline 3.1. The scale of the earth, the biosphere, and "Gaia": 4.5 billion years.

residue that all the planets, including our earth, were born. We have seen that as clouds of matter shrink, gravity tends to spin them and flatten them into disks. The solar nebula, the cloud of gas and dust from which our solar system formed, was no exception. As the Sun formed during a period of about 100,000 years, its gravitational pull drew most of the matter in the solar nebula to the center. But wisps of dust and gas, held at a distance by centrifugal forces, orbited the Sun like the rings around the large, gassy planets of Saturn, Jupiter, Uranus, and Neptune. We can be sure of this because in the late 1990s, astronomers managed for the first time to see similar disks around newly formed stars in our suburb of the Milky Way. The solar nebula consisted almost entirely of hydrogen and helium (ca. 98 percent of its mass), with other elements sprinkled through it in minuscule amounts.

As the Sun lit up, the inner rings of the solar nebula were heated more than the outer rings. This heat drove the more volatile (gassy) materials away from the inner regions. But farther out, from about the orbit of the future Jupiter on, it was cold enough for volatile gases to become liquids or solids. As a result, the inner orbits contained much more rocky material, while the more volatile materials accumulated farther from the Sun. This explains why the inner planets are rocky, while the outer planets (from Jupiter outward) are dominated by substances such as hydrogen and helium that are gases on earth. It also explains why the outer planets are so big: Jupiter contains more than 300 times the mass of Earth (though it is still only ca. $\frac{1}{1000}$ the size of the Sun), and Saturn almost 100 times Earth's mass. (Pluto, which is much smaller even than our moon, is no longer reckoned to be a true planet but rather the largest of the surviving planetesimals.) Water (ice) is the most common of all simple chemical compounds, as it is formed from the two most abundant reactive elements, hydrogen and oxygen. So, planets that formed at distances where water was normally a solid were bound to be larger than those formed in regions where water existed as a gas and could easily be driven away. The larger mass of the outer planets also made it easier for them to capture elements such as hydrogen and helium that remain gases even at extremely low temperatures. To this day, the solar system is divided into two main populations of planets: an inner ring of smallish, rocky planets, with densities exceeding 3 grams per cubic centimeter, and an outer ring of huge but less dense planets, with densities of less than 2 grams per cubic centimeter.

Though temperatures and materials varied from orbit to orbit, within each orbit particles of matter collided with each other or were drawn together by gravity. Sometimes, they stuck together, held by electrostatic forces— the same forces that enable a piece of rubbed amber to lift scraps of paper.

In a mechanism first guessed at by the German philosopher Kant in 1755, and known to astronomers as *accretion*, small, soft blobs of rock were formed from these gentle collisions. These formed lumps the size of snowballs, which grew into objects such as meteorites and then into planetesimals. Like dodgem cars, the planetesimals followed chaotic orbits and frequently collided with each other. As they grew larger, the collisions became more violent. Within 100,000 years, there existed many small planetesimals up to 10 kilometers in diameter. Modern comets such as Halley's comet are mostly survivors from this early stage of the solar system's history, and thus they help us imagine what some of these early planetesimals may have looked like. However, the comets that survive today mostly fell into more eccentric or more remote orbits, partly under the gravitational tug of the emerging superplanet, Jupiter. As a result, they escaped being incorporated into planets. Billions of comets still orbit beyond the outer planets in the Oort Cloud, which begins beyond Neptune, at more than 35 times Earth's distance from the Sun. Most are tiny, but some, such as Chiron, may be up to 200 kilometers in diameter.

About 100,000 years after the formation of the Sun, the newly formed sun blasted away the remaining gas and dust particles in the inner orbits, in what is known as the *T Tauri wind*. This is a phenomenon commonly associated with young stars. Presumably, the T Tauri wind also swept away any young atmosphere on the planetesimal that would eventually form the earth. What remained in the inner orbits were the solid planetesimals too large to be affected by the solar wind. Gradually, in all orbits, the largest planetesimals lured smaller objects into their gravitational nets, until the largest in each orbit had swept up most of the remaining material within its reach. In this way, perhaps within a million years of the Sun's formation, there emerged some thirty proto-planets, similar in size to our moon or to Mars; each dominated a particular orbit, and all circled in the plane of the original solar disk. During the next hundred million years, these consolidated into the planetary system we see today.

The inner planets (Mercury, Venus, Earth, Mars, and the asteroids) were made mainly from silicates (compounds of silicon and oxygen), as well as metals and trapped gases. Earth, for example, is made up of oxygen (almost 50 percent) and smaller amounts of iron (19 percent), silicon (14 percent), magnesium (12.5 percent), and many other elements of the periodic table. Between Mars and Jupiter, the asteroids may be the remnants of a "failed" rocky planet, whose formation was disrupted by the strong gravitational pull of nearby Jupiter. Jupiter, the largest of the planets, probably formed quite rapidly, perhaps 50 million or more years before Earth.[1] It is almost large enough for nuclear reactions to have begun in its center. It is almost, but not quite, a small star. If Jupiter had been slightly larger, then the solar system might have had two suns, and its structure and history would have been very different. The planets would have orbited in much less stable patterns, and it seems unlikely that life could have emerged on any of them.

The disks of matter that exist around all the larger planets (most spectacularly in the case of Saturn) show that all were large enough to have formed with their own nebulae, just like embryonic stars. In fact, Jupiter's nebula was so similar to that of the Sun that its inner moons, Io and Europa, are quite rocky, while its outer moons are more gassy, presumably because radiation from the early planet drove gassy elements away.

Is our solar system unique, or are solar systems quite common? Until recently, astronomers had no direct means of detecting the existence of planets around even the closest stars. It seemed that the solar system might be unusual, perhaps even singular. However, in 1995 astronomers for the first time demonstrated the existence of a planet orbiting another star, by careful measurements of slight wobbles in the star's motion. In the next six years, almost seventy other planets were identified in the same way. In May 1998, the Hubble Space Telescope took what seems to be the first photograph of a planet. It was huge—three times the size of Jupiter—and seems to have been ejected by a system of binary stars in the constellation of Taurus.[2] Astronomers have also photographed the accretion disks of embryonic solar systems. Such evidence suggests that solar systems may be quite common, though their exact structures may vary greatly. If only 10 percent of stars form with attendant planets, as recent evidence suggests, then even within our own galaxy there could be billions of stars with solar systems of some kind. This means that the astronomical niche in which we exist, though unusual on the scale of the universe, is not rare. Just within the Milky Way, there may be millions of planetary systems capable, in principle, of supporting life of some kind. Does this mean that life is common in the universe? We will return to this question below, and also in chapter 4, when we examine how life itself first appeared on Earth.

THE EARLY EARTH: MELTDOWN AND COOLING

Accretion was a chaotic and violent process, and it became more so as planetesimals grew in size and gravitational pull. Within each orbit, collisions between planetesimals generated immense heat and energy. How violent these processes were is suggested by the odd tilt and spin of many of the planets, which indicate that each of the planets was, like a billiard ball, struck at some stage by another large body of some kind. Visual evidence of these processes can be seen by looking at the surface of the Moon. Because the Moon has no atmosphere, its surface is not subject to erosion, so it retains the marks of its early history. And its face is deeply scarred by millions of meteoric impacts, as you can see on a clear night with simply a pair of binoculars. For perhaps a billion years, the earth's history was quite as violent, until the earth itself had swept up most of the remaining material in its orbit. The violence of this early, "Hadean" epoch of the earth's history explains why so little evidence survives from the period (see table A1, in appendix 1). After about a billion years, collisions became less frequent. Some planetesimals have survived, of course, to the present day. So collisions still occur, and some have played a crucial role in the earth's history. But such collisions are much rarer than they were in the Hadean epoch.

The early earth didn't have much of an atmosphere. Before it grew to full size, its gravitational pull was insufficient to prevent gases from drifting off into space, while the solar wind had already driven away much of the gaseous material from the inner orbits of the solar system. So we must imagine the early earth as a mixture of rocky materials, metals, and trapped gases, subjected to constant bombardment by smaller planetesimals and without much of an atmosphere. The early earth would indeed have seemed a hellish place to humans.

As it began to reach full size, the earth heated up, partly because of collisions with other planetesimals and partly because of increasing internal pressures as it grew in size. In addition, the early solar system contained abundant radioactive materials, which may have formed in the supernova explosion that occurred not long before the creation of the Sun. Much of this heat has been retained to the present day, though over time, much has also leaked to the surface from the earth's well-insulated core. As the earth heated up, its interior melted. Within the molten interior, different elements were sorted out by density, in a process known as *differentiation*. By about 40 million years after the formation of the solar system, most of the heavier metallic elements in the early earth, such as iron and nickel, had sunk through the hot sludge to the center, giving the earth a core dominated by iron. This metallic core gives the earth its characteristic magnetic field. The earth's magnetic field has played an extremely important role in the history of our planet: by deflecting the many high-energy particles streaming through space, it shielded the delicate chemical processes that eventually generated life here.

As heavy materials headed for the center of the earth, lighter silicates drifted upward, in a process analogous to those in a modern blast furnace. The denser silicates formed the earth's mantle, a region almost 3,000 kilometers thick between the core and the crust. With the help of the cometary bombardment, whose many impacts scarred and heated the earth's surface, the lightest silicates rose to the surface, where they cooled more rapidly than the better-insulated materials in the earth's interior. These lighter materials, such as the rocks we call *granites*, formed a layer of continental crust about 35 kilometers thick. Relative to the earth as a whole, this is as thin as an eggshell. Seafloor crust (formed largely from volcanic basalts) is even thinner, at about 7 kilometers. From the surface of the earth to its core is a distance of almost 6,400 kilometers; thus, even continental crust reaches only about $\frac{1}{200}$ of the way to the earth's core. Much of the early continental crust has remained on the earth's surface to the present day. The oldest portions of continental crust, found today in parts of Canada, Australia, South Africa, and Greenland, appear to be about 3.8 billion years old.

The lightest materials of all, including gases such as hydrogen and helium, bubbled through the earth's interior to the surface. So we can imagine the surface of the early earth as a massive volcanic field. And we can judge pretty well what gases bubbled up to that surface by analyzing the mixture of gases emitted by volcanoes today. These include hydrogen (H), helium (He), methane (CH_4), water vapor (H_2O), nitrogen (N), ammonia (NH_3), and hydrogen sulfide (H_2S). Other materials, including large amounts of water vapor, were brought in by cometary bombardments. Much of the hydrogen and helium escaped; but once the earth was fully formed, it was large enough for its gravitational field to hold most of the remaining gases, and these formed the earth's first stable atmosphere. Much of the methane and hydrogen sulfide was converted into carbon dioxide (CO_2), which soon dominated the early atmosphere. In a carbon dioxide atmosphere, the sky would have seemed red rather than blue to us. However, as the earth cooled, water vapor that had accumulated in its atmosphere fell in torrential rains lasting millions of years. These downpours created the earliest oceans. The first oceans must have existed before 3.5 billion years ago, for we know that living organisms existed by that date; their presence suggests that temperatures at the earth's surface had fallen below 100°C. The early oceans dissolved much of the carbon dioxide in the atmosphere. To a human eye the sky would have seemed to gradually turn blue.

The fact that water exists in liquid form on the earth's surface is of fundamental importance to us, for it means that Earth's temperatures were suitable for the appearance of the complex and fragile molecules that made up the earliest life forms. Why Earth's temperatures are so benign for life remains uncertain. Perhaps in all solar systems there is a narrow band—just far enough from the local sun not to boil, but close enough to get some warmth—within which life can emerge. Yet we know that atmospheres do not evolve according to simple and predictable rules. On Venus, the early atmosphere may have been similar to Earth's; but a thick blanket of clouds and the greater amounts of sunlight it received caused a runaway greenhouse effect, with the result that the surface of Venus is now hot enough to boil lead. Venus has in effect been sterilized. Because it is smaller and has a weaker gravitational pull, Mars has hardly any atmosphere today, though it may have had more in the past. Perhaps Earth proved suitable for life because of a rare combination of circumstances, suggesting that even if the universe contains billions of planets, few may be hospitable to life.[3] However, as we will see in chapter 5, once life formed, living organisms started making themselves at home, shaping the atmosphere and surface of the earth to make it more suitable for life.

Many of the ingredients of the early atmosphere (including much of its water), together with many of the organic chemicals that formed the first life forms, may have been brought to Earth by the comets that bombarded its surface for the first billion years of its life.[4] This constant bombardment may also explain the creation of the Moon, perhaps 50 to 100 million years after the formation of the solar system. Study of moon rocks has shown that this satellite is less dense than the earth, and contains much less iron. The discrepancy can be explained by supposing that the earth was struck a glancing blow by a proto-planet, perhaps the size of Mars, after the process of differentiation had been completed. The collision would have gouged out material from the mantle and crust of the early earth, but not from its iron-rich core. This debris would then have orbited the earth, like Saturn's rings, until eventually gathering into a single object by accretion, to create the Moon.

So, about one billion years after the formation of the solar system, the early earth had a moon; a hot iron core; a hot, semiliquid interior, the mantle; a thin but solid crust; seas; and an atmosphere dominated by nitrogen, carbon dioxide, and water vapor. To us it would have seemed a hot, dangerous, and unpleasant place, drowned in a constant acidic rain and covered, periodically, by huge seas of lava caused by collisions with comets or asteroids. But it contained all the ingredients necessary for the earliest forms of life to evolve and thrive. Above all, it contained water, because it was far enough from the Sun for water vapor to liquefy, but close enough to prevent it all from freezing.

EVIDENCE ABOUT THE EARLY EARTH

How can we know so much about the early earth? There is, of course, an element of speculation in the account I have given, but it is also based on a lot of hard data. Two types of information are so important that they are worth discussing in more detail.

We can bore only a tiny way into the earth; to study its deep interior, we must use indirect methods. Fortunately, methods of describing the earth's interior have developed as a by-product of the study of earthquakes. Geologists study earthquakes by using seismographs, which are instruments that measure sudden vibrations of the earth caused by violent shocks. By placing seismographs at different parts of the earth's surface, they can map these vibrations with great accuracy, defining their point of origin and their power and form. It is also possible to map how these vibrations travel through the earth's interior. These exercises have shown that waves travel in different ways through different types of material; from this evidence, it is possible to map out the different layers of which the earth is composed (see figure 3.1).

Even more remarkable are the techniques that enable us to attach absolute dates to events many millions or even billions of years in the past. Indeed, this capacity to offer precise absolute dates for events in the remote past—including those in the early history of the earth—is one of the most extraordinary features of the modern creation myth (see appendix 1).

In the past, remote dates had to be estimated using whatever techniques were at hand.[5] Genealogical records provided one of the most important of all ways of dating the past. In seventeenth-century Europe, biblical scholars used the genealogical lists in the Old Testament to calculate when God made the world. In the late eighteenth century, geologists learned how to determine relative dates for many major geological events in the distant past by studying the different layers in which particular fossils or rock types were found. Relative dates cannot quite tell us when an animal lived or a particular rock was laid down, but they can tell us the order in which events occurred. Paleontologists became connoisseurs of particular fossils that could be used as precise markers of relative ages. In the hands of an expert, a particular type of trilobite or the odd, saw-toothed marks left by ancient creatures known as graptolites can prove that rocks from quite different parts of the world were laid down at about the same time. Such methods were used to construct the earliest versions of the geological timescale, which tells us the rough order in which different rock layers and different types of organisms appeared (see Table A1). By the nineteenth century, even these rough techniques were suggesting that the earth had to be much more than 6,000 years old. However, most scientists believed the earth had existed for at most a few hundred million years.

Relative dating has become more and more refined, and remains a pow-

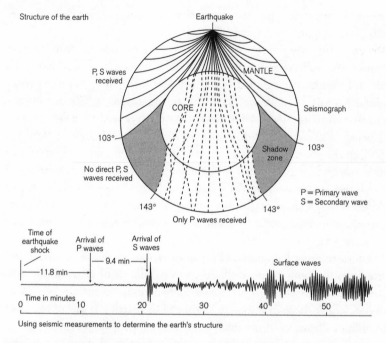

Using seismic measurements to determine the earth's structure

Figure 3.1.   The structure of the earth's interior. We do not yet have the ability
to penetrate deep within the earth. But we can use *seismic waves,* the vibrations
generated by earthquakes, to determine what's there. There are three kinds of
seismic waves: primary waves, secondary waves, and surface waves. Each moves
at a different speed and is affected in different ways by the material through which
it passes. So, by analyzing the speed of arrival of different types of waves at different points on the earth's surface, it is possible to tell a lot about the internal
structure of the earth. The graph shows the seismographic record of the earthquake at the top of the upper diagram, as recorded by the seismograph. Adapted
from Cesare Emiliani, *The Scientific Companion: Exploring the Physical World
with Facts, Figures, and Formulas,* 2nd ed. (New York: John Wiley, 1995), p. 174:
adapted from Arthur N. Strahler, *The Earth Sciences,* 2nd ed. (New York: Harper
and Row, 1971), p. 397, fig. 23.22; p. 395, fig 23.17.

erful method for dating rocks. But the most important revolution in dating
occurred in the twentieth century with the emergence of what are known
as *radiometric* dating techniques. In many situations, these enable us to tell,
with surprising precision, exactly when a particular object was formed. So,
using these methods, we can determine absolute as well as relative dates for
many events that occurred long before humans existed. Radiometric dating
techniques are described in more detail in appendix 1.

The dates we use to construct the modern account of the earth's creation are based mainly on the analysis of material still drifting through the solar system. Material on the earth's surface, or even deep within the earth, has been recycled so often that it can tell us little about the earliest stages of the earth's formation. The oldest datable rocks on Earth (from Greenland) are about 3.8 billion years old, which is perhaps 800 million years after the formation of the planet. To find out when the earth and the solar system were formed, we must use materials that have remained unchanged since the early days of the solar system's life. Meteorites (particularly the type known as *chondrites*) fit the bill nicely, as they seem to consist of debris from the solar nebula within which our solar system formed. This means that they formed early in the history of our solar system; and they have changed little since their creation. It is thus not surprising that radiometric dating techniques regularly yield dates of about 4.56 billion years for meteorites. The oldest moon rocks yield similar dates. The closeness of these dates and the absence of anything older in the solar system suggest that the solar system itself was created about 4.56 billion years ago.

THE ORIGINS OF MODERN GEOLOGY

How did the hot, early earth evolve into today's earth, with its blue skies, its oxygen-rich atmosphere, its mountains, continents, and oceans?

Before the 1960s, geography and geology were already well-developed fields of study, and they had accumulated much hard evidence about the ways in which landforms and oceans were constructed. But they lacked a central, organizing idea that could help explain the transition from the hostile early earth to the earth we know today. In the late 1960s, with the appearance and widespread acceptance of the theory of plate tectonics, earth sciences acquired a central idea or paradigm as powerful as that of the big bang theory in modern astronomy. Then, for the first time, it became possible to tell a coherent, scientific story of the earth's history.

The modern tradition of geology evolved in Europe, so it was greatly influenced by the creation myths of Christianity. But as we have seen, the belief that the earth had been created by God about 6,000 years ago began to be threatened as early as the seventeenth century. A Danish scientist, Nicholas Steno, first argued that fossils were the remains of organisms that had once lived on earth. He also argued that mountains were built up over long periods by familiar geological processes such as volcanic activity. These claims had significant consequences. For example, they suggested that fishlike fossils found high in the Alps might indeed be the remains of ancient forms of fish. Any nonmiraculous explanation of such facts had to suppose that the Alps had risen up from lands that had once been under water. And it was hard to imagine how such processes could have been compressed into a mere 6,000 years without imagining a series of catastrophic events in between. Some geologists, taking the biblical flood as their model, did indeed argue that the earth's history had included many catastrophic events. And such theories made it possible, at least in some circles, to defend the biblical chronology until the nineteenth century.

But geologists became more and more skeptical. In the eighteenth century, some began systematically to map different layers of rock. The nineteenth-century geographer Charles Lyell first stated clearly what came to be known as the principle of *uniformitarianism*. This was the principle, already hinted at by Steno, that the earth was created not in a series of catastrophes but over huge periods of time by the same slow geological forces that we see at work today. These included processes such as vulcanism (volcanic activity), which could raise land above its existing levels, and erosion, which slowly swept the material of highlands down to the lowlands and eventually into the sea. Lyell argued that most features of the modern world could be explained as the result of these opposite processes, the one building mountains up while the other tended to wear them down. And in a fundamental work, *Principles of Geology* (1830), he drew out the clear implication of this theory: the earth had existed for millions rather than thousands of years.

By the late nineteenth century, the conventional wisdom was that the earth had existed for at least 20 and perhaps as many as 100 million years. These figures were estimated by William Thompson (Lord Kelvin), by assuming that Earth and the Sun had once been molten balls of matter, which had gradually cooled. On this reading, the crucial factor in the earth's history was its gradual cooling over millions of years. As the earth cooled, the present configuration of the lands and seas had emerged, shaped by vulcanism and erosion. Not until the discovery of radioactivity early in the twentieth century, and the discovery by Marie Curie that radioactive materials produce heat, was it realized that the Sun and Earth might have sources of heat within themselves. This suggested that they were cooling much more slowly than Lord Kelvin had imagined, and were probably much older than his influential estimate.

WEGENER AND THE MODERN THEORY OF PLATE TECTONICS

In the meantime, an odd observation, first made in the seventeenth century, had prompted a number of thinkers to suggest a rather different way of describing the earth's history. The first modern maps of the world were produced in the century after Europeans began to travel to the Americas and the Pacific. As the English philosopher Francis Bacon pointed out in 1620, it was easy to see from these maps that the continents looked like pieces of a jigsaw puzzle. This similarity was most striking when the west coast of Africa was matched up with the east coast of South America. With a little imagination it was possible to suppose that at one time all the continents had fitted together. What could explain this odd fit?

The idea that the continents really had drifted apart was given a thorough scientific basis in a book called *The Origin of Continents and Oceans*, written in 1915 by a German geographer, Alfred Wegener. Wegener assembled a huge amount of evidence suggesting that at one time the continents had been joined together. He showed that the fit between the continents was much more impressive if, instead of matching them at their present-day water lines, he matched them at their continental shelves. Further, he showed that many modern-day geological features seemed to continue from one continent to another. For example, he described a series of rock formations, known as the *Gondwana sequence,* all formed, apparently, by glacial activity. The sequence reached from the north of Africa, through to West Africa, then to South America, through Antarctica, and into Australia. Wegener argued that these features had been laid down as each region had moved over the South Pole. In other words, the continents had not always been fixed in their present positions, but had, as it were, "drifted" across the surface of the earth. As a result, Wegener's idea came to be known as the hypothesis of *continental drift.*

Wegener's evidence was impressive, but he could not explain *how* blocks of land the size of Africa or Asia or the Americas could have moved across the surface of the earth. Partly because of this, in 1928 his theory was officially rejected by the influential American Association of Petroleum Geologists. For the next forty years, most geologists regarded his theory as no more than an interesting hypothesis, and they looked for more conventional explanations of the anomalies Wegener had explored. It was not until after the Second World War that it became possible to explain how and why the continents might move across the face of the earth. But once such an explanation was available, Wegener's ideas became respectable again. Indeed, with modern additions, they now form the central organizing idea of modern geology: the theory of plate tectonics.

The modern theory of plate tectonics originated from technologies developed during the Second World War. New forms of warfare encouraged the development of sonar to detect submarines. But sonar also made it possible to map the seafloor more thoroughly than ever before. As oceanographers began to examine the bottom of the sea in detail, some strange features emerged. One was a chain of high subterranean mountains that ran through the center of the Atlantic and through other seas as well. At the center of these suboceanic ridges were lines of volcanoes, from which lava seeped out onto the neighboring seabed.

Studies of the magnetic fields on the seabed near the suboceanic ridges revealed a further oddity. While rocks close to the ridges generally had a normal magnetic orientation, bands farther away often had a polarity opposite to that of the modern earth, with their north pole to the south and vice versa. Farther out, the polarity was reversed again, and so on, creating a series of bands with alternating magnetic polarity. Geologists eventually realized that the polarity of the earth itself seems to switch every few hundred thousand years, and this suggested that the different bands had been laid down in different periods. As other, more precise dating techniques were applied to the seafloor, it became clear that the youngest seafloor lay closest to the midocean ridges, while bands farther away became successively older. The oldest areas of seafloor were those farthest from the midocean ridges. These turned out to be at most about 200 million years old—much younger than the oldest parts of the continental crust, some of which are almost 4 billion years old.

In the 1960s, starting with the work of an American geologist, Harry Hess, a coherent explanation of all these anomalies began to emerge. Lava, seeping up through cracks that ran through most of the major ocean systems, was creating new seafloor. Such regions are known as *spreading margins*. As new oceanic crust was formed, it reared up in huge ridges of basalt, but it also acted like a wedge, driving apart seafloor that already existed. As a result, some oceans, such as the Atlantic, appeared to be widening. Modern satellite observations have shown that the Atlantic is getting about 3 centimeters wider every year; it's growing at about the same rate as our fingernails. This suggests that the Atlantic Ocean was born about 150 million years ago, as parts of what is now North America began to split away from what is now West Eurasia.

This evidence did not mean that the earth was expanding, for geologists also realized that there were areas of the earth, such as the western coast of South America, where seafloor was being sucked back into the interior. These are known as *subduction margins*. Here, tectonic plates collide, pushed together by seafloor spreading elsewhere in the world and jamming seafloor crust up against plates of continental crust. Oceanic crust, which consists mainly of volcanic basalts, is heavier than the granitic material that dominates continental crust. So, when an oceanic plate collides with a continental plate, the lighter continental plate usually rides over the oceanic plate. The oceanic plate dives beneath the continental crust and is eventually pulled down into the interior. (This constant recycling explains why oceanic crust is normally so much younger than continental crust.) Slabs of descending oceanic crust grind against the continental plates above them as well as the material below them, creating enormous heat and pressure. In South America, this heat, combined with the motions of both oceanic and continental crust, generates the volcanic activity that has created the Andes.

In some areas, regions of continental crust are forced together in what are known as *collision margins*. The most striking example is in northern India, where the plate that contains the Indian subcontinent has been forced up against the Asian plate. In such regions, both plates buckle up and huge mountain ranges (here, the Himalayas) are formed. Finally, there are regions in which plates seem to slide past each other, such as the San Andreas fault in California. Most plate movements cause earthquakes, because the friction between plates and the material beneath them ensures that plate movements are rarely smooth: they normally come as sudden slippages after a prolonged buildup of pressure. So in principle it is possible to map the edge of the various tectonic plates by mapping the regions of most intense earthquake activity.

Detailed mapping of regions where different portions of crust meet has shown that the uppermost layer of the earth (the lithosphere) consists of a number of rigid plates, like a cracked eggshell. There are eight large plates and seven smaller ones, as well as smaller slivers of material. These move over a layer of softer materials just below them, the asthenosphere, which is between 100 and 200 kilometers thick. The plates are driven by movements in the asthenosphere and also by the pressure of materials squeezed up from even deeper in the earth through the cracks between (and sometimes within) the plates. Like the scum on the surface of a slowly cooking soup, the more rigid material of the plates buckles, cracks, and moves because of the currents of softer, hotter, and more malleable materials underneath. In other words, it is the heat of the earth's interior that provides the power needed to move great plates of matter about the surface of the earth. That heat, in its turn, is generated largely by radioactive materials within the earth, which had been formed in the supernova explosion that occurred just before the creation of our solar system. Here was the geological motor that Wegener was unable to find: he could not possibly have anticipated that the continents were being pushed around the earth by the remnant energy from a supernova that exploded more than 4.6 billion years ago. And that takes us back, once again, to gravity, for it was gravitational forces that first constructed and then destroyed the star that died in that supernova explosion.

The theory of plate tectonics provided a unifying idea for many different aspects of geology. It can help explain mountain building, volcanic activity, and the many geological anomalies explored by geographers such as Wegener. And it shows how, in principle, it might be possible to construct a history of the earth's surface, showing what the surface of the earth looked like at different stages of its history. Meanwhile, the use of more accurate mapping systems, such as GPS (the Global Positioning System), has made it possible to measure the movement of tectonic plates with great accuracy.

A SHORT HISTORY OF THE EARTH AND THE ATMOSPHERE

The theory of plate tectonics, combined with what we know of the formation of the earth, means that today we can offer a reasonably coherent history of our earth.

The Hadean phase of the earth's history lasted from the planet's formation, 4.56 billion years ago, until about 600 million years later.[6] During that period, the earth's surface was hot, volcanic, and unstable. It was also subjected to a constant bombardment from comets and other surviving planetesimals.

By 3.8 billion years ago, at the beginning of what geologists call the *Archean era*, we know that continental crust had appeared on the earth's surface, because some crust as old as this is still present. It is also probable that seas existed. The earth's atmosphere was probably dominated by gases such as carbon dioxide, nitrogen, and hydrogen sulfide, much of it brought in by comets. There was little free oxygen, for oxygen is highly reactive and therefore combined with other elements to form chemical compounds. The earliest portions of continental crust may have moved, but we cannot be sure that plate tectonics worked in exactly the same way as today. With an atmosphere, and plenty of water about, processes of erosion and surface change probably occurred as fast as they do today. Rapid erosion and constant bombardment explain why the early earth's surface was made over many times, a process that left few traces for us today; our knowledge of the very earliest stages of the earth's history thus remains sketchy.

The earliest fragments of continental crust probably formed tiny and short-lived micro-continents. These may have been surrounded by seas with many small volcanic islands, as well as subterranean volcanoes. By about 3 billion years ago, some of these micro-continents must have fused into larger plates, for plates this old can be found at the heart of modern continents, including Africa, North America, and parts of Australia. But it is only for the last half billion years that we can actually begin to reconstruct the arrangement of these plates on the surface of the earth.

Modern geology has built up an increasingly sophisticated picture of tectonic movements during the past few hundred million years. These movements have been discovered largely by studying the magnetic orientation of modern rocks whose ages are known. From this, it is possible to estimate roughly where these rocks were when they first formed. Such studies seem to reveal a simple pattern of dispersal and convergence. About 250 million years ago, most of the continental plates were joined into a supercontinent, which Wegener had christened "Pangaea." It was surrounded by a single, large sea, known as Panthalassa. By about 200 million years ago, Pangaea began breaking up into two large continents. Laurasia, in the north, contained most of modern Asia, Europe, and North America; Gondwanaland, in the south, contained most of modern South America, Antarctica, Africa, Australia, and India. Then, both Laurasia and Gondwanaland began to fragment. Now, we may be in the early stages of a reconvergence, as Africa and India move north to join Eurasia. Recent evidence suggests that some 500 million years before the existence of Pangaea, there existed an even earlier supercontinent, now known as Rodinia.[7] But at present, this is as far back as we can trace modern processes of plate tectonics (see map 3.1).

This history is a vital part of the modern creation myth, for the exact configuration of continents and seas at different eras of the earth's history has played a crucial role in the way that life-forms evolved and atmospheres and climates operated, as we will see in chapter 5. In these and other ways, the history of the earth shaped the evolution of living organisms. The next two chapters will explore how living organisms made a home for themselves on the changing earth, and how the earth itself changed as it was clothed in a thin membrane of life.

SUMMARY

The Sun and the solar system were created at the same time, about 4.56 billion years ago, during the gravitational collapse of a cloud of matter. The Sun formed at the center of this cloud and absorbed most of the material it contained. However, the sprinkling of matter left outside the Sun orbited the new star in a flat disk. Within each orbit, clumps of matter formed through collisions and gravitational attraction, until eventually there appeared a single planetary body in each orbit. Because the more volatile elements had been driven away from the central regions by the solar wind, the inner planets tended to be more rocky and the outer planets more gaseous.

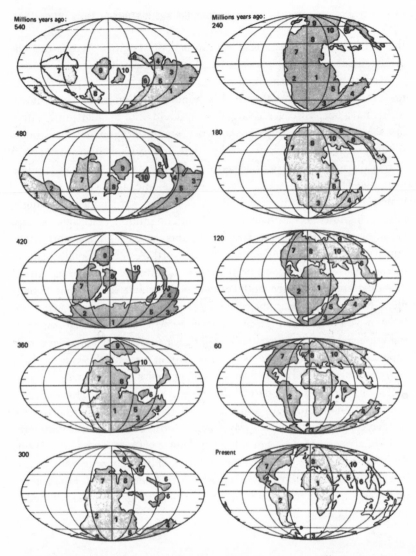

Map 3.1. The changing earth: tectonic movements over 540 million years. From Cesare Emiliani, *The Scientific Companion: Exploring the Physical World with Facts, Figures, and Formulas*, 2nd ed. (New York: John Wiley, 1995), p. 190: from Emiliani, *Dictionary of the Physical Sciences: Terms, Formulas, Data* (Oxford: Oxford University Press, 1987), p. 48, reproduced with permission of Oxford University Press, Oxford, England.

Soon after it formed, the early earth melted; heavy materials sank to the core, and lighter materials rose to the surface. By about 4 billion years ago, the earth's interior structure was similar to its structure today. However, the earth's surface and its atmosphere underwent a long process of change before they were as we observe them today. Since the emergence of the theory of plate tectonics in the 1960s, it has become clear that the continental plates have moved slowly over the earth's surface, gradually changing the configuration of continents and seas.

FURTHER READING

There are a number of good histories of the earth, including Peter Cattermole and Patrick Moore, *The Story of the Earth* (1986), and J. D. Macdougall, *A Short History of Planet Earth* (1996). The books by Preston Cloud (*Cosmos, Earth, and Man* [1978]; *Oasis in Space* [1988]) are classics, though some of their details may now be dated. Armand Delsemme, *Our Cosmic Origins* (1998), and Cesare Emiliani, *The Scientific Companion* (2nd ed., 1995), summarize many of the more technical details of this story, while Steven Stanley's *Earth and Life through Time* (1986) shows how closely linked are the histories of the earth and of life on earth. James Lovelock's several books on the Gaia hypothesis also describe the history of the earth and of life on earth as two closely intertwined stories. Isaac Asimov offers readable summaries, though some are beginning to date. Ross Taylor's short essay "The Solar System: An Environment for Life?" (2002) conveys well how contingent events made each planet unique.

PART II

LIFE ON EARTH

4

THE ORIGINS OF LIFE AND THE THEORY OF EVOLUTION

LIFE: A NEW LEVEL OF COMPLEXITY

"What is life?" The physicist Erwin Schrödinger asked this question in a famous series of lectures given in Dublin in 1943. Schrödinger's answer was remarkably prescient, for he wrote before we had any real understanding of the genetic basis of life. He argued that we should be able to explain life as scientifically as we can explain physics or chemistry. But he also understood that we cannot define life simply by referring to a checklist. Like all complex entities, living organisms manage significant flows of energy and matter, so they must have some form of metabolism. They take in and excrete energy and nutrients. They also reproduce, again like many other complex but nonliving entities from tornadoes to crystals. So neither metabolism nor reproduction alone can provide a satisfactory definition of life; it is the way they work together to create a new level of complexity that is critical. Schrödinger therefore suggested another way of defining what is distinctive about life. Life is not just complex—it is significantly *more* complex than anything else in the universe; and the level of orderliness achieved by living organisms is remarkable, given the general tendency of the universe toward disorder. "The unfolding of events in the life cycle of an organism exhibits an admirable regularity and orderliness, unrivalled by anything we meet with in inanimate matter."[1]

Stars can climb the thermodynamic down escalator (see appendix 2), but living organisms climb it with greater agility. Indeed, Eric Chaisson has argued that the level of complexity achieved by living organisms can be measured, roughly but quite objectively, by estimating the density of the energy flows that sustain them against the destructive pressure of the second

law of thermodynamics.[2] Table 4.1 gives Chaisson's approximations of these
energy flows. The right-hand column, which measures the amount of free
energy passing through a given mass in a given amount of time, appears to
indicate that living organisms can handle far denser flows of energy than
stars without breaking down. And this ability is what lets them climb farther and faster up the thermodynamic down escalator. In Schrödinger's famous phrase, each living organism seems to have an astonishing capacity
for "continually sucking orderliness from its environment."[3] The simplest
structures have also been around longest, and the more complex structures
have appeared more recently, which suggests that creating them was a more
difficult evolutionary task. Finally, it is also clear that the more complex entities in the bottom half of the table are more fragile. While stars and planets may live for many billions of years, even the longest-living organisms
(at least those we know of) can live for only a few thousand years, and most
live for only a few days or years. That the most complex structures break
down so fast is a measure of the difficulty of managing particularly dense
energy flows: this is the price living organisms pay for their aggressive challenge to the second law of thermodynamics. Thus in dealing with life we
are dealing with a new level of order and complexity, a new capacity to control and organize free energy that is achieved at the price of greater fragility.
As Martin Rees has written, "A star is simpler than an insect."[4] But a star
also lives much longer.

Chemical processes may have generated life elsewhere in the universe,
though at present we do not know if this is true. But we do know that life
appeared on Earth within 600 million years of the planet's creation. By geological standards, and given the harsh conditions of the early earth, that
was quick. And from the moment life appeared, living organisms have multiplied and changed in a dazzling and apparently endless cascade of new life-
forms, each finely tuned to handle the particular energies and resources in
its immediate environment. Unlike stars or crystals, which are general, all-
purpose antientropy machines, living organisms can adapt constantly to new
terrain and new challenges in their more flexible guerrilla war on entropy.
Collectively, living organisms explore their environment in ways that have
no parallel in the inanimate world. And what they find is new sources of
energy and new ways of organizing themselves so as to survive the hurricane of energy flowing through them. Not all of these changes lead to greater
complexity, but some do. This is why life has such an astonishing capacity
to conjure up new types of complexity.

What is the source of the energy differential that sustains these complex
entities? The answer to that question is blessedly clear: the ultimate source

TABLE 4.1. SOME ESTIMATED FREE ENERGY RATE DENSITIES

Generic Structure	Free Energy Rate Density $(erg\ s^{-1}\ g^{-1})$
Galaxies (e.g., Milky Way)	1
Stars (e.g., Sun)	2
Planets (e.g., Earth)	75
Plants (biosphere)	900
Animals (e.g., human body)	20,000
Brains (e.g., human cranium)	150,000
Society (e.g., modern human culture)	500,000

SOURCE: Eric J. Chaisson, *Cosmic Evolution: The Rise of Complexity in Nature* (Cambridge, Mass.: Harvard University Press, 2001), p. 139.

is gravity. We have seen that gravity can create stars—objects of great density and high temperatures. But the universe as a whole is extremely cold; its average temperature is about 3°C above absolute zero, the temperature of the cosmic background radiation. So stars are embedded within the cold universe like billions of tiny points of light and heat. It is no accident that we live huddled near a star, for here we can tap the huge energy flows pouring into space from the nuclear furnace at the Sun's core.

As for the *rules* of complexity in living organisms, these are different from those that dominate at the astronomical scale. Individual organisms (at least as we know them) flourish at much smaller scales than stars or planets. At those smaller scales, gravity counts; but other forces count for more. Life is shaped largely by electromagnetism and the nuclear forces that control how atoms work. These are the forces that determine how atoms are assembled and how they combine into larger and more complex molecules.

But at the biological level of complexity, new rules appear as well. Living organisms operate according to distinctive and more open-ended rules of change, which are superimposed on the simpler and more deterministic rules of physics and chemistry. The rules of biology are made possible by the high degree of precision with which living organisms reproduce. Handling large energy flows is such a delicate task that it requires extremely · precise mechanisms; the rule book for creating and re-creating such structures has to be complex, exact, and accurate. A system of reproduction that could copy these mechanisms only approximately would soon lose the required exactness (though one that copied them perfectly would rule out any possibility of change). Thus, a high level of metabolic precision requires a high (but not perfect) level of reproductive precision. This is why large organisms like ourselves need more genetic information than do bacteria. It is also why most of those studying the origins of life have concentrated on the origins of the genetic code, the intricate molecular "software" that explains why living organisms reproduce with a precision not matched by any other complex entities.

All in all, the shift from chemistry to life counts as one of the great transitions in the history of the universe. Complex organisms, replicating according to new and exquisitely precise blueprints, introduced new types of historical change—certainly on this planet, and perhaps in many other parts of the universe. As chemicals combine to form living organisms, emergent properties appear that we cannot explain simply by studying the chemicals from which organisms are constructed. So, to understand living things, we need a new paradigm, one that takes us beyond the rules of nuclear physics, chemistry, or geology and into the realm of biology. This chapter will discuss the basic rules by which living organisms change, and some current ideas about the origins of life on Earth. It will focus on the ideas of Charles Darwin, who first described clearly what is distinctive about the biological rules of change. The next chapter will survey the history of life on earth.

DARWIN AND THE THEORY OF EVOLUTION

Many societies have tried to explain life by assuming that there is a creator spirit or god who somehow breathed life into inanimate matter. Modern science regards this explanation as the easy way out, because theories that depend on deities can be made to explain more or less anything and are not subject to objective verification. Instead, modern science tries to explain the creation of life as a consequence of inanimate forces and processes, just as it approaches the creation of our sun and earth.

The fundamental idea used in modern biology to explain both the development and the origins of life is that of "evolution" by "natural selection." The theory was first presented systematically in Charles Darwin's book On the Origin of Species, which appeared in 1859.[5] Darwin rarely used the term evolution, perhaps because it seems to imply some sort of mystical force that drives biological change in particular directions and thus would contradict his own view of biological change as a more open-ended process. However, Herbert Spencer, who did the most to popularize the term, saw biological change as a movement from "lower" to "higher" life-forms, as a form of progress. This is unfortunate, because such an approach introduces arbitrary and subjective value judgments into our understanding of the history of living organisms. But despite the associations carried by *evolution*, we will continue to use the term simply because it remains the word most commonly applied to Darwin's theories of biological change.

Darwin argued that species are not fixed entities. They are constantly changing, and the way they change is governed by some simple rules. A species is a large collection of individual organisms that are similar enough to interbreed, but are not quite identical to each other. Species are defined by those features that individuals share rather than by the minor ways in which they differ from each other. However, over long periods of time, random variations in the features of individuals may cause the average features of an entire species to alter. Its average height might change, for example, or the average size of individual brains may grow. Such minor changes, accumulating over thousands of generations, must eventually transform the average features of the entire species. To understand how species change, we therefore must understand how and why the features of some individuals become more common, while those of others dwindle and disappear.

Darwin knew that in most populations only a minority of individuals survive to adulthood and produce offspring. Yet the future of the species can be shaped only by those individuals that do survive and reproduce. So later generations of life are the offspring only of the survivors. (Evolution, like history, apparently is written by the winners.) But what determines which individuals reproduce, and which do not? Pure chance may play a role here, of course. But in the long run, he argued, the individuals most likely to survive and reproduce are those that have had the good luck to inherit from their parents features that make them slightly better adapted to their environment. They will then pass these same features on to most of their offspring. Over time, these features will become more and more common because those individuals that do *not* possess them will produce fewer healthy offspring, until their lineages die out. Over thousands of generations, many small changes of this kind will ensure that the species as a whole appears to change or evolve in ways that make it better adapted to its environment. In this metaphorical sense, we can say that the environment "naturally" selects certain features and discards others, just as animal breeders "artificially" select some individuals to breed and not others. And it is in this metaphorical sense that species appear over time to "adapt" to their natural environments.

Adaptation is such an important notion in modern biology that it is worth defining more carefully. It refers to the fact that all living organisms seem to be exquisitely fitted for the environments in which they live. Indeed, so perfect is the fit between organisms and their environments that many of

Darwin's opponents argued, as some still argue, that organs such as the human eye or the elephant's trunk must have been designed by a benign creator. Darwin tried to show that blind processes could do the job equally well. Adaptation helps explain the great variety of living organisms, for there is a huge variation in environments to which organisms can adapt themselves. To describe these different environments, biologists and ecologists use the notions of *habitats* and *niches*. Habitats are simply the geographical environments in which species live. The idea of a niche is more complex, as it includes the *way* they live as well. The word *niche* is derived from the Latin word meaning "nest." In architecture, a niche is a recess or alcove in a wall in which a statue or other object can be placed. In biology and ecology, a niche is the particular way of living for which an organism seems to have been sculpted or adapted by evolutionary processes. The niche of a woodpecker is defined by the way it finds edible insects in certain trees; many single-celled bacteria find attractive niches in the guts of larger animals, including ourselves. But of course environments also change—and as they do so, while old niches may close down, new niches may open up elsewhere. As environments are varied and changeable, organisms have to keep adapting if they are to survive. This is why evolution never ends. Because there is no fixed standard of perfection or "progress," adaptation is an endless process.

Modern biologists use the idea of evolution to explain the colossal variety of life-forms on earth. They also use it to try to explain the initial emergence of life on earth, for it seems that nonliving substances may also have evolved by some simplified form of natural selection. And, given a favorable environment and enough time, they eventually evolved into living organisms. The idea of evolution is so basic to the modern understanding of what life is and how it changes that we must begin our account of the history of life on earth by describing the theory in more detail and seeing how it, too, evolved from older attempts to explain the emergence of life in all its variety.

ORIGINS OF THE MODERN THEORY OF EVOLUTION

We have seen that in the seventeenth and eighteenth centuries, some European scientists began to doubt the creation myth of the Judeo-Christian Bible. The Bible seemed to say that species were created by God, about 6,000 years ago, and that they remained essentially as God had created them. This belief was held even by the Swedish botanist Carl Linnaeus, the eighteenth-century founder of modern systems of *taxonomy*, or biological classification. Yet even in the seventeenth and eighteenth centuries, there were difficulties with this account. For example, many fossils suggested the existence of strange creatures never mentioned in the Bible or in historical records. Some, which appeared to be sea creatures, were found high up in mountains that had taken millions of years to create, while others were found buried deep within rocks. Surely, this location suggested, they must have been buried many millions of years ago.

Then there was the fact, known to every farmer, that species of dogs, cats, cattle, and sheep are not as fixed as they might seem. Indeed, by careful choice of mates, breeders of pigeons or dogs can produce some very strange creatures. Darwin was fascinated by the activities of pigeon breeders, and he was a member of two London pigeon clubs. Here, he describes some of the varieties he saw, all apparently bred from the common rock pigeon:

> Compare the English carrier and the short-faced tumbler, and see the wonderful difference in their beaks, entailing corresponding differences in their skulls. The carrier, more especially the male bird, is also remarkable from the wonderful development of the carunculated skin about the head, and this is accompanied by greatly elongated eyelids, very large external orifices to the nostrils, and a wide gape of mouth. The short-faced tumbler has a beak in outline almost like that of a finch; and the common tumbler has the singular and strictly inherited habit of flying at a great height in a compact flock, and tumbling in the air head over heels. The runt is a bird of great size, with long, massive beak and large feet; some of the sub-breeds of runts have very long necks, others very long wings and tails, others singularly short tails. . . . The pouter has a much elongated body, wings, and legs; and its enormously developed crop, which it glories in inflating, may well excite astonishment and even laughter.[6]

Did these exotic creatures belong to the same species that God had originally created? Or were they entirely new species? If they were new, then apparently God was continually tinkering with life—and such tinkering seemed to imply that his original creation may have been less than perfect. As Europeans traveled more widely in the centuries after Columbus, they also became aware that there were many more species than were mentioned in the Bible. The many contrasts in animal and plant life found in the Pacific, the Americas, and Eurasia posed a great challenge to Christian theologians. Had God created all these species? If so, why in such profusion? And why had he distributed them in such a curious and arbitrary way across the globe? Why were there no kangaroos in England, and no pandas in Australia?

By the late eighteenth century, some biologists were considering the possibility that living organisms changed over time by natural mechanisms of some kind, as it seemed messy to suppose that God was continually tinkering with his creation. Perhaps such a mechanism could account for why there were so many species and subspecies, and why so many were not described in the biblical account of creation. The trouble was that no one could explain *how* or *why* species changed.

In the early nineteenth century, Darwin's uncle, Erasmus Darwin, suggested that species evolved so as to adapt better to their environments. This made sense, because all living species do, indeed, seem to fit their environments with great precision. But, like all biologists of his time, he had no clear idea *how* they became so well adapted. In a book first published in 1809, the French naturalist Jean-Baptiste Lamarck suggested a possible mechanism. Perhaps minor changes acquired during a creature's lifetime could somehow be passed on to its descendants. For example, he argued, the ancestors of giraffes may have stretched to browse on leaves high up in trees. Those that stretched hardest may have passed on their long necks to their offspring. Gradually, long necks would have become more and more common, until eventually they became the main distinguishing feature of an entire species. Unfortunately, any animal breeder could tell what was wrong with this theory. *Acquired* characteristics—that is, qualities acquired through a particular lifestyle or the exertions of a particular individual—are not normally passed on to offspring. Only *inherited* characteristics are transmitted in this way. Time we spend in the gym does not guarantee that our children will be fit. A fattened pig will not necessarily produce fat offspring; but a pig whose ancestors were fat *is* likely to produce fat offspring.

Lamarck's mechanism didn't work. But if the genetic makeup of organisms was determined by their *past* (by what they inherited from their parents), how was it possible for them to adapt to *present-day* conditions? This was the conundrum that Darwin solved. From childhood, Darwin was fascinated by animals, and by his twenties he was already an expert naturalist. Like most naturalists of the day, he understood that species are malleable. He also understood that humans were quite capable of altering species through artificial selection. What he didn't know yet was why species also changed *without* human intervention. What, apart from a god or a human being, could allow some individuals to reproduce and condemn others to genetic extinction?

In 1831, his abilities as a naturalist and some fortunate family connections secured him a position as the naturalist on a ship called the *Beagle*, which was embarking on an expedition around the globe. Darwin was staggered by the variety of species he encountered on this trip, and by the extremely subtle variations he noted between closely related species. He also saw clear fossil evidence of evolution in creatures such as armadillos. In

South America, he saw fossil animals that were similar to living animals, but with slight differences. But it was in the Galápagos Islands, off the Pacific coast of Chile, that he discovered the clues that led him eventually to his theory of evolution. There, on a number of recently formed volcanic islands, he found several species of finches that seemed closely related to finches from the American mainland. Yet they differed slightly from island to island. Their beaks, for example, showed minor variations that ensured that each species was adapted with exquisite precision to the particular plants and animals that flourished on its home island.

Here was clear evidence of the sort of adaptation that Erasmus Darwin had written about. Clearly, species could in some sense "adapt" to changing environments. But how did they do it? In about 1838, Darwin read the work of Thomas Malthus, the pioneer of modern population studies, and this seems to have suggested the central idea of his theory. Malthus noted that in most species, including humans, many individuals (sometimes a large majority) do not survive to produce offspring. It was immediately obvious to Darwin that only those individuals that reproduced could have any influence on the nature of later generations. So it was important to explain why some individuals survived and others did not. When he studied the activities of pigeon breeders, the answer was clear. Breeders artificially selected some individuals and allowed only them to reproduce. In Darwin's time, this was already a highly developed art:

> That most skilful breeder, Sir John Selbright, used to say, with respect to pigeons, that "he would produce any given feather in three years, but it would take him six years to obtain head and beak." In Saxony the importance of the principle of selection in regard to merino sheep is so fully recognized, that men follow it as a trade: the sheep are placed on a table and are studied, like a picture by a connoisseur; this is done three times at intervals of months, and the sheep are each time marked and classed, so that the very best may ultimately be selected for breeding.[7]

But what did the selecting in the natural world? What "chose" some individuals to reproduce and condemned others to genetic extinction? What difference was there between those individuals that reproduced and those that did not?

The answer, he suggested, was "fitness." In a statistical sense, it had to be true that the individuals that survived and had healthy offspring were fractionally healthier than those that did not. They reproduced because they were healthy enough to survive longer than others and to find a healthy mate. In individual cases, of course, luck may have played a role. (If lightning strikes you dead, it really doesn't matter how "fit" you were.) But with large numbers, and over large periods of time, fitness must have played the crucial role. On average, those individuals that survived to adulthood and reproduced must have been slightly healthier, slightly better adapted to their environments, than the nonsurvivors. So it wasn't really that species *adapted;* it was the other way around. Those individuals who happened to be better adapted by pure chance were the ones most likely to survive and shape future generations of their species.

Darwin understood that this random statistical process of sorting, which occurs in all forms of life generation after generation, could alter species as effectively as any human breeder if repeated with sufficient frequency. Over and over again, millions of times in each generation, the environment eliminated some individuals, while allowing others to survive. Later generations inherited *only* the qualities of the survivors; as a result, over time the entire species began to resemble the survivors more than the nonsurvivors. In a metaphorical sense, then, the environment was acting like a human animal breeder. And this was why Darwin called the mechanism "natural selection," in contrast to the "artificial selection" performed by those who bred animals.

In this way, Darwin showed that purely statistical and totally mindless processes, repeated over and over again, could explain how species changed in ways that seemed to make them constantly adapt to changing environments. To understand his argument fully, it is vital to understand the random nature of many of these processes. Individuals vary from their parents in minor but essentially random ways. They do not, in any sense, "try" to adapt. It is not the individuals that "evolve," but the average features of the species.

Darwin argued that these mechanisms, repeated over long periods of time, can also explain how distinct species arose, for it is clear that populations of a single species scattered over a large area, and shaped by slightly different environments, are likely to evolve in slightly different ways. The Galápagos finches were a clear example to him of an early stage in the creation of distinct species. Over time, Darwin argued, such processes could explain all the variety of living organisms to be found on earth. As an admirer of the geologist Charles Lyell, Darwin was sure that the earth had existed for a very long time—long enough, perhaps, for such minute changes to create the huge variety of species existing today.

These were astonishing conclusions, for they implied something utterly revolutionary: all the beautiful and complex organisms on Earth, from amoebae to elephants to hummingbirds to human beings, can be created by blind, repetitive processes. Unconscious processes can create not just stars and galaxies, it seemed, but even life itself.[8] Such reasoning seemed to deprive God himself of any reason for existence, which is why Darwin's theory has met, and still meets today, such profound resistance.

Here is how Darwin himself described the workings of natural selection:

> There is no obvious reason why the principles which have acted so efficiently under domestication should not have acted under nature . . . [since] more individuals are born than can possibly survive. A grain in the balance will determine which individual shall live and which shall die,—which variety or species shall increase in number, and which shall decrease, or finally become extinct. . . . Owing to this struggle for life, any variation, however slight and from whatever cause proceeding, if it be in any degree profitable to an individual or any species, in its infinitely complex relations to other organic beings and to external nature, will tend to the preservation of that individual, and will generally be inherited by its offspring. The offspring also, will thus have a better chance of surviving, for of the many individuals of any species which are periodically born, but a small number can survive. I have called this principle, by which each slight variation, if useful, is preserved, by the term of Natural Selection[.][9]

EVIDENCE OF EVOLUTION THROUGH NATURAL SELECTION

In *On the Origin of Species* Darwin stated, as clearly as possible, the arguments for his theory of evolution. He also tried to deal with some possible objections to it. He had few illusions about the difficulty of the task he faced. Most of his readers were traditional Christians. They believed that God had created distinct species, and the thought that species might have emerged through blind processes shocked them. So it was to this audience that Darwin directed most of his arguments.

He was able to show from the fossil record that species seem to change over time. But this was perhaps the least powerful argument available to him, for the fossil record consists of a series of individual, fossilized "photos"; his opponents could easily argue that each was a distinct species created separately by God, and now extinct. What Darwin had to show was the existence of *transitional* species. Some fossils did appear to be halfway between existing animal types. The most famous fossil of this kind was the birdlike dinosaur known as archaeopteryx, which lived about 150 million years ago. Archaeopteryx seemed to be half reptile and half bird. The first fossil specimen was found in 1862, just after the publication of *The Origin of Species,* and Darwin was able to comment on it in later editions. It was exactly the sort of discovery that Darwin's theory had predicted. However, it was also easy for opponents to point out that the fossil record was extremely imperfect. There were vast gaps in all fossil lineages, so that the fossil record on its own could never offer a completely satisfactory proof of the workings of natural selection.

Darwin also offered many other types of evidence for his theory. He argued that where modern species shared many common features, this was a sign that they had evolved from common ancestors. Oddly, the evidence for this claim was most persuasive where apparently useless features survived from remote ancestors by a sort of biological conservatism. Whales have finger bones, which have no obvious adaptive function today. But their existence does make sense if we assume that modern whales are descended from land animals that once *did* have a use for hands and fingers. Indeed, modern whales may be distantly related to hippopotami, a species of contemporary mammals that seems to be in the early stages of adapting to an aquatic lifestyle. The theory of natural selection could explain such phenomena easily, for it suggested that organisms evolve in tiny steps, preserving much from the past, even if some of those features cease to be of any use. This argument was particularly persuasive because traditional theories found such evidence difficult to explain. What reason could a god have for preserving such useless organs, rather than redesigning each species from scratch?

Darwin was also able to show that the geographical distribution of species was more consistent with his theory than with the theory of God the Creator. Why should a god not have placed particular species in all those regions of the earth for which they were adapted? Why weren't all deserts full of camels? Why, instead, did naturalists find that most species in a particular region were closely related—so that, for example, in Australia there were mouselike creatures that were more closely related to kangaroos than to the mice of Europe? Darwin's answer, of course, was that marsupials lived in Australia because that was where their ancestors had lived. That was where they had evolved.

Particularly distasteful to Darwin's critics was another implication of his theory: humans might be closely related to apes (an implication that is alive and well today; see chapter 6). To some people, the idea is still distasteful; the anthropologist Yves Coppens remembers his grandmother saying to him: "*You* may have descended from a monkey, but *I* certainly didn't!"[10] But in Darwin's time, his theory faced many other difficulties. For example, most geologists believed the earth to be no more than 100 million years old. Darwin understood that natural selection needed huge periods of time in which to generate the immense variety of creatures present on Earth, and he conceded that 100 million years was probably not long enough. He believed that evolution worked extremely slowly. Indeed, he was convinced it worked so slowly that it could never be observed directly, and thus all the evidence for evolution would have to be indirect. Furthermore, nineteenth-century biologists had no real understanding of how inheritance worked. For Darwin's theory to work, the mechanism of inheritance had to be very accurate (otherwise no stable species could exist), but not too accurate (otherwise there would never be change). It was important that the qualities of parents be passed on to their offspring, but it was also important that there should be slight variations that might either enhance or threaten the health of particular individuals. But because no one understood fully how the mechanism of heredity worked, it was not certain that heredity worked in exactly the way required. Darwin himself could only suggest that the qualities of parents were "blended," like two colors mixed together. But this seemed to guarantee that variations from the norm would be eliminated in each generation, even if they were beneficial, an outcome that would have made natural selection impossible. The absence of a clear understanding of inheritance was to undermine the credibility of Darwin's theory for more than half a century.

Most of the difficulties Darwin faced were to be resolved in the twentieth century. Religious resistance to the theory of evolution diminished. Meanwhile, new types of evidence emerged to bolster the theory, and gaps in it were filled in. In important respects, Darwin's theory has been modified and improved. As a result, Darwin's central ideas have become the fundamental organizing principle of modern scientific accounts of the history of life on earth.

One reason why his theory is now so widely adopted is that in the twentieth century it proved possible to see evolution at work directly. It is easiest to watch evolution when studying small species that breed rapidly, such as fruit flies. We have also seen evolution at work when new forms of bacteria have appeared in response to the use of antibiotics (as further discussed below).

The fossil record is also much richer than it was in Darwin's time, and new discoveries have created a fuller account of evolution over long periods. This account can never completely prove Darwin's theories, but it remains perfectly consistent with them. Modern dating techniques, by pushing the age of the earth back from 100 million years to over 4 billion years, have also provided a time span forty times as long for Darwinian processes to work in. Finally, twentieth-century biologists came to understand how inheritance works, and their account is fully consistent with Darwin's theory. Gregor Mendel, a contemporary of Darwin's, had already figured out the basic principles of heredity, though his work was ignored until the twentieth century. He showed that although sexually reproducing organisms inherit traits (or *genes*) from both parents, they inherit them in discrete packages—one from this parent, and one from that. He also showed that in many instances, only one of these traits is expressed in the offspring. If your parents have blue and brown eyes, this doesn't mean that your eyes will be a muddy blue; you will inherit one color or the other. So inheritance does not automatically lead to the dilution of traits that Darwin had feared. Particular genes may not be passed on to all offspring; but if they are, they are passed on intact. We also understand exactly how genes are transmitted from generation to generation. Deoxyribonucleic acid, or DNA, transmits genetic information from an organism to its offspring with great accuracy, so that species have great stability. But it is not perfect. As DNA copies itself, it makes, on average, one error for every billion bits of genetic information, the equivalent of a typist making one error in half a million pages. This allows for the small variations necessary if evolution is to occur.[11] The explanation of DNA's structure by Francis Crick and James Watson, in 1953, was therefore a crucial stage in the consolidation of Darwin's theory as the central idea of modern biology.

Modern microbiology has also proved another hunch of Darwin's: all organisms on earth are related. All living organisms, from the simplest bacteria to the largest modern mammals, contain cells that use the same basic chemical processes and pathways. And they all use the same genetic code. In this sense, all living organisms are related. This means one of two things. Either life evolved only once or life evolved more than once, but only one of these experiments has survived to the present day, while all other lineages were eventually wiped out. In either case, all organisms living today, from humans to bananas to sea squirts and amoebae, are descended from the same (bacterial) ancestor.

In minor ways, Darwin's theory has been modified. For example, Darwin seems to have believed that all evolutionary changes occurred because they enhanced the survival chances of the individuals that carried them. But it is now clear that many genetic changes occur at random. There are large amounts of genetic material (perhaps as much as 97 percent of the human genome, for example) that have no impact on the makeup of the adult individual, so changes in these areas will not directly affect an individual's chance of survival. The general principle seems to be this: random changes that have no impact on the survival chances of individual members of a species can lead to slow, and essentially random, changes in the genetic structure (or *genotype*) of an entire species. However, such "neutral" changes may become significant in the future, if, as sometimes happens, some of these inactive genes are reactivated.

Darwin also seems to have believed that evolution occurred at a steady pace. It is now clear that this is not always true. In periods of climatic or environmental stability, species may change slowly. But if environments or climates change more rapidly, species can evolve and diversify very quickly. This is precisely how modern bacteria have evolved in response to the challenge of modern antibiotics. Where antibiotics are widely used, those individual bacteria least affected by antibiotics suddenly become much more likely to produce healthy offspring. Within a few generations, their genes will tend to dominate the species. In this way, there have appeared new species of bacteria that seem more or less immune to the action of antibiotics. It now appears that this rhythm is normal in evolution. During the earth's history, there have been both periods of extremely rapid evolutionary change and periods of relative biological stability. Evolution works in fits and starts, according to the modern theory of "punctuated evolution," which was proposed by Niles Eldridge and Stephen Jay Gould in 1972.

THE BEGINNINGS OF LIFE ON EARTH

Modern Darwinian theory can explain how modern organisms have evolved from the simple life-forms present on the early earth (see the next chapter). But can it take the next step and explain the origin of the earliest life-forms? Is it possible to offer a purely scientific explanation of the emergence of life from nonliving matter?

The notion that life might appear spontaneously has been taken seriously by scientists from at least the time of classical Greece.[12] And with good reason. After all, maggots seem to appear, apparently from nowhere, on the carcasses of dead animals. In the seventeenth century, studies using the newly invented microscope demonstrated that the air was full of tiny organisms, making it possible to explain the apparently spontaneous growth of organisms such as maggots by supposing that their eggs traveled through the air and settled on decaying meat. But this still left open the possibility that tiny microorganisms might be generated spontaneously, perhaps by some form of "life force" floating in the air.

In a remarkably simple experiment, the French biologist Louis Pasteur seemed to finally disprove the idea that life could be generated spontaneously. All life is based on organic molecules—that is, on molecules that use carbon. Because carbon can bond to itself in complex ways, it can form more complex and varied molecules than any other element. By the nineteenth century, many experiments had shown that if soups rich in organic materials were boiled to kill all living organisms, and then placed in airtight containers, no life would appear. However, some argued that this was because the containers excluded an all-pervasive life force. In 1862, Pasteur devised an ingenious experiment to test this idea. He boiled a broth of organic materials, and then placed it not in a sealed container but in a swan-necked container, open to the air. On the one hand, if a life force existed, it would surely be able to enter and use the organic materials to generate new organisms. On the other hand, spores or microorganisms floating in the air would not be able to travel up the neck of the retort. Pasteur's broths have remained sterile for more than a century, and can still be seen today in Paris. His experiment seemed to prove, finally, that life could not be generated spontaneously and that there was no life force floating through the air. Life could be generated only from life.

This cleared up one mystery, only to create another. If life could not be generated from nonlife, then how had life appeared on the early earth? Paleontologists knew that life seemed to appear quite suddenly in the so-called Cambrian era, which we now know began less than 600 million years ago. How could this sudden burst of life be explained? Were biologists forced back to the idea of a creator deity? Many nineteenth-century biologists felt they were, because any purely scientific explanation of the origins of life had to suppose that life could be generated from nonliving material, and Pasteur seemed to have shown that this was impossible. Another possibility was to argue that organic molecules were somehow quite different from inorganic molecules. Perhaps they were different in their origins and their capacity to generate life. Was there something special about carbon? This theory was disproved when chemists showed in the mid–nineteenth century that many organic molecules could be synthesized from inorganic chemicals in the laboratory.

Not until the twentieth century did a more plausible scientific theory of the origins of life emerge. This account, pioneered in the 1920s by Alexander Oparin and J. B. S. Haldane, uses the basic ideas of evolutionary theory to explain not just the evolution of life on earth but also its initial appearance. Its key idea is that evolution works to some extent even among complex but nonliving chemicals. Thus even some chemicals may be capable of evolving if, like crystals, they can stabilize themselves and create reasonably accurate copies of themselves. Once this happens, then those chemicals that produce the most stable "offspring" (i.e., the offspring best adapted to their environment) will tend to multiply more rapidly than those whose descendants survive less well. Such a process would be similar to Darwinian evolution. As such chemicals became better adapted to their environment, they might also become more complex until, eventually, we might start thinking of them as living organisms. Biologists refer to these processes as *chemical evolution*.

But precisely how chemical evolution generated the first living organisms remains unclear. To understand these difficulties, we must break the problem into several levels. First, we need to explain how the basic raw materials of life were created: the chemical level. Second, we need to explain how these simple organic materials were assembled into more complex structures. Finally, we need to explain the origins of the precise mechanisms of reproduction encoded in the DNA that is present in all living organisms today. At present, we have reasonably good answers to the first question; we have plausible answers to the second question; and we are still puzzled by the third question.

The first task now seems surprisingly easy. Living organisms are constructed, for the most part, from compounds of carbon and hydrogen. Carbon is critical because of its astonishing flexibility. Add hydrogen, nitrogen, oxygen, phosphorus, and sulfur and we can account for 99 percent of the dry weight of all living organisms.[13] It turns out that where the conditions are right and these chemicals are abundant, it is easy to construct simple organic molecules, including amino acids (the building blocks of proteins, the basic structural materials of all organisms) and nucleotides (the building blocks of the genetic code).[14] Pasteur's experiment had seemed to show that such molecules could not form spontaneously. We now know why: today's atmosphere, with its large amounts of free oxygen, offers a peculiarly hostile environment for simple organic molecules. Oxygen is extremely reactive, and when it reacts it generates heat (we normally become aware of its destructive power in fires). Oxygen is particularly destructive of organic molecules such as those in wood or paper. Wherever possible, it breaks them down into water and carbon dioxide in a form of slow burning.

However, Oparin and Haldane pointed out there may have been little free oxygen in the atmosphere of the early earth. Perhaps, then, life appeared long before the Cambrian era, in an oxygen-free atmosphere that allowed simple organic molecules to survive long enough to engage in the complex, slow-motion chemistry needed for chemical evolution to take place. In 1952 this possibility was tested in a famous, and remarkably simple, laboratory experiment conducted by two American scientists: Harold Urey and his graduate student, Stanley Miller. They created a model of the early atmosphere by filling a large, closed retort with methane, water, and ammonia. They warmed the mixture, and provided shots of free energy by passing electric currents through it, thereby simulating the lightning that undoubtedly flashed through the skies of the early earth. After seven days, they found a dark-red sludge in their retort. This contained several of the twenty most important amino acids. Amino acids are simple organic molecules (containing about twenty to forty atoms) that link up in different patterns to form the proteins that dominate the chemistry and the structure of all living organisms. By rerunning the experiment under slightly different conditions and in slightly different pseudo-atmospheres, scientists have shown that all twenty basic amino acids can be manufactured in this way. Here we have the basis for the construction of proteins, the fundamental building blocks of life. The Miller-Urey experiment also created, in smaller quantities, other important organic molecules, including sugars and the main components of nucleotides, the molecules from which the genetic code is constructed.

Some claimed that Miller and Urey had come close to creating living organisms. It is now clear that this was not so. Many difficult steps lie between the creation of simple organic molecules and the creation of life. In any case, the atmosphere of the early earth probably contained less ammonia and methane than the two chemists had supposed, and more carbon dioxide and nitrogen. Such an atmosphere would have been less fertile in simple organic molecules. But the Miller-Urey experiment remains important nevertheless. What it showed is that the creation of many of the basic chemical building blocks of life may not have been too hard on the early earth.

Since this experiment, amino acids, simple nucleotides, and even the phospholipids from which cell membranes are constructed have turned up in many surprising environments, both on Earth and in space. Amino acids have been identified in dust clouds in interstellar space. So have vast quantities of water and alcohols, which are vital to the manufacture of phospholipids. We know that both water vapor and many simple organic molecules are also present within meteorites and comets. The presence in space of water, as well as many different types of simple organic molecules, suggests that the entire solar system has been bombarded throughout its history with life's raw materials, either through violent impacts or through the constant drift of cosmic dust onto planetary surfaces. Indeed, it now seems possible that several bodies within the solar system—including Mars; Venus; Jupiter's planets, Europa and Callisto; and Saturn's moon, Titan—may have had (and in some cases may still have) liquid water, so they could have evolved simple organic molecules, even if they are now sterile (as both Venus and Mars seem to be). It is also possible that at least in their early life, the planets swapped organic material, as debris chipped off the planets by meteoritic impacts floated between them. In 1996, for example, it was claimed that a meteorite found in Antarctica, which had arrived about 13,000 years ago, contained trapped gases corresponding to the mix of gases in Mars's thin atmosphere. If this claim is correct (and many scientists doubt that it is), it may mean the meteorite came from Mars. If we discover living organisms on Mars, we will have to take seriously the possibility that they are related to us.[15] All in all, it now appears that the basic chemicals from which life is formed were abundant in the early solar system.

The second task is more difficult. It is to explain how these simple chemicals, containing no more than a few tens of molecules at the most, were assembled into the vast and complex structures necessary for life to exist. Even viruses contain up to 10 billion atoms organized in highly specific patterns, while the complex cells of plants and animals each contain between 1 and 100 trillion (i.e., 10^{12} to 10^{14}) atoms. At present, there is little certainty about where or how this great leap in size and complexity was achieved. Yet it was this change that created true life from organic chemicals. Currently, there seem to be three possible answers to the question of where life first began. It may have been created first in space, or on the surface of planets, or (and this is the latest possibility) deep within planets.

For many years, Fred Hoyle and Chandra Wickramasinghe have argued that Earth was seeded with life from outside. This theory is known as Panspermia. If organisms were first created on planets somewhere else in the universe, then of course we have simply shifted the problem to another planet, and are still left to explain how life could have been constructed there rather than here. Alternatively, it has been suggested that extremely simple organisms were assembled in space. We know that chemical processes occur in interstellar space, and that some simple organisms are robust enough to survive periods of space travel. Yet at present it seems unlikely that life itself could have originated in space, where both energy and raw materials are in short supply, thereby ensuring that chemical processes are normally very slow. Besides, many of the chemical reactions vital to life seem to require water in liquid form, and this cannot be found in space.

Planets—where conditions are more complex, free energy is more abundant, liquid water can collect, and chemicals occur in greater density and profusion—offer a more promising theater for biogenesis. Until recently, most biologists assumed that if life originated on Earth, it must have appeared on the surface. As early as 1871, Darwin suggested that life might have begun in "some warm little pond, with all sorts of ammonia and phosphoric salts, light, heat, electricity, etc."[16] Ever since Darwin's time, biologists have tried to figure out how natural chemical and physical processes could have assembled these chemicals into simple organisms somewhere in the seas or on the seashores of the early earth.

Water plays a crucial role in all these theories. Amino acids and nucleotides, once formed, could have been protected to some extent as long as they stayed in water. Both molecules can form naturally into long chains, though this process requires drier conditions. It is possible that such chains were formed in shallow coastal pools where dissolved molecules would have been periodically dried out and then dissolved again. Under the right conditions, chains of amino acids form proteins, while chains of nucleotides form nucleic acids. So, after millions of years, the seas of the early earth could have been full of simple organic chemicals, which could have joined together into more complex patterns. A. G. Cairns-Smith has suggested that in shallow water, tiny crystals of clay may have provided a template for the formation of more complex molecules.[17] Here, electrostatic forces may have held atoms in complex patterns governed by the molecular structure of the clay itself, until eventually they began to link together in new ways. It is even possible that crystals of clay, sitting inside early cells, played some of the roles of modern DNA by providing templates that could be used over and over again to produce the chemicals used in the metabolism of their host cells.

However they were created, early organic molecules could have formed a weak organic "soup" of simple proteins, nucleic acids, and other organic molecules. Such molecules have a natural tendency to form simple membranes or "skins" made of phospholipids and to form into small globules, "like drops of oil in a vinaigrette sauce."[18] Some of these molecules can also absorb chemicals through their skins in a process analogous to eating, which provides them with the energy needed to expand and absorb more chemicals. Furthermore, when they grow too large and ungainly, such molecules simply split into parts, each of which then goes its own way, just as large water drops on an oily surface can split into smaller droplets. So, along the shores and in the warm seas of the earth 4 billion years ago, there may already have existed organic molecules that duplicated many of the activities of life. They formed into cell-like globules with an outer skin; they "ate" other chemicals; and they could split into separate globules as if reproducing.

All these theories are plausible, but none can explain all the steps leading from nonlife to life. And there are problems with the "warm pond" theory, including the fact that the early atmosphere may not have been as favorable to organic evolution as Miller and Urey had supposed—particularly if, as some evidence suggests, the earliest living organisms appeared before

3.8 billion years ago, when the earth's surface was still being bombarded regularly by extraterrestrial material. Recent research has offered some promising new approaches to the problem, as it has revealed the existence of previously unknown forms of bacteria, the archaebacteria, that evolved well below the surface of Earth.[19] Like all prokaryotes (the simplest type of single-celled organisms), archaebacteria have no nucleus. But unlike most prokaryotes, which extract energy either from sunlight or from other cells, archaebacteria "feed" on chemical energy produced within the earth. They can extract energy from iron, sulfur, hydrogen, and many other unlikely chemicals that are buried in rocks or dissolved in seawater. Archaebacteria can live comfortably deep below the surface of the earth, even under extremely high temperatures and pressures, so their existence raises the possibility that life originated not on but well below the surface of planets. In the 1990s, archaebacteria were found living inside rocks more than a kilometer below the earth's surface; and they have been found living at temperatures well above boiling point in volcanic vents on the seafloor, as well as in porous rocks below the seabed. In 2001, they were found in a huge region that researchers described as a "lost city" on the seabed, where heat was generated not by volcanic activity but by chemical reactions associated with the exposure of a green rock known as *olivine* to seawater. Such areas may have been common early in the earth's history.[20] But archaebacteria also live in huge quantities at the earth's surface. They have been studied, for example, in the hot springs of Yellowstone National Park in Wyoming. Finally, the extreme habitats they occupy suggest the possibility that life may exist, or may once have existed, on neighboring planets and moons, for similar habitats may well exist elsewhere in our solar system.

There are several reasons for supposing that archaebacteria offer a better model than most other modern organisms for the earliest forms of life on Earth. Archaebacteria live in environments that have changed little since the Hadean era. And their ability to live well below the surface means that they would have been less affected by the meteoritic impacts that were common early in the earth's history and may have periodically wiped out life near the surface. They would also have been protected from changes in the earth's atmosphere, and from the heavy doses of ultraviolet radiation that bombarded the early earth before the appearance of an ozone layer. Though the habitats of heat-loving archaebacteria may seem forbidding to us, they may have been the best place for early organisms to establish themselves. These environments contained plenty of chemical food to produce organic material of the kind generated in the Miller-Urey experiments. Particularly around thermal vents, they also contained plenty of energy to drive multiple chemical reactions. Studies of the genetic material in archaebacteria also suggest that they have evolved far more slowly than most other surviving organisms. And, perhaps most striking, the oldest organisms of all, whether archaebacteria or ordinary bacteria, all appear to be heat-resistant. This suggests that however we classify the earliest organisms on earth, they were probably heat-loving organisms that evolved in the highly productive environments around deep ocean vents. If these arguments are right, then life may have appeared first beneath the surface of the earth and its seas, before producing new species that could survive in the cooler environments near or at the surface.[21]

The third task, to explain the origins of the genetic code, is even trickier than the previous two. In a sense this is the most fundamental problem of all, for the key to all modern forms of life appears to be a division of labor between nucleotides, which store and read the instructions for making an organism (the *genome*), and the proteins that use those instructions to construct each individual organism. Crudely speaking, nucleotides handle replication, while proteins handle metabolism. The distinction is almost exactly analogous to the distinction between hardware and software in computing. So, which evolved first, metabolism (chemical activity) or replication (the genetic code)—or did they evolve together?[22]

Oparin's theory implied that metabolism came first and accurate mechanisms of replication evolved later. This idea is intuitively plausible, on the grounds that hardware can exist without software but not the other way around. In Oparin's theory, the earliest organisms were bags of chemicals that could reproduce in a rough-and-ready way, and even "evolve." And they may have evolved in quite complex ways. Many people find it difficult to understand how random processes of this kind can generate complexity. However, evolutionary processes, even of a crude sort, are not in fact entirely random. Of all the chemical experiments conducted randomly in the early earth, some would have created more stable by-products than others. So the process of assembling early molecules did not necessarily begin from scratch each time. On the contrary, every time a relatively stable molecule was produced, it was likely to survive, thereby becoming, in turn, the basis for further experiments. As Cesare Emiliani points out, the odds of a monkey typing the entire Bible by tapping away randomly for millions of years are almost infinitely low. But if a rule is added saying that each time a correct letter is typed it is locked into position, then the odds change radically, and we can expect a Bible to be produced within a decade.[23] To put it slightly differently, we can say that the organic chemicals of the early earth were already subject to the laws of evolution. Most chemicals vanished, but those well adapted to their environment were likely to get "locked in." The few that survived long enough had offspring, and all later generations were their descendants. In this brutal way, the environment "selected" those chemicals best able to survive and reproduce.

And there are many reasons to suppose that chemical evolution of this kind may be quite a powerful mechanism of change. For example, some scientists speculate that there exist deep mechanisms that encourage organization where we might expect pure randomness.[24] In certain types of chemical reactions, a particular chemical may catalyze, or stimulate, its own production, in a process known as *autocatalysis.* Bring together a sufficient number of chemicals of this type and, eventually, there is likely to be a runaway reaction, rather like bringing together a critical mass of uranium in a bomb. Once the critical mass has been reached, these chemical chain reactions can form extremely complex structures very quickly. If this (essentially mathematical) logic is correct, then it suggests that the construction of the large and complex chemicals that were the earliest life-forms may be a natural tendency of organic chemistry. If so, then wherever in the universe conditions allow large amounts of organic chemicals to form and interact, life may be a near certainty.

But all forms of evolution require reasonably accurate mechanisms for replication; otherwise, even the most successful traits will be lost over time. So even theories that put metabolism first have to explain what mechanisms of replication existed in the early days. And this is not easy. DNA, the key to the genetic code in all living organisms today (except for a few viruses), is a fantastically complex molecule, containing billions of atoms. Untwisted, a single molecule of human DNA would be almost two meters long. The atoms of DNA are arranged in precise patterns that, like a piece of software, contain all the information needed to create a living organism. Every cell of our body contains a complete set of these instructions, though it uses only a tiny portion of its DNA instruction manual. These instructions are selected and triggered by the particular environment in which it finds itself. Thus, different parts of the code are used by brain cells and bone cells.

A DNA molecule consists of two long chains of nucleotides that are linked together by rungs, like a ladder. The ladder is then twisted into a long helix, like a spiral staircase. Each section of the ladder has attached to it one of four simple clusters of atoms known as *bases.* Each base can link with only one of the other four, and thus each rung of the ladder consists of two linked bases in a strict order. Adenine (A) links only with thymine (T) and cytosine (C) with guanine (G). The order in which these bases appear on each side contains the code for manufacturing the proteins from which organisms are constructed. Each group of three bases codes for a particular amino acid. Special molecules periodically unzip a part of the DNA spiral, and read off the order of bases in groups of three. Elsewhere in the cell, these amino acids are assembled into chains to make the thousands of proteins that drive chemical reactions and form structures within the cell. DNA can also replicate itself. First, the entire double helix splits down the middle like a zipper as the two bases that make up each rung of the ladder separate. Then, each base attracts a counterpart from the surrounding environment, A joining with T and C with G, until each half of the original helix has built up an entirely new, complementary chain. In this way, a single molecule of DNA can form two new molecules, each more or less identical to the parent molecule.

Explaining how this complex, elaborate, and elegant mechanism was constructed is one of the most challenging tasks facing modern biological theory. One problem is that DNA appears to be helpless on its own. Like any software, it is useless without hardware. So it is difficult to imagine how it could have evolved independently. But there are also problems with the notion that metabolism (the "hardware") evolved first—in particular, it is hard to see how rough-and-ready evolutionary processes could generate a high level of complexity. If cells reproduce without much precision, then the mechanism Emiliani describes simply doesn't work as well. Even if a complex organism evolves, the blueprint for it is likely to get blurred in later generations. For this reason, many who have struggled with these issues have insisted that life cannot achieve significantly new levels of complexity without a capacity to replicate more precisely. And that leads us back to the argument, despite its difficulties, that perhaps the genetic code came before complex metabolism.

What gave such theories a boost in the 1960s was the discovery that DNA's close cousin, RNA, is less helpless than DNA. RNA is software and can, like DNA, code information. But it exists in only a single strand, which means that it can also fold up like a protein and engage in metabolic activities. So, it can play both of the roles of life; it can reproduce itself and provide the set of instructions for reproducing. It can be both hardware and software. Perhaps the first molecules that replicated accurately enough to have some form of "heredity" were made from RNA. Indeed, some viruses even today use RNA instead of DNA as the basis for their genome.

The discovery that RNA can act as hardware as well as software gave rise to theories suggesting that RNA was the earliest form of life. In these theories, associated with the work of Manfred Eigen and Leslie Orgel, the genetic code evolved first, before complex metabolism, even before cells.[25] Unfortunately, RNA copies itself less accurately than DNA, and this creates real problems. A system of replication that is good but not quite good enough may be the worst of all possible worlds, because it may be bad enough to accumulate errors and good enough to transmit those errors faithfully to later generations. It has been shown that such a system may lead to breakdown more rapidly than the sloppier forms of replication required in "metabolism first" models of the origins of life. (Manfred Eigen, the great champion of RNA, has described this problem, ruefully, as the "error catastrophe.")[26]

Freeman Dyson has suggested that perhaps these two theories can be combined.[27] Perhaps metabolism did indeed come first, and cells without a precise mechanisms for replication dominated life on earth for many millions of years, managing, despite their limitations, to evolve many metabolic processes that still occur inside modern cells. One of these processes, present in all living organisms today, involves the storing of energy inside a molecule known as ATP (adenosine triphosphate). And this, as it happens, is closely related to another molecule that is a crucial component of RNA. So perhaps RNA evolved within such cells, which might have provided a more benign environment than the outside world for their evolution. Eventually, acting as a sort of parasite, RNA may have hijacked the reproductive mechanisms of the cell until, in an early form of symbiosis, the cell and its parasitic RNA reached a compromise under which the cell focused on metabolism and the RNA on reproduction. With more accurate mechanisms of replication available (if the "error catastrophe" could somehow be avoided), the RNA in such cells might eventually have evolved into DNA, which is a close relative of RNA.

Or perhaps life as we know it emerged through a symbiosis between two rather different types of organisms, one of which was good at metabolism and the other at coding. Something like this division of labor still exists today between bacteria and the many different viruslike entities that float between them (see chapter 5). Bacteria often use free-floating bits of software similar to viruses for their own purposes, while entities like viruses exploit the metabolic powers of bacteria and other organisms to reproduce. We can imagine a very early world in which metabolizers used viruslike organs to steady their own mechanisms of replication, while virus-type organisms used bacteria to do their metabolism for them, until eventually the two merged to form single organisms.

If none of these theories is totally persuasive, we should not be surprised. There doesn't yet exist a complete theory of the origins of life. In explaining the origins of the genetic code, the key to the emergence of really complex organisms, we are still in difficult territory. However, progress has been rapid in recent decades, and ongoing research holds out the promise that a more satisfactory story may appear within the next decade or two.

SUMMARY

Darwin's theory of evolution, as modified in the twentieth century, provides the fundamental organizing idea of the modern life sciences. Darwin argued that slight random variations within species explain why some individuals are more likely to reproduce than others. Those individuals that are slightly better adapted to their environment are slightly more likely to survive into adulthood and produce healthy offspring, so they are more likely to pass their genes on to later generations. In this way, by what he called "natural selection," species slowly change, and over time entirely new species can be formed.

Such processes may also have shaped the organic chemicals that floated in space, on the earth's surface, and below its surface in the early days of the earth's history. As the more "successful" and stable of these forms survived, over billions of generations more and more stable and complex organic chemicals appeared, through a chemical form of natural selection. In this way, the earliest living organisms were created, within less than a billion years of the formation of our earth. These organisms were the ancestors of all modern life-forms.

FURTHER READING

Darwin is one of the few founders of modern scientific thought whose own writings are worth reading. *On the Origin of Species* (1859; reprint, 1968) is still readable today; and Steve Jones has offered a modern update in *Almost Like a Whale* (2000). Armand Delsemme's *Our Cosmic Origins* (1998) offers a quick introduction, as do many good modern textbooks on biology and evolution. In recent years, biology has generated a superb literature for the general, nonspecialist reader. The writings of Stephen Jay Gould on evolution are always worth reading, even when he does not take a mainstream position. Daniel Dennett's *Darwin's Dangerous Idea* (1995) is a modern classic, while Ernst Mayr (*One Long Argument* [1991]) and John Maynard Smith (*The Theory of Evolution* [3rd ed., 1975]) offer slightly older accounts. There are several good books on modern attempts to explain the origins of life. Paul Davies, *The Fifth Miracle* (1999), is one of the most recent and most accessible. Erwin Schrödinger, *What Is Life?* (1944; reprint, 1992), is still worth reading, and is updated in Freeman Dyson's *Origins of Life* (2nd ed., 1999). A. G. Cairns-Smith's *Seven Clues to the Origin of Life* (1985) and Robert Shapiro's *Origins* (1986) are good, accessible discussions. The writings of Lynn Margulis and Dorion Sagan offer an immensely readable introduction to a view of life that emphasizes the role of bacteria (the "microcosmos"; see especially *Microcosmos* [1987]). Eric Chaisson (*Cosmic Evolution* [2001], *The Life Era* [1987], *Universe* [1988]), Stuart Kauffman (*At Home in the Universe* [1995]), and Roger Lewin (*Complexity* [1993], *Human Evolution* [4th ed., 1999]) discuss notions of complexity and their role in modern discussions of life.

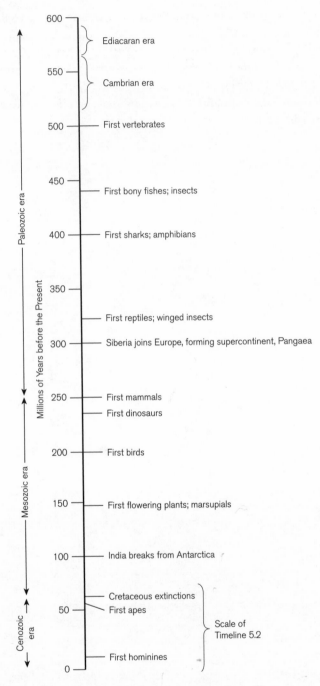

Timeline 5.1. The scale of multicellular organisms: 600 million years.

5

THE EVOLUTION OF LIFE AND THE BIOSPHERE

DIVERSITY AND COMPLEXITY

Once life had appeared on Earth, natural selection ensured that living organisms would multiply and diversify as long as they could find new niches to fill in a changing world. This chapter will describe the main changes in the history of life on Earth. How did evolution generate the variety of organisms present today? What are the main stages in the history of life on Earth? Though many details of this story remain obscure, its broad outlines are now remarkably clear.

After almost 4 billion years of evolution, most living organisms are still simple and small. Bacteria rule, as they always did, and few bacteria are more than a hundredth of a millimeter in diameter. Unlike stars (whose complexity does not necessarily increase with size), living organisms seem to get more complex as they get larger. So the predominance of bacteria is in accordance with the general rule that simpler entities are easier to create and sustain than complex entities, as well as being more durable and more numerous. The vast majority of living organisms belong to what Lynn Margulis and Dorion Sagan have called the "microcosmos."[1] This is why Stephen Jay Gould has argued that even if the appearance of life marks the emergence of new forms of complexity, the history of life on Earth is not merely a story of entities becoming complex. The simplest genetic recipes still work well, so there is no particular evolutionary virtue in complexity.[2] Indeed, in some cases organisms have evolved toward greater simplicity: snakes have lost their legs, moles have lost their eyes, and viruses have lost even the ability to reproduce independently. _

Nevertheless, it is also true that natural selection has experimented endlessly with new ways of living, and in the course of this 4-billion-year-long experiment, it has given rise to organisms more complex than those living on the early earth. They have appeared even though there seems to be no active drive toward greater complexity, and even though complex organisms may not be terribly important in the overall scheme of things. As John Maynard Smith and Eörs Szathmáry point out, "The theory of evolution by natural selection does not predict that organisms will get more complex.... Yet some lineages have become more complex."[3] A world of "macrocosmic" organisms eventually appeared within and alongside the microcosmic world; as large organisms ourselves, we tend to pay attention to this process— just as our history of the universe has concentrated on one obscure planet orbiting one obscure star, simply because that planet happens to be our home.

The story of increasing biological complexity can be told as a series of major transitions. These include the origin of life itself, the appearance of eukaryotic cells, sexual reproduction, the construction of multicellular organisms such as ourselves, and the appearance of organisms that join together in social groups.[4] At each stage, molecules, cells, and individuals were linked together in larger structures—like businesses during corporate mergers—and evolution had to find new ways for them to communicate and cooperate with each other. Explaining how complexity arose through natural selection means explaining why it occasionally proved advantageous (in Darwinian terms) for replicating molecules to cooperate in larger and larger entities, until there appeared towering organisms like us—huge structures made up of billions of closely cooperating cells.

Despite our size (we are to a single-celled organism what the Empire State Building is to a human), we shouldn't exaggerate our complexity. One familiar way of measuring increasing complexity is by estimating the number of genes it takes to construct different types of organisms. But this calculation, it appears, does not flatter us as much as we once thought. Humans do not have the 60,000 to 80,000 genes we once believed were necessary to construct us but half that number, about 30,000. Roundworms have two-thirds as many genes as us (ca. 19,000), and fruit flies just under half (ca. 13,000); even *Escherichia coli*, a bacterium that inhabits our gut, may have as many as 4,000 genes. So, though constructing large organisms is tougher than constructing small organisms, the difference is not as great as we once imagined. Our biological relatives include amoebae and roundworms as well as chimps.

THE ARCHEAN ERA: THE AGE OF BACTERIA

The most important evidence for the history of life on Earth comes from the fossil record, which tells us a great deal about the last 700 million years.

But this is less than one-fifth of the period during which life has existed here. The fossil record can tell us less about earlier periods—during which living organisms consisted of single cells, all living in the sea—because these early organisms lacked hard parts that could form fossils. However, paleontologists have learned how to find and analyze the tiny "microfossils" of bacteria, and the oldest of these date back 3.5 billion years, close to the earliest signs of life on Earth. In recent years, biologists have also made increasing use of techniques for studying and comparing the genetic material of different modern species. This work can show evolutionary links between modern species that cannot be detected from the fossil record alone.

In the conventional planetary chronologies, the Hadean era is the era of Earth's formation, and lasts until about 4 billion years ago; the Archean age is the earliest era of life on Earth, lasting from ca. 4 billion to ca. 2 to 2.5 billion years ago. The first life-forms on Earth evolved early in that age, and they did so in the presence of water. They may have been archaebacteria, which evolved in hot volcanic vents in or below the seabed. Or they may have been other forms of bacteria, if recent research is right in suggesting that both archaebacteria and eukaryotic organisms evolved from earlier and simpler organisms, the so-called eubacteria.[5]

Either way, life appeared early. Living organisms probably existed by 3.8 billion years ago, for rocks of this age from Greenland contain a level of the C^{12} isotope that is normally associated with the presence of life. Life was certainly present by 3.5 billion years ago, the date of rocks from South Africa and Western Australia that seem to contain microfossils of bacteria similar to modern cyanobacteria (i.e., blue-green algae).[6] These were similar to the organisms that turn stagnant water green today. Their presence suggests that the seas of the early earth were then already full of life. The speed with which life appeared on Earth has encouraged many biologists to think that given appropriate conditions anywhere in the universe, life may appear rapidly and naturally. So life, far from being rare, may exist throughout the universe. As Paul Davies has recently argued, the universe itself, at least in the current phase of its evolution, seems to be a "bio-friendly" place.[7]

But that "friendliness" has its limits. Any complex structure requires a constant flow of energy if it is to survive. So one of the fundamental tasks for all organisms is to find sources of nutrients and energy—a task that is not always easy. The solutions found by the earliest organisms on Earth had a profound effect on the history of life here and also shaped the planet itself.

The earliest organisms may have extracted their energy from chemicals below the earth's surface. They "ate" chemicals. If the earliest organisms were archaebacteria, they probably extracted the energy they needed from chemical vents deep within the seas. But quite early some organisms learned to acquire energy by eating other organisms. In this way, there emerged a clear distinction between primary producers, which extract their energy from the nonliving environment, and organisms higher up the food chain that feed on other living organisms, including the primary producers. If these had been the only ways of extracting energy, then the history of life on Earth would have been limited by the energy supplied from the earth's molten core and available to organisms living deep within the sea. But by at least 3.5 billion years ago, some organisms were living near the surface of the seas, where they learned to feed on sunlight. And the Sun is a richer source of energy than the heat engine at the center of the earth. The cells of cyanobacteria contained molecules of chlorophyll, which enabled them to process sunlight in the fundamental chemical reaction known as *photosynthesis*.

Photosynthesis is so important for life on earth that it is worth making the effort to understand how it works.[8] Molecules consist of atoms linked together by chemical bonds. However, creating chemical bonds requires energy, and some of that energy can be released by breaking the bonds again. So chemicals can be thought of as stores of energy. Living organisms get at the energy stored in organic molecules such as glucose by breaking their chemical bonds. As it also takes energy to break chemical bonds, the trick is to do it in such a way that more energy is released than is used to break the bond. This is the work of *enzymes*. Enzymes are molecules (mainly proteins) whose shape enables them to destabilize particular energy-containing molecules with very little effort. By doing so, they release much more energy than they expend. The principle of using a small amount of energy to release a much larger amount is one we are all familiar with—we do it whenever we use a match to light a fire. However, this whole process requires an initial input of energy to create the chemical bonds that act as stores of energy. This is where photosynthesis comes in. In photosynthesis, chlorophyll (in the presence of water and carbon dioxide) uses the energy of light to set up a tiny electric current. This current drives a complex chain of reactions that form molecules of substances such as glucose that can store energy. In this way, plants use the energy of sunlight to create tiny energy packages in their own bodies, which they unwrap when they need more energy. Of course, other organisms can also use the stored energy by eating plants. When we eat an apple, we are breaking into and using the energy stored by the apple tree. When we burn coal, we are releasing energy stored by trees of the Carboniferous era, 300 million years ago. In this way, large amounts of energy derived from sunlight can be stored in quite small packages. It is easy to forget that a cup of gasoline, which contains the energy stored by organisms many millions of years ago, can drive a truck up a hill. It is also easy to forget that without the constant inflow of energy from sunlight, the whole biosphere would run out of energy.

Using the complex chemical reactions of photosynthesis, living organisms began to harvest the colossal bounty of sunlight. Fueled by sunlight, life could flourish in ways that would have been unthinkable in a sunless world. Much of the rest of the history of life on Earth is governed by the different ways in which sunlight was captured, distributed, and divided among the different species that shared the planet. Human history is part of this story, for humans have found increasingly powerful ways of harvesting sunlight, through foraging, farming, and the use of fossil fuels.

Cyanobacteria are the remote ancestors of modern plants, and they are among the most important primary producers in the world today. Many cyanobacteria secrete a sticky slime, which enables them to stick together in mats. Over time, these form large, mushroom-shaped objects called *stromatolites*, with a thin upper layer of living bacteria atop a thickening layer of their congealed ancestors. Stromatolites still form in a few environments today (one of the most famous such places is Shark Bay in Western Australia), but fossil stromatolites are common from as early as 3 billion years ago. They are a reminder that many of the earliest life-forms were so successful that they still survive today, with remarkably little obvious change.

So we should resist the temptation to think that the Archean world was boring in comparison with ours. Properly understood, this world was as varied and exotic as our own. Margulis and Sagan offer a dramatic description:

> Shrunk to microscopic perspective, a fantastic landscape of bobbing purple, aquamarine, red, and yellow spheres would come into view. Inside the violet spheres of *Thiocapsa*, suspended yellow globules of sulfur would emit bubbles of skunky gas. Colonies of ensheathed viscous organisms would stretch to the horizon. One end stuck to rocks, the other ends of some bacteria would insinuate themselves inside tiny cracks and begin to penetrate the rock itself. Long skinny filaments would leave the pack of their brethren, gliding by slowly, searching for a better place in the sun. Squiggling bacterial whips shaped like corkscrews or fusili pasta would dart by. Multicellular filaments and tacky, textilelike crowds of bacterial cells would wave with the currents, coating pebbles with brilliant shades of red, pink, yellow, and green. Showers of spores, blown by breezes, would splash and crash against the vast frontier of low-lying muds and waters.[9]

In the microcosmos, genetic information can float around in the form of bits and pieces of DNA or RNA known variously as *replicons, plasmids, phages,* or *viruses.* These objects exist right at the borderline between life and nonlife, for most consist of little more than genetic information looking for a body to do their "living" for them. Bacteria can make use of these fragments of genetic information at any stage of their life, not just at reproduction. And they do so to supplement the limited range of metabolic skills available in their small genetic libraries.[10] Replicons may never become incorporated into the permanent genetic store of each bacterium, but, like borrowed software, they can be used by their host before moving on elsewhere. So bacteria can share genetic information with a flexibility not available to larger organisms, a feature that may help explain their astonishing variety and adaptability. Despite its tiny genetic library, each bacterium has access to a global genetic data bank that is off-limits to larger organisms such as us (or was off-limits, before the era of genetic engineering). As Margulis and Sagan put it: "For the macrocosmic size, energy, and complex bodies we enjoy, we trade genetic flexibility."[11] We will see later that symbolic language may have returned, at least to our species, some of the adaptive flexibility that bacteria enjoy through their ability to freely swap genetic material, by permitting us to swap information instead of genes.

By many measures, bacteria remain the dominant forms of life on earth today. Eventually, though, some single-celled organisms began to join together in more tightly organized structures. These constitute the first steps toward the creation of multicellular organisms such as ourselves. The beginnings of multicellularity occurred during the Proterozoic era.

THE PROTEROZOIC ERA: NEW FORMS OF COMPLEXITY

Early forms of photosynthesis pried hydrogen loose from hydrogen sulfide to store the energy of sunlight. Eventually, though, some forms of cyanobacteria learned to pry hydrogen loose from the much tougher bonds of water molecules in a more efficient form of photosynthesis, one of whose by-products was free oxygen. Over millions of years, this new, and more powerful, metabolic technology began to transform the early atmosphere by pumping into it huge amounts of free oxygen, a gas that was poison to most early life-forms.

At first, free oxygen was quickly reabsorbed in chemical reactions, such as rusting, that bound it to iron. (The presence of huge bands of rust from the early Proterozoic era is one of the reasons we know about the increase in free oxygen.) However, from about 2.5 billion years ago on, free oxygen was being produced too fast to be soaked up in this way, and it began to appear in the atmosphere. By 2 billion years ago, free oxygen may have accounted for 3 percent of the gases in the atmosphere; in the last billion years, the level has risen to about 21 percent.[12] If there were much more oxygen around, we'd self-ignite if we rubbed our hands together too hard!

The appearance of an oxygen-rich atmosphere was one of the greatest of all revolutions in the history of life on earth. Margulis and Sagan describe the change as "the oxygen holocaust."[13] Because oxygen is so reactive, its presence kept the atmosphere in a continuous chemical imbalance, creating a new level of chemical tension that could drive more powerful forms of evolutionary change. Here, fueled indirectly by the Sun, was a new source of free energy that could be used to construct more complex life-forms. As James Lovelock has written: "[Free oxygen] provides the chemical potential difference wide enough for birds to fly and for us to run and keep warm in winter; perhaps also to think. The present level of oxygen tension is to the contemporary biosphere what the high-voltage electricity supply is to our twentieth-century way of life. Things can go on without it, but the potentialities are substantially reduced."[14]

The life-forms that dominated the earth until less than 2 billion years ago were simple, single-celled organisms that lived in the sea. Biologists call such organisms *prokaryotes*. The DNA of prokaryotes floats freely within the cell. At reproduction, the cell splits, and each half receives an identical copy of the DNA of the parent cell. Thus, in prokaryotic organisms the offspring are normally clones of the parents. However, as we have seen, they can exchange genetic information "horizontally" with their neighbors, as well as "vertically" with their parents and offspring, an ability that allows for types of evolution not available to more complex organisms.[15] This partly explains why prokaryotes were so successful in discovering and exploiting many of the basic chemical processes on which life depends even today. They altered the earth's surface as well as its atmosphere; in the words of Margulis and Sagan, "The Age of Bacteria transformed the earth from a cratered moonlike terrain of volcanic glassy rocks into the fertile planet in which we make our home."[16]

However, until living organisms had access to the powerful batteries of an oxygen-rich atmosphere, there was a limit to their complexity and size. Free oxygen is extremely damaging to simple organic materials, which is why life could not have appeared in an oxygen-rich atmosphere. But after 2 billion years of evolution, life was robust and flexible enough to survive the appearance of this new pollutant. Though many species must have perished, those that managed to survive in an oxygen-rich atmosphere would have flourished, because oxygen can supply much more energy than most other forms of "food." In addition, free oxygen, floating high in the atmosphere, eventually created the ozone layer. Though only a few millimeters thick and about 30 kilometers above the earth's surface, this layer of three-atom oxygen molecules (O^3) shielded the earth from much ultraviolet radiation and thus made it easier for life to spread on land as well as in the sea. In these ways, the appearance of oxygen steered evolution into new pathways.

These changes may explain the appearance of distinctively new life-forms, known as *eukaryotes,* about 1.7 billion years ago.[17] Their arrival marks a clear increase in the genetic complexity of organisms, so it counts as one of the major transitions in the history of life on earth.[18] While most prokaryotes are tiny, somewhere between 1 and 10 thousandths of a millimeter in size, eukaryotic cells are normally much larger. Most are between 10 and 100 thousandths of a millimeter across, which means that the largest can just be seen with the naked eye. They are also more complicated and contain much more (roughly 1,000 times more) DNA than most prokaryotes, though much of this extra genetic information seems not to be used. Finally, eukaryotes thrived in an oxygen-rich atmosphere, because they found ways of exploiting this new energy source. On a paleontological timescale, their appearance is also quite sudden. Margulis and Sagan cite the analogy of the astronomer Chet Raymo: "The difference between the new cells and the old prokaryotes in the fossil record looks as drastic as if the Wright Brothers' Kitty Hawk flying machine had been followed a week later by the Concorde jet"[19] (see figure 5.1).

Because eukaryotes contain so much more genetic information than prokaryotes, and have access to more powerful energy sources, they have more metabolic tricks up their sleeve and can give rise to more complex organisms. By extracting energy from oxygen, eukaryotes were indirectly exploiting the efforts of photosynthesizing organisms such as cyanobacteria, which were constantly pumping fresh oxygen into the atmosphere. Eukaryotes have more flexible and adaptable membranes than prokaryotes, and these allow them to exchange energy, foods, and wastes with their environment with more precision. Eukryotes also protect their delicate genetic machinery in a special internal compartment known as a *nucleus.* Finally, their insides are more complex, for they contain internal organs, or *organelles.*

Lynn Margulis has shown that eukaryotes probably evolved through the joining together of different types of prokaryotes and their genetic material in a form of symbiosis—the process by which independent organisms evolve so as to become more dependent on each other. Symbiosis is extremely common, and it illustrates one of the most complex aspects of evolutionary change: the fact that competition and cooperation are so closely

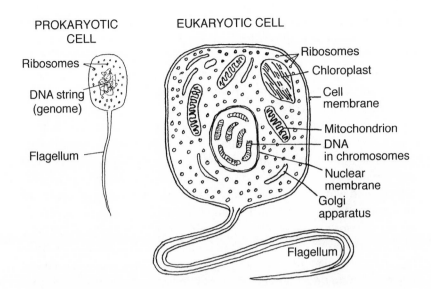

PROKARYOTIC CELL

Ribosomes

DNA string (genome)

Flagellum

EUKARYOTIC CELL

Ribosomes
Chloroplast
Cell membrane
Mitochondrion
DNA in chromosomes
Nuclear membrane
Golgi apparatus
Flagellum

Figure 5.1. Prokaryotic and eukaryotic cells compared. Eukaryotic cells are larger and more complex than prokaryotic cells. In all cells, ribosomes assemble proteins, using instructions from the DNA. Flagella, which allow movement, are present in many (but not all) cells. But eukaryotic cells also contain other structures (or organelles) that are not present in Prokaryota. Eukaryota keep their DNA in a special area (the nucleus), where it is protected by a special membrane, and often organized into special parcels called *chromosomes*. They also have mitochondria, which convert food into chemical energy; and many have chloroplasts, which convert light into chemical energy in the process known as *photosynthesis*. Finally, Eukaryota also have a cytoskeleton, a complex structure made up of protein rods and tubes, that organizes the different organelles within the cell. From Armand Delsemme, *Our Cosmic Origins: From the Big Bang to the Emergence of Life and Intelligence* (Cambridge: Cambridge University Press, 1998), p. 164. Reprinted with the permission of Cambridge University Press.

intertwined. In evolution, as in commerce, not all games let the winner take all. Often, the winning move for a particular organism requires cooperating with other organisms. Biologists identify several different kinds of symbiosis. *Parasitism* is a relationship in which one species benefits at the expense of another. Robins laying eggs in the nests of other birds are acting like parasites. But parasitism is not predation (in which the victim loses *everything*). If the relationship is to endure and benefit the parasite, the host must be kept alive, at least for a time; otherwise, the parasite gains little. In the relationship known as *commensalism*, two species live together and one benefits while the other seems to suffer no harm. *Mutualism* is a relationship in which both species seem to benefit from the partnership. Most flowering plants depend on insects or birds for pollination, but to attract their pollinators they "offer" nectar or food of some kind. Human agriculture involves a form of mutualism between humans and domesticated species of animals and plants. For example, humans eat maize (the cereal commonly known in North America as *corn*), and in some regions they will starve if the crop fails. But maize benefits from this relationship, because humans protect their plants, helping them to reproduce and flourish. Indeed, so dependent are modern strains of maize on this relationship that they can no longer reproduce without the help of humans. This is true symbiosis: a relationship in which one or both partners can no longer survive outside of the symbiotic relationship. Such relationships are surprisingly common in the animal world because where both partners gain something, the relationship may prove more stable than if one partner is gaining very little. This is why virulent disease bacteria so often evolve in ways that do less and less harm to their "hosts." The most familiar examples for humans are "childhood" diseases such as chicken pox, which have all evolved from more virulent species that did so much damage to their hosts that they sometimes killed them off.

In extreme cases, mutualism may lead to the creation of a single organism from two species that had been independent. In a sense, eukaryotes are therefore the first "multicelled" organisms. With the appearance of eukaryotes, as Margulis and Sagan note, "Life had moved another step, beyond the networking of free genetic transfer to the synergy of symbiosis. Separate organisms blended together, creating new wholes that were greater than the sum of their parts."[20] We know this step was taken, because the internal organelles of eukaryotes seem to have developed from once-independent prokaryotic organisms that may originally have acted like parasites. Eukaryotic organelles include thousands of tiny ribosomes, in which different proteins are manufactured according to recipes contained in the DNA. They also include mitochondria, which specialize in extracting energy from oxygen in chemicals "eaten" by the cell, and lysosomes, which destroy harmful intruders. The whiplike flagella, which are present in most eukaryotes, allow them to move; thus eukaroytes, unlike prokaryotes, can move to congenial environments rather than just surviving wherever they drift. Presumably, motion revolutionized many evolutionary processes: "As the steam engine sped up the cycles of industrial production, including the manufacture of more steam engines, spirochete partnerships may have initiated a burst of development, increasing the number and diversity of symbiotic life forms."[21] Eukaryotic photosynthesizers also contain chloroplasts, or packets of chlorophyll. Indeed, the earliest eukaryotic organisms were probably green algae. The fact that some of these organelles, including mitochondria and chloroplasts, still contain their own DNA (mitochondria contain about a dozen genes) is one of the reasons for thinking that eukaryotes evolved through the symbiotic union of once-distinct organisms.

Eukaryotes reproduce in more complex ways than do prokaryotes. While most prokaryotes produce identical copies of themselves, eukaryotes normally reproduce only after merging the genetic material between two different parent individuals. The DNA from two adults merges randomly to produce new strands of DNA, which contain a mix of the genes of both parents. As a result, individuals vary more among eukaryotes than among prokaryotes. They are no longer clones of their parents. This innovation, the first step toward sexual reproduction, had a profound impact on the pace of evolutionary change, for it gave natural selection a greater variety of bodies to choose from in each generation. The evolutionary acceleration triggered by the emergence of eukaryotic organisms and of sexual reproduction explains why, in the last billion years, life has flourished in entirely new ways, creating the profusion of large life-forms that inhabit the modern earth. Sexual reproduction counts, along with the appearance of eukaryotic organisms, as one of the major turning points in the history of life on earth.[22]

THE CAMBRIAN EXPLOSION: FROM THE MICROCOSM TO THE MACROCOSM

The evolution of eukaryotic cells was part of a suite of evolutionary changes that were as significant as any since the first emergence of life on earth.

Sexual reproduction accelerated evolutionary change. Increasing amounts of free oxygen and the evolution of breathing (i.e., the ability to extract energy from oxygen) made more energy available for more exotic and powerful forms of metabolism. And, perhaps most important of all, eukaryotic organisms began to combine into teams that eventually formed the first multicellular organisms. Taken together, these changes help explain the "Cambrian explosion"—the relatively sudden proliferation of larger, more complex, and more energetic life-forms that began almost 600 million years ago.

The first multicellular organisms (as opposed to mere colonies of organisms, such as stromatolites) may have evolved as early as 2 billion years ago.[23] But they became common only in the last billion years. Before such organisms could flourish, there were some serious problems to overcome. Most important of all, large numbers of cells had to be able to communicate and cooperate with each other in new ways.

This was not easy. To understand how it happened, it is important to distinguish between different types of biological cooperation. The first is symbiosis, which we have already encountered. A second type of cooperation arises from the formation of societies or colonies made up of many individuals of the same species. Sometimes these animal societies are held together quite loosely. We know that early organisms, including cyanobacteria, gathered in huge colonies, for stromatolites are formed from such colonies. However, though they gave protection to individuals, these colonies are not yet examples of symbiosis, as the individual organisms can still survive on their own when necessary. Some modern sponges appear to be single organisms, but in fact they can be passed through a sieve. They will break into a pulp, then reconstitute themselves as the individual cells regather. Equally extraordinary is a type of amoeba that feeds on bacteria. Joël de Rosnay explains:

> If you deprive it of food and water, it emits a distress hormone. Other amoebas rush to the rescue and gather into a colony about a thousand strong—a thousand "individuals," as it were, moving like a slug in search of nourishment. If they don't find it, they stop moving, put up a spore-producing stalk, and remain there indefinitely, just like that, as long as it's dry. But if you add water, the spores germinate and give rise to independent myxamoebas, which head off in different directions.[24]

Such entities consist, essentially, of millions of separate individuals huddling together in a sort of crowd, capable if necessary of working together in a team.

In so-called social animals, such as ants or termites, the dependence between individuals is greater. In ant or termite colonies, many individuals are sterile. Cooperation of this kind poses a serious problem to evolutionary theory, for what evolutionary advantage could there be for an organism that has no offspring? Why would genes evolve that seem to commit suicide in this way? The solution seems to be that natural selection works in such communities as long as the cooperating organisms are closely related. Genes that encourage an individual organism to enhance the reproductive chances of close relatives can indirectly enhance their own survival chances. For example, a sterile worker ant may share 50 percent of its genetic material with other ants in the colony that can reproduce. By helping them to reproduce, it is encouraging the survival of many of its own genes. Indeed, it can be shown mathematically that in certain situations the survival of particular genes may be maximized not just by having lots of offspring but also by helping relatives to have lots of offspring. But those relatives must be close. (The geneticist J. B. S. Haldane once commented that he would lay down his life for two brothers or eight cousins; he was referring to the fact that he shared half his genetic material with his brothers, but only one-eighth with his cousins.)[25] Only through arguments of this kind is it possible to show how Darwinian mechanisms of natural selection can give rise to individuals that cooperate with other members of their species, even if doing so reduces their immediate reproductive chances.

Multicellular organisms are an extreme example of this type of cooperation. Organisms such as ourselves consist of hundreds of billions of individual cells, yet only a tiny number, the so-called germ cells, have any chance of reproducing. Why do the bone, blood, and liver cells put up with this? The answer seems to be that all the cells contain the same genetic material. They are clones, so they have identical DNA molecules. By cooperating, they maximize the survival chances of their shared genetic blueprint. In genetic terms, they have a common "interest" in ensuring the survival and reproduction of the entire organism, and therefore of the tiny number of germ cells. In such organisms, billions of individual cells cooperate so closely that we no longer think of them as separate organisms, but consider them to be parts of a single, complex, multicellular organism.

So, before multicellular organisms could evolve, there had to exist a mechanism that allowed a single germ cell (a fertile egg) to multiply into many different kinds of genetically identical cells in the adult form. What happens is that each cell inherits the same genetic material; but as the organism develops, external factors switch on different genes in different cells, leading different cells to develop differently. Once set, these genetic switches can then be passed on to further cells, so that a single brain cell may multiply by cloning many identical daughter cells.[26] In the same way, muscle cells produce more muscle cells, bone cells produce more bone cells, and so on. This secondary form of heredity, by which only some of the total genome contained in the DNA is expressed in each cell, is characteristic of the way cells develop in all multicellular organisms.

The first extensive fossil evidence of multicellular organisms dates from the Ediacaran era, ca. 590 million years ago. But the fossil record of multicellular organisms really becomes abundant during the Cambrian era, from ca. 570 million years ago. Quite suddenly, in geological terms, there appeared organisms with protective shells, made from secretions of calcium carbonate. Their shells have survived exceptionally well as fossils. The worldwide appearance of shells marks the beginning of the Cambrian era, but it is hard to tell whether this signals a real flourishing of multicellular organisms or merely the appearance of organisms more likely to be preserved as fossils.

The largest multicellular organisms, such as trees or humans, may contain as many as 100 billion cells—as many cells as there are stars in the Milky Way. Humans have as many as 250 or more distinct types of cells, which are created and controlled by the activity of about 30,000 genes. At the other extreme are simpler multicellular organisms, such as fruit flies, which have only about sixty different types of cell. The hydra, an invertebrate that consists of little more than a translucent tube about 30 millimeters long, has only sixty different types of cell.[27] Clearly, the evolution of multicellularity implied a significant increase in the complexity of living organisms. (As always, we must be careful not to assume that "more complex" means "better.")

Few multicellular species survive in the fossil record for more than a few million years. As a result, the multicellular species that exist today are a tiny proportion of the many different species that have evolved in the past 600 million years. However, some of the larger *types* of species that have evolved since the Cambrian explosion have proved remarkably durable. It is as if there emerged standard general patterns, on which evolution kept weaving minor variations.

To understand the history of these different types of multicellular organisms, we need a system of classification, a *taxonomy*. Biologists classify species into many different groups and subgroups. The smallest unit of classification is a species. A single species consists of individual organisms that are so similar biologically that they can, in principle, interbreed with each other, but not with members of other species. Modern humans constitute a single species. According to one well-known definition, a species consists of "groups of actually or potentially interbreeding natural populations, which are reproductively isolated from other such groups."[28] Similar species are grouped together in genera, related genera are grouped into families and superfamilies, and these, in turn, are classified within orders, classes, phyla, and, finally, kingdoms and even superkingdoms.

Currently, biologists differ about the best ways of classifying organisms at the highest levels. Linnaeus, the pioneer of modern systems of classification, grouped all organisms into two large systems: plants and animals. However, as they made more use of microscopes, biologists became aware of a vast array of single-celled organisms that did not fit into these two categories. In the middle of the nineteenth century, the German biologist Ernst Haeckel suggested that all single-celled organisms be classified within a separate kingdom of Protista. Then, in the 1930s, biologists realized that there was a fundamental difference between cells with nuclei and those without. As a result, they began to divide all organisms into two distinct kingdoms, the Prokaryota (organisms whose cells had no nuclei) and the Eukaroyta (organisms whose cells had nuclei). In some systems, the Eukaryota also include all multicellular organisms. In the second half of the twentieth century, powerful arguments emerged for the creation of separate kingdoms for fungi and for viruses (which are so simplified that they cannot even reproduce without hijacking the metabolic systems of other organisms). In the 1990s, Carl Woese proposed a new large classification to distinguish between the archaea and other forms of bacteria. Like all prokaryotes, archaea do not have nuclei; but unlike other prokaryotes they take in energy neither from sunlight nor from oxygen but from other chemicals.

Table 5.1 describes one contemporary system of classification at the highest levels. This recognizes two superkingdoms, the Prokaryota and Eukaryota. Within these, it recognizes five kingdoms: Monera (the only kingdom within the Prokaryota), Protista (single-celled eukaryotic organisms), and Plantae, Fungi, and Animalia (all multicelled Eukaryota). Using this system, we can say, for example, that modern humans belong to the superkingdom of Eukaryota; to the kingdom of animals (Animalia); to the phylum Chordata, or backboned animals; to the class of mammals, or Mammalia; to the order of primates; to the superfamily Hominoidea (which includes humans and apes); to the family Hominidae (which includes humans, gorillas, and chimpanzees); to the subfamily Homininae (which includes humans and their ancestors of the last 4 or 5 million years); to the genus *Homo*; and to the species *sapiens*.[29]

Multicellular organisms divided quite quickly into three great kingdoms: plants (organisms that derive their energy from photosynthesis), animals (organisms that consume other organisms), and fungi (organisms that digest other organisms externally before absorbing nutrients from them). In the Ediacaran era, from ca. 590 to 570 million years ago, an astonishing range of multicellular organisms appear, some quite similar to organisms such as sponges, sea worms, corals, and mollusks that still exist today. But some species were quite different from anything that exists today. This is also true of many Cambrian organisms such as those excavated in the "Burgess shale" in British Columbia, which date from ca. 520 million years ago. This was clearly a period of genetic experimentation. From it there emerged, through adaptation or perhaps (as Stephen Jay Gould has argued) more through the sheer exuberance of evolutionary change, a number of basic patterns for the organization of multicellular plants and animals.[30] Many of these patterns have survived to the present day.

The discovery of fossil spores from the Ordovician era (510–440 million years ago) suggests that plants were the first multicelled organisms to leave the sea and colonize the dry land. For multicelled organisms, colonizing the land was like settling another planet. Above all, the process required spe-

TABLE 5.1. THE FIVE-KINGDOM SCHEME OF CLASSIFICATION

Superkingdom	Kingdom	Members
Prokaryota (single-cell organisms without nuclei)	Monera	bacteria, blue-green algae, and archaebacteria (sometimes put in a separate kingdom)
Eukaryota (organisms whose cells have nuclei and organelles)	Protista (mostly single-celled)	protophyta, protozoa, and slime fungi
	Plantae (multicell, contain chlorophyll and photosynthesize, normally cannot move)	algae, bryophytes (mosses, liverworts, and horn-worts), ferns, psilo-phytes, lycopodiophytes, conifers, gnetophytes, ginkgophytes, cycads, and flowering plants
	Fungi (multicell, no chlorophyll, obtain energy by decom-posing organic remains, normally cannot move)	yeasts, toadstools, and mushrooms
	Animalia (multicell, no chlorophyll, obtain energy from other organisms, often mobile)	protozoans, sponges, corals, flatworms, tapeworms, arthropods, mollusks, lampshells, annelids, bryozoans, echinoderms, hemichor-dates, and chordates, including the vertebrates

cial protective equipment to prevent the organism from drying out and col-lapsing. The wet insides had to be protected by an insulating layer of some kind; indeed, all land animals still carry small surrogate seas within them-selves, and it is there that their young are fertilized and begin to grow. Once on land, and without the buoyancy of water, bodies also needed more in-ternal rigidity, a problem most often solved by using calcium compounds secreted from their cells to make skeletons. Special ways of feeding, breath-ing, and reproducing had to be developed as well. The land was, as Margulis and Sagan succinctly describe it, a hellish environment "of torturous sun, biting wind, and decreased buoyancy."[31] The earliest dry land colonizers were similar to modern liverworts or ferns. The first seed-bearing trees appeared during the Devonian period (410–360 million years ago). These formed huge forests from which most modern reserves of coal are derived. (Coal, like oil and natural gas, the other major "fossil fuels," is literally formed from the fossilized remains of once-living organisms.)

Animals have taken multicellularity further than have plants. Their cells are more specialized, and communicate with each other more efficiently. Animals have also made a specialty of mobility and complex behavior. But this is not necessarily a cause for pride; rather, it is a sign of the all-pervasiveness of symbiotic relations in evolution, for plants don't just feed animals but also exploit the mobility of animals to spread their own seeds. They don't need brains or legs—they use ours![32] The first animals to move on land were probably arthropods, a bit like giant insects. We know of their presence from the Silurian period (440–410 million years ago). They included creatures similar to modern scorpions but as large as human beings. Arthropods, such as modern lobsters, carry their skeletons on the outside. In contrast, vertebrates, the group of animals that includes modern human beings, all have internal skeletons. The earliest vertebrates evolved in the sea, during the Ordovician period, between 510 and 440 million years ago, from wormlike ancestors. They included early forms of fish and sharks. All vertebrates had a backbone, limbs, and a nervous system whose parts were concentrated at one end, the head. This thicket of nerves at the end of the backbone was the first brain. Eventually it was to become the seat of consciousness, for at some point in the evolution of vertebrate brains (we do not know when), there appeared the capacity not just to react to stimuli but also to *feel* them, to be in some sense aware of them.

With the appearance of the earliest nervous systems, we can perhaps claim that simple forms of "consciousness" had evolved, if we accept Nicholas Humphrey's argument that consciousness is the capacity to feel sensations, even where there is no systematic thought or self-awareness. "To be conscious is essentially to have sensations: that is, to have affect-laden mental representations of something happening here and now to me."[33] But consciousness clearly has degrees or grades, and these depend on how the brain represents and experiences the external world. Terrence Deacon has suggested that the experiences Humphrey refers to here should really be described as *sentience*, while the word *consciousness* should be applied to the way in which organisms "represent aspects of the world to themselves."[34] He argues that all creatures with nervous systems can construct internal representations of the external world that allow them to react in more complex ways to external changes. Thus, a bear will come to feel that there are similarities between all animals that look like bears. It may also learn to feel a close correlation between the onset of winter and a desire to hibernate. These felt representations of the external world may exist in all animals with nervous systems, though their diversity, the number of links between them, and the power of the sensations associated with them may increase in larger-brained species. But Deacon further argues that only humans can think in symbols—that is, in purely arbitrary signs that link together many different types of representation and can create a quite distinct internal world of their own.[35] So, it seems likely that the inner world of humans shares much with that of the earliest organisms with brains, even if the glare of our symbolic thinking normally casts a mental shadow over these more direct and universal forms of sensation. The sensations that lurk in the background of our consciousness are probably shared with all organisms that are "aware" in this minimalist sense.

Though the earliest forms of consciousness evolved in the sea, consciousness was to flourish most spectacularly on land. Vertebrates (animals with backbones) first colonized the land in the late Devonian period, though the earliest steps on land may have taken place in the Silurian. Modern land vertebrates are still quite similar in their basic layout. All have four limbs, each with five digits even when, as in snakes, the limbs and digits have almost shriveled away. Such similarities suggest that all of them—amphibians, reptiles, birds, and mammals—are descended from the earliest colonizers of the land. The first amphibians evolved from fish that could breathe oxygen and whose fins could be used to move on land, like modern lungfish. However, amphibians have to lay their eggs in water, which normally confines them to seashores, rivers, or ponds. Reptiles evolved eggs with hard shells, just as trees evolved seeds with tough skins, so both types of organism can reproduce on dry land. The earliest reptiles appeared about 320 million years ago, during the Carboniferous era (360–290 million years ago). But evidence is growing that they began to flourish after the "great dying," the massive extinctions of ca. 250 million years ago (at the end of the Permian period), which were perhaps caused, like the later Cretaceous extinctions, by a huge asteroid impact.[36] To modern humans, the most spectacular of the ancient reptiles were the dinosaurs. These first appeared during the Triassic period (250–210 million years ago) and flourished until the end of the Cretaceous period, ca. 65 million years ago.[37]

Modern evidence suggests that the many different species of dinosaurs died out quite suddenly after an asteroid collided with the earth, creating a huge dust cloud.[38] The Cretaceous impact may have created the 200-

kilometer-wide Chixculub crater in the north of Mexico's Yucatán Peninsula. For many months, temperatures on the earth's surface would have dropped as the Sun's light was obscured. But then, as layers of dust insulated the earth, creating a sort of greenhouse effect, temperatures would have risen once more. These wild temperature fluctuations were enough to wipe out many species of warm-adapted creatures. Modern birds, with their insulating layer of feathers, may be descendants of the few species of dinosaurs that survived the catastrophe of the late Cretaceous period.

MAMMALS AND PRIMATES

Unlike reptiles, mammals are warm-blooded and furry. Their young are nourished before birth inside the bodies of females and after birth from modified sweat glands that produce milk (see timeline 5.2).

Mammals first appeared during the Triassic, at about the same time as the earliest dinosaurs. However, they remained limited in variety, number, and size during the era of the dinosaurs. Typically, mammals were small night creatures—similar in size, perhaps, to modern shrews. Their small bodies, and the fact that many lived in underground burrows, may have helped them survive the catastrophe that wiped out so many types of dinosaurs. After the disappearance of most dinosaurs, mammals flourished in a spectacular radiation of new species. They soon filled the many ecological niches vacated by dinosaurs. Mammal browsers appeared, alongside mammal carnivores, mammal insect eaters, and mammal tree dwellers. The asteroid impact of the late Cretaceous period counts, therefore, as a crucial event in the prehistory of our own species. If the asteroid had been on a slightly different trajectory, say a few minutes faster or slower, mammals would have remained limited in numbers and variety, and our own species could not possibly have evolved.

The crisis of the late Cretaceous is a reminder of the capriciousness and open-endedness of evolutionary change. Evolution has no preplanned direction. There was no inner necessity about the way in which life evolved on Earth. It may have been likely that photosynthesizing organisms would develop, or that, eventually, multicellular organisms would evolve.[39] In this limited sense, it was likely that larger and more complex organisms would at some point appear. But there was no necessity about evolution taking the particular pathways it took on Earth.

Humans belong to the group of mammals known as *primates*. Today, the primates include about 200 species of monkeys, lemurs, and apes. Most have been tree dwellers. This particular niche encouraged the evolution of limbs with opposable thumbs (that could grasp); of eyes that could see

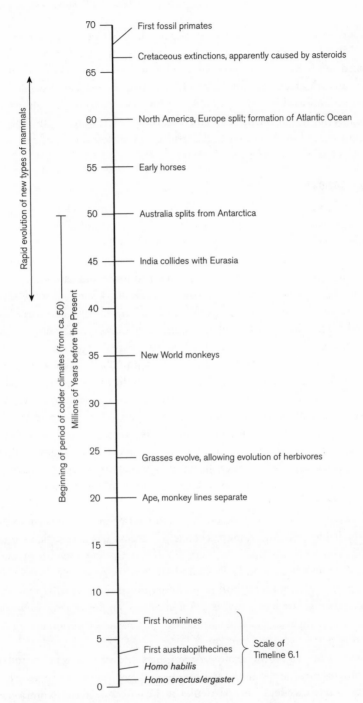

Timeline 5.2.   The scale of mammalian radiations: 70 million years.

stereoscopically, in order to judge distances accurately; and of brains that could control the complex movements of limbs and process the complex information from eyes. The earliest primates appeared soon after the extinction of the dinosaurs, and the group rapidly diversified, unconscious beneficiaries of the fluky crash landing at Chixculub.

Humans belong to the superfamily of primates known as the Hominoidea (hominoids). This includes human beings and the apes (chimpanzees, gorillas, gibbons, and orangutans) among living creatures, as well as many extinct species. The fossil record suggests that the earliest hominoids appeared roughly 25 million years ago in Africa. That our evolutionary ancestors had lived in Africa is an idea that even Darwin held, though he had far less evidence than we have today. His reasoning was simple: in the modern world, the animals most like ourselves all live in Africa. Chimpanzees and gorillas, he argued, are more similar to modern humans than are Asian orangutans. For his contemporaries, though, discriminating between chimps and orangutans was missing the point; it was the thought that we might be related to apes of *any* kind that was shocking and insulting. But the idea was not as novel as it may have seemed, and many communities have thought of primates as closely related to humans. A Portuguese missionary, Father Alvares, who worked in Sierra Leone early in the seventeenth century, reported that "there are heathen that claim to be descendants of this animal [the *dari*, or, in modern language, the chimpanzee], and when they see it they have great compassion: they never harm it or strike it, because they consider it the soul of their forefathers, and they think themselves of high parentage. They say they are of the animal's family, and all that believe they are descended from it call themselves *Amienu*."[40]

According to the classification adopted here, the Hominoidea are divided into three main groups: the Hominidae, the Pongidae (orangutans), and the Hylobatidae (gibbons). The Hominidae, in turn, include two main groups: the Gorillinae (gorillas and chimpanzees) and the Homininae. The Homininae are the only primates who have customarily walked on two legs. Molecular dating techniques suggest that the hominine line diverged from the Gorillinae between 5 and 7 million years ago. Modern humans are the only living members of the hominines, but the group also includes many extinct species, among which are our immediate ancestors (see chapter 6).

EVOLUTION AND THE EARTH'S HISTORY: "GAIA"

I have told the history of life on earth and of the earth itself as if these were different stories. In fact, they are closely linked. The evolution of new life-forms transformed the earth's atmosphere by pumping huge amounts of

free oxygen into it. The dead bodies of millions of plants and animals created the carboniferous rocks and the huge deposits of fossil fuels that power
modern industries, and in these ways they transformed the earth's geology.
Meanwhile, archaebacteria excavated and mined the regions beneath the
seabed.

It is possible that the impact of life on our planet has been even more
far-reaching. James Lovelock has argued that forms of cooperation between
living organisms are much more extensive than we normally recognize. Indeed, he has argued that in some sense living organisms constitute a single,
earthwide system; he calls it "Gaia," after the Greek goddess of the earth.
Gaia acts as a huge self-regulating superorganism, which automatically
maintains an environment suitable for life at the earth's surface.

> The Gaia hypothesis, when we introduced it in the 1970s, supposed
> that the atmosphere, the oceans, the climate, and the crust of the Earth
> are regulated at a state comfortable for life because of the behavior
> of living organisms. Specifically, the Gaia hypothesis said that the
> temperature, oxidation state, acidity, and certain aspects of the rocks
> and waters are at any time kept constant, and that this homeostasis
> is maintained by active feedback processes operated automatically
> and unconsciously by the biota.[41]

As one illustration of these mechanisms, Lovelock points out that though
the heat emitted by the Sun has almost certainly increased by perhaps 40
percent during the last 4.6 billion years, the temperatures at the earth's surface appear to have been maintained at roughly 15°C, or within the range
suitable for life to evolve and flourish. What mechanisms could maintain
such a stable global thermostat? Algal blooms may provide an example of
these feedback processes, linking the living and nonliving environments.
Many algae generate a gas called dimethyl sulfide (DMS). When it reacts
with oxygen high in the atmosphere, DMS creates minute particles around
which water vapor condenses. In effect, by producing large amounts of DMS,
algae create clouds. Massive cloud cover reduces the amount of sunlight that
reaches the surface, cooling the earth's surface and thus reducing the numbers of algae near the surface. As a result, the amount of DMS generated
declines, cloud cover begins to decline, and the amount of sunlight reaching the surface increases. So algae create a sort of worldwide thermostat,
which maintains the earth's surface within a limited range of temperatures
by constantly adjusting the amount of cloud cover. Lovelock's theory holds
that the biosphere (the totality of life on Earth) is held in a state of rough
equilibrium by many interlocking negative feedback loops of this kind.[42]
One reason why Lovelock's theory has been greeted with skepticism is

that it is hard to explain why particular species should have evolved in ways that benefit the biosphere as a whole. The theory of natural selection encourages us to think of competition rather than cooperation as the dominant force in evolution, because there are many individuals and few niches. So the existence of cooperation between species always requires special explanation. Within multicellular organisms, genetic similarity seems to explain cooperation. And we have already seen that there are many forms of symbiosis in which both species benefit. But the idea of cooperation on a global scale is harder to justify. Why should algae have evolved the capacity to emit DMS unless it is "adaptive" for the species that have evolved this ability, unless it helps them reproduce their own genes? Lovelock has always insisted that there must be a Darwinian logic behind such processes, but explaining that logic is not easy. In specific instances, we can sometimes see how the benefits to a particular species coincide with those to the biosphere as a whole. It has been suggested, for example, that some algae may have the ability to ride high in rising air currents before descending again in rain. Because it disperses their offspring widely, this process offers a clear evolutionary advantage to the species as a whole. The release of DMS may help in the process in several ways. When reacting with oxygen, DMS can generate heat that may help create updrafts to carry bacteria aloft. Once high in the clouds, the water vapor and ice crystals that form around the byproducts of this reaction can protect the algae from drying out in the upper atmosphere. The same ice crystals may also help carry them back to earth. Such arguments, like the "hidden hand" of Adam Smith's economic theory, may help explain how competition between individuals and species led to outcomes that were, on balance, beneficial to most life-forms. The seeding of clouds may be just one of myriad ways in which life itself, particularly bacterial life, helps keep the biosphere in a state suitable for the survival of life in general.[43]

There is a further possibility, which is that cooperation is far more natural in the bacterial realm than in the realm of large organisms because bacteria exchange genetic information more freely than do large organisms. As we have seen, bacteria can exchange replicons almost as easily as modern humans exchange coins. This means that in the bacterial world, at least, natural selection shapes entire teams of bacteria that work together. As Margulis and Sagan put it:

> Its minimal number of genes leaving it deficient in metabolic abilities, a bacterium is necessarily a team player. A bacterium never functions as a single individual in nature. Instead, in any given ecological niche, teams of several kinds of bacteria live together, responding to and

reforming the environment, aiding each other with complementary enzymes. . . . Intricately meshed in this way, bacteria occupy and drastically alter their environments. In huge and changing numbers, they perform tasks of which individually they are incapable.[44]

If this argument is correct, it may be that bacterial cooperation extends even to the planetary scale—in which case it would be much easier to explain the forms of cooperation that Lovelock has detected within "Gaia." It would certainly make sense of the fact that bacteria seem to play the crucial roles in maintaining the viability of the biosphere.

Whether the Gaia hypothesis is literally true or not, it is a powerful and inspiring idea. *"Si non è vero, è ben trovato"* (if it isn't true, it was well invented). Besides, it is abundantly clear that living organisms really have transformed the earth's surface. But the reverse is also true. Geological changes have shaped evolution. In eras in which most of the continents were joined together, there was less biodiversity than in eras in which the continents were scattered more widely over the face of the planet, because there was less ecological variety. Today, the continents are widely scattered, so life has been exceptionally diverse in the earth's recent history (until the actions of our own species began to reduce that diversity in the past few centuries). By altering the number and variety of available niches, the rearrangement of the continents by plate tectonics may explain why in the past 500 million years there have been at least five periods of sharply declining biodiversity—periods in which perhaps 75 percent or more of all species may have vanished. Of these, the most catastrophic occurred in the late Permian period, ca. 250 million years ago, as the supercontinent of Pangaea formed. In this period, it appears that more than 95 percent of all marine species may have become extinct.[45] The evolution of primates occurred during a period of accelerated evolutionary change caused both by the meteoritic impact of the late Cretaceous and by the existence of an exceptional number of distinct ecological niches in a geologically complex world.

The precise configuration of continents and seas at any time can also profoundly affect climatic patterns. Indeed, our own species evolved in a period of unusually rapid climatic and ecological change. Climates became cooler during the Miocene epoch, from 23 to 5.2 million years ago. Reduced evaporation from the oceans meant that climates generally also became drier, so that forests shrank while steppe and desert regions spread. These changes were partly due to rearrangement of the earth's continental masses, as the Atlantic sea widened, and as Africa and India both drifted north to collide, respectively, with the western and eastern parts of the Eurasian landmass. When warm, equatorial water can circulate freely to the poles, it keeps the earth's climates warm. In the current era, the presence of Antarctica over the South Pole prevents warm water from warming the South Pole, while the circle of continents around the North Pole limits the movement of equatorial waters northward. This conjunction, which blocks off the circulation of warm water at both poles, may be unique in the planet's history. The trend toward cooler, drier climates has accelerated during the Pliocene epoch, from 5.2 to 1.6 million years ago, and into the Pleistocene, the epoch in which our hominine ancestors evolved. About 6 million years ago the Mediterranean became a semienclosed inland sea locking up some 6 percent of the salt of the world's oceans. With lower concentrations of salt, the remaining oceans froze more easily and the ice cap of the Antarctic began to expand rapidly, causing a sharp fall in world temperatures. About 3.5 to 2.5 million years ago, ice sheets began to form in the northern hemisphere and in Antarctica, and by 900,000 years ago there were already large sheets in the far north. The world had entered the "ice ages" (see figure 5.2).[46]

Specialists in climatic history can now measure the most recent changes in global climates with great precision. Oxygen has three isotopes (i.e., atoms with different numbers of neutrons in their nuclei). Because ice sheets and water absorb them in different amounts, the ratio of these isotopes remaining in the oceans varies depending on the amount of water locked up in ice. By finding fossil organisms that have taken in oxygen, and measuring the ratios of these isotopes in their remains, scientists can estimate the size and extent of the world's ice sheets when these organisms were alive. Such calculations show that within the longer, cooling trend, there have also been shorter cycles of warmer and colder periods. These are partly caused by changes in the earth's tilt and orbit. It is now clear that the frequency of these shorter cycles has changed in the last 5 million years. A complete cycle, from a warm interglacial period to a longer glacial period and back again, lasted about 40,000 years up to ca. 2.8 million years ago. From then until ca. 1 million years ago, they lasted about 70,000 years, and in the last million years the dominant cycles have lasted about 100,000 years.[47] The current pattern seems to consist of brief interglacials, or warm periods, lasting about 10,000 years, and much longer cool periods, with short periods of extreme cold preceding quite rapid transitions to a new interglacial. The most recent ice age began about 100,000 years ago, and lasted until about 10,000 years ago. So, for the last 10,000 years, the earth has been in a warm, interglacial phase of these cycles.

The significance of both the long-term cooling of the earth's climates and of these shorter-term cycles for the study of hominine evolution is that they created unstable ecological conditions. All land organisms had to adapt to

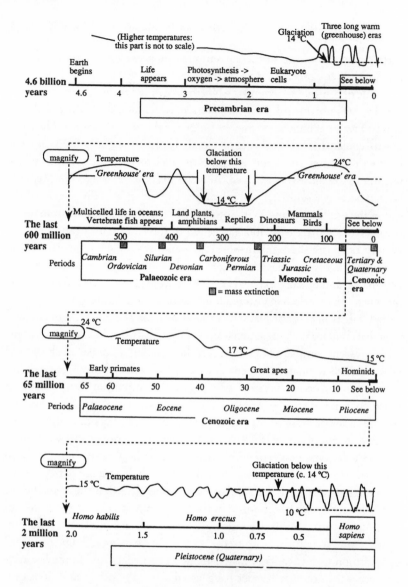

Figure 5.2. Temperature fluctuations at different timescales. There is nothing new about global warming; average surface temperatures on earth have fluctuated on several different scales throughout the earth's history. As these graphs show, our species evolved in an era of cooling temperatures. But temperature changes appear to have become more erratic in the past million years or so, the period known as the *ice ages*. From A. J. McMichael, *Planetary Overload: Global Environmental Change and the Health of the Human Species* (Cambridge: Cambridge University Press, 1993), p. 27. Reprinted with the permission of Cambridge University Press.

periodic changes in climate and vegetation, and that necessity undoubtedly accelerated the pace of evolutionary change. Modern humans are one product of this period of accelerated change.

INDIVIDUAL SPECIES AND THEIR HISTORIES

Today, there may be between 10 and 100 million different species on the earth. Each consists of many individual organisms that can, in principle, interbreed. Species may be divided into many regional populations, which are separated geographically. A *population* consists of those members of a species who actually encounter each other and may breed, while the species consists of all individuals who are biologically similar enough to interbreed, even if, in practice, most will never meet.

The rest of this book will focus on the history of just one of these many species, our own. But first it may be worth describing some of the more general features of species histories. Ever since Darwin, it has been clear that species are not eternal. They evolve from other species; they exist, sometimes for many millions of years; then they either become extinct or evolve into one or more other species. In this sense, each species has its own history, even though most of these histories will never be recorded. During its lifetime, each species may undergo minor changes of many different kinds. Regional variations may appear, but biologists will continue to classify the individuals as members of a single species as long as they can continue to interbreed and produce fertile offspring. For example, all modern breeds of dogs can interbreed, despite their great variety of shapes, sizes, and temperaments (which is the product of artificial selection). This is why domestic dogs are regarded as members of a single biological species.

We can describe a species' history largely in terms of its population history. If a new species succeeds in establishing itself, then it has found a niche within the community of other species, a way of extracting enough resources from the environment so that individual members of the species can successfully survive and reproduce. Migration into new regions, or minor innovations in lifeways or genetic endowment, may enable a species to widen that niche, or even to exploit new niches or new regions. When this happens, the population of a species may grow. And its growth often follows a characteristic pattern that we can summarize in a formula: migration, innovation, growth, overexploitation, decline, and stabilization (MIGODS).[48] The initial innovation leads to rapid population growth. Eventually, too many individuals are produced, until at least one critical resource (such as food, water, or space) becomes so scarce that further growth is impossible.[49] This is the stage of overexploitation. It is followed by sometimes catastrophic

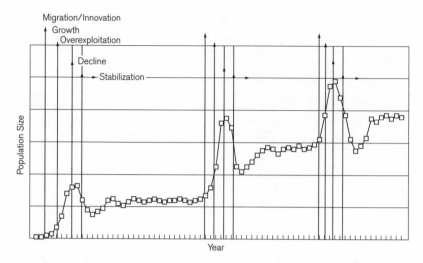

Figure 5.3.   Basic rhythm of population growth. A schematic representation
of typical patterns of population growth. Eventually, this pattern ends for most
species in a period of decline leading to extinction.

population decline, though the decline may be less drastic if population
growth slowed as the species reached the maximum sustainable level. Finally, populations may rise again as the species adapts in subtler ways to
the opportunities and limits of its environment and reaches the stage of stabilization. In its existence, a particular species may pass through this cycle
many times. But eventually there will come a time when a phase of decline
is not followed by a phase of stabilization—perhaps because the environment changes, or because other species transform the environment in ways
that prove fatal. The species will become extinct, though it may leave descendants that are different enough to be classified as members of a new
species.

This rhythm is described graphically in figure 5.3, which charts the population growth of an imaginary species. It provides a way of describing the
characteristic rhythms of a species' history, and of its relationship with other
species and with the biosphere as a whole. It also provides some ideas that
may help us identify some of the more important similarities and differences between our own history and the histories of other species.

SUMMARY

During almost 4 billion years, evolutionary processes have generated all the
biological diversity apparent on the modern earth. Indeed, the species alive

today are only a tiny sample of the total number of species that have evolved during the earth's history.

For more than 3 billion years, life consisted only of single-celled organisms. However, even in the world of bacteria there was change. Cells acquired the ability to secure energy from sunlight and eventually from oxygen. Eukaryotic cells acquired internal organelles. And from ca. 600 million years ago, some cells joined together to form multicellular organisms, the first nonmicroscopic organisms on earth. Since the Cambrian explosion, trees, flowers, fish, amphibians, reptiles, and primates have all evolved. Many other evolutionary experiments may also have flourished and vanished without leaving any traces.

As life evolved so did the earth itself, and the two processes were interrelated at many points. Living organisms created carboniferous rocks and an oxygen-rich atmosphere. At the same time, the processes of plate tectonics slowly shaped and reshaped the earth's surface and its climatic patterns in ways that accelerated or slowed the rate of evolutionary change, while violent events such as meteoritic impacts and volcanic eruptions occasionally diverted the course of evolution in particular regions. The biosphere and the earth have evolved together as part of a complex, interlinked system.

Within this constantly changing system, the basic ecological units are particular species. Each has its own history, which is governed by relations with other species. The history of each species is shaped mainly by that species' particular niche, the way it extracts resources (including food) from its surroundings. Over time, the niche of a species may alter in more or less subtle ways, and these alterations may affect the population of the species. The history of each species is shaped largely by these fluctuations in numbers, which are related in turn to changes in the environment and to the way each species exploits its environment. The characteristic ways in which populations change suggest a way of approaching the history of living species in general and our own species in particular.

FURTHER READING

Despite its title, John Maynard Smith and Eörs Szathmáry's *The Origins of Life* (1999) is a history of life on Earth, constructed around the central idea of the evolution of complexity. Hubert Reeves et al., *Origins: Cosmos, Earth, and Mankind* (1998), follows similar themes. In *Life's Grandeur* (1996 [U.S. title, *Full House*]), Stephen Jay Gould has criticized the idea that life has, in any fundamental sense, become more complex over time, while in *Wonderful Life* (1989) he stresses the serendipitous nature of evolution. Malcolm Walter's *The Search for Life on Mars* (1999) is extremely helpful on the earliest fossil evidence of life on Earth. There are short surveys of the history of life in Armand Delsemme, *Our Cosmic Origins* (1998), and longer accounts can be found in Richard Fortey, *Life: An Unauthorised Biography* (1998), and Steven Stanley, *Earth and Life through Time* (1986). Lynn Margulis and Dorion Sagan powerfully remind us of the importance of bacterial life on earth in *Microcosmos* (1987) and *What Is Life?* (1995); and the writings of James Lovelock argue for the crucial role of bacteria in regulating the environment for "Gaia." Paul Ehrlich's *The Machinery of Nature* (1986) is a good introduction to ecological issues. Tim Flannery's *The Future Eaters* (1995) and *The Eternal Frontier* (2001) offer superb ecological histories of the lands in and around Australia and of North America, respectively, on geological timescales.

PART III

EARLY HUMAN HISTORY
Many Worlds

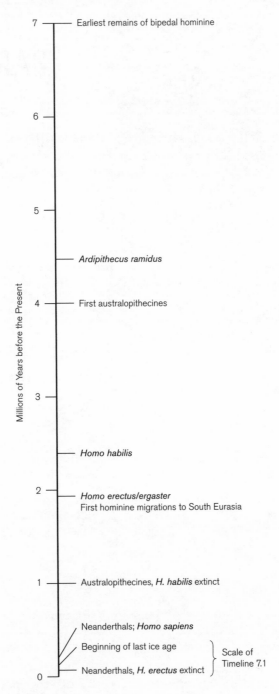

Timeline 6.1.   The scale of human evolution: 7 million years.

6

THE EVOLUTION OF HUMANS

The rest of this book will be concerned mainly with the history of just one species, *Homo sapiens.* There are two justifications for narrowing the focus in this way. The first is that we—the author and readers of this book—belong to this species. To know ourselves, we must know the history of *Homo sapiens.* The second reason, less obvious and less parochial, is that the history of our species is significant at some surprisingly large scales.

When we try to explain the *appearance* of human beings, we face once again the paradox of beginnings. How can something utterly new appear? We are animals, we evolved according to Darwinian rules as did any other living organism, and we are remarkably similar to closely related species, such as the other hominoids (the great apes). Yet we are also radically different from even our closest relatives. Somehow or other, our species has moved beyond the Darwinian rules. And this is why our impact on the earth has been far greater than that of any other large organism.

How can we explain both what unites us to other animals and what divides us from them?

HUMAN HISTORY: A NEW LEVEL OF COMPLEXITY

We have seen similar transitions before. Human history marks the sudden and unexpected emergence of a new level of complexity, as did the first appearance of stars, of life on Earth, or of multicelled organisms. We have seen that complex entities are rarer than less complex entities, they are more fragile, and, because they have to climb faster up entropy's down escalator (see appendix 2), they have to manage denser energy flows. We have also seen that transitions to greater complexity come about through the creation of new forms of interdependence, as entities that once existed more or less independently are incorporated within new and larger structures. Finally, we have seen that as new levels of complexity have appeared, they seem to operate according to new rules ("emergent properties," in the jargon of complexity theory).

Human history also marks the emergence of a new level of complexity on Earth.[1] As with earlier transitions, human history links once-independent entities into larger patterns of interdependence; and this process is associated with greater energy flows that have a profound transformative effect. From the perspective of the twenty-first century, we can measure some of these changes. Humans, acting together, have learned how to manage increasingly large energy flows. Though the spectacular implications of these changes have become apparent only in the past two centuries, their roots lie deep in the Paleolithic era, or "Old Stone Age."[2]

Table 6.1 shows how humans have learned to extract from their environment more than just the energy needed to survive and reproduce. They have shown an entirely new capacity for "ecological innovation." From early in human history, skills such as the management of fire increased the amount of energy available per capita. In the past 10,000 years, agriculture has increased the food energy humans can extract from a given area, while the domestication of large herbivores in the past 6,000 years increased the amount of energy available for traction power as well. In the past two centuries, the use of fossil fuels has multiplied per capita energy use many times over. As the total number of humans has also increased from perhaps a few hundred thousand in the Paleolithic to a few million 10,000 years ago and more than 6 billion today (see figure 6.1), the total amount of energy controlled by our species has multiplied by at least 50,000 times. This is a staggering amount of energy to be under the command of a single species, and it helps explain why our species has had such an impact on the entire biosphere. A powerful way of measuring this impact is to estimate how much of the energy supplied to the biosphere from sunlight is co-opted for use by humans. Net primary productivity (NPP) is that portion of energy from sunlight that enters the food chain through photosynthesis and is turned into plant material. This, in turn, feeds most other organisms. NPP can thus be used as a rough measure of the biosphere's energy "income." Modern calculations suggest that our species is currently co-opting for its own use at least 25 percent, and by some measures 40 percent, of all the NPP available to land-based species. Paul Ehrlich sums up the story told by these remarkable figures: "One of many millions of species, *Homo sapiens* is now co-opting about a quarter of all the products of photosynthesis for its own use."[3]

Increasing human control of energy has shaped human history and the

TABLE 6.1. HUMAN PER CAPITA ENERGY CONSUMPTION
IN HISTORICAL PERSPECTIVE (UNITS OF ENERGY = 1,000 CALORIES PER DAY)

	Food (incl. animal feed)	Home and Commerce	Industry and Agriculture	Transport	Total per Capita	World Population (mill.)	Total
Techno. Soc. (now)	10	66	91	63	230	6,000	1,380,000
Indust. Soc. (1850 CE)	7	32	24	14	77	1,600	123,200
Adv. Agric. (1000 BP)	6	12	7	1	26	250	6,500
Early Agric. (5000 BP)	4	4	4		12	50	600
Hunters (10,000 BP)	3	2			5	6	30
Proto-humans	2				2	n.a.	n.a.

SOURCE: I. G. Simmons, *Changing the Face of the Earth: Culture, Environment, History,* 2nd ed. (Oxford: Blackwell, 1996), p. 27.

histories of many other species as well. It has also enabled humans to multiply at an accelerating rate. Tables 6.2 and 6.3 and figure 6.1 summarize human population growth over the past 100,000 years. As human numbers
have grown, so has the ecological range of our species; by 10,000 years ago,
and perhaps as early as 30,000 years ago, humans could be found living in
all continents apart from Antarctica. In the Paleolithic era, human history
was characterized mainly by the increasing *range* of human settlement. In
the past 10,000 years, the increasing *density* of human settlement has been
the main shaper of human social evolution, as humans learned how to live
in larger and larger communities, from villages to towns, cities, and states.

    Resources used by humans are, by definition, unavailable to other species.
So, as human numbers have risen, other species have felt the pinch. Domesticates such as sheep and cattle, and unintentional domesticates from
cockroaches to rats, have flourished. But many more species have fared less
well, and an alarming number have died out. That process also began in the
Paleolithic, when human activity helped drive to extinction close relatives
such as Neanderthals, as well as many other large species, including mammoths in Siberia, horses and giant sloths in the Americas, and giant wom-

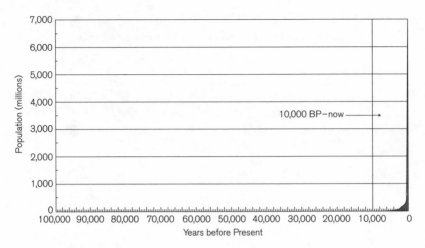

Figure 6.1.   Populations of *Homo sapiens,* 100,000 BP (before present) to now. Based on table 6.2.

bats and kangaroos in Australia. Today, rates of extinction caused by human activity have accelerated. Currently, 1,096 of 4,629 mammal species (24 percent) are thought to be "threatened," as are 1,107 of 9,627 bird species (11 percent), 253 of 6,900 reptile species (4 percent), 124 of 4,522 amphibian species (3 percent), 734 of 25,000 fish species (3 percent), and 25,971 of 270,000 higher plant species (10 percent).[4] And because the pace of extinctions is accelerating, we can expect many more species to vanish in the near future. These figures provide a powerful measure of the planetary impact of human history, for paleontologists have surveyed rates of extinction over much of the past 600 million years, and the current rates appear similar to those of the five or six most drastic extinction eras during that time span.[5] This means that the impact of human history will be visible on scales of at least a billion years. In other words, if interstellar paleontologists visit this planet in one billion years' time and try to decipher the history of the planet using the tools of contemporary human paleontologists, they will identify a major extinction event that coincides with the presence of our species.

These figures also help us gauge the uniqueness of human history. No other large animal species has multiplied like humans, or occupied such a wide range, or controlled such vast ecological resources. (Once again, the possible exceptions are species such as cattle or rabbits that have multiplied as part of the human ecological team.) Our history is utterly different even from that of our closest relatives, the chimps. Though they are extremely

TABLE 6.2. WORLD POPULATIONS AND GROWTH RATES,
100,000 BP TO NOW

Date BP (years)	Estimated World Population	Rate of Growth Each Century since Previous Date (%)	Implied Doubling Time (years)	Source of Figure
100,000	10,000	—	—	Stringer, 150
30,000	500,000	0.56	12,403	Livi-Bacci, 31
10,000	6,000,000	1.25	5,580	Livi-Bacci, 31
5000	50,000,000	4.33	1,635	Biraben
3000	120,000,000	4.47	1,583	Biraben
2000	250,000,000	7.62	944	Livi-Bacci, 31
1000	250,000,000	0.00	∞	Livi-Bacci, 31
800	400,000,000	26.49	295	Livi-Bacci, 31
600	375,000,000	−3.18	n.a.	Livi-Bacci, 31
400	578,000,000	24.15	320	Livi-Bacci, 31
300	680,000,000	17.65	427	Livi-Bacci, 31
200	954,000,000	40.29	205	Livi-Bacci, 31
100	1,634,000,000	71.28	129	Livi-Bacci, 31
50	2,530,000,000	139.74	79	Livi-Bacci, 31
0	6,000,000,000	462.42	40	Livi-Bacci, 31

SOURCES: J. R. Biraben, "Essai sur l'évolution du nombre des hommes," *Population* 34 (1979):
13–25; Massimo Livi-Bacci, *A Concise History of World Population,* trans. Carl Ipsen (Oxford:
Blackwell, 1992); and Chris Stringer and Robin McKie, *African Exodus* (London: Cape, 1996).

close to us genetically, physically, socially, and intellectually, we have no evidence that their numbers, the range they occupy, or their technologies have
changed greatly during the past 100,000 years. Indeed, that is precisely why
humans can be said to have had a "history," while the very idea of chimps
having one seems slightly bizarre. Most animal species don't have histories
as we usually use the word; once they have evolved, they tend to remain
within their original niche until they vanish from the fossil record. Whole
families or orders of species, such as the dinosaurs or mammals, can be said
to have histories, because different species within these groups can evolve
in many different ways and thus the numbers, the ranges, and the ecological "technologies" of whole families of animals do change. But the same is

TABLE 6.3. GROWTH RATES IN DIFFERENT HISTORICAL ERAS

Era	Start (years BP)	End (years BP)	Population at Start (millions)	Population at End (millions)	Rate of Growth Each Century (%)	Implied Doubling Time (years)
Late Pal. Era	100,000	10,000	0.01	6	0.71	9,752
Mid. Pal.	100,000	30,000	0.01	0.5	0.56	12,403
Upper Pal.	30,000	10,000	0.5	6	1.25	5,579
Agrarian Era	10,000	1000	6	250	4.23	1,673
Early Ag.	10,000	5000	6	50	4.33	1,635
Era of Ag. Civs.	5000	1000	50	250	4.11	1,723
Exc. 1st mill. CE	5000	2000	50	250	5.51	1,292
Modern Era	1000	0	250	6000	37.41	218
Early Mod. Era	1,000	200	250	950	18.16	415
Indust. Era	200	0	950	6000	151.31	75

SOURCE: Table 6.2.

not normally true of single species. Humans have multiplied and diversified their behaviors in ways that are characteristic not of single species but of entire families or orders of animals—and they have done so in an astonishingly short period.

Clearly, a fundamental threshold of some kind was crossed with the appearance of our species. Human history marks the appearance of new rules of historical change. So, to focus on human history is not just a matter of genealogical vanity. The appearance of our species marks a significant turning point in the history of our planet. As A. J. McMichael writes: "Each species is an experiment of Nature. Only one such experiment, *Homo sapiens*, has evolved in a way that has enabled its biological adaptation to be complemented by a capacity for cumulative cultural adaptation. This unprecedented combination of the usual biologically-based drive for short-term gain (food, territory and sexual consummation) with an intellectual capacity to satisfy that drive via increasingly complex cultural practices is what distinguishes the human 'experiment.'"[6]

EXPLAINING THE APPEARANCE OF HUMANS

Over the years, many "prime movers" have been proposed to explain the transition to humanity. These range from bipedalism, which freed our dexterous hands for toolmaking (Darwin's preferred answer), to hunting and meat eating, to large brains, to human languages. The explanation that follows focuses on the importance of human language, but allocates supporting roles to some of these other factors.

A somewhat abstract explanation is implicit in the preceding paragraphs. All species adapt to their environments, but most have only one or two adaptive tricks in their repertoire. In contrast, humans seem to constantly develop new ecological tricks, new ways of extracting resources from their environments. In the jargon of economists, humans seem to have a highly developed capacity for "innovation." And they innovated not on the Darwinian scale of hundreds of thousands or millions of years, but on a scale ranging from thousands of years to decades and even less time. Our challenge is to explain how, when, and why human beings acquired their new level of ecological creativity. If we can explain this vastly enhanced capacity, we will have gone a long way toward explaining what is distinctive about human history.

We have seen that the emergence of new forms of complexity always involves the creation of large structures within which previously independent entities are locked into new forms of interdependence and new rules of cooperation.[7] Following this hint, we should expect to find that the transition to human history is primarily marked not by a change in the nature of humans as individuals but rather by a change in the way individuals relate to each other. This suggests that we should focus not just on changes in the genes, the physiology, or the brains of earlier forms of humans but also on changes in the ways our ancestors interacted.

Like many other transitions of this kind, the emergence of our species was quite sudden. On the paleontological scale, it was an almost instantaneous event. This means that we should expect to find a single trigger. In star formation, temperatures rise over long periods until suddenly a trigger is released when hydrogen starts to fuse. So with human evolution: adaptive skills that may have evolved over many millions of years were suddenly transformed when a threshold of some kind was crossed. How can we describe this threshold? It clearly has something to do with an enhanced capacity to learn. Many animals learn, from flatworms to toads. But most of what most animals learn is lost when they die. Of course, some teaching goes on. Chimp mothers teach their children to crack nuts or fish for termites by demonstrating how to do it. And the infants, in time, may teach *their* children. But we know of no animal that can describe what to do in the abstract— no animal that could explain how to fish for termites without giving a demonstration, or give an account of a pathway without walking along it; and we certainly know of no animal that could describe abstract entities such as gods or quarks or pink elephants. The past and future, too, are abstractions, for only the present can be experienced directly; thus animals without symbolic language may lack the ability humans have to deliberately think about the past and imagine the future. These are severe limitations. For many years, the primatologist Shirley Strum observed a troop of baboons in Kenya that she named the "Pumphouse Gang." Compared to other troops, they were virtuoso hunters; they often ate meat as frequently as once a day. But they hunted most successfully when led by one particular male. And they had no way of storing his skills or knowledge once he was gone.[8]

Human language, however, allows more precise and efficient transmission of knowledge from brain to brain. That means that humans can share information with great precision, creating a common pool of ecological and technical knowledge, which in turn means that for humans, the benefits of cooperation increasingly tend to outweigh the benefits of competition. (John Mears has referred to humans as "highly networked creatures.")[9] Furthermore, the ecological knowledge contributed to that pool by each individual can survive long after his or her death. So knowledge and skills can accumulate nongenetically from generation to generation, and each individual has access to the stored knowledge of many previous generations. Thus what is distinctive about humans is that they can learn *collectively*. Cellular thinking (thinking that focuses on the individual) makes it hard to see this; but in explaining the distinctiveness of humans, we must learn to compare individual chimps not with individual humans (where the differences are significant but not transformative) but with entire *groups* of humans. We won't understand the difference if we compare individual human brains with individual chimp brains; we will begin to comprehend it only if we compare individual chimp brains with the huge, collective brains created by millions of humans over many generations.

The possibility of learning *collectively* changes everything. McMichael writes:

> The advent of cumulative culture is an unprecedented occurrence in nature. It acts like compound interest, allowing successive generations to start progressively further along the road of cultural and technological development. By traveling that road, the human species has, in general, become increasingly distanced from its ecological roots. The transmission of knowledge, ideas and technique between generations has given humans an extra, and completely unprecedented, capacity for surviving in unfamiliar environments and for creating new environments that meet immediate needs and wants.[10]

Collective learning is what gives humans a history, because it means that the ecological skills available to humans have changed over time. And there is a clear directionality to this process. Over time, processes of collective learning ensure that humans as a species will get better at extracting resources from the environment, and their increasing ecological skills ensure that, over time, human populations will increase. Generalizations about collective learning cannot predict the exact timing or geography of such processes, of course, nor how far they are likely to proceed, nor their detailed consequences; but such generalizations can tell us something about the long-term shape of human history on large timescales.

To get a feeling for the power of collective learning, it is enough to imagine life as it might be if we had to learn everything from scratch, receiving little more from family or community than hints about appropriate social behavior and eating habits, which is more or less the intellectual heritage of young chimps. How many of the artifacts around us (each of which embodies stored knowledge) could we invent or construct in a lifetime? Asking such questions is a powerful reminder of the extraordinary extent to which our lives as individuals depend on the accumulated knowledge of millions of other humans over many generations. Humans as individuals are not that much cleverer than chimps or Neanderthals; but as a species we are vastly more creative because our knowledge is shared within and between generations. All in all, collective learning is so powerful an adaptive mechanism that one might argue it plays a role in human history analogous to that of natural selection in the histories of other organisms.

Why can humans learn collectively? Because of the distinctive nature of human language. Human language is more "open" than nonhuman forms of communication. It is open grammatically because its strict rules of grammar allow us to generate a nearly infinite number of meanings from a small number of linguistic elements, such as words. It is also open semantically—that is, it can convey a wider range of meanings—because it can refer not just to what is in front of us but also to entities that are not present, and even to entities that could never be present. By using symbols, we can gather large amounts of information stored in our memories into single blocks; then we can construct even larger conceptual structures with these symbolic building blocks. Symbols enable us to abstract from the concreteness of things—to refer, as it were, to the distilled "essence" of what is around us. But they can also refer to other symbols. So they can condense and store huge amounts of information, just as the symbolic tokens we call *money* offer a compact and efficient way of storing and exchanging abstract values.[11] Symbolic languages let us store and share information that may have been accumulated during thousands or millions of lifetimes. All in all, symbolic language is a vastly more powerful data mover than are any presymbolic forms of communication. As Terrence Deacon has argued, presymbolic forms of communication "can only refer to something else by virtue of a concrete part-whole link with it, even if this has no more basis than just habitual coincidence. Although there is a vast universe of objects and relationships susceptible to nonsymbolic representation, indeed, anything that can be present to the senses, this does not include abstract or otherwise intangible objects of reference."[12]

If this argument is on the right track, it suggests that to understand the evolution of modern humans, we need to explain the emergence of symbolic language. But it is important to note straightaway that there was nothing inevitable about this process. Unlike star formation, which was statistically predictable, given what we know of the workings of gravity and the strong and weak nuclear forces, biological change is more random and open-ended, which is why living organisms are much more varied than stars. The elements that eventually combined into our species came together erratically and haphazardly, and there was never any certainty that they would assemble themselves in this particular way. As late as 100,000 years ago, well after our species had appeared, human populations may have fallen to as few as 10,000 adults, which means that our species was as close to extinction as mountain gorillas are today.[13] This statistic is a powerful reminder not merely of the haphazardness of evolutionary processes but also of the fragility of complex entities. The appearance of human beings on Earth was an extremely chancy business.

EVIDENCE AND ARGUMENTS: CONSTRUCTING THE STORY OF HUMAN EVOLUTION

Creating a coherent and plausible account of human evolution has been one of the great achievements of twentieth-century science. But how has this story been constructed? Before we look more closely at the story of human evolution, we must examine the types of evidence and the sorts of arguments that have been used to assemble it.

The fossil evidence includes the bones of ancestral species as well as the remains they left behind: their tools, scraps of food, and the marks they made on bones or rocks. Modern paleontologists can glean a remarkable amount of information from a bone. A jawbone can do more than identify a species; the patterns of toothwear can tell us about an animal's normal diet, and that can tell us about the environments it lived in and the way it exploited them. A skull can tell us about the intellectual capacities of a species. And the lower part of a skull can often tell us whether a species walked on two legs or four;

with bipedal animals, the spine enters the skull from below, whereas in quadrupeds it enters from behind. A toe bone on its own indicates how an animal walked: if the big toe is separated from the other toes (as it is in most primates), we can be sure the foot was still used for grasping and not yet specialized for walking. Often, a few bones are all we find. But the bonanza of a more complete skeleton, such as Lucy (40 percent of whose skeleton was found by Don Johanson in Ethiopia in the early 1970s), tells us much more. Lucy and the remains found near her were about 3 to 3.5 million years old, and they offer detailed evidence of the physiology of at least one species of early humans from that era.

The remains of human activity are equally important. Most important of all have been discoveries of stone tools, partly because tools made of less durable materials—bark, bamboo, and so on—rarely survive. Microscopic analysis of the cutting edges of stone tools can tell us what they cut; analysis of where the stone came from can tell us whether their makers actively sought out particular raw materials from other areas; reconstruction of flakes from the sites where stone tools were made can tell us much about how they were made; and techniques of toolmaking can give us valuable hints about how our ancestors thought. Analysis of the bones of other animals from early human sites can tell us much about whether or not our ancestors ate meat, as well as how they hunted. For example, careful analysis of cut marks on bones has sometimes found signs of human butchering overlaid on the tooth marks of carnivorous animals. Presumably this means that early humans scavenged animals first killed by other predators. All these types of material evidence can also be dated, more or less accurately, using the increasing number of modern dating techniques. (On radiometric dating techniques, see appendix 1.)

But the fossil record is spotty; and until very recently, we had no fossils at all for the crucial period, from about 4 to 7 million years ago, when the hominines (the lineage leading to our own species) diverged from the lineage that leads to modern chimps. So other forms of evidence have to fill in the gaps. One of the most important in recent decades has been provided by molecular dating. As we saw in chapter 4, much evolutionary change is random. This is particularly true of those parts of a species' genome that do not directly affect its survival chances, including the large amounts of "junk DNA" and the DNA contained in the mitochondria of all human cells. Genetic change in these parts of the genome is "neutral"—it doesn't affect the developed organism. Change in junk DNA is thus like the shuffling of a vast deck of cards. Fortunately, random processes of this kind are subject to general statistical laws. If you take a new deck of cards arranged by suit and number and shuffle it a few times, a statistician can estimate roughly how often it has been shuffled by determining how much the pack differs from its original condition. The larger the number of cards and shuffles, the more precise and reliable such estimates can be.

In an article first published in 1967, two biochemists working in the United States, Vincent Sarich and Alan Wilson, argued that much genetic change is subject to similar rules.[14] Thus, if we take two modern species and calculate the differences between their DNA sequences, we can estimate pretty well when their two lines diverged from a common ancestor. In this way, the evolution of DNA can provide a sort of genetic clock. The idea was ridiculed at first, partly because many took it as a fundamental tenet of natural selection that all evolutionary change was adaptive, which should have implied that changes did not occur in statistically predictable ways. However, it is now agreed that much change is indeed random—and in any case, the results of such dating methods have turned out to fit remarkably well with many other types of evidence. Genetic comparisons of this kind are now made and used routinely to understand the relationship between different species, though some problems remain. For example, it is clear that not all genetic change occurs with the regularity necessary if it is to be used as a clock. But these methods can be extremely valuable for many purposes, particularly in the study of human evolution.[15]

The first thing Sarich and Wilson showed was that, genetically speaking, we are closer to chimpanzees than was once thought. In the 1970s, it was widely believed that the two lines leading to humans and apes had diverged at least 15 million, and perhaps as much as 30 million years ago—a comfortable distance for those unhappy at the thought of close kinship with chimps. Yet the DNA of modern humans differs from that of our nearest living relatives by only about 1.6 percent. That is to say, 98.4 percent of our DNA is identical to that of modern chimpanzees. This means that all the variation between our history and that of chimps must be explained with reference to the 1.6 percent of our genetic material that is different from that of chimps. Comparisons of the rate of genetic change of mammals were possible because it was known that mammal species had diverged rapidly from each other about 65 million years ago, when the dinosaurs were driven to extinction. But it turned out that humans and chimps differed from each other only by about 10 percent as much as the differences between major groups of mammals, which suggested that they had diverged from each other approximately 5 to 7 million years ago. This implies that at the time of that divergence, there lived an animal that was the ancestor of both modern humans and modern chimps, though it would have looked different from either of these living species. The thinness of the fossil record in this period means that we can say little about this ancestral being.[16] But we can be sure that such an animal existed—otherwise we would not exist! Similar arguments suggest that humans and gorillas had a common ancestor 8 to 10 million years ago, and humans and orangutans about 13 to 16 million years ago.

We also know a lot about the environments in which our hominine ancestors evolved, based on analyses of changing climates and plant and animal remains. In the past few million years, global climates were dominated by the erratic and unpredictable climatic changes of the ice ages (see chapter 5). These changes altered habitats and environments, which favored species that were highly adaptable and able to use a wider variety of ecological niches. Generalist or "weedy" species that can adapt well to ecological disruption, such as modern humans, may have been typical products of the ice ages.[17]

In combination, these various techniques enable us to describe the physical evolution of hominines and the environments they lived in, but describing behavior is much trickier. Fossils can tell us something about lifeways; but to go further, we have to rely on modern analogies with other species that may have lived similarly. Researchers in recent decades, beginning with Jane Goodall and Dian Fossey, have studied the lives of great apes in the wild, and we now know a lot about how they live and about their social, sexual, and political relations.[18] Such studies can suggest how early hominines may have lived; but they can also mislead us, for different types of ape, and even different communities of a given type of ape, can live in different ways. For example, *Pan troglodytes*, the most familiar species of chimp, lives in communities dominated by closely related males, which are joined by females from other communities. Males form hierarchies, but these are changeable, and females may mate with several males, circumstances that make the sexual and political life of such communities extremely complex. Gorillas, in contrast, usually live in smaller groups of several females, with one or perhaps two males. Orangutans are for the most part solitary, coming together only to mate. So deciding exactly what analogies with primate societies can tell us about the societies of early hominines is not easy.

The same is true of the other analogy that has been extremely influential in studies of hominine evolution: the analogy with modern "foraging" societies.[19] Anthropologists constantly remind paleontologists that modern foraging societies are very modern—all have been influenced in some way by modern society. So building theories about hominine or early human social structure on these analogies may be risky. Nevertheless, because the technologies and the social structures of modern foraging societies are certainly closer to those of early humans than are those of a modern urban community, the anthropologists' warnings are routinely ignored. Modern studies, such as those of the San peoples of southern Africa, have helped us construct plausible models of how early hominines and humans hunted, how males and females related to each other, and what sort of power games may have been played. Perhaps most important, they have reminded us that societies that seem simple to modern city dwellers were, in their own ways, both complex and sophisticated. After all, it was no mean task to live successfully for thousands of years, in the deserts of southern Africa or Australia, or in the Siberian tundra, using Stone Age technologies.

Finally, modern understanding of how other species have evolved has been used to construct models of how humans may have evolved. For example, it is quite common to find that a new species evolves that is remarkably like the immature individuals in the species from which it has evolved. This process is called *neoteny*, and it occurs through minor alterations in the genetic switches that control the life cycle of a species. Such changes can launch a cascade of secondary and tertiary effects that cause significant evolutionary change. It has been argued that in many respects, humans are more like young chimps than like adult chimps; this similarity implies that we may have evolved, in part, through some form of neoteny, while modern chimps may have remained more like the adults of our common ancestor. Equally, modern evolutionary research has shown that evolution often occurs in fits and starts. If a new niche appears, perhaps as a result of climatic change, it is often filled quite rapidly ("rapidly" in evolutionary terms, which means in a few hundred thousand or even a few million years) with a large number of quite similar species, most of which may then get weeded out, leaving only one or two surviving lines. This process is known as an *adaptive radiation*, and each radiation seems to be associated, roughly speaking, with a particular ecological trick. Among our ancestral species, as we will see, there seems to have been several adaptive radiations, each of which, we can now recognize, added something new to the package that became us.[20]

All these types of evidence have been used to construct the modern account of how humans evolved. That account is far from perfect, but it is far richer and is based on far more evidence than the accounts of even ten years ago.

PRIMATE AND HOMININE RADIATIONS

I have argued that the evolution of symbolic language may mark the critical threshold that leads to human history. But symbolic language could not have made such a difference if our ancestors had not had other qualities that enabled them to exploit the advantages that it bestowed. Among the most important of these preadaptations are sociability, preexisting linguistic skills, bipedalism and dexterous hands, meat eating and hunting, a long period of childhood learning, and large forebrains. Here, we will try to trace the haphazard processes by which these various elements evolved and combined in the package of features that makes up our species.

Primate Heritage

We share many of the features named above with other primates.[21] Most primates have been tree dwellers. Animals that live in trees have to be able to see well, or they fall out. So all primates have good, stereoscopic vision. Smell is less important, much less important than it is for dogs, which is why most primates have smallish snouts and flattish faces. Visual information, in a complex, three-dimensional environment, requires considerable processing, so most primates have largish brains in comparison to their body size, and the primate line as a whole has been characterized by increasing relative brain size. Larger brains usually imply longer lives—perhaps because they imply greater dependence on learning, and learning improves with age (in principle). Tree dwelling also requires dexterity, so most primates have hands and feet that can grip and manipulate objects well. In practice, this means that their thumbs and big toes can be opposed to their other fingers and toes. Tree dwelling also encourages a greater specialization of labor between front and back limbs than is normal for ground-dwelling species. Though most primates can grip with both their feet and hands, the hindlimbs tend to specialize in locomotion while the forelimbs specialize in gripping.

Humans belong to a particular group of Old World primates known as the Hominoidea. This includes humans and the apes—chimpanzees, gorillas, orangutans, and gibbons—as well as all their now-extinct ancestors. The oldest fossils classified within this superfamily of organisms are just over 20 million years old, which means they appear early in what geologists call the Miocene era (ca. 23–5.2 million years ago). These remains belong to a species known as *Proconsul*.[22] Though the hominoids probably evolved in Africa, from as early as 18 million years ago hominoid remains also turn up in the southern parts of the Eurasian landmass, from France to Indonesia. The hominoids were a diverse group, and for a time they may have been more numerous than other species of Old World monkeys. Their migrations provide a typical example of an adaptive radiation.

The fossil record is not well-enough defined for us to be sure which were the evolutionary tricks that best define the hominoids, though increasing size, greater manual dexterity, larger brains, and a willingness to move away from tree cover may be among them. These are all features that we share with the remaining members of this superfamily of primates.

Bipedalism and the First Hominines

The Homininae are a subfamily of the family of Hominidae, the great apes. The hominines include only our own immediate ancestors. Their story begins at the transition from the Miocene to the Pliocene eras, between 5 and 6 million years ago. The construction of that story begins with the realization, based on molecular dating techniques, that about 6 million years ago, there existed somewhere in Africa an animal that was the ancestor of both modern chimps and modern humans. Since then, in a series of adaptive radiations, a large number of different species of hominines has appeared—perhaps as many as twenty or thirty. Whereas thirty years ago the difficulty was to find *any* hominine remains, today the difficulty is to decide which of the many species we now know of lies on the line that evolved into modern humans.

For a modern paleontologist, the holy grail is to find the remains of the species from which both chimps and humans are descended. And it is possible that this species, or something close to it, has already turned up. In 2000, a team of French and Kenyan archaeologists, working north of Nairobi, found the remains of a creature about 6 million years old, which was promptly dubbed "Millennium Man" in the press.[23] But its true status remains uncertain. Its appearance is sufficiently apelike that many paleontologists have placed it on the chimp rather than the hominine side of the great divide between our two species. Similar criticisms have been leveled at another possible candidate for the oldest hominine, *Ardipithecus ramidus kadabba,* some remains of which were found by an American team of archaeologists in the Great Rift Valley of Ethiopia, as reported in the journal *Nature* in July 2001.[24] These remains have been dated to between 5.2 and 5.8 million years ago. They include a toe bone, whose shape suggests that this creature walked on two legs. At present, most paleontologists are agreed that the decisive feature distinguishing hominines from apes is bipedalism: all known species of hominines are bipedal, while no known species of apes are (though chimps can stand for short periods).[25] So the determination of whether these early specimens were really bipedal or not will be crucial; for now, the evidence is equivocal.

Debate over the significance of these finds is complicated by the fact that no one is quite sure why bipedalism evolved, though there have been many theories.[26] Some focus on the role of climatic change. Twenty million years ago, the African continent was relatively flat, and its equatorial regions were covered fairly evenly with tropical forest. But beginning about 15 million years ago, the African tectonic plate began to tear in half. Tectonic activity along the Great Rift Valley has created a chain of highlands and rift valleys running north and south along the eastern part of the continent. By splitting open the earth's crust, the rift valleys have provided a happy hunting ground for fossil hunters. But it was the mountains that may explain the presence of hominine fossils here, for they cast a huge rain shadow over the eastern parts of the continent, making it drier than the lands to the west. Yves Coppens has argued that this aridity drove some species into less-forested landscapes, where they had to move greater distances between stands of trees in order to find the types of foods to which they were accustomed. This might have encouraged the evolution of a more upright stance, for the knuckle-walking characteristic of chimps is not a good way of traveling over long distances. Unfortunately for this promising theory, some of the most recent early hominine fossils, including those of *Ardipithecus ramidus kadabba*, have turned up in environments that were probably forested.[27]

Perhaps bipedalism enabled hominines to see potential predators from a greater distance in open country. Or perhaps it was more energy-efficient than the knuckle-walking typical of chimps, and enabled early hominines to search for food over larger areas. Or perhaps walking upright in unshaded environments provided some protection from the midday sun by limiting the area of skin exposed to direct sunlight. These and other pressures would have favored those individuals that found it easiest to walk upright. (The last argument may also explain why hominines, at some point in their evolution, became less hairy than the other great apes.) Comparisons with chimps are suggestive, for as Coppens points out, chimps try to stand in three situations: "to see farther, to defend themselves or to launch an attack—since standing up frees their hands and allows them to throw stones—and to carry food to their offspring."[28]

Whatever the causes of bipedalism, the fossil evidence, thin as it is, shows that within 2 million years, a number of bipedal species had appeared. These include the species known as *Ardipithecus ramidus ramidus*, whose remains were found in Ethiopia in 1994 and dated to ca. 4.4 million years ago. These early hominine species constitute the first major adaptive radiation in the history of hominines, and their success is probably associated with the advantages of bipedalism, whatever they were.

Australopithecines

The next two hominine radiations are associated with a group of species that paleontologists refer to as *australopithecines.*

All australopithecines were bipedal. We know this from the structure of the pelvis, the relative length of arms and legs, and the entry point of the spine into the skull (from below rather than from behind). The oldest of several species of australopithecines known at present is *Australopithecus anamensis,* a species whose remains were found in the Lake Turkana region in northern Kenya in 1995. These have been dated to ca. 4.2 million years BP.[29] The best-known australopithecine fragments were found in Ethiopia in the 1970s by the American paleontologist Don Johanson. He found 40 percent of the skeleton of a bipedal female that he christened Lucy (reportedly after the song "Lucy in the Sky with Diamonds"). Lucy was about 1.1 meters tall, though other remains found nearby were up to 1.5 meters tall. All these remains were between 3 and 3.7 million years old; they are normally classified as members of the species *Australopithecus afarensis,* so named after the Afar valley in Ethiopia, where they were found.[30] In 1998, in South Africa, an even more complete australopithecine skeleton was found along with its skull. This has been dated to between 2.5 and 3.5 million BP. The famous Laetoli footprints found by Mary Leakey may have been even older, for they date from at least 3.5 to 3.7 million years BP. These were made by three australopithecines, two of whom walked side by side, while a third walked in the steps of the leader. They apparently held hands as they walked through what may still have been hot volcanic ash. These astonishing footprints confirm directly what other fossil remains suggest indirectly: the oldest-known hominines were bipedal. In 1995, archaeologists working in Chad, well to the west of the Great Rift Valley, discovered the remains of a new species, *Australopithecus bahrelghazali,* which seems to have lived between 3 and 3.5 million years ago. Clearly, australopithecines lived on both sides of the Great Rift Valley. The several hundred individual australopithecines whose remains have been found in this century thus occupied a large area, reaching from Ethiopia to Chad to South Africa.

Though australopithecines walked on two legs, close study of their anatomy and particularly their hands has shown that they remained well adapted to life in the trees, and their walking was not yet as efficient as that of modern humans. Even more important, they had small brains, ranging in size from ca. 380 to 450 cubic centimeters. This contrasts with the 300 to 400 cubic centimeters of modern chimps, and an average brain size of 1,350

Figure 6.2.   A reconstruction of Lucy, an australopithecine who lived in the Hadar valley in what is today Ethiopia about 3.2 million years ago. Lucy was about 1.1 meters tall and had a brain about the size of that of a modern chimpanzee. From G. Burenhult, general ed., *The First Humans*, vol. 1 of *The Illustrated History of Humankind*, 5 vols. (San Francisco: HarperSanFrancisco, 1993). Copyright © 1993 by Weldon Owen Pty., Ltd./Bra Bocker AB. Reprinted by permission of HarperCollins Publishers, Inc.

cubic centimeters for modern humans. The first distinguishing feature of the hominine line was not braininess but bipedalism (see figure 6.2).

There is strong reason to think that our own lineage can be traced back to early forms of australopithecines. But there also appeared, in a quite distinct radiation, a second group of australopithecines that, in the language of paleontology, were more "robust"-looking than *afarensis*. These existed between 3 and perhaps as late as 1 million years ago and are sometimes assigned to a separate genus, *Paranthropus*. What was distinctive about them, and marks them out as a different evolutionary line from us, is that they evolved exceptionally strong jaws for the grinding of tough, fibrous plant foods. They therefore had heavyset skulls with flamboyant crests that provided anchors for powerful chewing muscles.

What can we say about the lifeways of australopithecines? If we start with diets, it seems likely that most australopithecines depended mainly on the types of foods their ancestors had eaten in forested environments. Their teeth are adapted for grinding the shells of tough or fibrous fruits, leaves, and other plants. However, they probably ate meat occasionally, for direct observations have shown that most living primates are occasional carnivores and eat as much meat as they can.[31] Either they hunted small or weak animals (including other primates) or they scavenged meat from animals that had died naturally or been killed by other carnivores. But in the main, australopithecine diets were vegetarian.

Analogies with modern primates occupying similar niches suggest that australopithecines probably lived in small family groups that traveled together, with individuals foraging separately for their own food. There is no evidence that they had more linguistic ability than modern chimps. This does not mean that there was no politics or no communication. As in many modern primate societies, males and females probably formed hierarchies of dominance and spent a lot of time dealing with, and presumably thinking about, group politics. Like modern chimps, australopithecines may have communicated through gestures, sounds, and activities such as grooming. But neither chimps nor australopithecines had the vocal apparatus or the intellectual ability necessary to precisely communicate abstract information.

Studies of the societies of primates closely related to modern humans offer contradictory suggestions about the nature of the earliest hominine societies. Genetically, we are closest to chimpanzees, and members of the best-known chimp species, *Pan troglodytes*, live in bands linked by closely related males. Males stay with their natal groups, while females move away from their natal groups. But most australopithecine species, unlike chimps, appear to have been quite sexually dimorphic (i.e., males were much larger than females). This suggests that in some respects australopithecine "societies" may have been closer to those of gorillas.[32] Among gorillas, males are large because they compete with each other for access to females, which ensures that the largest males produce the most offspring. The result is a social world in which a dominant male and perhaps one other, younger male will travel with several females and their children in groups of up to about twenty individuals. Perhaps we should envisage a world somewhere between these structures. This may have been a world of bands somewhat smaller than those of modern chimps, in which closely related males competed for dominance and for access to females. Perhaps dominant males gained access to several females, but that access was not exclusive. Australopithecines may have lived in a world whose males engaged in more or less constant competition with each other to attract female partners. Yet at the reproductive core of this world of competitive but closely related males, there existed smaller but closer units, consisting of mothers and their children, as in most modern primate groups. Chimpanzee mothers are known to form durable and clearly affectionate relations with their children, while males show little interest in the tasks of child rearing and no sense of paternity. All in all, there is little to suggest that australopithecines were radically different in their physiology or lifeways from the ape species of today.

Tool Use and Meat Eating: Homo habilis

For hominid specialists, Olduvai Gorge, a 50-kilometer-wide canyon in the Serengeti Plain in northern Tanzania that is part of the African Rift Valley, is a special place: finds from here have provided the best evidence that our species evolved in Africa. Here, in 1960, Jonathan Leakey—the son of Louis Leakey, one of the pioneers of modern studies of human evolution—found a hominine fossil about 1.4 meters tall. Louis Leakey claimed that it belonged to the same genus as human beings *(Homo)* and therefore christened it *Homo habilis,* or "handy man." This made it the oldest species of the genus that includes modern humans.

Though many anthropologists felt the remains simply belonged to an unusually gracile form of australopithecine, two factors encouraged Leakey to think that this species was more "human." First, associated with *Homo habilis* he found the earliest evidence for the systematic manufacture and use of stone tools. The skills involved in these activities seemed significantly more complex than those evident among earlier hominines. Second, the brains of *habilis* were a lot bigger than those of the australopithecines, ranging from 600 to 800 cubic centimeters. *Homo habilis* seemed to be a tool-using, learning animal, like modern humans; so perhaps the appearance of the new species, about 2.3 million years ago, marked the real beginnings of human history. Modern anthropologists have retained Leakey's nomenclature, and there is no doubt that *habilis* shows distinctive features, some of which may have been triggered by ecological changes caused by cooler, drier climates beginning about 2.5 million years ago. For example, stone tools

made by *habilis* show signs of "handedness," which implies a division between left and right sides of the brain; that, in turn, may be a necessary precondition for improved linguistic skills.[33] Nevertheless, recent research into
the growing number of *habilis* remains and sites has shown that the gulf
between them and modern humans in intellectual ability and lifeways is
wider than Leakey supposed.[34]

Part of the reason for these shifts in attitudes toward *habilis* is that modern paleontologists are less impressed than Leakey was by signs of tool use.
We now know that many animals use tools of some kind, and chimpanzees
use tools in more ways than any other animals apart from humans. For example, chimps have been observed inserting sticks into termite mounds, then
quickly pulling them out and eating the termites that are still clinging to
them; some even use rocks to crack open nuts. However, *habilis* seems to
have used tools in new ways that required more planning and foresight. Paleontologists describe their stone tools as Oldowan, a name taken from the
Olduvai Gorge in which so many have been found. These tools have a quite
distinctive form that persists in the archaeological record for almost 2 million years, until almost 250,000 years ago (see figure 6.3). They consist
mainly of large stones, often river cobbles of tough basalt or quartzite, from
which smaller chips have been removed by striking with a "hammer" stone,
to create one or two cutting edges.

Making such tools requires considerable planning and experience, much
more than is needed to make the simple tools used by chimpanzees. Modern experiments in stone knapping have shown that the original stones need
to be chosen carefully, and struck with precision. In fact, making stone tools
requires precisely those skills that are the forte of the prefrontal cortex, that
part of the brain that was to expand most significantly in human evolution.
It is possible that tool use evolved through a process known as Baldwinian
adaptation (named after the nineteenth-century American psychologist who
first described it systematically). This is a form of evolutionary change that
appears to combine Darwinian and cultural elements, because behavioral
changes lead to changes in an animal's lifeways, thereby creating new selective pressures that lead, over time, to genetic changes. For example, species
that learn new behaviors that let them live in cold climates may eventually
adapt genetically to their new environments by evolving furry coats (as did
mammoths or woolly rhinos). Among humans, groups that herded domestic animals eventually acquired, over many generations, an enhanced capacity to digest milk, as rare mutations that prolong the production of the
milk-digesting enzyme lactase into adult life became more common. Perhaps,
in a similar way, hominine individuals that were most skilled at making and

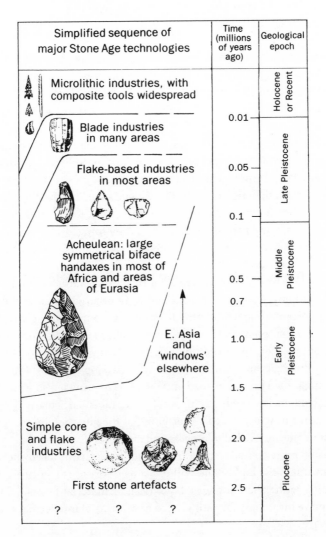

Simplified sequence of major Stone Age technologies	Time (millions of years ago)	Geological epoch
Microlithic industries, with composite tools widespread		Holocene or Recent
Blade industries in many areas	0.01	Late Pleistocene
Flake-based industries in most areas	0.05 / 0.1	Late Pleistocene
Acheulean: large symmetrical biface handaxes in most of Africa and areas of Eurasia	0.5 / 0.7	Middle Pleistocene
E. Asia and 'windows' elsewhere	1.0 / 1.5	Early Pleistocene
Simple core and flake industries	2.0	Pliocene
First stone artefacts	2.5	Pliocene
? ? ?		

Figure 6.3. The evolution of stone tools over 2.5 million years. From Steven Jones, Robert Martin, and David Pilbeam, eds., *Cambridge Encyclopedia of Human Evolution* (Cambridge: Cambridge University Press, 1992), p. 357. Reprinted with the permission of Cambridge University Press.

using tools gained such selective advantages that they had more offspring than others, so that their intellectual skills were soon incorporated into the genetic makeup of the entire species. If so, then tool use may have been both a cause and an effect of brain growth, in a process of positive feedback.

What were the stone tools used for? Modern experiments have shown that Oldowan choppers could be used successfully to break up bones or to work more crudely in wood. But the chips struck off them were probably more important than the cores, for these made small, sharp flakes that could be used for butchering and carving. So we can imagine *habilis* individuals and groups carrying pebbles with them as they foraged, and striking flakes off those pebbles when necessary. Microscopic examination of their edges has shown that Oldowan stone tools had many uses. Perhaps their most important role was in making available a richer and more varied diet. They could be employed to get at tubers that were otherwise inaccessible. Even more important, choppers and the flakes made from them could be used to scare away other predators from killed animals, to get at the marrow bones of large animals, and to butcher their carcasses. However they got their meat, *habilis* individuals ate more of it than did australopithecines, as dental evidence suggests. This richer food stuff may have provided some of the extra metabolic energy needed to support larger brains, particularly if, as seems likely, meat eating permitted a shortening of the gut, thereby reducing the amount of energy needed to process and digest food. Meat eating may also have led to more complex social lives, for it has recently been shown that chimps value meat highly and will use it as a sort of currency—a way of bargaining with others for sexual, political, or material favors.[35] In short, eating more meat may have stimulated new forms of intellectual and social complexity.

But we should not exaggerate the importance of meat in *habilis* diets. It is no longer believed that these primates were more than occasional hunters, perhaps like some groups of modern chimps.[36] Study of *habilis* teeth suggests that they, too, lived mainly off fruits and plant foods, even if meat provided an occasional and highly valued supplementary food. Besides, their stone tools were remarkably simple in comparison with those of modern foragers and, while useful in foraging or scavenging, they would have been of little use in true hunting. Close examination of cut marks on bones at *habilis* sites shows that they butchered carcasses but did not always kill them, for the cuts they inflicted often lie *over* the tooth marks of other animals. They may have killed small animals, but they probably scavenged the meat of larger animals that had died of natural causes or had been killed by other animals.

Anatomical studies also suggest that *habilis* was not completely bipedal, and may have spent much time in trees. So we should imagine *habilis* groups of any number from five to thirty individuals foraging separately during the day, like modern primates or australopithecines, and perhaps coming together at night and taking refuge in trees. Their preferred ecological niche was still similar to that of the australopithecines, though the scavenging of meat was more important to them and they spent more of their time on the ground.

All in all, there is no clear evidence of the quantum leap in intelligence and social complexity that Louis Leakey assumed when he first encountered *Homo habilis*.

Larger Brains and Ranges: Homo ergaster *and* Homo erectus

Homo habilis lived in East Africa with several other species, including robust australopithecines *(Paranthropus)*. Indeed, in a pattern that is common in the early history of new adaptations such as bipedalism, the early history of hominines shows a great variety. Perhaps six or more different species of hominines lived at the time of *habilis*.

About 1.8 million years ago, at the transition from the Pliocene to the Pleistocene period on the geological timescale, there appeared a new hominine species, known to modern anthropologists as *Homo erectus* or *Homo ergaster*.[37] A spectacularly well-preserved sample of *ergaster*, dated to ca. 1.8 million years ago, was found at Nariokotome, in Kenya, in 1984. "Turkana boy," as this fossil is known, is the most complete of all hominine fossils. Turkana boy died while still an adolescent, but he was already more than 1.5 meters tall and had a skull of ca. 880 cubic centimeters, almost ⅓ larger than those of most *habilis* individuals.[38]

By one million years ago, in one of the more spectacular hominine radiations, various forms of *erectus/ergaster* displaced all other forms of hominines. *Homo ergaster* individuals were taller than *habilis* and had larger brains, ranging from 850 to 1,000 cubic centimeters. This brings them close to the range of brain sizes in modern humans. There are other signs that they were significantly closer to modern humans. From ca. 1.5 million years ago, they began to manufacture a new type of stone tool, known as Acheulian hand axes, whose production demanded more intellectual sophistication than Oldowan tools. They are shaped more precisely and more elegantly than Oldowan choppers. And they are shaped on all sides, to form pear-shaped "axes," normally with at least two cutting edges. Sometimes, Acheulian stone tools were finished with a bone hammer to produce a finer edge. Some *ergaster* populations may also have learned to use fire. This would have provided valuable protection, particularly in cave dwellings, and would also have made it possible to soften and clean meat by cooking it. However, if they did use fire, they did so unsystematically. There is no evidence, for example, that they used hearths.[39]

It is likely that *ergaster* had linguistic abilities superior to those of *habilis*, but how superior is hard to tell. Larger forebrains suggest an increased capacity to understand and process symbols, while a larynx that sat lower in the throat may have allowed greater vocal flexibility; as a result, vocal communication may have increased in importance in comparison with gestural communication. Still, there is little direct evidence of the rich capacity for symbolic activity that is apparent in the fossil evidence of modern humans, so it seems likely that symbolic communication, even if it existed in some form, had not yet had a revolutionary impact on either the behavior or the consciousness of *ergaster*.[40] Steven Mithen has made the interesting suggestion that *ergaster* individuals may still have used what language ability they had mainly in social situations.[41] There is no evidence that language was used to deal with technological problems, for once they appeared, the Acheulian axes of *ergaster* show little change over a million years. And, though *ergaster* diets probably included more meat than the diets of their *habilis* relatives, it is unlikely that even they engaged in systematic hunting of the kind we find among modern foragers.

The most important sign that there was an increase in the behavioral flexibility of these species is the fact that they included the first hominines to migrate out of East Africa and then out of Africa entirely and into Eurasia. By about 700,000 years ago, communities of *Homo erectus* lived in parts of southern Asia and had even entered Ice Age Europe. *Erectus* remains were first found in Indonesia in 1891, and perhaps the best-known finds were made in the 1920s at the Zhoukoudian cave, which is today just a suburban train ride from modern Beijing. All in all, *erectus* explored a wider range of niches than those used by *habilis*—"wider" both ecologically and geographically. In particular, they apparently managed to live in regions whose climates would have been too cold or too seasonal for *habilis*.

An increase in the niches available to a species is normally a sign of considerable demographic success, and it seems reasonable to assume that the numbers of hominines increased with the number of available niches. Though we do not know the numbers of any early species of hominines, they were probably similar to the populations of great apes before the twentieth century. There were perhaps a few tens of thousands or perhaps as many as one or two hundred thousand hominines at any one time, and their numbers probably grew as they migrated into western and northern Africa and then into southern Eurasia. But there is as yet no evidence for long-term growth in the population numbers even of *erectus*. So we should not exaggerate the significance of these migrations out of Africa. In southern Eurasia, *erectus* entered environments that were more seasonal than those of East Africa's savanna lands, but otherwise quite similar. And many other mammal species had made similar migrations, including earlier species of hominoids. Finally, it is striking that *erectus* did not manage to inhabit the cold heartland of northern Eurasia.[42] Nor is there any evidence that they made the sea crossing to Australia and Papua New Guinea.

Prehuman Hominines of the Past Million Years

During the past million years, several new types of hominines appeared in different parts of Africa and Eurasia. And everywhere the brains of these species expanded rapidly. Eventually, many had brains as large as 1,300 cubic centimeters, which puts them within the range of modern human brain sizes. Beginning about 200,000 years ago, there also appears, after a long period of little technological change, a new type of stone technology: the so-called Levallois or Mousterian tools. In these, a stone core, shaped like a tortoise shell, was prepared in such a way that several flakes could be struck from it with a single, precisely calculated blow. Presumably a more varied tool kit was tied to the exploration of new niches.

Why should hominine brains have grown so quickly? Explaining brain growth is harder than it may seem, for large brains are rare—and with good reason. The modern human brain is arguably the most complex single object we know. Indeed, E. O. Wilson has argued that the evolution of the human brain constitutes one of the four great turning points in the history of life on earth.[43] Each human brain contains perhaps 100 billion nerve cells, as many cells as there are stars in an average galaxy. These connect up with each other (on average, each neuron may be connected to 100 other neurons) to form networks of astonishing complexity that may contain 60,000 miles of linkages. Such a structure can compute in parallel. That means that although each computation may be slower than that of a modern computer, the total number of computations being carried out in a particular moment is much, much greater. While a fast modern computer may be able to complete one billion computations a second, even the brain of a fly at rest can handle at least a hundred times as many![44] Surely, evolving a biological computer as powerful as this must have been a good Darwinian move.

But though this argument is intuitively plausible, there is a serious problem with it. If brains are so obviously "adaptive," why have so few species evolved really large brains in comparison with their body sizes? The trouble is that brains are costly to maintain. The human brain uses 20 percent of the energy needed to support a human body, but accounts for only 3 percent of body weight. Bearing large-headed infants is also difficult and dangerous, particularly for a bipedal species, as bipedalism requires narrow rather than wide hips. In other words, growing big brains is a chancy evolutionary gamble. So we cannot just assume that big brains evolved because they were obviously advantageous. Instead, we have to find more specific explanations.

One answer may be that brains provided good radiators for species living out in the open. This answer is not as flippant as it sounds. But there may be subtler and better answers. Perhaps there were feedback loops, involving forms of Baldwinian evolution. Changes in one area (either genetic or behavioral) may have caused changes in other areas, which created new selective pressures that reinforced the original change. One such loop, as we have seen, may link tool use and brain size.

A second loop, which may have operated in tandem, links sociability and brain size. Even among chimps, it has been shown that the capacity to calculate social relations accurately can increase the reproductive chances of individuals. And such processes may set up relatively speedy feedback loops, as more socially skilled individuals mate more frequently, producing more offspring who, in turn, are likely to have greater social and political skills. Eventually, such processes could have encouraged the expansion of those parts of the brain best able to make complex social calculations.[45] However, larger brains made birth a more painful and difficult process. At some stage, this problem may have been solved by a change in the rate of infant development. Hominine babies were born at an earlier stage of maturity. But this solution meant that infants became more and more helpless and demanded more parenting. That increased the importance of mothers being surrounded by a supportive social group, including both males and females. This shift may be linked to the fact that humans, unlike most other great apes (except for orangutans), have lost the estrous cycle; as a result, they can be sexually active even when conception is impossible. The partial separation of sexuality from reproduction may have encouraged stronger pair-bonding between males and females, thereby increasing the role of males in parenting, a change that may also be linked to declining sexual dimorphism in humans.[46] Whatever the details of these complex processes (and the archaeological record is too ambiguous for any certainty), hominines had to become more social as their brains grew larger. But living in larger or more complex social groups requires, as we have seen, complex social skills; and by and large, those with the greatest social skills were most likely to find mates. Feedback cycles of this kind—with increasing brain size stimulating increased social complexity, which encouraged further expansion in brain size—may explain why, at certain periods in hominine evolution, human brains (and particularly the prefrontal cortex) have expanded rapidly.[47]

A further possibility is that brain growth occurred as a by-product of quite small changes in the developmental schedule of hominines. As we have seen, neoteny, or the evolution of species similar to the *juvenile* forms of the species from which they evolved, occurs because of slight rearrangements in the genetic codes governing the rate and timing of development, as a result of which most features of a species develop more slowly, except for its sexual maturity. Thus, adult humans have flat faces and are relatively hairless. Chimps also have these qualities, but only during their youth. As they age, their muzzles push outward and they become more hairy. Most important of all, modern humans maintain the rate of brain growth typical of juvenile chimps, but sustain these rates for longer periods. This means that they grow larger brains and maintain for a longer time the rapid learning pace of juveniles. In this way, small alterations in the genes that control developmental processes can have a huge impact on the adult form of neotenous species.

A final possibility is that rapid brain growth has something to do with the evolution of more sophisticated forms of language. As with tool use, language skills probably correlated closely with brain power, giving those individuals with slightly bigger brains a significant Darwinian advantage. This would have accelerated the evolution of even larger brains in one more evolutionary feedback loop. We will explore this line of argument more carefully in the next chapter.

Whatever the cause, we know that hominine brains grew quickly beginning about 500,000 years ago. These changes provide clear evidence of increased intellectual capacity, and perhaps of increased linguistic ability. But, frustratingly, there is still little evidence of revolutionary changes in hominine lifeways. The best-known of these later hominine species are the Neanderthals. The first Neanderthal fossils were found in 1856 in the Neander valley in Germany. Though Neanderthals were long assigned to the same species as modern humans (technically, they were known as *Homo sapiens neanderthalensis*), recent genetic tests, using remnant DNA from Neanderthal fossils, suggest that the human and Neanderthal lines diverged perhaps as much as 700,000 to 550,000 years ago.[48]

Neanderthals first appear in the archaeological record about 130,000 years ago, and they vanish from the record as recently as 25,000 years ago. Their brains were as large as, and perhaps even larger than, those of modern humans, but their bodies were tougher and stockier. They clearly had the ability to hunt, and this enabled them to occupy Ice Age landscapes that had not been inhabited by any earlier hominines—for example, in parts of modern Ukraine and southern Russia. However, their hunting methods were inefficient and unsystematic in comparison with those of modern foragers, or even humans of the upper Paleolithic era. Their stone tools, usually described as Mousterian, are more complex than those of *erectus*, but show far less variety and precision than the stone tools of modern humans. There are hints of Neanderthal art or burial ritual, both of which might have signaled an increased use of symbolic communication (but the evidence is ambiguous). And there is little sign of great social complexity. Like earlier hominines, Neanderthals seem to have lived primarily in simple family groups that had limited contact with each other. There is no evidence that Neanderthals could have had the same impact on the planet as modern humans.

SUMMARY

This is a frustrating conclusion. We have seen that the evolution of modern humans was a revolutionary event within the history of the earth. And we can see all the elements of modern humanity being assembled over several million years. Hominines evolved larger brains, which gave them an increased behavioral flexibility and perhaps also the beginnings of a capacity for symbolic language. They learned to use tools in more complex ways than any other primate, which gave them access to a more varied diet. Taken together, these changes apparently enabled *Homo erectus* to explore a wider range of habitats than any other closely related species. Yet there is not any clear evidence in the fossil record of revolutionary changes in the behaviors even of later hominine species before ca. 250,000 BP. We have not yet left the realm of natural selection, in which genetic change eclipses cultural change. It is hard to imagine how any earlier species of hominine could have transformed the world as our own species has done. This is true even of Neanderthals, a species remarkably close to us genetically, with brains as large and perhaps even larger than ours. What is it, then, that is revolutionary about modern humans and human history? And in what way did the changes described in this chapter prepare the way for their revolutionary ecological impact? The next chapter will offer some tentative answers.

FURTHER READING

There are many good popular books on human evolution, but the field is changing so quickly that books can date rapidly. One of the best texts is Roger Lewin's *Human Evolution* (4th ed., 1999), while *The Cambridge Encyclopedia of Human Evolution* (1992), edited by Steve Jones et al., is a superb reference work. Two of the major figures in the field, Richard Leakey and Donald Johanson, have both written accessible books on the subject (Leakey, *The Origin of Humankind* [1994]; Johanson and Maitland A. Edey, *Lucy* [1981]). Jared Diamond's *The Rise and Fall of the Third Chimpanzee* (1991) is a punchy survey of the field, and Paul Ehrlich's *Human Natures* (2000) is another recent general survey. Other general surveys include Göran Burenhult, ed., *The Illustrated History of Humankind* (5 vols., 1993–94); Brian Fagan, *People of the Earth* (10th ed., 2001), a widely used text; Robert Foley, *Humans before Humanity* (1995); Ian Tattersall, *Becoming Human* (1998); Robert Wenke, *Patterns in Prehistory* (3rd ed., 1990); and Peter Bogucki, *The Origins of Human Society* (1999). Clive Gamble's *Timewalkers* (1995) is one of the best general surveys of the Paleolithic era. On the evolution of consciousness and thought, Steven Mithen, *The Prehistory of the Mind* (1996); Terrence Deacon, *The Symbolic Species* (1997); Steven Pinker, *The Language Instinct* (1994) and *How the Mind Works* (1997); William Calvin, *The Ascent of Mind* (1991) and *How Brains Think* (1998); and Nicholas Humphrey, *A History of the Mind* (1992) are all valuable, though this remains territory in which there is more speculation than hard proof. Craig Stanford's books, *The Hunting Apes* (1999) and *Significant Others* (2001), give a good insight into what modern primatology has to offer to the story of human evolution. In *Nonzero* (2000), Robert Wright discusses the crucial role of non-zero-sum games in human history.

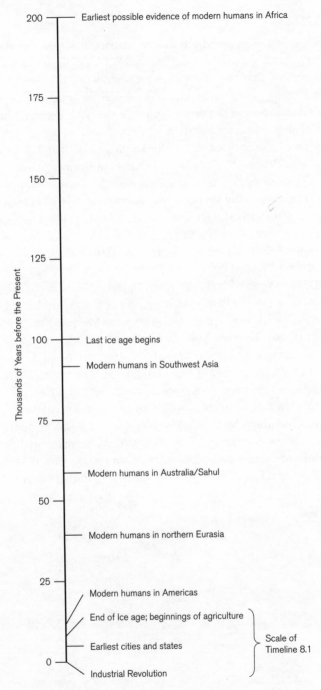

Timeline 7.1. The scale of human history: 200,000 years.

7

THE BEGINNINGS OF HUMAN HISTORY

THE EVOLUTION OF HUMAN LANGUAGE

Many features contributed to the unique evolutionary package that is our species. But the previous chapter argued that the most critical was the appearance of symbolic language, which released the new and uniquely potent adaptive mechanism of collective learning. So, to understand when human history really began, we have to understand when and how humans acquired their aptitude for symbolic language.

This is murky territory, for language leaves no direct signs in the fossil record; our attempts to understand the evolution of human language depend on ambiguous hints in the fossil record, padded out with a heavy wadding of theory. Not surprisingly, experts disagree even on the fundamental question of when human language first appeared. Henry Plotkin writes:

> Some put it as recently as 100,000 years ago or less, a few put it back beyond two million years from the present, and the majority go for somewhere in the region of 200,000 to 250,000 years ago. It is extremely unlikely to have occurred instantaneously, if one defines instantaneously as either a single miraculous mutation or a period of time less than about 1,000 years. . . . It most likely was smeared out over tens of thousands, perhaps a few hundred thousand years.[1]

Currently, it is common to suppose, building on the insights of the linguist Noam Chomsky, that language, like some other distinctive human abilities, depends on the evolution of particular "modules" or "organs" within the brain that contain the programs for particular skills. Human brains, it is argued, have a generalized computing capacity that is extremely powerful. But they also contain specialized modules for language and for many other skills, perhaps including social skills, technological skills, and ecological or environmental knowledge. Such theories are tempting, particularly in the case of language. Human infants acquire language with a speed and fluency that is incompatible with any process of learning by trial and error and that has no parallel among our closest relatives, the chimps. In some sense, it seems, the human capacity for language must be hardwired into our brains, and it must have been wired in quite recently, in evolutionary terms. If so, those interested in hominine evolution must try to explain how a language module evolved.[2]

Steven Mithen has proposed that a number of once discrete brain modules, some of which may have been present in the earliest hominines, merged quite suddenly—perhaps within the last hundred thousand years—in a sort of linguistic "big bang."[3] But exactly how this might have happened remains unclear. There are other difficulties with the "Swiss army knife" view of the human brain. Human brains are certainly different in significant ways from those of apes (not just in their size), but it has proved impossible to locate a distinct "language" module. Language skills appear to be distributed through many different parts of the mind, and their location differs even between individuals. Language seems to be a product of networks of interactions between different parts of the brain, rather than the work of any one language area.[4]

In *The Symbolic Species*, Terrence Deacon has offered an account of the evolution of human language that does *not* rely on the idea of specialized modules. His argument begins with the use of symbols, the most distinctive feature of human languages. Representations of the external world can exist in three distinct forms. The simplest two depend on the detection of similarities (which Deacon calls "icons") or correlations ("indices") between events and things.[5] Iconic similarities enable organisms as simple as bacteria to react in one way to all manifestations of warmth or light, and in another way to cold or darkness. On the other hand, Pavlov's dogs learned that there was a correlation between eating and the sound of bells because the two regularly occurred together. As a result, they linked the two phenomena despite the absence of any iconic similarity. Both these ways of learning depend on one-to-one correspondences between internal and external events. However, "symbols," the third form of representation, refer not just to the outer world but also to whole collections of icons and indices, so they can be used to create much more complicated inner maps of reality.

But symbolic thinking is tricky. It can be done only if iconic and indexical forms of representation can be held, as it were, in the background, while other parts of the mind distill their conceptual essence into a symbolic form of some kind. According to Deacon, "The problem for symbol discovery is to shift attention from the concrete to the abstract; from separate indexical links between signs and objects to an organized set of relations between signs. In order to bring the logic of token-token relationships to the fore, a high degree of redundancy is important" (p. 402; and see chap. 3, passim). This intellectual maneuver requires a lot of computing power. Deacon's argument makes clear the size of the hurdle that had to be surmounted before symbolic thinking was possible, and this helps explain why symbolic modes of representation are apparently confined to modern human beings, with their exceptionally large brains.

Large brains are not enough, however. Symbolic language also requires many other intellectual and physiological skills. These include a capacity to quickly make and process symbolic gestures or sounds and to understand rapid sequences of symbolic sounds uttered by others. How and why could such a coherent and complex set of skills develop together in the comparatively short period of a few million years? Deacon's answer is that they emerged through a process of co-evolution during which hominines evolved to take increasing advantage of rudimentary forms of symbolic communication, while languages themselves evolved to accommodate, with increasing delicacy and precision, the changing abilities and peculiarities of the hominine brain. Such changes probably involved some type of Baldwinian evolution, in which slight behavioral modifications gave a significant reproductive advantage to those individuals most skilled at these new behaviors. That advantage, in turn, would create powerful selective pressures in favor of these skills; in this way, what started out as a purely behavioral development may eventually have been inscribed both in the genetic code of our species and in the deep structures of human languages.[6] Rudimentary forms of symbolic communication may have appeared first as a result of minor behavioral changes similar to those observable in modern chimpanzees in experimental situations. But once they became habitual, these new forms of communication may have created new selective pressures by enhancing the reproductive chances of those individuals who were, for genetic reasons, most adept at them.

This discussion suggests that the initial steps toward a symbolic language probably began a long time ago to allow time for the evolution of the many behavioral and genetic changes that have made modern language possible. It also suggests that the first steps required brains little different from those of modern chimps. But these initial steps were probably followed by evolutionary changes whose most evident feature (at least in the fossil record)

would have been expansion in the size and importance of the prefrontal cortex, the front part of the brain. Finally, it is only at a later stage in human
evolution that we should look for direct evidence of efficient symbolic communication. Deacon's account of the extreme difficulty of symbolic communication suggests that once that threshold was crossed, we might expect a
sudden change in the quality and nature of human communication—
something along the lines of Steven Mithen's linguistic big bang.

The first steps toward symbolic language may have involved a combination of gestures and sounds. Under experimental conditions, chimpanzees
can be taught to use signs symbolically, even if their capacity to symbolize
remains limited, and australopithecines may have been as competent lin-
guistically as modern chimps.[7] But if we could observe australopithecines
communicating with each other, we might still be uncertain whether this
was really "language." Deacon explains:

> The first symbolic systems were almost certainly not full-blown lan-
> guages, to say the least. We would probably not even recognize them
> as languages if we encountered them today, though we would recognize
> them as different in striking ways from the communication of other
> species. In their earliest forms, it is likely that they lacked both the
> efficiency and the flexibility that we attribute to modern language. . . .
> The first symbol learners probably still carried on most of their social
> communication through call-and-display behaviors much like those of
> modern apes and monkeys. Symbolic communication was likely only
> a small part of social communication. (p. 378)

If this reconstruction is correct, it suggests that australopithecines had a limited ability to live in a symbolic realm, which may have permitted a modest degree of abstract thought and perhaps even a degree of *self*-conscious-
ness. However, for the most part, we should assume that australopithecines,
like most animals with brains, lived in an experiential world dominated by
the sensations of the present moment rather than in a psychic world like
that of modern humans, within which we can often conjure up what is *not*
present, including the past and future.[8]

Studies of *Homo habilis* skulls show that their brains were not merely
larger than australopithecine brains; they were also organized differently.
In particular, there are hints of the division of labor between left and right
sides that, in modern humans, is reflected in "handedness." This feature, in
common with increased brain size, may reflect selection for improved symbolic ability, since the segregation of functions to different parts of the brain
may have increased the ability of the brain to process different types of in-

formation in parallel.[9] Deacon suggests that other skills related to language may have been present in *habilis* and later hominines:

> *Homo habilis* and *Homo erectus* would have had greater motor control [than australopithecines] and probably also exhibited some intermediate degree of laryngeal descent as well [thereby increasing the variety of sounds that could be made]. *Homo erectus'* speech might have been somewhat less distinctive as well as slower than modern speech, and the speech of *Homo habilis* would have been even more limited. So, although their speech would not have had either the speed, range, or flexibility of today, it would have at least possessed many of the consonantal features also found in modern speech. (p. 358)

But we should not exaggerate these skills. The relatively high larynx of all early hominines suggests that they could not produce the range of sounds (particularly vowels) used by modern humans. If they spoke, they probably did so with a limited vocabulary of words dominated by consonants. Gestures may still have carried most of the burden of communication. Because they lacked the ability to manipulate symbols with the speed or dexterity of modern humans, their communication would have been limited and slow by modern standards. Most important of all, we do not yet see in the archaeological record any signs of the significantly enhanced adaptive ability associated with collective learning.

It is during the past 500,000 years or so that we begin to find evidence of a more decisive shift toward symbolic language, combined with increased adaptive creativity. Neanderthals had brains as large as humans (see figure 7.1), but studies of the base of Neanderthal skulls suggest that they, too, lacked the capacity to manipulate sounds in the complex ways demanded by modern human languages. And this, combined with the absence of any other unequivocal evidence for extensive symbolic activity among Neanderthals, leads us to believe that Neanderthals did not use a fully developed form of language, though their presence in parts of Ice Age Eurasia indicates that they did have an enhanced capacity to adapt to new environments. However, the rapid growth in brain size among several distinct species of humans in the past 500,000 years suggests that a process of rapid co-evolution was taking place, in which several distinct abilities crucial to symbolic language were evolving together and quite swiftly. These may have included the descent of the larynx (necessary to make possible more complex manipulation of sounds), increasing lateral specialization within the brain, and increasing ability to control the breath and to recognize and analyze sounds rapidly and precisely.[10]

Figure 7.1.   Neanderthal and human skulls. The skull on the left is Neanderthal (from La Ferrassie); the skull on the right is of a modern human (from Cro-Magnon). Modern genetic evidence suggests that humans and Neanderthals are less closely related than was once thought. From Chris Stringer and Clive Gamble, *In Search of the Neanderthals* (London: Thames and Hudson, 1993), p. 185.

WHEN DOES HUMAN HISTORY BEGIN?

When do we first get evidence for the existence of humans that not only *looked* like modern humans but also *behaved* and *communicated* with each other like modern humans? This is one of the most important questions that a historian can ask, for it is really a question about the beginnings of human history.

In recent years, two rather different answers have been available. The first is now a minority position, though it is still defended vigorously by some scholars, including Milford Wolpoff and Alan Thorne. They argue that humans evolved slowly toward their modern forms throughout Afro-Eurasia, over almost a million years. Thus all the hominine remains found throughout Afro-Eurasia in the past million years should be treated as examples of a single, evolving species with regional variants, some of whose features, including skin color and facial characteristics, survive to the present day. In this view, regional populations continued to interbreed, so they always remained part of a single species.[11] If this account is accurate, we must conclude that human history is perhaps a million years old, though its most distinctive features do not become apparent until more recently. There are several difficulties with this approach, however. Above all, the great variety of fossil remains from the past million years, the huge area they cover, and the probability that few individuals traveled large distances make it difficult to see how we can regard these remains as evidence of a single, evolving species.

A second view, which is currently more popular, is that modern humans appeared more abruptly, somewhere in Africa, between 100,000 and 250,000 years ago.[12] The crucial evidence for this conclusion is genetic, though it is also compatible with recent fossil finds. Studies of the genetic material of modern humans show that we vary far less than do neighboring populations of gorillas. This suggests that our species is very young—perhaps only 200,000 years old. If we had been around much longer, there would have been time for much more genetic variety to accumulate both within and between regional populations. Furthermore, *most* of the genetic variety within modern humans occurs within African populations, which suggests that this is where humans have lived longest. Presumably, then, Africa is where modern humans *(Homo sapiens)* first appeared. Indeed, this theory suggests that for at least half of our history, modern humans lived exclusively in Africa.

This account of the relatively abrupt appearance of our species fits well with what we know of typical patterns of evolution. Modern humans, like many hominine species, may have evolved by a process known to biologists as *allopatric speciation*. When populations of a species range over a large area, it is common for some groups to become isolated. They may enter a valley or cross a mountain or a river that then cuts them off from other members of their species. If they cease to interbreed with other populations of their species, they will soon begin to diverge genetically from the parent population. If the isolated population is small, and if the ecological conditions of its new home are very different from its old home, it may diverge rapidly, because selective pressures are strong and favorable genetic changes can spread more swiftly in small populations. Besides, a small population is unlikely, for purely statistical reasons, to be entirely typical of the parent population, and in it such deviations can multiply quickly. (This is known as the *founder effect*.) For all these reasons, new species often evolve rapidly in small populations living at the edge of the range of a parent species. If this is how our species evolved, then all modern humans are descended from a small and isolated group of ancestors who lived in Africa between 100,000 and 200,000 years ago. If they lived in southern Africa, this would indeed place them at the extreme edge of the range of hominine populations of the Middle Paleolithic (the era from 200,000 to 50,000 years ago).

But there is a problem with this theory, too, for most of its supporters agree that evidence of distinctively modern *behaviors,* including human *language,* does not appear before the Upper Paleolithic, which began about 50,000 years ago. Archaeological evidence from Eurasia and Australia suggests that some quite decisive changes occurred in human behavior about 50,000 years ago. The markers that archaeologists have taken as signs of modern human behavior are of four main types. First are new ecological adaptations, such as the entry into new types of environment. Second are new technologies, such as the appearance of small, precisely made, and sometimes standardized blades that may have been hafted, as well as the use of new materials such as bone, all of which would presumably have enhanced the capacity to enter new environments. Third are indications of greater social and economic organization, which show up in evidence for networks of exchange extending over large distances, improved ability to hunt large animals, and evidence of an increased capacity for organization and planning. Fourth, and in some ways most important of all, are indirect signs of symbolic activity, such as the appearance of artistic activity of various kinds, which would have accompanied the use of symbolic language. On the basis of evidence for all these types of change, a number of archaeologists and prehistorians have argued that there was a "revolution of the Upper Paleolithic": a late, and remarkably sudden, flowering of human creative activity, beginning ca. 50,000 years ago, which marks the true beginning of human history.

But why the apparent gap between the appearance of modern humans and the appearance of modern behaviors? This has remained a tantalizing puzzle. It has tempted some scholars to suppose that critical changes may have taken place in the wiring of human brains within the last 100,000 years; in that case, the *real* beginning of human history should be put later than the genetic evidence might suggest. Recently, however, two American paleontologists, Sally McBrearty and Alison Brooks, have proposed an elegant resolution of these difficulties, based largely on a close analysis of the archaeological evidence from Africa. Their account dovetails neatly with the account of language origins offered in the previous section, as it seems to demonstrate how a process of genetic evolution of the kind familiar to biologists was transformed, about 250,000 years ago, into a process of cultural evolution of the kind familiar to historians. The next section will be based largely on their revised account of early human history in Africa.[13]

In "The Revolution That Wasn't," McBrearty and Brooks have shown that the abrupt changes apparent in the Eurasian and Australian evidence

are not seen in evidence from Africa. Here, they argue, evidence of fully human behavior can be found much earlier than the Upper Paleolithic, perhaps from as early as 250,000 years ago, but it appears piecemeal and gradually. Evidence for the use of small blade tools, some of them hafted, as well
as for the use of grindstones and pigments appears very early, while evidence for other innovative technologies—including fishing, forms of mining, long-distance exchanges of goods, the use of bone tools, and migrations
into new environments—also can be seen earlier than in Eurasia. Neither
cultural nor anatomical changes appear in a "big bang"; instead, they evolve
more fitfully.

> There was no "human revolution" in Africa. Rather, . . . novel features
> accrued stepwise. Distinct elements of the social, economic, and subsis-
> tence bases changed at different rates and appeared at different times
> and places. We describe evidence from the African MSA [Middle Stone
> Age, ca. 250,000–50,000 BP] to support the contention that both human
> anatomy and human behavior were intermittently transformed from
> an archaic to a more modern pattern over a period of more than
> 200,000 years. (p. 458)

Instead of a revolution of the Upper Paleolithic, what is apparent in Africa
is a slow process of change that seems to reflect the "fitful expansion of a
shared body of knowledge" over many small groups and large areas (p. 531).
And this, they argue, is just what should be expected if modern humans lived
in small groups and developed these skills community by community.

Furthermore, they argue, the earliest of these changes coincide with the
appearance of a new hominine species, recently dubbed *Homo helmei*, which
is so close to modern humans that it may prove necessary to reclassify its
members as belonging to our own species, *H. sapiens*. Remains that are
firmly attributable to *H. sapiens* are certainly present in Africa by 130,000
years ago, and perhaps as early as 190,000 years ago, but there is no sharp
discontinuity between the two species (p. 455). All in all, they maintain that
in Africa, unlike Eurasia, the genetic evidence and behavioral evidence combine to offer a coherent account of how our species originated and began to
display the ecological creativity that is unique to our species.

> Both *H. helmei* and early members of *H. sapiens* are associated with
> MSA technology, and thus it is clear that the main behavioral shift
> leading to modernity lies at the Acheulian-MSA boundary about
> 250–300 ka [thousand years], not at the MSA-LSA [Later Stone Age]
> boundary at 50–40 ka as many assume. We have shown here that many
> sophisticated behaviors are present in the MSA. This implies increased
> cognitive abilities with the appearance of *H. helmei*, and behavioral

similarities and a close phylogenetic relationship between *H. helmei* and *H. sapiens*. It could be argued that the specimens referred here to *H. helmei* are more correctly attributed to *H. sapiens*, and that *H. helmei* should be sunk into *H. sapiens*. If that is the case, our species has a time depth of ca. 250–300 ka, and its origin coincides with the appearance of MSA technology. (p. 529)

If McBrearty and Brooks are right, we can say that human history began somewhere between 300,000 and 250,000 years ago in Africa.

AFRICAN ORIGINS: THE FIRST 200,000 YEARS

Before about 100,000 years ago, humans were confined to Africa; but within Africa they pioneered new technologies and lifeways and they occupied new environments, including those of the forests and deserts. Only after ca. 60,000 years ago do humans begin traveling into regions that no earlier hominines had settled, including Australia (which required the ability to cross a significant body of water), Ice Age Siberia (which required the ability to adapt to extremely cold conditions), and eventually the Americas.

Evidence for the earliest (and longest) phases of human history in Africa is tantalizingly thin. In principle, we know that once language appeared, each community had its own history that was rich in epic stories, great names, disasters, and triumphs. But because we cannot see these histories, we have to portray the large trends, forgetting about the details that mattered to individuals. There is little we can do about this except to periodically make the imaginative effort to remember that each community *did* have its own detailed history, which was as vivid and live to community members as any history constructed today on the basis of written sources.

These generalizations are true of the entire period of human history traditionally referred to as *prehistory* because of the absence of written sources. But they apply with particular force to the earliest eras of human history. Less archaeological work has been done in Africa than in Europe, dating is tricky, and, as always, trying to explain behavior on the basis of archaeological evidence is difficult. Besides, we should expect that in those early days, processes of collective learning would have worked extremely slowly; we should not be looking yet for spectacular displays of technological virtuosity. As McBrearty and Brooks note, "Early modern human populations in late Middle Pleistocene Africa were relatively small and dispersed, change was episodic, and contact among groups intermittent. This resulted in a stepwise progress, a gradual assembling of the modern human adaptation" (p. 529).

Despite these difficulties, McBrearty and Brooks make a strong case for

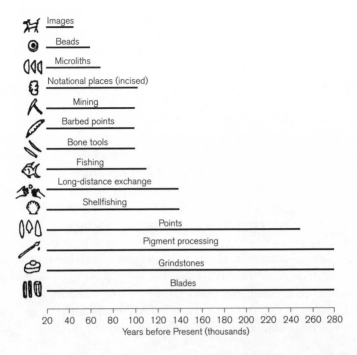

Figure 7.2.   Behavioral innovations of the Middle Stone Age in Africa (duration in thousands of years). Adapted with permission from Sally McBrearty and Alison S. Brooks, "The Revolution That Wasn't: A New Interpretation of the Origin of Modern Human Behaviour," *Journal of Human Evolution* 39 (2000): 530.

the appearance in Africa, 250,000 years ago, of all the crucial changes once regarded as evidence for a revolution of the Upper Paleolithic (see figure 7.2). The earliest and clearest signs of new behaviors can be seen in changing stone technologies. Most striking is the disappearance, after 250,000 BP, of the Acheulian stone technologies associated with various forms of *H. ergaster*. In their place there appear new and more delicate types of stone tools. Some may have been hafted so they could be used as spears or projectiles, an innovation that would have permitted safer and more precise hunting of large animals. Traces of the gums used by modern hunters to hold blades in place have been found on at least one early blade, and many early stone blades are shaped in ways that are consistent with hafting.[14] In addition, there are signs of the use of small-scale resources such as fish and shellfish. These are technologies that do not appear outside of Africa until after ca. 50,000 BP.

Humans were also adapting to new environments, in particular to desert and forest regions that no earlier hominines had used.[15] Evidence of new forms of social organization and of local "cultures" appears in the quite distinct stylistic patterns found in stone implements. There is also evidence of complex patterns of exchange, sometimes over several hundred kilometers. Such behavior suggests that though most of the time humans lived in family groups joined together in small bands, they had occasional friendly contacts with other groups—sometimes over large distances. The creation of such networks (Robert Wright describes them as "giant regional brains")[16] marks a radical break with what we know of the social systems of living great apes. It is tempting to interpret it as indirect evidence of improved forms of communication. More direct signs of modern linguistic skills appear in the form of ornamental objects, as well as grindstones apparently used to grind pigments. Both have been found in Africa well before the Upper Paleolithic. These provide the clearest evidence for the existence of symbolic activity, symbolic thought, and therefore symbolic language.

None of these scraps of evidence is unambiguous, but taken together they help us piece together the earliest stages of the process of collective learning that has culminated, 250,000 years later, in the world we know today. And they suggest that this process was linked directly with the appearance of new species of hominines capable of using symbolic language.

SOME RULES OF COLLECTIVE LEARNING

Symbolic language enabled humans, unlike other closely related species, to share information and to learn together. How did this pooling and sharing of knowledge generate the long-term changes that distinguish the history of humans from that of closely related species? In exploring what is distinctive about human history, we will need to focus, above all, on those factors that determined the pace and geography of processes of collective learning. Why was ecological innovation slower in some eras and faster in others? Why was it slower in some regions and faster in others? If, as I have argued earlier, collective learning is the most important distinguishing feature of human history, we clearly need to keep a close eye on these questions.

In practice, of course, processes of collective learning were as unpredictable as any creative process. But some general rules are worth noticing at the outset, as these will suggest which changes were most likely to accelerate or retard the accumulation of ecologically significant knowledge—the types of knowledge that, over time, have given humans their unique power to manipulate the material world. Two factors stand out: the volume and variety of the information being pooled, and the efficiency and speed with which information is shared.

The first critical factor is the size of information networks, or the number of communities and individuals that can share information.[17] Intuitively, we should expect the potential synergy of a network of information exchanges to increase at an accelerating rate as the number and diversity of people exchanging information increase.[18] It may be easiest to understand this rule in terms of a model network in which there are a number of nodes (*vertices*, in graph theory; *people* or *communities*, for our purposes), but the total intellectual synergy is proportionate to the number of possible links between those nodes (*edges*, in graph theory). Then the math is easy. The number of possible links between 2 nodes is 1, the number between 3 is 3, and the number between 4 is 6; in general, if the number of nodes is n, the total number of links is $(n \times (n-1))/2$. In reality, not all connections are made. But the important point is that the number of *possible* connections (and thereby the *potential* informational synergy of the entire network) increases faster than the number of nodes, and the difference between the two rates increases as the number of nodes increases. So, as networks expand in size, their potential intellectual synergy increases much faster: "larger and denser populations equal faster technological advance."[19]

The variety of information being pooled may be as important as the sheer volume. Neighboring communities living similar lifeways may be able to help each other fine-tune technologies and skills, but they are unlikely to introduce radically new ideas. Fundamentally new forms of information are likely to be shared only where communities living different lifeways come into significant contact. To be sure, differences in lifeways often act as a barrier to contact; but sometimes, as in some forms of trade, they do not. Indeed, where dissimilar groups belong to the same information networks, we are most likely to find processes of collective learning leading to significant changes in technologies and lifeways.

This abstract model suggests that it is important to try to describe the size and variety of information networks—the regions over which information can be exchanged. It also suggests another important principle: as the size and variety of information networks grow, we should expect to find not just an accumulation of new knowledge but an *acceleration* in the accumulation of new knowledge. And at the most general level, this is exactly what we observe over long periods of human history.

The second critical factor is the efficiency with which information is exchanged. To define the size of a region within which information may be exchanged is one thing. But within that region the speed and regularity of exchanges may vary greatly. The efficiency of information exchanges reflects, above all, the nature and regularity of contacts and exchanges between different communities. And these may be shaped by social conventions, geographical factors, and technologies of communication and transportation. Within a single information network, processes of collective learning may be more or less powerful in different regions; it is thus possible to imagine regions in which more information is pooled, in greater variety and in greater concentrations, than in other regions.

These arguments suggest a useful general principle: the size, diversity, and efficiency of information networks should be an important large-scale determinant of rates of ecological innovation. In subsequent chapters, we will attempt to track the changing synergy of processes of collective learning by examining the size and variety of information networks in different parts of the world, as well as the varying efficiency with which information was pooled within those networks.

In the Paleolithic era, the existence of small groups that had limited contact with each other ensured that exchanges of ecological information worked sluggishly. In a single lifetime, each individual was unlikely to encounter more than a few hundred individuals, and most of that lifetime would have been spent in the company of no more than the ten to thirty individuals who belonged to the same family. The amount of information that could be exchanged in such networks was clearly limited; these limitations help explain what seems to us the glacial slowness of technological change in the Paleolithic era, even though by hominine standards, technological change was actually rapid.

Other factors would also have slowed the pace of change. Societies made up of many small communities tend to display great linguistic diversity. In Aboriginal Australia, a population of several hundred thousand people may have had 200 different languages. Though related to each other, these languages were distinct and only close neighbors could communicate easily with each other. In California as late as 1750, at least 64 and perhaps 80 different languages were spoken, and in Papua New Guinea even today, there are almost 850 living languages.[20] Cultural differences would also have limited the exchange of ecological and other forms of information, as would the large distances between neighboring groups in a world in which each group needed a large territory to support itself. All in all, it should be no surprise to find that new technologies and new adaptations evolved slowly in the Paleolithic. And they emerged locally, so the earliest human societies were probably extremely varied: each group made its own adaptive experiments in relative isolation, and opportunities to pool technological discoveries remained limited.

PALEOLITHIC LIFEWAYS

Anyone trying to determine how the earliest humans lived must depend on a lot of guesswork. And studies of modern foraging communities suggest that lifeways varied greatly in their details from group to group. Still, there are some broad generalizations we can make with considerable confidence.[21] The small number of fossil remains, combined with what we know from observations of modern foragers, makes it certain that the number of early humans was small and that they lived in small communities. How small we really cannot know. But it seems a reasonable guess that for some time, human populations were similar to those of modern chimps, with perhaps significant fluctuations both up and down.

We can be sure that groups were small because all modern foraging technologies require large areas of land to support small populations. In early Holocene Europe, for example, foraging lifeways could support population densities of up to one person to every 10 square kilometers, while early forms of farming could support between fifty and one hundred people in the same area.[22] We have no reason to think that Paleolithic communities were any more efficient in this respect. Modern foragers are mostly nomadic, moving to different parts of their home territory in different parts of the year. Their diet normally depends largely on gathered foods, including plants, nuts, tubers, and small animals of various kinds. In addition, most hunt larger animals and highly value their meat, even though catching it is uncertain; as a result, smaller, more reliable forms of food usually make up the basis of their diet. Living a foraging life requires immense knowledge about available resources, about the migration patterns of birds and animals, and about the life cycles of particular plants, so it would be a mistake to underestimate the ecological skills of such communities.

How well did people live in the Paleolithic? A modern city dweller transported into this world would not find things easy, but the once-popular assumption that the lives of foragers were intrinsically harsh is exaggerated. It is probably equally true that a citizen of Paleolithic Siberia transported suddenly into the twenty-first century would find life hard today, if in different ways. In a deliberately provocative essay published in 1972, the anthropologist Marshall Sahlins describes the world of the Stone Age as "the original affluent society." He argues that an affluent society is "one in which all the people's material wants are easily satisfied," and he suggests that by some standards, Stone Age societies met this criterion better than do modern industrialized societies.[23] He points out that affluence can be achieved either by producing more goods to fulfill more desires or by limiting one's

desires to what is available (the "Zen road to affluence"). Using modern anthropological data to gain some insight into the life experience of Stone Age
societies, he accepts that levels of material consumption were undoubtedly
low among Stone Age peoples. Indeed, nomadism, by its very nature, dis-
courages the accumulation of material goods, for the need to carry what one
owns limits any desire to accumulate material possessions. Studies suggest
that modern nomadic societies may also deliberately check population
growth using many different methods, including prolonged breast-feeding
of children (which inhibits ovulation) and more brutal techniques, such as
the abandonment of excess children or of older members no longer capable
of moving with the rest of the community. In all these ways, foraging communities may have limited their needs.

Nevertheless, Sahlins argues that normal levels of consumption in such
communities were more than adequate to supply basic needs. Capable of exploiting an extremely wide range of foodstuffs, foragers in all but the harsh-
est of regions rarely suffered from serious shortages. And nomadism in small
groups provided variety and freedom from the diseases characteristic of
larger, sedentary communities. Even more strikingly, attempts by anthropologists to assess how much time modern foragers spend "working" for a
living suggest that far from toiling desperately just to stay alive, they work
less than most wage earners or household workers in modern industrial societies. Studies of traditional communities from Arnhem land showed that
"people do not work hard. The average length of time per person per day
put into the appropriation and preparation of food was four or five hours.
Moreover, they do not work continuously. The subsistence quest was highly
intermittent. It would stop for the time being when the people had procured
enough for the time being, which left them plenty of time to spare."[24] Here,
there was plenty of what we are tempted to call "leisure" time. Researchers
studying other modern communities of foragers have come to similar conclusions. And, given that today's foragers have generally been driven away
from regions of greatest abundance, there is little doubt that those of the
Upper Paleolithic would, if anything, have spent an even smaller proportion of time working. There have been many attempts to sketch changes in
work patterns as societies have increased in size from the Paleolithic to the
modern day. In summary, these suggest that daily work time for adult males
and females has increased, on average, from ca. 6 hours in foraging societies
to ca. 6.75 hours among horticulturalists to ca. 9 hours among intensive
farmers, falling to slightly less than 9 hours for modern industrialized urban dwellers. Total time spent on "housekeeping" has increased as dwellings
have become more permanent, containing more goods, but the proportion

of housekeeping done by men has decreased as societies have become larger. On the other hand, time spent making and repairing household goods has decreased as households have begun to acquire more goods from outside specialists.[25]

All in all, Sahlins concludes that Stone Age society was a world of abundance, in the sense that most basic desires could be satisfied with a minimum of stress and effort. It may be that Sahlins's article was a deliberate overstatement, intended to counter a traditional view of human history that could see only progress in the transition from forager to farmer to industrial worker. There is little reason to think that life expectancies in Stone Age societies were much above 30 or 40 years; and undoubtedly many people died in ways that could now have been avoided. But there is no getting around the basic paradox that Sahlins highlighted: the increasing "productivity" of human societies has created societies in which more is desired, and less free time is available to enjoy what is available. Rising productivity levels have supported larger populations, but it would be difficult to prove that they have generated increased levels of human contentment. Humans collectively have got better and better at extracting resources from the environment, but we cannot automatically equate this change with "betterment" or "progress."

The earliest humans probably lived, like most hominines, in family groups of ten to twenty related individuals who traveled together. The family was the community in which most people lived most of the time. Because (being human) they talked to each other, we can also be pretty sure that they *thought* of those closest to them as "family" or "kin." All primates live in groups that we can loosely think of as families. But only with the appearance of symbolic language was it possible to share ideas about family and kin. This means that kinship (whether based on ties of blood or ties of convention such as marriage) became the fundamental organizing principle of human social networks in early human history. In his simple but influential model of social structures, Eric Wolf has suggested that "kin-ordered" societies constitute a major type of human community, one that survives in many different forms even in the modern world.[26] But family groups rarely lived in total isolation. Like modern families, each was normally part of a network of related communities that periodically met with each other, particularly when supplies of food were plentiful enough to feed large numbers. At such meetings (known in Australia as *corroborees*), groups probably swapped information and even individuals with other groups that included at least some close relatives. Within these networks, a sense of kinship could define who you were, who you could trust, and who you had to be wary of.

Modern analogies suggest that the Paleolithic sense of kinship was embedded in a distinctively Paleolithic set of economic relations. We can perhaps understand these relationships by imagining a social equivalent of the law of gravity. Humans are intensely social creatures; thus every individual exerts a gentle gravitational pull on every other individual, which is why humans live in groups. But each group also tugs gently at the ideas, the goods, and the people inside neighboring groups. We have seen that even modern chimps (who are also extremely social) exchange valued goods such as meat as a way of cementing relations within their community. Among humans, exchanges of information, goods, and favors of many kinds provide the social gravity that holds close-knit groups such as families together. These exchanges should be thought of not as trade in a modern sense but rather as a form of gift-giving. In the Christian world, Christmas is a modern survivor of such exchanges, in which the gifts themselves (think of socks, ties, and cheap perfumes) are less important than the social relations they symbolize. In such a context, gifts are exchanged not mainly for economic advantage but primarily to maintain good relations. Anthropologists refer to the principle behind such exchanges as *reciprocity*.[27] Reciprocity depends on building up good relations through gift-giving as a sort of insurance for the future. Robert Wright quotes an account of Eskimo life that makes the point well: "the best place for [an Eskimo] to store his surplus is in someone else's stomach."[28]

The opposite of reciprocity is vengeance. Where reciprocity failed to prevent conflict, individuals or families took vengeance for wrongs done to them. After all, in small, stateless communities, if individuals or families did not exact justice, no one else would do it for them. The anthropologist Richard Lee reports a modern example, which hints at what capital punishment may have meant in the Paleolithic world:

> /Twi had killed three other people, when the community, in a rare
> move of unanimity, ambushed and fatally wounded him in full day-
> light. As he lay dying, all the men fired at him with poisoned arrows
> until, in the words of one informant, "he looked like a porcupine."
> Then, after he was dead, all the women as well as the men approached
> his body and stabbed him with spears, symbolically sharing the respon-
> sibility for his death.[29]

Large-scale warfare, like large-scale trade, was probably rare in the Paleolithic era. For the most part, exchanges of gifts (and also of the negative gifts of violence and insult) remained personal and "familiar." Nevertheless, these exchanges played a fundamental role in survival, creating systems of knowledge, alliance, and mutual assistance that embraced many distinct family groups and covered huge areas. And we can be sure that group violence did occur even in Paleolithic societies, as it does within modern families, as well as among modern nonhuman primates.[30]

Though we cannot be certain, it is likely that social networks were thought of as extending into the nonhuman world. Symbolic language makes it possible to imagine, and to share what is imagined. Such sharing lies at the base of all forms of religious thought. Modern studies of small-group religions suggest that the earliest human communities thought of the entire cosmos as bound into webs of kinship. Totemic thought—the belief that particular families or lineages are related to particular animal species and may return to life in animal form—reflects a sense of close kinship with the animal world that seems to be pervasive in small communities, even today. The supernatural world may also have been seen as a distinct but accessible realm—almost like a separate tribal territory, with whose occupants one could negotiate, fight, or intermarry. This was a realm into which people could travel, certainly at death and often even in life. And when they did so, rituals and symbols of kinship provided a sort of passport between worlds. Modern shamans plead with, negotiate with, and even "marry" supernatural beings in order to pacify them or secure their favor. Above all, they give gifts of food or sacrificed animals to please or pacify the gods, so that reciprocal gift-giving shapes relations with the world of spirits as well as humans. The relationship between kinship thinking and religion survives even in the great modern religions, which often describe transcendental beings as parents or ancestors, to whom one must give gifts or "sacrifices" as marks of respect. But in relatively egalitarian communities, it seems that the world of the gods, too, was thought of as egalitarian and individualistic. Christopher Chase-Dunn and Thomas D. Hall report that in northern California before European colonization,

> there was little hierarchy among the many powers and beings. Many groups believed that Coyote, the trickster, had created the universe. No families or lineages had special relationships with deities or sacred ancestors. Rather, it was the job of each individual to seek out and establish relations with those spiritual forces that were to become his or her special ally. An individual who obtained a great deal of this kind of "power" was more likely to become a shaman, but each person constructed his or her own relationship to the spiritual world. This kind of religious cosmology is quite resistant to claims of seniority or hierarchy.[31]

In at least one respect, though, it is likely that Paleolithic thinking about the world was very different from the thinking typical of later eras of human history: it was much more specific. People dealt not with "the gods" in general but with this spirit and that magical force, just as their technologies were not generalized but highly specific, concerned with this particular herd of deer, or that particular forest or shoreline. And this characteristic may be why, as far as we can tell, the religions and cosmologies of the Paleolithic world were attached so strongly to particular places.[32] Because Paleolithic communities were so small, their thinking about the world lacked the distinctive modern concern with universality and generality. It was particular places that mattered most of all; such places were the source of everything that mattered. Something of this sensibility may be captured in what Hobbles Danaiyarri, from Yarralin in Australia's Northern Territory, once said to Deborah Bird Rose: "Everything come up out of ground—language, people, emu, kangaroo, grass. That's Law."[33]

"EXTENSIFICATION": MIGRATIONS OF THE UPPER PALEOLITHIC AND THEIR IMPACTS

The small size of Paleolithic groups and the limited exchanges between them ensured that ecological knowledge accumulated slowly, slowly enough that it is often assumed (wrongly) that there was no technological evolution at all during this period. In fact, though it is not easy for us to see the details, we can be certain that a huge amount of ecological knowledge was accumulating within Paleolithic communities. Indeed, looking back from modern times, it is easier for us than for contemporaries to see that change *was* occurring, for most of the changes that stand out in retrospect (as opposed to the births, deaths, and other life events that mattered at the time) occurred on scales too large to be noticeable within a single life span.[34] Over many thousands of years, the size and diversity of environments occupied by humans within Africa increased. It did so by a process that we can usefully describe using the ugly word *extensification*, whose complement is the more familiar notion of *intensification*. Extensification means an increase in the range of humans without any parallel increase in the average size or density of human communities, and consequently with little increase in the complexity of human societies. It involves the gradual movement of small groups into new lands, usually adjacent to and similar to those they have left. Humans moved in this way in part because they had the adaptive flexibility to do so, whereas related species such as chimps lacked the ability to move far beyond the habitats in which they had evolved. As for their motivation to migrate, that may have ranged from social conflicts within the home group to local overpopulation. But it is important to note that extensification leaves the average group size unchanged, even as it may lead to a slow expansion in the range and the total number of modern humans.

So, though humans had to constantly make minor adaptations to new habitats, in the course of which they developed the new technologies necessary to live in environments as diverse as tropical forest and arctic tundra, the synergy of collective learning did not increase greatly.

Whatever the cause, and however slow such changes might appear to modern eyes, when repeated many times, over perhaps seven to eight thousand generations and 250,000 years, they eventually led modern humans to settle all continents apart from Antarctica. Beginning ca. 100,000 BP, evidence is found of the presence of modern humans outside of Africa. The first evidence is the appearance of modern human skulls in the Middle East that date to about 100,000 years ago. This means that modern humans lived in the Middle East at the same time as Neanderthals. In this region, at least, members of the two species may even have met each other.[35] Like earlier hominines, modern humans would have found it easy to migrate eastward or westward around the Mediterranean or toward Asia, for in southern Eurasia they found environments quite similar to those in Africa.

The first migrations into very different environments were into the continent of Sahul (which included modern Australia and Papua New Guinea), and into the Ice Age steppes and tundra lands of northern Eurasia (see maps 7.1 and 7.2). No earlier hominines had made these migrations; so they provide decisive evidence for the increasing ecological creativity of modern humans. The difficulty of occupying colder northern latitudes is evident in the long time it took modern humans to move from the Middle East to Europe and Inner Eurasia. Modern humans first appear in these regions from ca. 40,000 years ago. Humans were in Ukraine by 40,000 to 30,000 years ago, and had probably settled parts of northern Siberia by 25,000 years ago. Eventually, communities living in eastern Siberia crossed into the Americas—perhaps using boats, or perhaps crossing the land bridge across Beringia that was exposed during the colder parts of the last ice age. We know that humans had entered the Americas by about 13,000 years ago, but there are hints that they may have arrived earlier, possibly as early as 30,000 years ago.

Meanwhile, some humans had made the first significant sea crossing from what is today Indonesia into Sahul. As late as the 1960s, there was no firm evidence of settlement in Australia before 10,000 years ago. But since then, dates for the settlement of Sahul by modern humans have been pushed back in time. Humans had certainly arrived by 40,000 years ago, and they may have arrived earlier. Recent evidence, examined with the new dating technique of thermoluminescence, suggests a date of almost 60,000 years for the occupation of the Malakunanja rock shelter in Arnhem land in Northern Australia, while a skeleton found at Lake Mungo in New South Wales

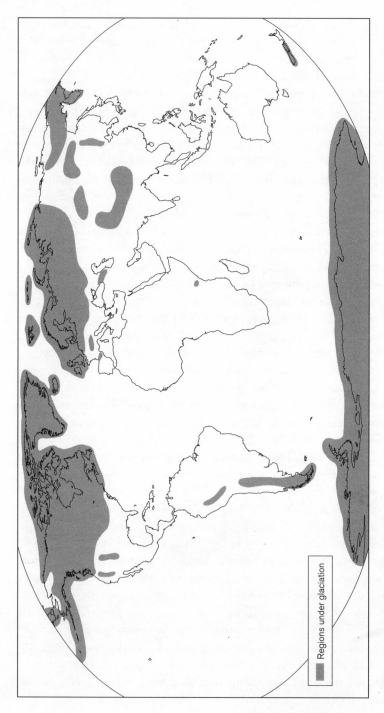

Map 7.1. Extent of glaciation during the ice ages. Data from Neil Roberts, *The Holocene: An Environmental History*, 2nd ed. (Oxford: Blackwell, 1998). p. 89.

Regions under glaciation

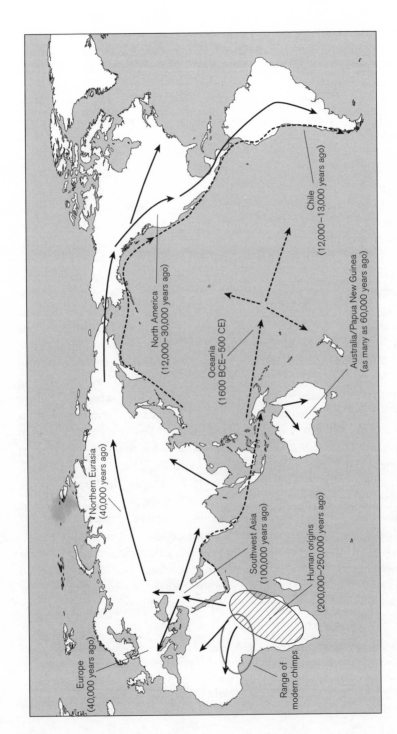

Map 7.2.   Migrations of *Homo sapiens* from 100,000 BP.

Europe
(40,000 years ago)

Northern Eurasia
(40,000 years ago)

North America
(12,000–30,000 years ago)

Chile
(12,000–13,000 years ago)

Oceania
(1600 BCE–500 CE)

Australia/Papua New Guinea
(as many as 60,000 years ago)

Southwest Asia
(100,000 years ago)

Human origins
(200,000–250,000 years ago)

Range of
modern chimps

in 1974 has recently been dated at 56,000 to 68,000 BP.[36] These are significant dates, because no earlier hominines had managed to settle Sahul. And the reason is clear. Even during the last ice age, when sea levels were lower than today, the journey to Sahul required a sea crossing of at least 65 kilometers. At other times, the distance was at least 100 kilometers. Any humans traveling to Sahul from Timor or the Sula group of islands had to be superb sailors. And they had to be careful planners, for populations that drifted to Sahul by chance would not have been large enough to form permanent colonies. So, settling Sahul required technologies that we do not find in any previous hominine species (see map 7.2). Careful analysis of genetic variations in modern populations confirms the story of migration that is apparent in the fossil record. They show that East Asian and Australian populations diverged more than 50,000 years ago, and Amerindian populations diverged from North Asian populations 15,000 to 35,000 years ago.[37]

As humans moved into these new environments, they had to develop new technologies. Improved control of fire may have been one of the most important of all technological developments in the later Paleolithic. We have seen that some communities of *H. ergaster/erectus* may have used fire, but only in limited ways. Modern humans put fire to many more productive uses. It provided warmth and some protection against predators. It was also used for cooking, a development that made it possible to process and use foods that might otherwise have been unusable: heating softened the fibers in meats and destroyed the toxins that had evolved as a form of protection in many plant species, from tubers to legumes.[38] Fire could also be used to shape entire landscapes, and as a supplement to hunting and gathering. In a famous article, the Australian archaeologist Rhys Jones referred to such techniques as "fire stick farming."[39] Fire stick "farmers" deliberately set fire to bushland in regular cycles. In part, their aim was to prevent buildups of combustible material that could lead to hotter and more dangerous fires. But by clearing away underbrush, fire stick farming also encouraged the growth of new plants that, in turn, attracted browsers that could be hunted. Recent research suggests that such techniques may have been used as early as 45,000 years ago.[40] But, at least in temperate zones, they have been used more or less continuously ever since, with a profound effect on entire biota. As Stephen Pyne writes:

> hardly any plant community in the temperate zone has escaped fire's selective action, and, thanks to the radiation of *Homo sapiens* throughout the world, fire has been introduced to nearly every landscape on earth. Many biotas have consequently so adapted themselves to fire that, as with biotas frequented by floods and hurricanes, adaptation

has become symbiosis. Such ecosystems do not merely tolerate fire, but often encourage it and even require it. In many environments fire is the most effective form of decomposition, the dominant selective force for determining the relative distribution of certain species, and the means for effective nutrient recycling and even the recycling of whole communities.[41]

In some form, the practice can be found in many different parts of the world at the end of the Paleolithic era, and more recently.[42] Captain Cook saw the smoke of bush fires while sailing off the Australian coast in the eighteenth century, and Magellan saw huge plumes of smoke off Tierra del Fuego. Modern anthropological research has also revealed a long history of fire use in North America.[43] According to I. G. Simmons,

> the Beaver Indians of northern Alberta had a sophisticated and delicately tuned approach to creating fire. Certain patches of vegetation were burned deliberately in order to maximize their value as resources. Openings or clearings ("yards") were created within a forest area and maintained by burning; grass fringes of streams, wetlands, trails and ridges ("corridors") were similarly created and maintained, for they were both areas where hunted species of animal would either collect or traverse, or both. Fires were also set along traplines, around lakes and ponds, and within large areas of dead fallen trees which otherwise had no resource value; indeed, they were a danger since if ignited in summer they might start off a crown fire, whereas the Indian groups controlled time and place so as to produce only surface fires. So the yards and corridors may well have existed alongside a natural fire-produced mosaic, or could have used natural patterning as a starting point and maintained a version of it.[44]

So pervasive is the use of fire that the Dutch sociologist Johan Goudsblom has argued that it constitutes the first great technological transition in human history.[45]

In colder climates, improved hunting techniques were crucial, for while accessible plant foods were scarcer than farther south, there were huge herds of herbivores to be hunted on the Ice Age steppes of Russia, Siberia, and North America. The evidence for new forms of technological creativity is particularly abundant from eastern Europe. In this region, Upper Paleolithic innovations may have included some of the earliest forms of weaving and pottery, technologies that were once thought to have appeared first in the Neolithic era. Sites from the Moravian lowlands, dated to between 28,000 and 24,000 years ago, suggest the use of fired clay and also of weaving, probably to make nets and baskets, as well as simple forms of clothing.[46] There is also evidence in eastern Europe in the Upper Paleolithic for improvements in clothing, particularly in northern environments. At Sungir, near Vladimir in Russia, there is a burial dating to ca. 23,000 BP; it contains the remains of a boy and a girl, who wore clothing covered with beads. The position of the beads suggests that the clothes, made from hides and furs, were carefully tailored and well fitting. The girl's grave is the more elaborate. It contained more than 5,000 beads, many ivory lances, and other carved ivory ornaments. The boy's grave also contained many beads, as well as a belt made from 250 carved fox teeth, a bracelet, a pendant, several spears, and a mammoth figure carved from ivory. Many Upper Paleolithic sites also contain bone needles.[47]

Dwellings became more specialized. There is particularly striking evidence of systematic and well-planned building from what is today Ukraine and southwest Russia.[48] Perhaps most astonishing of all, communities in some regions exploited local resources so efficiently that they became less nomadic. The clearest evidence for the existence of Upper Paleolithic "villages" also comes from Ukraine, where Olga Soffer has studied almost thirty Upper Paleolithic sites. Many have mammoth bones and pits for the storage of frozen meat. Linked to these are other, less permanent sites—on high ground, away from the river valleys—which may have been temporary summer hunting camps. The earliest mammoth bone dwellings date from ca. 20,000 BP, but similar dwellings are present at many sites in the Dnieper basin, usually near river valleys. At Mezhirich, on the river Dnieper, there are large concentrations of mammoth bones, along with carefully prepared hearths and many bone or ivory ornaments. Mammoth bones provided a frame for dwellings partly dug into the ground, and covered over with skins. There were about five dwellings, each about 80 square meters in area and housing up to ten people. The builders used mammoth bones not just for scaffolding but also as "tent pegs," in preference to wood, which rots more easily. They forced them deep into the ground and cut sockets into which they inserted wooden poles. They also used mammoth bones as fuel, after splintering them.[49] These settlements were probably winter camps for groups of perhaps thirty people, who may have occupied them for as long as nine months each year. The relative permanence of these settlements is reflected in the care with which they were built. At the Kostenki 21 site, there were several dwellings along 200 meters of the Don river shore, set 10 to 15 meters apart. One dwelling, near marshland, had an area paved with limestone slabs to avoid the damp. There are also objects that seem to have ritual importance, such as the two musk ox skulls found at Kostenki. Perhaps these were the site of annual gatherings or ritual activities that affirmed the unity of related groups.[50] The inhabitants of these Ice Age villages lived off frozen stocks of meat, kept in storage pits and thawed out by fire. The meat, most of which came from gregarious herbivores such as mammoth or bison, was hunted in summer and autumn, when the animals were at their fittest. Each year, some of the inhabitants moved out to temporary summer camps for the hunting season. On returning, they stored meat in pits whose depth suggests they were dug from the top layer of permafrost as it thawed during the brief summers.[51]

The skills needed to survive in such environments were social as well as technological. In harsh environments, knowledge is as crucial as tools; modern anthropological studies suggest that knowledge was highly valued, and carefully codified and stored in stories, rituals, songs, paintings, and dances. In the Upper Paleolithic there are many hints that information and prestige goods of various kinds were being exchanged—sometimes over huge areas. This does not mean that such exchanges were regular, but it does mean that information could spread extensively, though slowly and fitfully. The astonishing Venus figurines, which appear from the Pyrenees to the river Don at the coldest period of the last ice age, about 20,000 years ago, are a spectacular example of such diffusion. Even more astonishing are the similarities between the cave paintings of southwestern Europe and those of western Mongolia toward the end of the Upper Paleolithic.[52] In Sahul, too, there is evidence that goods and ideas could be exchanged over vast areas. The Wilgie Mia ochre mine in Western Australia has been excavated for thousands of years, using technologies including wooden scaffolding, heavy stones to bash the rocks, and fire-hardened wedges to extract the ochre embedded within the rocks. The mine's red ochre, which may have represented the blood of a Dreamtime being, was traded from Western Australia right across the continent to distant Queensland.[53]

Technologies that gave early humans access to more and more diverse environments, enabling them to settle on all the world's major landmasses, imply an increase in the total number of humans. But estimating how human populations grew in the Paleolithic is extremely tricky. Most calculations depend on little more than careful guesswork. And there is a danger, which should be admitted at the outset, that in any deductions from such figures we will merely rediscover the assumptions behind the original guesses. Nevertheless, *if* such estimates are accurate, even within a wide margin of error, they suggest some clear and important conclusions. Though early human populations were undoubtedly small and probably fluctuated significantly, we have seen that the range of humans expanded markedly within Africa over 150,000 years and more. This expansion in range suggests that the total number of early humans also increased. As noted in chapter 6, genetic evidence hints that the number of modern humans may have fallen dangerously low (perhaps to 10,000 adults) about 100,000 years ago, at the beginning of the last ice age.[54] However, the migration of some modern humans out of Africa—first into the Middle East and then, beginning ca. 50,000 years ago, into central and northern parts of the Eurasian landmass, as well as to East Asia and Australia—must imply a significant increase in human numbers after that date. The harsh conditions of the later stages of the last ice age may have slowed growth, but the spread of humans into entirely new environments such as Siberia and the Americas presumably had the opposite effect, at least on a global scale. One indirect sign of population growth is the increasing number of settlement sites from the Upper Paleolithic: in the lands from north of the Black Sea to the northern ice sheets, only six Neanderthal sites have been found, but more than 500 sites date from the period after 50,000 years ago.[55] The Italian demographer Massimo Livi-Bacci proposes a global figure of "several hundred" thousand for Upper Paleolithic populations of ca. 30,000 years ago, and a figure of ca. 6 million at the end of the last ice age, almost 12,000 years ago (see tables 6.2 and 6.3).[56]

If we take these three figures—10,000 at the beginning of the last ice age, a guess of ca. 500,000 early in the Upper Paleolithic, and another guess of 6 million at the end of the last ice age, 10,000 years ago—we can calculate some approximate growth rates for early human populations. Taken at face value, these figures imply that human populations multiplied by a factor of ca. 1.006 every century from 100,000 to 30,000 BP, a rate that yields a doubling time of ca. 12,500 years. In the period from 30,000 to 10,000 BP, world populations grew at a factor of ca. 1.013 every century, yielding a doubling time of ca. 5,600 years.

These growth rates are rapid by comparison with those of any other large mammal. Yet they are slow by the standards of later human history. Table 6.3 shows that on these figures, the average doubling time for populations in the agrarian era fell to about one-sixth of what it was in the late Paleolithic era. In the modern era, the average doubling time has fallen again, to about one-eighth of what it was in the agrarian era. One way of getting a general feeling for the difference between these eras is by estimating average population densities. The total land surface of the earth (including Antarctica) is ca. 148 million square kilometers. Dividing world populations at different eras into this figure, we get a notional average population density of 1 person for every 25 square kilometers in 10,000 BP. In 5000 BP, the same average area would have contained ca. 8 people; in 2000 BP, ca. 42 people; in 1800 CE, ca. 160 people; and today, ca. 1,013 people. This is just

one way of saying that since the end of the Paleolithic era, world populations have multiplied a thousand times, from 6 million to about 6 billion.
As this chapter has shown, this astonishing change began deep in the Paleolithic era, with the first migrations into new terrain within Africa.

THE HUMAN IMPACT ON THE BIOSPHERE

Though they may seem crude to modern humans, the technological skills
that made this expansion possible imply a marked increase in human ecological control. That increase was enough to have a significant impact on Paleolithic environments. Fire stick farming offers the most spectacular example, for it seems that the regular firing of landscapes over many thousands
of years could transform large areas, sometimes in fundamental ways.[57] In
Australia, fire-loving species such as eucalypti multiplied under a regime of
fire stick farming while other species declined; thus the eucalyptus-domi-
nated landscapes that European migrants took for the "natural" landscape
of Australia were, in fact, as much a human artifact as the landscaped gardens of eighteenth-century Britain.

Another important way in which Paleolithic communities began to
shape their surroundings was by driving other species to extinction. Improved hunting techniques and the increasing use of fire may both have
played a role here, as did the spread of humans into new environments. Particularly threatened were many large species, or megafauna: large mammals,
reptiles, and birds that reproduced slowly and were therefore more vulnerable to sudden population declines. Mammoths, woolly rhinoceros, and giant Irish elk vanished in northern and inner Eurasia; horses, elephants, giant
armadillos, and sloths vanished in North America.[58] In Australia many genera of large marsupials vanished, including the *Diprotodon*, a wombat-like
creature about 2 meters high (see figure 7.3). And they seem to have vanished within 10,000 years of the first arrival of humans.[59] As Darwin's col-
laborator, Alfred Wallace, noted as early as 1876, the extinctions occurred
with varying degrees of intensity in much of the world, from the Pacific to
Eurasia to the Americas: "We live in a zoologically impoverished world, from
which all the hugest, and fiercest, and strangest forms have recently disappeared; and it is, no doubt, a much better world for us now that they have
gone. Yet it is surely a marvelous fact, and one that has hardly been sufficiently dwelt upon, this sudden dying out of so many large Mammalia, not
in one place only but over half the land surface of the globe."[60]

Scientists have long debated the relative importance in these extinctions
of climatic change and human overhunting. Both may have played a role,
but as we begin to date the extinctions more precisely, the evidence is

Figure 7.3.   Extinct (and dwarfed) Australian megafauna: shadow drawings of Australia's lost and dwarfed fauna. The human hunter at left gives some idea of their size. From Tim Flannery, *The Future Eaters: An Ecological History of the Australasian Lands and People* (Chatswood, N.S.W.: Reed, 1995), p. 119; courtesy of Peter Murray.

mounting that the main extinctions, certainly in newly colonized regions such as Siberia, Australia, and the Americas, coincide with the arrival of humans.[61] This is where the extinctions were most severe. Australia and the Americas may have lost 70 to 80 percent of all mammal species over 44 kilograms in weight; in Europe, about 40 percent of megafauna disappeared, and in Africa only ca. 14 percent.[62] In recent times, too, species were particularly vulnerable in environments such as the Pacific islands, whose animals had no previous experience of dealing with humans. The absence of any sign of similar rates of extinction in previous periods of rapid climatic change during the Pleistocene also supports the claim that human activity is implicated. Whatever the cause, the removal of most large mammal species from Australia and the Americas was to prove momentous. By eliminating several species that might eventually have been domesticable, it may have slowed or prevented the emergence of agriculture in these huge regions, as well as depriving them of a major potential energy source.[63]

There is a sad and striking end to the story of Paleolithic extinctions. Those species driven to extinction by the spread of modern humans probably included the last remaining hominines who were not members of our species. Neanderthals, as we have seen, had brains as large as those of modern humans, and they were creative enough to settle in cold regions of modern Russia and Europe that no earlier hominines had occupied. But they apparently lacked the technological creativity of modern humans, presumably because they lacked a developed symbolic language. In the Middle East, modern humans were present at the same time as Neanderthals; moreover, in this region modern humans seem to have used tools similar to those of their Neanderthal neighbors. But the two species used similar tools in different ways. Studies of the bones of prey species left by modern humans show that most animals were taken either in summer or winter, while those from Neanderthal sites were taken throughout the year. In other words, modern humans were probably moving around more, and taking prey more selectively, while Neanderthals were occupying the same site year-round. These subtle differences may point to more profound differences between the two groups. The greater mobility of modern humans suggests that different groups had more contact with each other, and may have shared information more widely, while Neanderthal groups and individuals remained more isolated from each other. Among modern foraging communities, particularly in colder regions (similar, perhaps, to those of the Middle East during the last ice age), information sharing between different groups can be vital to survival. At the same time, groups that are more self-sufficient and less mobile may be more vulnerable to sudden ecological crises. Such groups, with their less efficient hunting methods, may also have to expend more physical energy in order to survive. That need may explain why Neanderthals seem to have been so stocky; their hunting relied more on individual strength than on collective cunning.[64]

Over time, these differences told, as modern humans spread more widely and eventually migrated into regions occupied by Neanderthals. One such region may have been the south of France, which probably had the densest populations of any region of Upper Paleolithic Europe late in the last ice age (which may be why it also contains 80 percent of Europe's cave art).[65] In France, there is evidence that Neanderthal communities survived during most of the last ice age and may have tried to borrow some of the new technologies of their neighbors. But they had little success with it. The last Neanderthals perished somewhere in southwest Europe, 25,000–30,000 years ago. It is just possible that a similar story was played out at about the same time at the eastern end of the Eurasian landmass as well, as evidence has emerged that other hominine populations may have survived there as late as Neanderthals, vanishing perhaps 50,000 or even 27,000 years ago.[66]

Even in the Paleolithic, the ecological virtuosity of modern humans had both destructive and creative sides. The migrations of Paleolithic humans, their cave art, and their technological skills rightly win our admiration; but the elimination of so many other large animals, including the only surviving species of hominines, is a powerful reminder of a more deadly side to human history.

SUMMARY

Recent research suggests that modern humans, equipped with a symbolic language and the capacity for collective learning, appeared in Africa about 250,000 years ago. Gradually, community by community, humans evolved new technologies and learned to live in new environments. Beginning ca. 100,000 years ago, humans began to migrate out of Africa and into lands no earlier hominines had settled, lands whose occupation required entirely new ecological skills. The continent of Sahul was occupied between 60,000 and 40,000 years ago; Ice Age Russia and Siberia were occupied from ca. 30,000 years ago onward; and the Americas were certainly occupied by migrants from Siberia by 13,000 years ago, and perhaps much earlier. As humans spread, they began, for the first time, to have a significant impact on the biosphere, transforming landscapes with fire and hunting a large number of Pleistocene megafauna to extinction. By the end of the last ice age, ca. 10,000 years ago, humans occupied all habitable parts of the world except the many islands of the Pacific. They had also driven the only other surviving hominines to extinction.

FURTHER READING

The early history of our own species is complex territory, and riddled with controversy. There are several good general surveys, including Peter Bogucki, *The Origins of Human Society* (1999); Göran Burenhult, ed., *The Illustrated History of Humankind* (5 vols., 1993–94); Roger Lewin, *Human Evolution* (4th ed., 1999); Ian Tattersall, *Becoming Human* (1998); Richard Klein, *The Human Career* (1999); Luigi Luca and Francesco Cavalli-Sforza, *The Great Human Diasporas* (1995); Chris Stringer and Robin McKie, *African Exodus* (1996); and Robert Wenke, *Patterns in Prehistory* (3rd ed., 1990). This chapter relies heavily on a superb recent article by Sally McBrearty and Alison Brooks, "The Revolution That Wasn't" (2000), but it is too early to know if this account will achieve general recognition. The early history of language is equally controversial. Aspects of the current debates on the subject are covered in books by Terrence Deacon (*The Symbolic Species* [1997]), Steven Mithen (*The Prehistory of the Mind* [1996]), Henry Plotkin (*Evolution in Mind* [1997]), John Maynard Smith and Eörs Szathmáry (*The Origins of Life* [1999]), and Steven Pinker (*The Language Instinct* [1994]). Clive Gamble's *Timewalkers* (1995) is one of the best recent surveys of Paleolithic history, with a strong focus on changing social relations and networks. Tim Flannery's *The Future Eaters* (1995) is a superb, if controversial, book on early ecological impacts of humans in Sahul; his more recent work, *The Eternal Frontier* (2001), discusses the ecological history of North America. The work of Olga Soffer (see the articles listed in the bibliography) is fundamental for understanding the settlement of Ice Age Russia. *The Cambridge Encyclopedia of Human Evolution* (1992), edited by Steven Jones et al., is also useful for many details in this chapter.

PART IV

THE HOLOCENE
Few Worlds

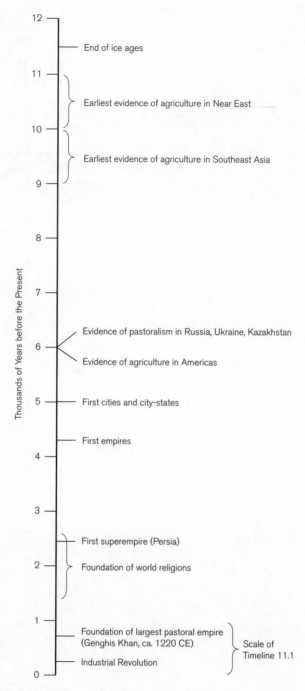

Timeline 8.1. The history of agrarian societies and urban civilizations: 5,000 years.

8

INTENSIFICATION AND THE ORIGINS OF AGRICULTURE

The Agricultural Revolution involved restructuring the food
economy, shifting from a nomadic life-style based on hunting
and gathering [foraging] to a settled life-style based on tilling the
soil. Although agriculture started as a supplement to hunting and
gathering, it eventually replaced it almost entirely. The Agricultural Revolution entailed clearing one tenth of the earth's land
surface of either grass or trees so it could be plowed. Unlike the
hunter-gatherer culture that had little effect on the earth, this new
farming culture literally transformed the surface of the earth.

In the geological timescale, the Pleistocene era ends and the Holocene era
begins ca. 11,500 years ago, at the end of the last ice age. From about this
time, human history sets out in a new direction. A threshold is crossed, with
a shift from extensive to intensive technologies. In the Paleolithic era, the
increasing ecological power of our species shows up in the exploration of
new environments during migrations that took humans around the world.
From the early Holocene, it takes the form of *intensification:* new technologies and lifeways that enabled humans to extract more resources from a
given area of land. As a result, though most of human history took place in
the Paleolithic era (chronologically speaking), most humans have lived during the last 10,000 years (see figure 8.1).

Loosely, we can refer to the new technologies of the early Holocene as
agriculture. They stimulated population growth and encouraged humans to
settle in the large, concentrated communities we call villages and towns.
Denser settlement encouraged more exchanges of ideas, and stimulated collective learning so that the pace of technological change accelerated. But
larger and denser settlements also created novel social and organizational
problems, whose solutions required both new social relationships and larger
and more complex social structures. Over thousands of years, and at vary-

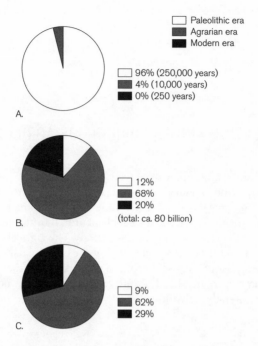

Figure 8.1.   Three eras of human history compared. The Paleolithic, agrarian, and modern
eras compared by (a) Duration (240,00, 10,000,
and 200 years, respectively). (b) The number
of humans that lived in each era (ca. 80 billion
humans have been born since our species first
appeared, according to the estimates of M. Livi-
Bacci, *A Concise History of World Population*
[Oxford: Blackwell, 1992], pp. 31, 33). (c) The
number of years lived in each era. Because life
expectancies have risen dramatically in the
modern era, it looms larger when measured by
the number of lives lived (estimates from Livi-
Bacci, *A Concise History of World Population*,
pp. 31, 33).

ing speeds, these changes spread through much of the world. They mark the
most fundamental change since the evolution of modern humans.

The dynamism of the Holocene era shows up most clearly as population
growth (see figure 8.2, and tables 6.2 and 6.3). We have seen already that in
European prehistory, even the earliest forms of farming could support perhaps 50 to 100 times as many people as foraging technologies could in a

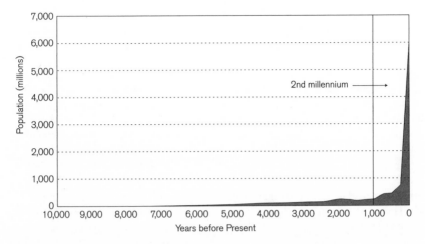

Figure 8.2.   Human populations, 10,000 BP to now. Based on table 6.2.

similar area.[1] This is why the transition to agriculture shows up as a clear upward inflection in the graph of world population growth. Of course, our figures for world populations in this period are approximate. Nevertheless, the multiplication of settlement sites in this period shows that populations really did grow more rapidly than in the Paleolithic era. The estimates in tables 6.2 and 6.3 suggest that world populations rose from ca. 6 million in 10,000 BP to about 50 million 5,000 years ago, which implies a six- to twelve-fold increase in 5,000 years.[2] On average, populations were doubling every 1,600 years, while in the Upper Paleolithic they doubled, on average, every 6,000 years. These changes mark the start of a new demographic epoch, one whose characteristic long-term growth rates were sustained for almost 10,000 years before accelerating even more decisively in the modern era.

In the 1930s, the Australian archaeologist V. Gordon Childe proposed that this suite of changes be called the "Neolithic Revolution." Archaeologists first used the term Neolithic (or New Stone Age) to describe distinctive polished stone tools that appear from about 10,000 years ago. But Childe insisted that the real significance of this period lay in something more revolutionary: the emergence of agriculture. Agriculture laid the foundations for all the most important developments of later human history. Today, many prehistorians resist Childe's term because they know that when examined closely, the changes turn out to have been gradual. Contemporaries could hardly have known they were living through a revolution. Nevertheless, Childe's notion of a Neolithic or agrarian revolution de-

TABLE 8.1. A PERIODIZATION OF HUMAN HISTORY

Name of Epoch	Rough Time Periods	Distinctive Features
Era 1: Many Worlds: The Paleolithic, and the beginnings of human history	300,000/250,000 BP– 10,000 BP	Earliest signs of human adaptive virtuosity; many small, loosely linked communities; population growth and extensification; humans enter new environments, settling most of the habitable world; extinction of all other surviving hominine species
Era 2: Few Worlds: The Holocene and the agrarian era	10,000 BP–500 BP	Intensification and dense, interconnected settlements; increasing variety of adaptations, new types of community, increasingly artificial environments, growing populations; three separate world zones, moving through similar trajectories at different speeds governed by different synergies of informational exchange
Era 3: One World: The modern era	500 BP–Now	Single, global system; collective learning at species level; sharp acceleration in extraction of resources; control of biospheric resources; extinction of other organisms

serves to survive, for on the scale of human history as a whole, the changes were both rapid and revolutionary (see table 8.1). During a mere 7,500 years, between 11,500 and 4,000 years ago, agricultural communities with domesticated plants and animals appeared in at least three quite separate regions of the world, and perhaps as many as seven. The lifeways pioneered in these regions of "pristine" agriculture then diffused, as agriculturalists migrated to new regions or as other communities incorporated the new techniques into their own lifeways, which may already have been half agricultural. Through a complex blend of migration, diffusion, and local invention and reinvention, agricultural lifeways spread, with many local variations, along existing or newly created networks of exchange to most of the world.

This chapter concentrates on what I will call the *early agrarian era*. This is the period of human history in which there existed agricultural communities, but no cities and no states. As we will see, its chronology varies from region to region. In some regions, it began 10,000 to 11,000 years ago, and ended about 5,000 to 6,000 years ago; in others, it began much later and survived into the twentieth century.

THE HOLOCENE PERIOD OF HUMAN HISTORY

The End of the Last Ice Age

The coldest phase of the last ice age occurred between about 25,000 and 18,000 years ago. Beginning 18,000 years ago, climates became warmer and wetter, sometimes quite suddenly, though there were also brief returns to glacial conditions (for example, between ca. 13,000 and 11,500 years ago). After ca. 11,500 years ago, climates have remained, for the most part, typical of the warm periods between ice ages known as *interglacials*, though there have been occasional periods of warmer or cooler weather. All of recorded human history has taken place within the Holocene interglacial.

As climates warmed, the ice sheets covering much of North America, northern Europe and Scandinavia, and eastern Siberia thinned and retreated. As the ice melted, sea levels rose, drowning coastal regions in much of the world. The change was most dramatic in northern latitudes, where lands freed from the weight of the great ice sheets literally floated upward.

Climatic changes transformed landscapes and vegetation.[3] Regions of desert and tundra contracted, while forests expanded. In Eurasia and North America, forests migrated into what had been the cold steppes of the Ice Ages, creating some of the largest forested zones in the world. Birch and pine migrated fastest and farthest, followed by deciduous species such as hazel, elm, and oak. In warmer regions of Africa and South America, where forests had largely disappeared, they returned to create tropical forests almost as extensive as the temperate forests of northern latitudes. Where forests spread, they displaced steppe species, such as the herds of mammoth, bison, and horses that had grazed the Ice Age steppes of Eurasia and North America. In place of those animals, they brought species such as boar, deer, and rabbit, together with a whole range of new plant foods such as nuts, berries, seeds, fruit, and fungi. For humans, these species were more difficult to exploit than the large herbivores hunted in northern latitudes during the ice ages. But in some regions these smaller prey species flourished in huge numbers as climates warmed, and their sheer quantity made them more attractive. Between 10,000 and 5500 BP, increased humidity turned what is now the Sahara Desert into a lush region of lakes and woodlands, whose inhabitants have left astonishing rock paintings of lifeways that are inconceivable in the dry Sahara of today.

While plants and animals had to adapt to climatic change, so did human beings. But they adapted in different ways in different parts of the world; thus human societies became more diverse during the Holocene.

Three Worlds

As sea levels rose early in the Holocene, the land bridges between Siberia and Alaska, Japan and China, Britain and Europe, and Australia, Papua New Guinea, and Tasmania were all drowned. Indonesia, a southern peninsula of Asia during the Ice Age, became an island archipelago, and the gap between Indonesia, Australia, and Papua New Guinea widened. With humans now settled throughout the world, this severing of ancient links threatened to divide humans into separate populations with separate histories. As Robert Wright nicely puts it: "The Old World and the New World were now two distinct petri dishes for cultural evolution."[4]

The separation was at no time complete. The arrival of the dingo in Australia perhaps 4,000 years ago, or of Indonesian trepang fishers in recent centuries, proves that Australia was never totally cut off from Indonesia and Asia. And Papua New Guinea certainly had contacts with Austronesian migrants to Indonesia from ca. 1600 BCE onward.[5] The narrowness of the gap across the Bering Straits and the brief settlement by Vikings of a colony in Newfoundland show that the Americas were never totally cut off from Eurasia. Moreover, the presence of South American sweet potatoes in Polynesia demonstrates that there must have been some contact between the Americas and the various communities that settled the Pacific during the past 3,000 years. Nevertheless, these contacts were so limited that it makes sense to think of human history during much of the Holocene era as taking place within three distinct world zones, with the addition of a fourth—the Pacific zone—in the past 4,000 years.[6] The major world zones of the Holocene were the Afro-Eurasian zone, which includes Africa and the entire Eurasian landmass, as well as offshore islands such as Britain and Japan; the Americas, from Alaska to Tierra del Fuego, as well as offshore islands such as those of the Caribbean; Australia and Papua New Guinea; and from ca. 4,000 years ago, the island societies of the Pacific (see maps 8.1 and 8.2).

Within each of these world zones, it was possible, in principle at least, for ideas, influences, technologies, languages, and even some goods to travel from one end to the other. There was always indirect contact between Papua New

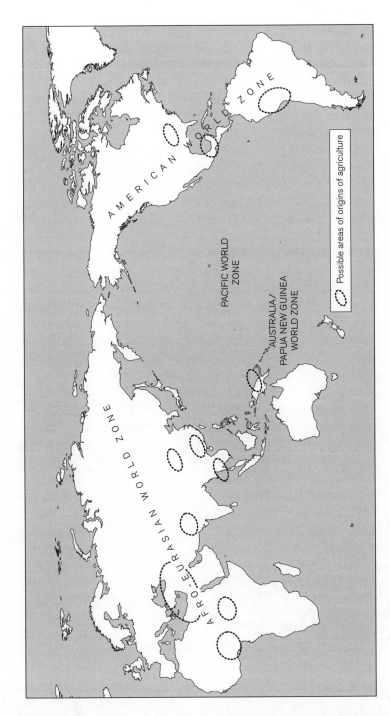

Map 8.1. World zones of the Holocene era.

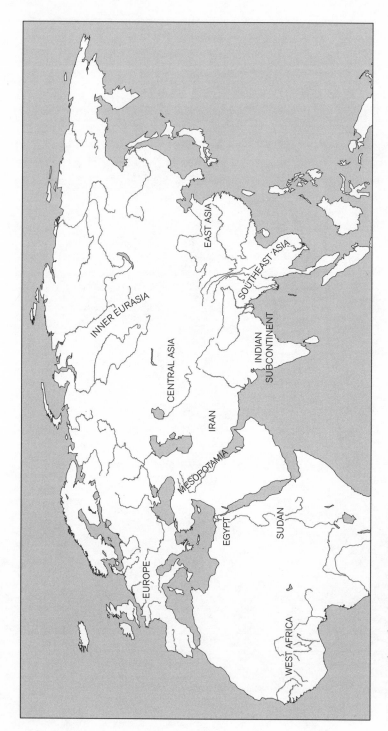

Map 8.2.   The Afro-Eurasian world zone.

Guinea and Australia across the island chain of the Torres Strait. Within Australia, prestige goods such as pearl shells from the northwest traveled in relays across much of the continent, while "baler shells" from Cape York in the far northeast were made into ornaments that were used in rituals and sorcery as far away as southern Australia and the deserts of western Australia.[7] The islands of Polynesia and Micronesia were settled in a series of linked migrations by communities whose similarities are apparent in language and in the archaeological remains of the so-called Lapita cultures.[8] In Afro-Eurasia, the Sahara was a region of steppe and savanna land until ca. 4,000 years ago, and thus sub-Saharan Africa was not as separate from the rest of the region as it has been since then. Technologies of pastoralism originated in both Inner Eurasia and Saharan Africa; from there they spread through the Eurasian steppes and to eastern Siberia in one direction and to the Middle East and East Africa in another. Indo-European languages spread to Sinkiang, India, and western Europe; Afro-Asiatic languages spread through much of Africa and also into the Middle East; and Turkic languages spread from Mongolia to Anatolia. In the Americas, early migrants traveled, generation by generation, from Alaska to Tierra del Fuego, creating, as the linguist Joseph Greenberg has shown, a coherent linguistic zone embracing all of South America and most of North America as well.[9]

It will be helpful to think of each of these regions as a separate world zone during much of the Holocene, because doing so can help us distinguish between universal and regional trends. There are astonishing parallels between the histories of these worlds, but there are also striking and important differences. Intensification of some kind occurred in all world zones, and the steady growth of human adaptive capacity that was already evident in the Paleolithic era continued right around the globe. But the pace of change varied, and so did the nature of the adaptations that emerged in each zone. Explaining both the similarities and the differences between these different histories will be a central task of the next three chapters.[10]

WHAT IS AGRICULTURE?

Of all forms of intensification in the early Holocene era, agriculture is by far the most important. But what is agriculture?

Like the fire stick "farmers" discussed in the previous chapter, agriculturalists systematically groom the environment to favor those plant and animal species they find most useful. But agriculture raises productivity to a particular degree, by grooming so intense that it eventually transforms favored species through an early form of artificial selection. It depends on the early form of genetic engineering known as *domestication*.

Domestication

Domestication is a symbiotic process in which one species, instead of just preying on another, protects the second species and encourages its reproduction, so as to create a more reliable source of food. We have seen that this pattern of co-evolution, leading from predation to symbiosis, is common in evolutionary history, and there is a good Darwinian logic to it. Because excessively brutal predation may kill off the prey species, the most effective predators (both large and small) are those that take their prey selectively, and even try to ensure the prey's survival as a species. Both species benefit from such a relationship. While the predator gains more control over an important food source, the prey species finds a protector happy to ensure its survival and reproduction—at a price. Sheep and maize would not be as abundant as they are if humans had not domesticated them. Domestication occurs among many different species. There are, for example, species of ants that treat aphids more or less like domestic cattle, in return for honeydew. With their antennae they stroke the captive aphids to stimulate the production of honeydew. In return for their honeydew, the ants take care of the aphids and ensure their successful reproduction.[11]

No sharp dividing line exists between predation and domestication. But in tight symbiotic relationships, both species change, behaviorally or genetically, until a point is reached at which one or both can no longer survive without the other. In human history, the genetic changes have occurred principally in the domesticated species. Humans have also changed genetically—for example, some have acquired an increased capacity to digest the raw milk of domesticated animals. But the most significant human adaptations have been behavioral and cultural. The greater speed of cultural change explains why symbioses with humans developed much faster than symbiotic relations between nonhumans.

Domestication describes that stage of the symbiosis at which at least one of the partners cannot survive on its own. In the case of agriculture, this means that domesticates can no longer survive or reproduce without human support, while many human communities can no longer live without their preferred domesticates. Domesticated sheep are too slow and stupid to survive in the wild, and modern maize, or Indian corn, cannot reproduce without human help, as its seeds cannot scatter freely.[12] In a recent account of agricultural origins, Bruce Smith defines domestication as "the human creation of a new form of plant or animal—one that is identifiably different from its wild ancestors and extant wild relatives."[13] The creation of new species of animals begins when humans start controlling how their animals reproduce and cut them off from contact with wild populations. With domesticated plants, it begins with harvesting, planting, and weeding, for these practices remove domesticates from genetic contact with neighboring populations and give them an artificial head start over their "wild" cousins. In both cases, human intervention places a barrier between the wild and the domesticated species. This encourages rapid genetic changes in a manner similar to allopatric speciation, though here it is humans, rather than migrations or geographical changes, that have created the genetic barriers between populations of the same species.

Once humans start separating a population from its wild relatives, it can evolve rapidly.[14] Certain changes are familiar to archaeologists. Domesticated seed plants often have tight clusters of seeds that are more firmly attached to the stem than those in wild varieties, because humans find it easiest to collect (and therefore to replant) thick concentrations of seeds; moreover, isolated seeds or those loosely attached to the stem are likely to fall off during harvesting, so they are unlikely to be replanted. Domesticated plants also tend to develop large seeds with thin skins, for similar reasons. Where plants are planted closely together and competition for sunlight is fierce, those seedlings that sprout first are most likely to survive; and these are likely to have thin skins and larger internal food stores, which give them an edge over their rivals. The fattest, fruitiest, and earliest sprouting plants are also more likely to be selected by humans for replanting. So, when looking for evidence of domestication, paleobotanists look for seeds that are larger and have thinner skins than the wild varieties, that are held together in clusters, and that are more tightly held to the stem by a strong rachis (i.e., connecting axis). Domesticated animals undergo analogous changes, though these are usually harder to detect in the archaeological record. Decreasing size is a common marker, whether caused by deliberate selection for more docile, manageable beasts or by poorer nutrition under conditions of domestication. Different herd composition is another marker. Often, females outnumber males in domestic herds, because males are culled early. The old are also more likely to be culled in domestic herds.

Agriculture is not a synonym for *domestication*. Many societies have adopted limited forms of domestication, either of plants or animals, without becoming dependent on them and without becoming sedentary. And while pastoralists are as dependent on domesticates as are agriculturalists, they rely principally on animal rather than plant domesticates. In addition, pastoralists, like foragers, are often nomadic. In contrast, agriculturists normally exploit both domesticated plants and animals, and most are sedentary. Though agriculturalists may still hunt and fish, the subsistence basis

of their communities comes from their domesticates. Finally, plant domesticates are usually more important than animal domesticates in agricultural
societies. This is a consequence of the basic ecological rule that organisms
lowest on the food chain transmit the energy of sunlight most efficiently.
At each step in the food chain, about 90 percent of that energy is lost; thus,
human lifeways depending mainly on plant-based foods can normally support larger population densities than lifeways (such as pastoralism) dependent mainly on animal-based foods. So plant domesticates account for most
of the demographic dynamism of the agricultural revolution.

As table 8.2 shows, the domestication of different species of plants and
animals continued throughout the Holocene, and occurred, apparently quite
independently, in several different parts of the world. However, these figures
simply reflect the earliest evidence of domestication. The step from domestication to lifeways based mainly on agriculture was rapid in some areas
(such as Southwest Asia, central Asia, and China) but slower elsewhere—
particularly in the Americas, where there is a gap of several thousand years
between the earliest signs of domestication and the earliest evidence of lifeways based mainly on agriculture.

Chronology and Geography of Early Domestication

Additional research may push the dates recorded in table 8.2 farther back
in time, perhaps by centuries, in some cases by millennia. Researchers may
also identify other centers of domestication that have been missed so far.
Likely candidates may appear in the tropics, particularly in Papua New
Guinea and Indonesia, and in the Amazonian rain forests (where the main
crops were manioc, potato, and peanut). In parts of Papua New Guinea, taro
was probably being cultivated some 9,000 years ago; by 5,000 to 6,000 years
ago, true agriculture was supporting permanent villages in forest clearings
throughout the country, using indigenous (or perhaps imported) species of
taro and yams as staples.[15]

Writing more than a century ago, Francis Galton suggested that the first
steps to domestication involved a sort of ecological "audition." Humans
probably "auditioned" numerous prey species, but many failed their try-
out because they lacked some crucial quality necessary to make them viable
domesticates. The failures included deer (which proved too skittish) as well
as acorns and hazelnuts (which proved less nutritious and harder to store
than domesticated cereals and legumes, though both continued to be used
as famine foods). The first species successfully domesticated by humans was
probably the wolf. Wolves were domesticated late in the Upper Paleolithic,

TABLE 8.2. FIRST RECORDED EVIDENCE OF DOMESTICATES

Date (1000 years BP)	Southwest Asia	Central/ East Asia	Africa	Americas
13–12			dog	
12–11	dog, goat, sheep			
11–10	emmer/einkorn wheat, barley, pea and lentil, pig			
10–9	rye, cattle			bottle gourd, squash
9–8	flax			chili pepper, avocado, beans
8–7		foxtail millet, bottle gourd, dog		maize, llama/alpaca
7–6	date palm, vine	water chestnut, common millet, mulberry, rice, water buffalo	finger millet	
6–5	olive, donkey	horse, cattle (zebu), onion	oil palm, sorghum	cotton
5–4	melon, leek, walnut	camel (Bactrian)	yam?, cowpea	peanut, sweet potato
4–3	camel (dromedary)	garlic	cat, pearl millet	guinea pig, manioc
3–2				potato, turkey
2–1				pineapple, tobacco

SOURCE: Adapted from Neil Roberts, *The Holocene: An Environmental History*, 2nd ed. (Oxford: Blackwell, 1998), p. 136.

and all breeds of modern domestic dogs are descended from these early do-mesticates.[16] But domesticated wolves did not have the transformative im-pact of later domesticated species, for instead of offering an alternative to foraging lifeways, they were used to help with the hunt.

The "Neolithic revolution" really begins with the domestication of a small number of seed plants. The earliest evidence for this change comes from Southwest Asia, along the narrow corridor linking Africa and Eurasia into the largest premodern exchange network on Earth. It is probably no accident that agriculture appeared first within the largest and oldest world zone, that of Afro-Eurasia. Nor is it an accident that it occurred in the corridor linking two very different regions, for "hub" regions of this kind (see chapter 10 for a fuller discussion) were clearinghouses for ecological information accumulated over huge areas. Another hub region, in Mesoamerica, linked North and South America; and here, too, agriculture appeared early.

The earliest agrarian sites in Afro-Eurasia are concentrated in a region known to archaeologists as the Fertile Crescent. This is an arc of mainly high land that runs northward through parts of modern Israel, Jordan, and Lebanon, then curves eastward along the border between Turkey and Syria before turning south along the Zagros Mountains on the Iraq–Iran border. Between 11,000 and 9000 BP, at least eight different species of plants were domesticated in this region. They include lentils, peas, chickpeas, bitter vetch, flax, and the cereals—emmer wheat, einkorn wheat, and barley. The three cereal crops all seem to have been domesticated in the region near Jericho, between 11,500 and 10,700 BP, probably by communities that had once harvested them in the wild.[17] Within these few centuries, all three cereals underwent the changes usually associated with domestication. They developed larger seeds and a tough stalk holding the seeds to the stem.

Sheep and goats were probably domesticated in the north of the Fertile Crescent by communities that had previously hunted them. However, on the whole, animals seem to have been domesticated slightly later than plants. Indeed, the presence of crops that could be used as fodder may have been a prerequisite for animal domestication in many areas. Pigs were domesticated in the north of the Fertile Crescent, along the modern border between Turkey and Syria.[18] Unlike sheep and goats, they compete with humans for food, which may be why they were domesticated later. Cattle were also domesticated later than sheep and goats. The earliest certain remains of domesticated cattle date to about 9300 BP.[19] The delay may be because their wild ancestor, the aurochs, was a dangerous beast. (We know this because wild aurochs survived until three centuries ago: the last was drawn in Poland in the early seventeenth century CE.) However, like sheep and goats, aurochs were also gregarious. This meant it was possible to control whole herds by taming or supplanting their leaders.[20] In cattle, as in sheep and goats, domestication soon led to genetic changes as animals with undesirable characteristics such as skittishness or aggression (or even intelligence!) were culled.

China was a second area of early domestication. Recent research has shown that it occurred earlier there than was once thought. Rice was probably domesticated in southern China along the Yangtze River, about 9,500 to 8,800 years ago, by foragers who had previously harvested wild rice. Millet was domesticated along the Yellow River in northern China by 8,000 years ago. Pigs may have been domesticated independently in the north. By the eighth millennium BP, both the millet-based systems of northern China and the rice-based systems of southern China were well established.

A third wave of domestication occurred between 6000 and 4000 BP. African forms of millet and sorghum were domesticated south of the Sahara beginning at least 4,000 years ago, and maybe considerably earlier. The different conditions, and the appearance of domesticates quite distinct from those of the Fertile Crescent, suggest that domestication in sub-Saharan Africa was little influenced by what had happened in Southwest Asia.

Recent research suggests that in the Americas, domestication occurred later than was once believed. Nowhere is there now firm evidence of thoroughgoing domestication before about 5500 BP. This is the date of the earliest sample of domesticated maize found so far, from the Tehuacán Valley of Mesoamerica, southeast of modern Mexico City. Maize is descended from a wild species known as teosinte; together with beans and varieties of squash, it was to become the most important of all American plant domesticates. South America was the only region of the Americas in which animal domesticates played a significant role. Here, guinea pigs, llamas, and alpacas were domesticated at least by ca. 4000 BP, at about the same time as quinoa and potatoes. Animal domesticates were less important in the Americas, because the most promising potential domesticates, including horses and camels, had become extinct there at the end of the last ice age, probably because of human overpredation. Indeed, it is possible that the long gap in American prehistory before early forms of domestication and settled agriculture can be explained in part by the small number of potential animal domesticates that survived the first wave of human migrations to the region.[21]

Domestication also occurred in the third world zone, in Papua New Guinea. Here, it occurred early, but its impact was more limited than in the other world zones.

Once it had appeared, agriculture did not sweep all before it. Indeed, from a modern perspective, what is striking about the period covered in this chapter is how slowly agriculture spread. Though some communities began to depend primarily on their domesticates and became true agriculturalists, many others preserved their traditional foraging lifeways while adopting one or two domesticates as supplements. In Papua New Guinea, populations of farmers coexisted with neighboring foragers until modern times. In the Americas, the slowness with which domestication spread is most apparent in the communities of eastern North America that domesticated sunflowers and gourds. Here, the lack of suitable potential domesticates may explain agriculture's slow progress. Although agricultural lifeways were well developed to their south by the time they began to plant and cultivate local domesticates, about 4,000 years ago, hunting and gathering remained important for almost another 3,000 years, because the local domesticates could not supply a complete nutritional package. When Mexican maize arrived ca. 1,800 years ago, it did not flourish. Not until the introduction about 1,100 years ago of new strains of maize that could cope with the northern winters, as well as Mexican beans and squash, did agriculture take off in this region.[22]

In northeastern Africa, along the Nile, the suite of domesticated animals and plants typical of the Fertile Crescent appeared after 9000 BP (of these, only barley was indigenous to Egypt), but it took several thousand years before agricultural villages spread widely. In Europe, domestication spread from the Fertile Crescent to the Balkans and the Mediterranean coast of Italy and France beginning about 9,000 years ago. From there it spread north into temperate zones with distinctive climates and ecologies, where methods of domestication had to be modified before they could succeed. At one time, it seemed possible to trace a clear "wave of advance" of agriculture through Europe between 6,000 and 8,000 years ago. However, more detailed studies have shown that though domestication did spread through Europe, its progress was slower and less triumphal than had first appeared. Agrarian communities settled particularly in regions of easily worked loess soils. But elsewhere, especially in the northwest and northeast of the subcontinent, they had only a limited impact for several millennia. Instead, local communities of foragers often adopted some agrarian techniques and maintained trading relations with agrarian communities, without themselves becoming true farmers. Domestication and the agrarian lifeways associated with it remained merely options or supplements to foraging lifeways; in many parts of the Neolithic world, foragers and agriculturalists were linked in regional networks of exchange.

Similar patterns can be seen in other areas that were influenced but not dominated by agriculture during the early agrarian era, from Russia west of the Urals to central Asia and northern Mexico.

THE ORIGINS OF AGRICULTURE

How can we explain the transition to agriculture?[23]

It may seem that the question is easy to answer. Processes of collective learning ensured that human communities would continue to explore ways of extracting resources from their environment, and eventually they were bound to stumble on agriculture. Besides, agriculture was so much more productive than most foraging lifeways that it is tempting to suppose that once it had been "invented," it was bound to spread fast. The earliest attempts to explain the Neolithic revolution did indeed make these assumptions, seeing agriculture as an invention that spread from a single epicenter because of its inherent superiority to all other human adaptations.

However, research in the twentieth century revealed two significant problems with such explanations. First, as we have seen, agriculture did not in fact spread from a single center. Instead, it appeared, apparently independently, in many different regions of the world, in all three world zones. How can we explain the near-simultaneity of these changes in parts of the world that seem to have had no contact with each other? As Mark Cohen has stressed, "The most striking fact about early agriculture, ... is precisely that it is such a universal event."[24]

Second, we can no longer assume that communities of foragers were bound to adopt agriculture once they learned about it. Indeed, we are no longer so confident that the appearance of agriculture can automatically be regarded as a sign of progress. To be sure, agriculture can support larger populations than foraging lifeways, and thus in the long run agricultural communities are likely to outcompete foraging communities when the two lifeways come into conflict. But it is also clear that many foraging communities have resisted adopting agricultural practices even when they knew about them. As members of a foraging community in the Kalahari Desert told a modern researcher, why would one want to work as a farmer when there are so many mongongo nuts available to eat? In the far north of Australia, particularly in Cape York, Aboriginal populations knew about farming, because islanders to their north practiced it. But they chose not to adopt it. In Russia and Ukraine, too, foragers and farmers coexisted, perhaps for several thousand years after farmers entered the region some 6,000 to 7,000 years ago.[25] Foragers saw agriculture as an option, but not an inevitability.

And their conservatism may have been perfectly rational. Evidence from skeletal remains shows that early agriculture bred new forms of disease and new forms of stress.[26] Farmers have less varied diets than foragers in warm climates, so they are more subject to periodic shortages; foragers can more easily switch to alternative sources of food. Famine is a paradoxical by-product of the agricultural revolution. Farming communities are also more subject to diseases carried by the rats, mice, bacteria, and viruses that flourish in moderately large sedentary communities. Even more important, genetic comparisons of modern disease bacteria suggest that in Afro-Eurasia, where livestock were domesticated, disease bacteria spread easily from herd animals such as cattle, chickens, and pigs to humans. The diseases exploited the fact that humans, too, became herd animals once they settled down to farm in village communities.[27] The most successful strains, and the ones that survived longest to become endemic, were those—including smallpox and flu—that infected their human hosts without killing them. A further sign of declining health in early agrarian communities may be that Neolithic skeletons seem to be shorter, on average, than those of Stone Age foraging societies; moreover, there is no evidence that life expectancies rose or infant mortality declined with the appearance of early forms of farming.[28] In both types of society, no more than 50 percent of all children born could expect to reach adulthood; those who did generally had a life expectancy of no more than about 25 to 30 years, though some individuals may have lived into their 50s and 60s.[29] All in all, it seems that the appearance of agriculture did more to depress standards of human welfare than to raise them. John Coatsworth writes: "Bioarchaeologists have linked the agricultural transition to a significant decline in nutrition and to increases in disease, mortality, overwork, and violence in areas where skeletal remains make it possible to compare human welfare before and after the change."[30]

Any adequate account of the origins of agriculture must explain both the chronology of early agriculture and the reasons why communities of foragers took up agricultural lifeways despite their apparent drawbacks. Why would one prefer a lifeway based on the painful cultivation, collection, and preparation of a small variety of grass seeds, when it was so much easier to gather plants or animals that were more varied, larger, and easier to prepare?

Explanatory "Prime Movers" in the Neolithic Revolution

Modern attempts to explain the Neolithic revolution date from the 1920s. The Russian geneticist N. I. Vavilov studied modern relatives of domesticated plants in the belief that the areas of their greatest genetic variety would prove to be where they originated, and probably where they were first cultivated. He identified eight likely "hearth areas" for early agriculture. Vavilov's list of hearth areas resembles modern lists of this kind, and the principle that studies of modern plants can tell us a lot about the early history of domesticates underpins the modern science of paleobotany. V. Gordon Childe argued that climatic change may have created "oases" of dense settlement whose inhabitants were forced to intensify their production methods in order to survive. In its general form, the position still has some plausibility, though the details of his original argument no longer stand. Robert

Braidwood undertook the first systematic archaeological investigations of early agriculture in Iraq, studying two villages, Karim Shahir and Jarmo: the first was occupied by foragers, the second by agriculturalists. And Richard MacNeish pioneered the study of early agriculture in the Americas in a series of expeditions, begun in the late 1940s, that were devoted to the early history of maize.[31]

Since these pioneering studies, an immense amount of research on the origins of agriculture has been undertaken. It is now reasonably clear what the main components are, but we do not yet know exactly how they are interwoven. The main factors include climatic change; various forms of intensification among foragers; population growth, which in some regions forced communities of foragers to exploit smaller territories and use them more intensively; increasing exchanges between communities; and, finally, the availability of potential domesticates. Any explanation must include some combination of these elements. The following account combines the insights of several closely related models and data from several different regions, though it best fits what we know of Mesopotamia and the Fertile Crescent. It argues that there were several distinct stages in the evolution of agriculture, which occurred, with minor modifications, in all regions of early domestication.[32]

As the reader will see, the simplified account offered here is different from the triumphalist accounts of the early twentieth century. Instead, like the story in Genesis, it describes a temptation, a fall, and an expulsion.

Cultural Preadaptations and Ecological Know-How   Most Upper Paleolithic communities knew many of the things that farmers need to know. Technologically, they were preadapted for agriculture. We assume this is so because modern foraging communities must have a sophisticated understanding of the plants and animals in their environment. They know the conditions under which favored species flourish, and they know how to nurture and encourage those species they favor—for example, by removing weeds or other rivals. Most small-scale societies understand that plants grow from seeds or cuttings, and that human activities can stimulate or inhibit growth.[33] Donald O. Henry has described the ecological know-how of Paleolithic peoples as one of the "necessary" conditions for the appearance of agriculture.[34]

It is also certain that significant forms of intensification appeared among foragers who had little or no contact with agriculture. Such communities are often referred to by anthropologists as "affluent foragers." The previous chapter has already described the astonishing mammoth-hunting cul-

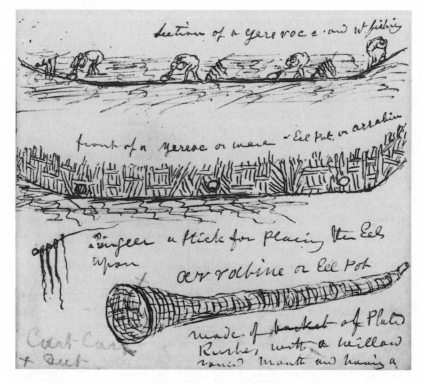

Figure 8.3.   Intensification in Australia: eel traps. An eel pot and trap sketched by G. A. Robinson in western Victoria, 1841, showing [*top*] the "front of a *yeroec* or weir" with "eel pot or *arrabine*," set into the holes in the weir; [*center*] "*lingeer* or stick for placing eels upon"; and [*bottom*] "*arrabine* or eel pot made of plaited rushes." From Josephine Flood, *Archaeology of the Dreamtime* (Sydney: Collins, 1983), p. 206: after George Augustus Robinson's 1841 journal, courtesy of the Mitchell Library, Sydney.

tures of late Paleolithic Ukraine and mentioned the dense populations of southern France, which lived by harvesting the large stocks of game and fish just south of the European ice sheets. Wherever we see foraging communities becoming more sedentary, we know they are using more intensive technologies, because to stay in one place for long periods they have to use its resources more intensively. But intensification of this kind becomes much more apparent early in the millennia after the end of the last ice age. In some form, intensification appears in all three world zones, and in all three it led to some degree of sedentism (i.e., the practice of establishing a permanent or semipermanent settlement). It is important to insist on this point, as it

Figure 8.4.   Intensification in Australia: stone houses. More than 140 stone structures have been found by Lake Condah, Victoria. From Josephine Flood, *Archaeology of the Dreamtime* (Sydney: Collins, 1983), p. 207; artist's impression by David White, *Age 29.1.81.*

is often thought that some parts of the world simply stagnated after agriculture had originated in a few favored regions.

In Australia there is now plenty of evidence of intensification, particularly in the past 5,000 years. Intensification enabled populations to increase, and in some areas it led to sedentism. Stone tools become much more varied during this period. New, smaller, and more finely made stone tools appear in many parts of Australia, including small points in Central Australia that may have been used as spear tips (there is no evidence of bows and arrows in Australia). Some stone points were so beautifully made that they were treated as ritual objects and traded over hundreds of miles. Elsewhere, backed blades were made, probably to be set in rows in weapons such as the terrifying "death spears," whose serrated edges ensured that the wounds they inflicted were almost always fatal.[35] The dingo, a semidomesticated dog, appeared about 4,000 years ago, and as it is closest to modern species of Indian dogs, it probably came from across the Indian Ocean rather than from Indonesia.[36]

New techniques meant new ways of extracting resources. In the Australian state of Victoria, elaborate eel traps were built, in some cases incorporating canals up to 300 meters long (see figure 8.3). Josephine Flood describes them:

> Traps were built across the stone races and canals; nets or eel pots were
> set in apertures in the stone walls, which were often constructed in a
> V-shape. The eel pots were made from strips of bark or plaited rushes
> with a willow hoop at the mouth. The tapered shape allowed men
> standing behind the weirs to grab the eels as they emerged through
> the narrow end of the pot. The fishermen killed the eels by biting
> them on the back of the head.[37]

So many eels could be stored and harvested in these traps that they attracted substantial and relatively permanent settlements (see figure 8.4). The survival of clusters of low stone huts (in one case, up to 146 at a single site) confirms the reports of early European visitors that there were Aboriginal villages in the area.[38] These communities lived off the many local species of game, from emu to kangaroos, as well as local vegetable foods such as daisy yam tubers, ferns, and convolvuluses.

In coastal regions of Australia, shell fishhooks were an innovation that gave access to new sources of food and enabled population growth. Some communities began to harvest plants such as yam, fruits, and grains in ways that suggest an incipient agriculture. The techniques for harvesting yams used then (and today) encouraged regrowth; fruit seeds were deliberately planted in refuse heaps to create fruit groves. In some of the more arid areas of Central Australia, European travelers observed communities harvesting wild millet with stone knives and storing it in large hayricks. Grindstones used on seeds have been found dating back as far as 15,000 years in some regions, demonstrating that these practices were very old.[39]

Similar changes occurred in many parts of the world toward the end of the Paleolithic and early in the Holocene. In Mesoamerica, there are signs from as early as 9,000 to 10,000 years ago of the intensive exploitation of several species that would later become agricultural staples, including early forms of maize, beans, and squashes. Some coastal communities in Mesoamerica enjoyed such abundant marine resources that they became largely sedentary from perhaps 5,000 years ago.[40] In the Baltic region of northwestern Eurasia, signs of intensification also appear soon after the end of the last ice age. Brian Fagan notes:

> The Mesolithic peoples who lived along the shores of the newly
> emerging Baltic Sea developed an astonishing range of fish spears,
> nets, harpoons, and traps, many of them preserved in waterlogged

sites. Spears and arrows were tipped with tiny stone, bone, or antler barbs. Ground-edged tools were used for woodworking and processing forest plants. Large canoes, some of them dugouts hollowed from tree trunks, were in evidence.[41]

These were stable and largely sedentary communities of affluent foragers. They relied on hunting, fishing, and the collecting of plant foods. Some Baltic settlements were huge. Archaeologists have found year-round sites in which as many as 100 people may have lived. Some of these sites were occupied continuously from ca. 3000 to 1500 BCE.[42]

Early evidence of affluent foragers has also been found along the Nile valley, in southern Egypt and Sudan. Near Aswan, 18,000 years ago, communities hunted large game, fished (which presumably meant they were reasonably sedentary), and also ground flour from wild grasses; at a nearby site, dated to ca. 15,000 years ago, there are stone blades covered with a sheen that indicates they were used for harvesting wild grains.[43] But the best-known affluent foragers of this era are the Natufian communities that appeared about 14,000 BP along the eastern Mediterranean coast in parts of modern Israel, Jordan, Lebanon, and Syria and survived for more than 2,000 years. Sediment cores have shown that the Natufian community of Ain Mallaha, which flourished ca. 13,000 years ago in the upper Jordan valley, had access to wild cereals and acorns, as well as lakeside resources such as fish, turtle, shellfish, and lake birds, which they caught using nets and hooks.[44] Natufian communities also hunted gazelle. With such abundant resources available nearby, they began to settle down in villages that were often six or seven times as large as any earlier settlements in the region, containing up to 150 people.

In all these areas, communities of foragers were pioneering new techniques, some of which involved more careful tending of plant and animal resources. Occasionally, these new techniques allowed entire communities to become more or less sedentary. These changes mark important steps on the path to agriculture.

As human technologies changed, they began to affect nearby species, particularly those exploited most intensively. For example, foragers might bring favored plant species back to base camps where, over several years, their seeds would form stands of plants ready for consumption by later generations of foragers. Such practices can exert powerful selective pressures, for it is obvious that over time, those fruit that taste best are most likely to be seeded around human campsites, while wild populations may remain less "tasty."[45] Over time, such intense manipulation of particular plant populations can lead to significant genetic changes.

Genetic Preadaptations and Potential Domesticates   Some species were more amenable to selective manipulation than others. Indeed, some potential domesticates appear in retrospect to have been preadapted for domestication. And this fact constitutes the second "necessary condition" for agriculture listed by Henry. Further, as Vavilov argued, the distribution of these potential domesticates offers a useful explanation for the geography and the "style" of domestication in different regions. Of the many wild species auditioned by humans as potential domesticates, few passed the test, and in some areas none did. Indeed, the availability or nonavailability of nutritious and easily domesticable plants and animals may have been a crucial determinant of the geography of early agriculture, and therefore a crucial determinant of much later human history.[46] Of several hundred thousand plant species, only a few hundred have been domesticated successfully, and most of these are of marginal importance when compared to the dozen major crops that provide most of the world's food today.

The qualities humans sought in potential domesticates were hardiness, nutritional value, adaptability, and the ability to breed under varying conditions. Animals had to be sociable; able to live and breed in large, compact herds; and characterized by social hierarchies that predisposed them to follow leaders, whether animal or human. The nature of the available domesticates may also help explain the chronology of early domestication. Jared Diamond has argued persuasively that the potential domesticates available in the Fertile Crescent were unusually varied, attractive, and easy to domesticate and that those features go a long way toward explaining why agriculture appeared first in this region. The ease of domestication of the region's main cereal crops can be demonstrated by the remarkably small change they have undergone from their wild state; wild barleys and wheats were abundant, nutritious, and easy to harvest and grow. In contrast, the domestication of maize was much trickier; teosinte had to be trained for several millennia before it could support large populations.[47] The lack of potential animal domesticates after the megafaunal extinctions of the early Holocene era also slowed the adoption of agriculture in Mesoamerica. There, only the dog and turkey were domesticated, and neither was as valuable as the main animal domesticates of the Fertile Crescent. The paucity of animal domesticates deprived American farmers of traction power, manure, and a rich source of protein. In Papua New Guinea, too, the nutritional limitations of local domesticates such as taro, which contains little protein, lessened the demographic impact of agriculture and restricted its spread.

The existence of potential domesticates and of much relevant ecological knowledge constitutes crucial preconditions for agriculture. But these factors cannot explain the timing or the motivation for the transition to fully developed agriculture.

Climatic Change, Population Pressure, and Exchanges   Given that agriculture emerged in several widely separated parts of the world within the relatively short span of a few millennia, it is tempting to look for global mechanisms that might have triggered change in different locations. Two possible triggers are climatic change and population pressure.

The climatic changes at the end of the last ice age were erratic and unpredictable. However, their most general effect was to raise average temperatures. Whatever their precise direction and nature, these changes must have stimulated cultural as well as genetic shifts throughout the world. As climates and environments altered, human communities had to experiment with new foods and new techniques. This was particularly true in areas such as the Eurasian steppes, where a combination of overhunting and global warming drove traditional prey species such as mammoth out of the lands they had once inhabited.

Climatic change also transformed environments. In some regions, warming climates increased the availability of both plant and animal foods. Henry argues that potential plant domesticates may have been rare before the end of the last ice age, pointing out that under colder conditions, ancestral forms of rice, cereals, and maize would have been isolated and confined to lowland regions. However, with the spread of warmer and wetter climates, they became more abundant and spread to upland regions. There, milder conditions encouraged them to produce seeds over a longer period, which made them more valuable for humans. This argument finds most support in the Fertile Crescent, where the spread of cereals can be traced by pollen studies. But it seems likely that the warmer, wetter conditions of the early Holocene increased the range and number of warmth-loving plants such as cereals in many parts of the world. Abundance was particularly great in regions with good supplies of water from rivers, lakes, or marshes, while regions with varied ecologies produced an increasing variety of plant and animal foods. In southeastern Turkey, as an experiment conducted by Jack Harlan showed in the 1970s, it was possible even under modern conditions, and in only three weeks, to harvest enough wild grain to support a family for an entire year. Increasing abundance of nutritious plant foods may in turn have attracted herbivores. And eventually, such "gardens of Eden" would have attracted humans, too. Where resources were particularly abundant, foraging communities may have become more settled, thereby perhaps taking a crucial step toward agriculture.

The second global factor is harder to identify in the archaeological record but equally hard to exclude from any account of agricultural origins: population pressure. The argument that population growth, far from being limited by available technologies (the familiar Malthusian argument), may create pressure for technological change in agriculture has been developed in the work of Ester Boserup, but Mark Cohen has done the most to explore its possibilities as an explanation for the origins of agriculture. His argument, essentially, is that population pressures encouraged individuals and groups to move to less densely settled regions. The eventual result was that population pressure by the early Holocene was so evenly distributed "that groups throughout the world would be forced to adopt agriculture within a few thousand years of one another."[48] There are several reasons to think that population pressures grew at the end of the last ice age, particularly in the Afro-Eurasian zone. The settlement of harsh environments such as tundra regions, a decline in the hunting of large animals (as many were hunted to extinction), and the increased use of smaller packets of food, such as shellfish and seeds, all hint at gentle population pressure. So does the increasing number of habitation sites.[49] But most important of all, we have seen already that humans had occupied all the earth's habitable continents by the beginning of the Holocene era, thereby eliminating any easy opportunities for further extensification. Given the foraging technologies of the Paleolithic, human populations in much of the world were already pressing at the limits of the earth's carrying capacity. Paul Bairoch observes, "According to the estimates of Hassan, the optimum carrying capacity of the world under hunting and gathering is some 8.6 million (5.6 million in tropical grasslands and only 0.5 million in temperate grasslands)."[50]

In particular regions, climatic change may have exaggerated these pressures, for as global temperatures rose, so did sea levels. In areas such as the Persian Gulf, this change undoubtedly forced coastal foragers to encroach on the territories of their neighbors. (One of the difficulties of testing this hypothesis is that most of the relevant sites are now under water.) The geography of Paleolithic migrations also highlights the existence of a few bottlenecks where population densities may have been exceptionally high. These were regions through which many peoples had to pass if they were to move to other lands. The region between Mesopotamia and the Nile was certainly such a region. By Paleolithic standards, it had relatively dense populations from as early as 80,000 or 90,000 years ago. Mesoamerica may constitute another such bottleneck, and so, probably, did the narrow stretch of habitable lands to the west of the Andes. It is harder to see whether such arguments can be applied to the Yellow or Yangtze valleys in China, but it may be that even there, localized abundance created bottlenecks that forced communities of foragers to live within ever smaller territories.

A third factor, closely linked to population growth, may also have encouraged sedentism: an increase in interregional exchanges. In foraging communities, the practice of gathering temporarily for the exchange of goods, rituals, and people in marriage has been widely documented. Foragers gathered where they could intensify food production, at least for a few weeks. Here is one description of such meetings, taken from the memoirs of a nineteenth-century British pastoralist in the Australian state of Victoria:

> At the periodical great meetings trading is carried on by the exchange of articles peculiar to distant parts of the country. A favourite place of meeting for the purpose of barter is a hill called Noorat, near Terang. In that locality the forest kangaroos are plentiful, and the skins of the young ones found there are considered superior to all others for making rugs. The Aborigines from the Geelong district bring the best stones for making axes, and a kind of wattle gum celebrated for its adhesiveness. This Geelong gum is so useful in fixing the handles of stone axes and the splinters of flint in spears, and for cementing the joints of bark buckets, that it is carried in large lumps all over the Western District. Greenstone for axes is obtained also from a quarry on Spring Creek, near Goodwood; and sandstone for grinding them is got from the salt creek near Lake Boloke. Obsidian or volcanic glass, for scraping and polishing weapons, is found near Dunkeld. . . . Marine shells . . . and freshwater mussel shells, are also the articles of exchange.[51]

Andrew Sherratt has suggested that exchanges of valued goods between foraging communities may have encouraged dense and perhaps even long-term settlement at the hubs of regional networks of exchange. Such exchanges were particularly intense in the early Holocene along the Levantine corridor, between Anatolia and the Red Sea; they might have stimulated communities already exploiting natural stands of cereals on well-watered highlands to try encouraging their growth in lowland regions crossed by flourishing "trade" routes. Indeed, he points out that in the 1960s, Jane Jacobs argued that large settlements such as Jericho may have appeared first, at points where exchanges were most intense; simple forms of agriculture may have emerged to support already existing settlements, with small villages appearing later.[52] Of course, the same exchanges would have also encouraged swapping of the ecological techniques needed for early types of farming.

In some regions, therefore, localized abundance, gentle population pressure, and increased exchanges may have conspired to encourage sedentism.

Sedentary communities had appeared even during the Upper Paleolithic; but in the absence of domestication, these experiments led to no permanent or widespread changes in technologies or lifeways. However, at the end of the last ice age, the presence of more abundant potential domesticates, and perhaps of increasing population pressure, ensured that such experiments would prove more common, more significant, and more lasting. The Natufian cultures of the Middle East offer a good example of these processes.

Population Growth, Intensification, and Specialization   Sedentism is not the same as agriculture, but it was probably a vital, unplanned step *toward* agriculture. In the Middle East, Natufian populations grew rapidly, and Natufian villages fissioned and spread throughout the eastern Levant after 14,000 BP. Population growth was almost certainly caused by sedentism, even if it was also, in some regions, a cause of sedentism. As the previous chapter showed, mobile communities of foragers have good reasons to limit population growth. But if they settle down, those limits to population growth can be relaxed. Babies do not need to be carried so much; grain-based diets (particularly if foods are cooked) make it possible to wean children earlier; birth intervals will shorten; and females will reach puberty earlier. All these factors would have accelerated population growth in less mobile communities.

Sedentism also tended to transform both the technologies of sedentary foragers and the genetic nature of the plants and animals on which they fed. Increased dependence on a small number of abundant and easily harvested food sources reduced people's familiarity with the wide range of species and techniques they had used when more nomadic. This was a Neolithic form of "de-skilling." But the same processes would also have increased specialized knowledge of particular favored species. Sedentary communities would have learned much more about the life cycles, diseases, and patterns of growth of the small number of species closest to their settlements. This lore would have added greatly to the already substantial ecological knowledge that foragers had of the life cycles of their prey species, and of how to protect and propagate them efficiently. The careful tending of these species would also have encouraged genetic changes that favored domestication, as poorer specimens were rejected. Finally, the clearance of land for permanent buildings would have created ideal conditions in which hardy species could thrive, particularly if humans used them regularly so that their seeds accumulated around human settlements.

Over time, sedentary communities of foragers would have found that their own numbers increased, that their knowledge of particular prey species increased, and that these same species began to change in ways that made them more and more useful.

The Trap of Sedentism   As the populations of sedentary communities increased, and as they became more dependent on a narrowing range of favored species and more skilled at raising the productivity of these species, both the possibility and the desirability of returning to nomadic lifeways diminished. This is a pattern we can describe as the trap of sedentism. Within just a few generations, sedentary foraging communities may have found that they had become committed to a sedentary lifeway as they lost ancient skills, and as population growth reduced the territories available to each community. As a Neolithic Malthus could have predicted, eventually population growth was bound to outstrip the abundant natural resources that had encouraged sedentism in the first place. Alternatively, periods of local climatic deterioration may have reduced the amount of naturally occurring foodstuffs available. In either case, after a few generations of sedentism, communities would have found themselves pressing at the ecological limits of regions whose resources had seemed abundant when they had first settled down. At this point, if the option of returning to more nomadic lifeways was no longer available (because neighboring regions were also overpopulated) or no longer seemed attractive (because the sedentary lifestyle had begun to seem normal), communities had little choice but to intensify further, putting more effort into raising the productivity of a few favored species.

This decision constituted the final step to fully developed forms of agriculture. These processes are clearest of all in evidence from Mesopotamia. Natufian communities suffered as climates deteriorated between 13,000 and 11,500 BP. There are signs of growing nutritional deficiencies, of increased female infanticide, and of increasing differences of rank, all of which may be responses to resource crises.[53] Some communities in the Fertile Crescent, particularly in more arid regions, responded by returning to more mobile lifeways. But in regions with abundant water supplies and surviving stands of wild cereals, some communities began to intensify production of particular foodstuffs, such as cereals. The crucial step was to plant seeds in soil cleared of other plants. Analogies with modern foraging societies and also with horticultural societies, in which women appear to do most of the agricultural work, suggest that it may have been women who pioneered these techniques, while men concentrated on hunting and other activities that took them away from their home villages.[54] At first, deliberate cultivation may have been a purely defensive step designed to aid survival in deteriorating conditions, for after 13,000 BP Natufian population levels seem to have declined sharply. However, it clearly worked, for there soon appeared communities that depended more and more on the use of domesticated species, first of plants and then animals. Many communities continued to treat domestication as a limited supplement to traditional foraging lifeways—but some did not. For them, domestication provided the basis for an entirely new lifeway.

The first true agricultural villages appeared in Southwest Asia only after ca. 10,500 BP. The village of Abu Hureyra, near the border between modern Turkey and Syria, illustrates how rapid this transition could be.[55] At about 10,500 BP a village was established here; its pit dwellings were made with reed roofs and wooden uprights. Their occupants used wild grains but also hunted gazelles. The gazelles would arrive regularly each spring, they were killed in large numbers, and their meat was stored. Thus these communities stored both meat and grains. They may have deliberately planted some grains, and perhaps they penned wild game such as gazelles. Agriculture and livestock herding then developed quite rapidly during several hundred years after ca. 10,500 BP. The population of the village grew to about 300 to 400 people. About 9,700 years ago, a new village appeared that covered a larger area; its inhabitants still depended on gazelle. But by ca. 9,000 years ago, in a rapid transition lasting perhaps only a century, they had become agriculturalists, with substantial herds of domesticated sheep and goats and with crops of cereals and pulses. They built simple rectangular houses from mud brick, with narrow lanes and courtyards.[56] By this time, similar villages had appeared in many parts of the Fertile Crescent (see map 8.3).

A General Explanation for Agricultural Origins?

This sequence—preadaptation; then increased sedentism encouraged by climatic changes, gentle population pressure, and increased exchanges; then intensification and further population growth, leading finally to fully committed agriculture—fits the Fertile Crescent quite well. But does it apply to other regions of early agriculture?

It used to be thought that domestication preceded sedentism in the Americas. This may well be true, in the sense that nomadic or seminomadic communities may have played a crucial role in the early stages of the domestication of crops such as maize. But recent revisions of the date of domestication in the Americas suggest that here, too, sedentism may have been crucial to the emergence of forms of agriculture that could lead to more fundamental transformations. Chinese data are too scanty to offer firm conclusions, but the same sequence seems perfectly possible there as well, and the same is true

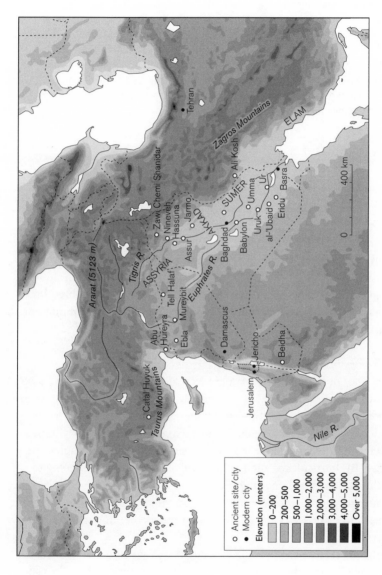

Map 8.3. Ancient Mesopotamia.

of early agriculture in sub-Saharan Africa.[57] In the best recent survey of the problem, Bruce Smith argues that

> in many regions of the world, experiments leading to seed plant domestication and, eventually, agriculture were carried out in a common set of conditions. The experimenters were hunter-gatherer [forager] societies that had settled by lakes, marshes, or rivers—locales so rich in wild resources that these societies could establish permanent settlements and rely to a considerable extent on local plant and animal communities. Thus a sedentary way of life, supported by the plentiful resources of an aquatic zone, seems to have been an important element in early experiments with domestication.[58]

EARLY AGRARIAN LIFEWAYS

How did people live in the earliest agrarian communities? To answer this question we must break with the strict chronology of this chapter, because early agrarian era societies were not confined to the period between 11,500 and 4,000 BP. In some regions, such as the highlands of Papua New Guinea, they survived well into the twentieth century; in many regions, including much of the Americas, semisedentary communities survived until just a century or two ago.[59] But the question is of great importance, as communities of independent farmers have been so widespread and have endured for so long that their lifeways and histories constitute an important though neglected chapter of human history.

Technologies: Horticulture, Not Agriculture

Technologies of the early agrarian era were different from those we associate with agriculture today. For this reason, it is common to refer to them as *horticulture*. By and large, these techniques were less productive than later technologies, and this may be one reason why the health of early agricultural communities was poorer in some respects than in most foraging communities. By horticulture, modern anthropologists mean technologies of plant cultivation that do not use plows or draft animals. In such societies, the main agricultural implement is little more than a sort of hoe or digging stick, used to plant seeds and to clear away weeds that might compete for nutrients from the soil.

Horticultural societies have survived in many regions of the world to the present day. Some regions and some crops may be better adapted to such technologies than to modern forms of plow agriculture, but horticulture is normally less productive. Digging sticks cannot turn over tough top soils, so horticulture can be practiced only in areas with fertile, easily worked soils, such as loesses. Besides, horticulturalists often do without the fertilizer provided by domestic animals. These limitations help explain why early forms of agriculture failed to spread into many regions that were farmed intensively in the later agrarian era. In areas such as modern Ukraine, early horticulturalists farmed lands on river terraces with loess soils, leaving upland regions between the rivers to nomadic foragers. Most early horticulturalists also continued to hunt and gather. Indeed, hunting, gathering, and fishing have remained important aspects of horticultural and agricultural lifeways to the present day.

Village Communities

The earliest agrarian era communities consisted of independent farming villages. Each constituted a largely self-sufficient society. Beyond them, there were no higher authorities, no states or regional chiefs, though networks of exchange (which were sometimes quite extensive) did exercise an impersonal influence on most communities.

Like villages of the Papua New Guinean highlands early in the twentieth century, early agrarian era villages varied greatly in size, from twenty to forty households up to several thousand. Some of these villages might have appeared to us as small towns. Permanent settlements appear to favor a different architecture from the temporary settlements of more mobile communities. Whereas communities of nomads tend to consist of circular "humpies" or windbreaks, the buildings in villages are made to last, and usually this means that they are square or rectangular. (However, in northern China, well-built round houses survived for a long time. Their remains can still be seen in the village of Pan-po outside Xian.) More permanent dwellings required a clarification of family arrangements, for they sharply pose the question, Who lives with whom? Thus, house sizes and designs suggest that the nuclear family may have acquired a sharper definition within these villages. There may also have emerged a clearer sense of "property," both of the individual household and of the village as a whole (see figure 8.5). When defensive walls appear in some regions, toward the end of the early agrarian era, we can be sure that villagers have begun to acquire a strong sense of family and village property.

The groups who lived in early agrarian era villages were larger than the families and bands that dominate the Old Stone Age. Kinship certainly remained the main principle that organized these communities before the appearance of large-scale chiefdoms and states. However, the nature of kinship thinking must have changed to accommodate the larger, more tightly organized and more permanent communities of these farming villages. Nu-

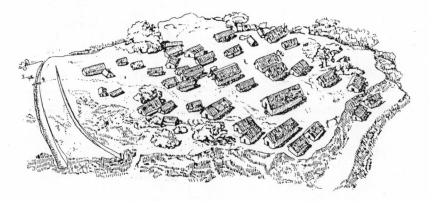

Figure 8.5.   Early agricultural villages from Ukraine. Reconstruction of Kolo-
miyshchina village, from the fourth millennium BCE. From Marija Gimbutas,
The Civilization of the Goddess: The World of Old Europe, ed. Joan Marler (San
Francisco: Harper and Row, 1991), p. 106.

clear families had to clarify their relations with each other and with the village as a whole, which meant creating more elaborate notions of kinship of
a kind familiar to modern students of village societies. This is why it is reasonable to assume that the main social structures of the early agrarian era
were similar to those that Elman Service has described as "tribes," as opposed
to the much simpler "bands," which rarely included more than fifty people,
and usually far fewer.[60] Because tribes may contain hundreds of people, they
need more elaborate ways of classifying the relationships between individuals and households. Often a sense of unity is maintained by assuming that
everyone is descended from the same ancestor.

Hierarchies or Equality?

Though individuals in most nomadic foraging communities may be ranked
clearly by gender and age, and there may well exist *personal* hierarchies
besides, in most other respects foraging societies have to be egalitarian. As
long as they are nomadic, they cannot create stored surpluses that could
generate significant distinctions in wealth. Agriculture required regular
storage of surpluses, and sustained much larger communities. It thereby
created the preconditions for a concentration of wealth and the appearance
of new forms of inequality. Indeed, there are signs that inequalities began
to appear as soon as foragers became more sedentary. Early Natufian communities may have consisted simply of small numbers of related families.
However, as Natufian communities grew in size, more complex relationships began to emerge as the problems of managing village activities and

controlling village conflicts became more complicated. The main problem sedentary communities face is that individuals can no longer easily deal with conflicts by just moving on or joining other groups. Agriculture ties individuals and whole groups much more firmly to a particular piece of land, and it therefore forces them on occasion to act collectively. For this and other reasons, larger communities find it necessary, for some purposes, to choose leaders. And choosing leaders inevitably means some form of hierarchy. Even in some Natufian burials, archaeologists have detected differences between a minority of individuals buried with ornamental objects, who were presumably of higher status, and a majority buried less grandly. The fact that even children are sometimes buried in grander style suggests that high rank could be inherited, so there may have existed a system of ranked lineages.

Similar pressures operated in all early agricultural villages. Yet during the early agrarian era, there were limits to how far such inequalities could go. Particularly in regions where agriculture was new and there was little competition for resources, communities remained quite egalitarian. In the early stages of the Tripolye culture in Ukraine, for example, houses do not vary greatly in size, and the objects they contain do not suggest great variations in wealth. It is evidence of this kind that led the Lithuanian-born American archaeologist Marija Gimbutas to argue that the entire early agrarian era may have been a period of relative equality between men and women and between different families.[61] There probably existed a clear division of labor by gender. In most farming communities, having children was essential for the survival of family units; and in worlds with high rates of infant mortality but without contraception or bottle-feeding, this meant that the lives of women were dominated by the bearing and rearing of children. But there is no reason to assume that such differences in gender roles necessarily implied systematic forms of gender inequality.

Relations with Other Societies

As we have seen, early agrarian era communities coexisted with foraging communities. They also traded with other agrarian communities. The large networks of exchange in the early Neolithic thus linked communities living in very different ways. The evidence for extensive systems of exchange is clearest from the Middle East—particularly from Anatolia, where the early town of Çatal Hüyük traded in obsidian, a volcanic glass used for making sharp blades.

Undoubtedly these contacts included conflict and raiding, and early agrarian societies may have engaged in semiritualized struggles of various kinds in which there were occasional casualties (as there are in the modern ritualized conflicts we call "sports"). But it is unlikely that such conflicts were highly organized or regular enough to be regarded as warfare. Most early agrarian era communities do not have large caches of weapons. Nor do they have fortifications as a matter of course. Even in Jericho, the oldest-known farming village, the walls, which were once believed to have been fortifications, are now thought to have been an early form of flood control.

AGRICULTURAL IMPACTS

With the appearance of farming, there began a fundamental change in the relation between humans and the natural world. Already in the Paleolithic era, human activity affected other organisms. But when humans engaged in agriculture for the first time, they began to remake the nonliving environment as well—its soils, rivers, and landscapes—to create new environments tailored specifically to their own needs.[62] Agriculture means altering natural processes in ways that benefit humans, so it means interfering in natural ecological cycles. By removing unwanted species (weeding), agriculturalists deliberately create artificial landscapes in which processes of succession, which might have returned the land to its previous state, are prevented. The land is deliberately kept free of many species, and therefore is maintained below its natural productivity level. In return, the productivity of those species favored by humans is increased, as they are given extra access to nutrients, water, and sunlight. But reducing plant cover also increases the rate of erosion, because plants hold soil together through their roots, create humus that binds soils together, and reduce the size and kinetic energy of raindrops as they hit the ground.[63] And erosion, together with intense cultivation of a small number of crops, can accelerate nutrient cycles, forcing humans to start deliberately maintaining soil fertility, whether by adding manure or ashes, by rotating crops, or by letting the land recuperate in fallow periods. Humans also continued to remake the organisms around them, not just by genetic engineering of domestic crops and animals but also by hunting down animals (such as wolves) that threatened them or their domesticates.

As humans began to reshape their environments in ways that suited them better, they may have experienced a growing sense of separation between the "natural" and "human" worlds. The sense of community between humans and their environment, which is still apparent in modern foraging communities, probably diminished in agricultural communities. It would have been replaced by a sense of alienation—a sense that the natural world was at best indifferent to humans, and at worst positively hostile.

However, in the early Holocene era, these changes affected only small parts of the world, and early agrarian technologies had a limited effect on the natural environment.[64] Only when agricultural techniques began to spread more widely did the human impact on the natural world become more significant.

SUMMARY

The end of the last ice age marks a fundamental turning point in human history. With the advent of agriculture, human societies began to acquire the demographic and technological dynamism that has driven historical change in recent millennia. Agriculture appeared in several different parts of the world in the millennia after the end of the last ice age. Explaining why communities of foragers took up agriculture is not easy, but the main steps seem reasonably clear. Most of the skills needed were already present in foraging societies. Present as well were a number of plant and animal species that were preadapted for domestication. Climatic changes spurred experimentation with new technologies and created new regions of abundance that encouraged sedentism, while sedentism, in its turn, encouraged local population growth. Eventually, as populations grew, sedentary communities had to either resume more traditional nomadic lifeways or intensify even further. Those that chose the second path created the first truly agricultural societies.

However, the advantages of early agricultural technologies were not so great that they spread rapidly or automatically. Instead, early agrarian era communities expanded slowly, as migrants colonized those areas suitable for the era's forms of horticulture. For many millennia, agrarian communities coexisted with neighboring populations of foragers. Most of the early agrarian era was therefore characterized by slow population growth (by modern standards), limited conflict, and limited ecological impacts. The early agrarian era was a relatively peaceful world of small village communities living among other communities that continued to practice foraging lifeways similar to those of the Upper Paleolithic. Historians have largely neglected this period of human history, so it is important to remember that it lasted as long as the subsequent era, which was dominated by the emergence of cities, states, and empires.

FURTHER READING

Bruce Smith, *The Emergence of Agriculture* (1995), and John Mears, "Agricultural Origins in Global Perspective" (2001), are good recent surveys of the huge literature on agricultural origins. Mark Cohen, *The Food Crisis in Prehistory* (1977), argues for the importance of population pressure in explaining agricultural origins; David Rindos, *Origins of Agriculture* (1984), describes the evolution of agriculture as a largely unconscious process of symbiosis. In *Guns, Germs, and Steel* (1998), Jared Diamond emphasizes the distribution of potential domesticates as a key to explaining the timing and geography of early agriculture. Donald Henry, *From Foraging to Agriculture* (1989), offers a detailed account of the Natufian cultures and their role in early Mesopotamian agriculture, while Richard MacNeish, *The Origins of Agriculture and Settled Life* (1992), offers a detailed survey of agricultural origins in the Americas. General surveys of lifeways in this period can be found in Göran Burenhult, ed., *The Illustrated History of Mankind* (5 vols., 1993–94), and Robert Wenke, *Patterns in Prehistory* (3rd ed., 1990); Marija Gimbutas's *The Civilization of the Goddess* (1991) offers a controversial account of early agrarian lifeways and gender relations, some of whose implications are summarized in Margaret Ehrenberg, *Women in Prehistory* (1989). Neil Roberts, *The Holocene* (1998); Clive Ponting, *A Green History of the World* (1992); and I. G. Simmons, *Changing the Face of the Earth* (1996) discuss some of the ecological implications of early forms of agriculture. Andrew Sherratt's "Reviving the Grand Narrative" (1995) argues for the importance of exchange networks in the origins of agriculture and in historical evolution in general. John Mulvaney and Johan Kamminga, *Prehistory of Australia* (1999), and Josephine Flood, *Archaeology of the Dreamtime* (1983), offer authoritative introductions to the history of Australia during the early Holocene.

9

FROM POWER OVER NATURE TO POWER OVER PEOPLE
CITIES, STATES, AND "CIVILIZATIONS"

SOCIAL COMPLEXITY

In the early universe, gravity took hold of clouds of atoms, and sculpted them into stars and galaxies. In the era described in this chapter, we will see how, by a sort of social gravity, cities and states were sculpted from scattered communities of farmers. As farming populations gathered in larger and denser communities, interactions between different groups increased and the social pressure rose until, in a striking parallel with star formation, new structures suddenly appeared, together with a new level of complexity. Like stars, cities and states reorganize and energize the smaller objects within their gravitational field.

The urbanized, state-organized, and often warring communities that were the products of these changes have been the main focus of modern historiography. So it has been all too easy for historians to forget how different these communities were from the small-scale and relatively nonhierarchical societies of the Paleolithic and the early agrarian eras. In fact, most of human history (chronologically speaking) has taken place in communities quite innocent of state power. Even in the villages of the early agrarian era, for most people, most of the time, the important relationships were personal, local, and fairly egalitarian. Most households were self-sufficient, and people dealt with each other as people rather than as the representatives of institutions.

Then, about 5,000 years ago, the first states appeared. Small city-states existed in southern Mesopotamia by ca. 3200 BCE (see map 9.1). By 3100 BCE a state had appeared in Egypt, where a regional official (named Menes or Narmer) united north and south into a single kingdom and founded the

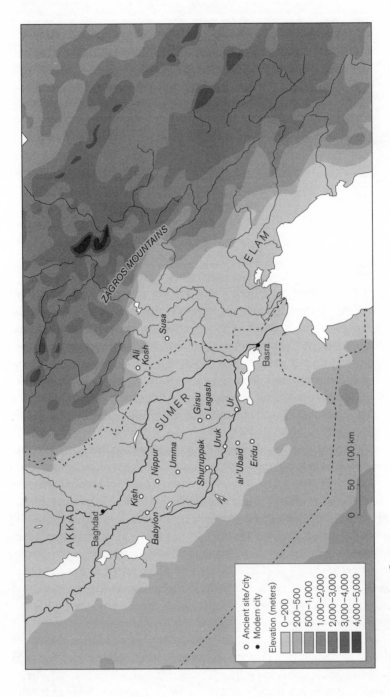

Map 9.1.   Ancient Sumer.

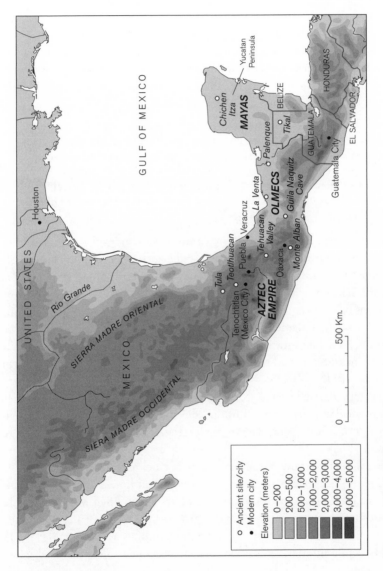

Map 9.2.   Ancient Mesoamerica.

first Egyptian dynasty. States also appeared in other regions of rising population density—in northern India and China by ca. 2000 BCE, and in
Mesoamerica by 1000 BCE (see map 9.2). The appearance of the first states
marks a critical transition from personal relations to impersonal power, and
from power over things to power over people.[1] The world of hierarchy, power,
and states is the one we know today. It is a world in which the wealth and
power of individuals and communities can vary tremendously, according to
their birth, their gender, and the ethnic group to which they belong. Marvin
Harris describes the change as the end of equality.

> For the first time there appeared on earth kings, dictators, high priests,
> emperors, prime ministers, presidents, governors, mayors, generals, ad-
> mirals, police chiefs, judges, lawyers, and jailers, along with dungeons,
> jails, penitentiaries, and concentration camps. Under the tutelage of the
> state, human beings learned for the first time how to bow, grovel, kneel,
> and kowtow. In many ways the rise of the state was the descent of the
> world from freedom to slavery.[2]

States were normally embedded within larger regions that included other
states and their hinterlands. I will describe such regions as *agrarian civilizations. Civilization* is often taken as a synonym for progress, but that is
not the sense in which the word is used here. Though there are clear differences between agrarian civilizations and other types of human community, I make no judgments about the intrinsic worth of any particular type
of society. I define agrarian civilizations as large societies based on agriculture, with states and all that that implies (literacy, warfare, etc.). The term
agrarian civilization may seem self-contradictory, for we associate civilization (a term derived from *civis*, the Latin word meaning "citizen") with states
and particularly their cities. But the adjective *agrarian* reminds us that all
premodern cities depended on rural hinterlands at the city's edge or in more
distant villages.

It will help to think of the emergence of cities and states, like the evolution of multicellular organisms, as a process that linked once independent
entities into larger unities. Figure 9.1 offers a simplified way of thinking
about some of the main stages in this process (and see table 9.1).[3] The transition discussed in this chapter can be thought of as a shift from level 4 to
level 5, while agrarian civilizations in general are normally organized at levels 5 and 6.

How can we explain this fundamental transition? Increasing population
density in farming regions provided the demographic and physical raw materials used to construct the first cities and states, and increasing congestion
provided much of the motivation for creating states.[4] But did local com-

TABLE 9.1. SCALES OF SOCIAL ORGANIZATION

Level	Type and Scale of Social Structure	Size (populations)
7	*The Modern Global System:* embraces all world societies in hierarchy of influence, wealth, power	6 billion+
6	*World Systems and Empires:* embrace large regions linked culturally, economically, and sometimes politically	100,000s to millions
5	*States/Nations/Cities/Supratribal Associations:* large, economically and militarily powerful systems, with state or near-state structures	1,000s–100,000s+
4	*Cultures/Tribes/Towns and Surrounding Villages:* linked reproductive groups, sometimes with single leadership, e.g., "big men" or "chiefs"	500–1,000s
3	*Reproductive Groups/Clusters of Villages:* related local groups whose members often intermarry, and who share a loose sense of kinship and culture	50–500 people
2	*Local or Subsistence Groups/Villages/Bands/ Camping Groups:* several parental groups that travel or live close together	8–50 people
1	*Parental or Family Groups:* mother and children, often with father, sharing a dwelling	2–8 people

munities willingly join together, or were they pushed together? The answer is probably a bit of both.

"Top-down" theories highlight the element of coercion, seeing states as institutions imposed on majorities by privileged and powerful minorities. This approach is common in Marxist theories of the state, which view states primarily as mechanisms for exploitation. While some individuals (mainly farmers) continued to extract resources from the natural world, as their ancestors had done before them, a new layer of rulers now appeared; they began to extract resources from their fellow humans by manipulating large networks of influence, wealth, and power. Human society became the "niche" in which elites foraged for the resources they needed. Society became multilayered, with a base level of those who exploited nature (the primary producers) and upper layers of those who exploited those who exploited nature. These changes created a new "food chain" within human society, and the emerging division of interests between elites and those they

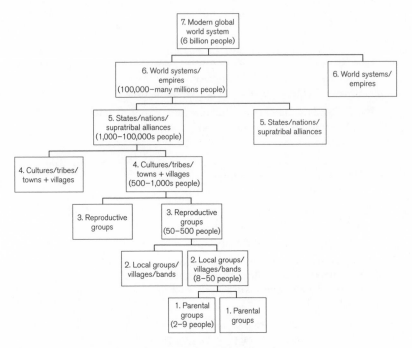

Figure 9.1.   Scales of social organization. A highly schematic attempt to capture the significant differences in scale of human social organization.

exploited is undoubtedly part of the explanation for the emergence of complex social structures.

But exploitation, like symbiosis, is never simple or unambiguous. Like predation in the nonhuman world, it can take more or less brutal forms. Lynn Margulis and Dorion Sagan observe, "In the long run, the most vicious predators, like the most dread disease-causing microbes, bring about their own ruin by killing their victims. Restrained predation—the attack that doesn't quite kill or does kill only slowly—is a recurring theme in evolution."[5] Within relationships of restrained predation, both partners may gain something, and exploitation may be softened by shared interests. In early states, including those of Mesopotamia, China, and Mesoamerica, exploitation could take extraordinarily brutal forms, including large-scale human sacrifices. But just as disease viruses often evolve less virulent strains that can exploit their prey without killing them, so human rulers eventually learned to protect the farmers they exploited (much as farmers protected their own herds of livestock). In this way, primary producers could become as dependent on the elites that ruled them as the elites were on primary producers. William McNeill has described these new relationships as a form of parasitism: "Disease germs are the most important microparasites humans have to deal with. Our only significant macroparasites are other men who, by specializing in violence, are able to secure a living without themselves producing the food and other commodities they consume."[6] Both elites and those they exploited had to adapt to the new, multilayered "ecology" that was emerging within human society, for the new structures transformed the intimate, ancient structures of village, household, and family.

"Bottom-up" theories of state formation stress that as societies become more complex, people find that they need state-like structures to survive. This process provides some striking analogies with the nonhuman world. Transitions to greater social complexity have occurred in the histories of many species, though not, apparently, among our closest relatives, the great apes. We have seen how single cells combined, first in loose structures such as stromatolites or sponges and eventually within multicelled organisms such as ourselves, in which there is a division of labor between different cells, each depending on the smooth functioning of the entire community. Multicelled organisms can also combine into larger communities. Some, like herds of antelope, are large but simple; others can be very complicated. Many species of social insects, such as ants, termites, and bees, live in dense communities, whose members are utterly dependent on the larger whole. Their environment (as in modern cities) consists mainly of other members of their own species and the structures they create. In the most complex communities, such as termite mounds, individuals become rigidly specialized, and special forms of communication and coordination are required if the community is to function effectively. Individuals communicate by sight, by touch, and by exchanging chemicals known as *pheromones*. Special routines evolve to deal with congestion, pollution, and conflicts between individuals. And hierarchies appear.

To us, these communities can look surprisingly like states, with their own caste systems and with their own methods of controlling and disciplining individuals. This is why human researchers find it so natural to talk of "queen bees" or "worker ants." As Lewis Thomas has written, ants "are so much like human beings as to be an embarrassment. They farm fungi, raise aphids as livestock, launch armies for wars, use chemical sprays to alarm and confuse their enemies, capture slaves. The families of weaver ants engage in child labor, holding their larvae like shuttles to spin out the threads that sew the leaves together for their fungus gardens. They exchange information cease-lessly. They do everything but watch television."[7] The parallels are indeed eerie, and they lend credibility to bottom-up theories of state formation. These see states as the solutions to problems experienced by all members of dense, congested communities. Humans also found that as they lived in larger and more complicated communities, they had to divide up tasks and knowledge; this development required new forms of communication, such as calendars to help people schedule their activities, or writing to help track the obligations and possessions of individuals. Individuals became more de-pendent on the group as a whole, and the group had to be organized in new ways as individuals swapped skills and resources. Yet as they began to co-ordinate the energies and skills of millions of individuals, these larger com-munities acquired an ecological power that no individual could match, though all stood to benefit from it in varying degrees. So the logic of state formation among humans parallels similar processes among the social in-sects. The major difference, just as we saw when examining the appearance of agriculture, is that humans adapted culturally while insects evolved ge-netically. This explains why the transition to social complexity could occur so much faster among humans.

Any attempt to explain state power will have to combine top-down and bottom-up theories, for they are in fact complementary. The rest of this chap-ter will attempt a systematic explanation of the emergence of state power, by which I mean the concentration, in the hands of a few people, of sub-stantial control over considerable human and material resources. This for-mulation leaves plenty of room for argument (over the word *substantial*, for example), but it helps focus attention on two crucial preconditions for large power structures. The first is the appearance of large accumulations of human, material, and intellectual resources; the second is the appearance of new ways of managing and controlling these resources.

INTENSIFICATION: NEW WAYS OF EXTRACTING RESOURCES FROM THE NATURAL WORLD

Shifting to new levels of complexity meant tapping and managing new en-ergy sources. The new energy sources were generated by more intensive technologies (the theme of the first half of this chapter). Constructing social structures that could manage these huge energy flows without breaking down was a complex task that eventually generated the coordinating mecha-nisms we know as *states* (the theme of the second half of this chapter).

Transitions to new levels of complexity often depend on positive feed-back mechanisms—cycles in which one change encourages another, which stimulates a third, which magnifies the first, and so on around the circle.

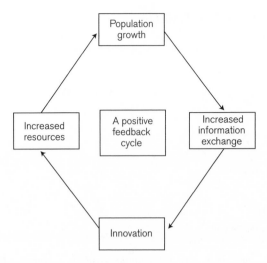

Figure 9.2.   Agriculture and population growth:
a positive feedback loop.

One of these causal chains played a fundamental role in the transition to larger and more complex social structures. It links population growth, collective learning, and technological innovation (see figure 9.2). Increasing the size and density of human communities stimulated processes of collective learning by increasing the size and variety of the networks within which information and goods could be exchanged. The intellectual synergies possible within these larger networks encouraged the development of new and more intensive technologies, which made it possible to support even larger human communities.[8] This feedback loop accelerated rates of innovation and growth, an outcome that helps explain why the emergence of agriculture counts as such a significant shifting of gears in human history. The pace of change may seem slow by modern standards, but it was fast by the standards of the Paleolithic era—and dazzling in comparison with the pace of genetic change in the nonhuman world.

In the millennia after the appearance of agriculture, there emerged within the Afro-Eurasian and American world zones several new technologies whose cumulative effect was to raise the productivity of technologies based on domestication. Here I will describe three of the more important changes, roughly in order of increasing intensification: shifting cultivation, the "secondary products revolution," and irrigation. Table 9.2 gives a general idea of the profound impact of different levels of intensity on food pro-

TABLE 9.2. ENERGY INPUT AND POPULATION DENSITY OVER TIME

	Energy Input (GJ/ha)	Food Harvest (GJ/ha)	Population Density (Persons/km²)
Foraging	0.001	0.003–0.006	0.01–0.9
Pastoralism	0.01	0.03–0.05	0.8–2.7
Shifting agriculture	0.04–1.5	10.0–25	10–60
Traditional farming	0.5–2	10–35	100–950
Modern agriculture	5–60	29–100	800–2,000

SOURCE: I. G. Simmons, *Environmental History: A Concise Introduction* (Oxford: Blackwell, 1993), p. 37.
NOTE: GJ/ha = gigajoules per hectare.

duction per hectare and on population densities, in different eras of human history.

Shifting Cultivation

Shifting or swidden cultivation is a seminomadic form of agriculture, still widely practiced today (mainly in forested environments). Indeed, it was swidden cultivation that enabled early agriculturalists to move beyond the easily worked soils preferred by the earliest farmers into forested environments, such as those of northern Eurasia. Because it normally uses fire to clear new areas for cultivation, swidden agriculture can be thought of as an adaptation of Paleolithic techniques of fire stick farming to the new technologies of the agricultural era.[9] It is a way of exploiting the nutrients stored in trees. Swidden farmers often clear an area of forest by felling or girdling the trees; they then burn the felled trees and grow crops between the remaining tree stumps in the highly fertile ash left behind. In Europe, early Neolithic cultivators grew cereals in clearings made with stone axes.[10] In newly created clearings, crops are not only nourished by the ash of felled trees but also freed from competition with other plants, so they grow exuberantly. But after three or four years, the soil's fertility is usually exhausted and it is necessary to move on. Where populations are small, entire communities can move in cycles of 20 to 50 years, which may be long enough to allow each patch of forest to regenerate between cycles. But as populations grow, the cycles inevitably shorten and the clearings become more permanent, a process that eventually created the forest-free agrarian landscapes more familiar in the modern world. In this way, shifting cultivation eventually led to massive deforestation. Overall, since the early Holocene, forests have declined by ca. 20 percent, from about 5 billion to 4 billion hectares. Until recently, the decline was more marked in temperate forests (32–35 percent) than in tropical forests (4–6 percent), but today, deforestation is most rapid in regions of tropical forest.[11]

The "Secondary Products Revolution"

Shifting cultivation has been practiced in varying degrees, in all the world zones. But the second major form of intensification occurred only in the Afro-Eurasian zone, because it depended on new ways of exploiting domestic livestock—and megafaunal extinctions in the Americas and Australia had left those zones hardly any potential domesticates to exploit.

As farmers entered the temperate zones of eastern and central Europe, they had to adapt their farming methods to colder and wetter climates. Andrew Sherratt has argued that between about 5000 and 3000 BCE, there occurred several important changes in farming methods that helped solve some of these problems.[12] He lumps these changes together under the label of the "secondary products revolution." The new techniques created a closer symbiosis with domestic livestock, and thus enabled humans to exploit their livestock more efficiently.

In the early agrarian era, domesticated animals were used mainly as sources of stored meat and hides. Though they had to be fed throughout their lifetime, they were used only once, when they were slaughtered. This inefficient method of exploitation may explain why animal domesticates were less important than plants in most early agrarian era communities. However, from ca. 5000 to 4000 BCE, farming communities in parts of Afro-Eurasia learned how to make good use of their animals even when they were alive by exploiting their secondary products—in particular, their milk and their wool. The farmers also learned to use animals as a new source of energy, particularly of traction power. Large animals such as horses, camels, or oxen soon became the most powerful sources of widely available mechanical energy. This was a revolutionary change, perhaps as dramatic in its way as the more recent fossil fuels revolution, for it provided the most significant new form of power since humans had first learned to make effective use of fire. Draft animals can deliver 500 to 700 watts of power, while humans, at best, can deliver about 75 watts.[13] The traction power of cattle or horses was used to carry people, as well as to pull carts and plows.

Horse- or ox-drawn plows were particularly important; because they could turn over soils more effectively than digging sticks, they could work tougher soils. Greater use of livestock also increased the amounts of dung available to maintain soil fertility. More efficient exploitation of livestock increased the productivity of farmers, while the increasing availability of dung as fertilizer and of plows to turn over the soil made it possible to farm less productive soils. Thus the new techniques enabled agriculture to spread into regions of heavier, clayey soils, such as those of northern Europe.

These changes also made it possible for the first time to settle arid steppe-lands, for they allowed some groups to live almost entirely off the produce of their livestock. The secondary products revolution turned domesticated herbivores into efficient machines for transforming grass into energy us-able by humans, just as, much later, the Industrial Revolution would find transformative new ways of extracting energy from coal. Pastoralists ex-ploited the new technologies by settling the huge areas of African and Eurasian grassland that had been too arid to farm. Because the most effec-tive way of exploiting grasslands was to let herds graze over large areas, pas-toralists often had to live a nomadic or seminomadic existence. So we often think of pastoralism as intrinsically nomadic, though in fact it need not be. Early forms of pastoralism probably appeared in the steppelands of south-eastern Russia and western Kazakhstan about 4000 BCE, but the militaris-tic and highly nomadic horse-riding pastoralism of later millennia devel-oped fully only in the first millennium BCE, after the development of new and improved forms of saddles. Pastoralism also evolved in Southwest Asia and in East Africa.

The secondary products revolution was a form of extensification, inso-far as it enabled human communities to settle regions that had previously been almost uninhabitable. But it also counts as a form of intensification, because it allowed *denser* settlement, and because the use of animal trac-tion improved transportation networks throughout Eurasia. In the long run, this revolution transformed communications, commerce, and warfare in the Afro-Eurasian zone, making it possible to move goods and soldiers more easily and rapidly over larger distances, whether in carts or chariots (from ca. 2000 BCE) or on horseback. In Eurasia, pastoralists provided the links that drew the agrarian civilizations of China, India, and Mesopotamia into a sin-gle, trans-Eurasian network of exchanges. This ensured that technologies, religions, even disease immunities could be shared throughout the region. All in all, the technologies of the secondary products revolution guaranteed that the Afro-Eurasian zone would become the largest area of shared knowledge on earth.[14]

It has been argued that the secondary products revolution, particularly the development of plow agriculture, may also have played a significant role in the evolution of more hierarchical gender relations. In horticultural so-cieties, as we have seen, women normally carry out most agricultural work.

Yet in societies that farm with plows, the agricultural work is normally done by men. And it has been suggested that the male "takeover" of farming was an important step toward the emergence of less egalitarian gender relations. According to Margaret Ehrenberg, "Anthropologists have shown that in present-day societies a significant . . . correlation exists between plough agriculture and patrilineal descent and land ownership in the same way as there is a correlation between non-plough agriculture and the heavy involvement, and consequent enhanced status, of women."[15] Yet there are difficulties with this theory. One is that even if men spent more time on agricultural tasks in societies using plow agriculture, the role of women in productive and reproductive activities remained as fundamental as ever. Another is that many communities never transformed by the secondary products revolution, such as farming societies in the Americas, also developed strikingly patriarchal structures. So we should not link patriarchy too closely to the appearance of any one lifeway or technology. As I will argue later, institutionalized patriarchy probably arose alongside institutionalized hierarchies in general; it evolved hand in hand with (and overlapped with) slavery, class, tribute exaction, and the state.

Irrigation

Like shifting cultivation, irrigation of some kind was practiced in all world zones, though it had its greatest impact in Afro-Eurasia and, to a lesser extent, in the Americas. In many warmer lands, there is plenty of sunlight available for photosynthesis, but plant growth is limited by lack of rainfall. Irrigation is a way of making the water of rivers or swamps available for crop cultivation. It has been one of the most important of all forms of agricultural intensification and remains vital today, whether in suburban gardens or the great grain factories of the American Midwest. Early forms of irrigation were often extremely simple, involving little more than the deliberate diversion of stream water into fields of crops by digging small channels. In areas of abundant water, such as the Euphrates Delta in southern Mesopotamia, irrigation was often just a matter of diverting small portions of the flow from the many small rivers that flowed into the Euphrates. Given such techniques, farmers benefited from the region's rich alluvial soils, which were laid down by the two great rivers, the Tigris and Euphrates. Eventually, though, as farming communities grew and as new forms of organization appeared, irrigation works became more elaborate; large, carefully planned networks of canals were built, using the labor of thousands of people. In arid lands with fertile soils, such as the flatlands of Mesopotamia or the lands along the Yellow River in China, irrigation could raise agricultural productivity decisively, which is why irrigation has been one of the most revolutionary of all technological innovations.

Irrigation was also used in many other areas. In Papua New Guinea there is evidence of irrigation as early as 9,000 years ago. In southern China and other parts of Southeast Asia, rice farmers developed many techniques of terracing and irrigation to increase the productivity of their main crop. In Mesoamerica, too, sophisticated new forms of irrigation evolved later in the agrarian era. In the first millennium CE, Mayans drained and filled swamps with town refuse, so as to form productive and easily worked soils that could support rapid population growth. Improved strains of maize also raised productivity in Mesoamerica. However, there was no secondary products revolution, because of the absence of suitable large domesticates. This had a profound impact on American agriculture, and it may explain many differences in the historical trajectories of the Americas and Afro-Eurasia.[16]

Other Innovations

Many other innovations also appeared in agrarian regions—in textile making, pottery making, building, and metallurgy, to name a few areas. The earliest pottery may be from the Jomon culture in Japan and may date back to the beginning of the Holocene era. In Mesopotamia, the earliest evidence of pottery comes from about 6500 BCE. It was used for carrying liquids and for cooking as well as for storage of foodstuffs. Pottery was also made in South and Central America from perhaps as early as 3000 BCE. In both the Afro-Eurasian and American world zones, pottery making was a natural development for peoples who built homes from mud and cooked their food in ovens or over fires. Soft metals, such as gold, silver, and copper, were worked in many parts of the world during the early agrarian era, but they were used mainly for ornamentation. The earliest evidence of such metalwork in Mesopotamia comes from ca. 5500 BCE; the same metals were also worked later in the Americas. But the working of hard metals, which could be used for weapons or tools, was a later development, because their manufacture required higher temperatures and more efficient ovens. Hard metals were made from alloys, such as bronze (made from copper and tin or sometimes arsenic) or iron (which is hardest if combined with some carbon). They were only ever made in Afro-Eurasia. That they appeared nowhere else is surprising, for the skills needed to work hard metals were similar to those needed to fire pottery. Bronze was first worked in Sumer in the fourth millennium BCE, and by 2000 BCE bronze was also being worked in China. Hard irons were first produced perhaps in the Caucasus in the middle of the second millennium BCE, and spread throughout many regions of Afro-Eurasia in the first millennium BCE; for this reason, the first millennium is often called the Iron Age. The first true steels were probably produced in the Roman Empire.

Population Growth

More productive agrarian technologies encouraged population growth. But population growth itself counts as a form of intensification, for in the era before fossil fuels, the energy resources available to human societies came mostly from human or animal muscle power. More people and more cattle thus meant that more productive power was available, wherever social structures were efficient enough to control and coordinate the activities of large numbers of people and livestock.[17]

In the Fertile Crescent, where these processes have been studied most closely, the long-term spread of village communities can be traced in the increasing number of sites. After ca. 5000 BCE, villages spread from the Fertile Crescent down into the flat desert lands and the marshlands along the great rivers of the Mesopotamian plain. In the arid plains, farmers had to make more use of river water, through simple forms of irrigation. They could also exploit the large number of fish in the great rivers. As farming communities multiplied, spread, diversified their techniques, and improved their productivity, both the resources they produced and the populations they supported increased. As we have seen, world populations grew from perhaps 6 million to 50 million in the period between 10,000 and 5,000 years ago.

On the largest scale, the trend of accumulation is clear. But it is important to remember that on the scale of decades or centuries, accumulation was a chaotic and erratic process. Population densities might increase in one region, and then decline as a result of climatic change, overworking of the soils, or some other cause. As Robert Wenke puts it: "The whole history of early complexity, in fact, seems to be a messy 'boom-or-bust' cycle, with only a very general long-term over-all trend toward complexity."[18]

HIERARCHY: EMERGING INEQUALITIES IN WEALTH AND POWER

More productive technologies and larger, denser communities created the preconditions for the emergence of states.

Evidence of Emerging Inequality

As available resources increased, societies had to face, for the first time, the task of dealing with surpluses, whose control and distribution posed entirely novel problems. And, surprisingly quickly, their distribution became lopsided, so that gradients of power and wealth appeared. Surpluses began to sustain populations of privileged (and mainly male) specialists: artisans, traders, warriors, priests, scribes, and rulers.

It is worth noting how paradoxical this steepening of hierarchies is. For the increases in productivity associated with the agrarian revolution might, in principle, have raised the average material living standards of all members of society. The reality was different. Unlike water, which prefers to lie flat as it accumulates, material wealth in complex societies likes to pile itself up into huge pyramids. Offering some explanation for this curious but fundamental feature of complex societies will be one of the central tasks of this chapter. But the general principle can be stated straightaway. As populations became denser, people, like termites, found that they needed ways of organizing and coordinating their activities. But this meant conceding power to organizers who used that power in ways that benefited themselves as much as (and often more than) the communities they controlled. Inequality is what all top-down theories of state formation predict.

Archaeologists have many ways of tracking inequality. Even in the most sophisticated early agrarian era communities—such as the Anatolian town of Çatal Hüyük, which flourished from ca. 6250 to 5400 BCE, traded in obsidian over wide areas, and had a population of perhaps 4,000 to 6,000 people—there is little sign of significant differences in wealth. However, there are minor differences in the ways people are buried, and archaeologists argue that such differentiation shows one of the earliest responses to increasing population densities: the emergence of ranked clans. As communities increase in size, kinship thinking and the social mechanisms based on it are stretched to their limits. It is impossible to think of a community of 4,000 people as a single family. But it *is* possible to retain a loose sense of kinship by assuming that all members of a community share a common ancestor. (Whether the ancestor is mythical or real is unimportant.) Once this happens, the symbolic logic of kinship dictates that different lineages will trace their descent to different children of the ancestor, some to senior children and some to junior children. And in this way, whole lineages may be thought of as senior and junior, just as individuals within a family may be ranked by seniority. Ranked lineages arise naturally out of the ideology of kinship, for even in the most egalitarian kin-ordered communities, people were often ranked within families by age and seniority. So, kinship thinking naturally predisposed people to accept the authority of senior members of senior clans.

Archaeologists know there is inequality when houses vary in size and in the value of the objects they contain. Special objects or types of clothing can also hint at high status. Welfare and nutritional levels can also tell us much about hierarchies, for elite groups were almost invariably better fed than those they ruled. So bioarchaeologists often find differences in the height of members of different social groups. John Coatsworth writes: "Among ancient Mesoamericans, ruling elites of nobles, priests, and warriors controlled access to food, particularly scarce sources of protein.... In England in 1800 ... the adult male members of the titled nobility stood a full five inches taller than the population as a whole."[19]

Equally suggestive is the appearance of monumental architecture. Some large structures, such as Stonehenge, have no obvious utilitarian functions. They may have been used as ritual centers, and perhaps as astronomical observatories. Others, such as the ziggurats and pyramids of Mesopotamia, Egypt, or Mesoamerica, often contain burials, or perhaps palaces or temples, all of which indicate the presence of high-status individuals. Such structures have appeared in all societies in which states have later emerged, and in many that did not develop state structures. The most spectacular are undoubtedly the Egyptian pyramids, the first of which were built in the middle of the third millennium BCE. The appearance of such structures suggests that religious thinking was changing, too, as human communities became larger and more complex. Just as ranked hierarchies appeared among humans, so there began to appear elite gods, who required an appropriate degree of respect. For, as the sociologist Émile Durkheim first suggested, our thinking about the way the universe works often mirrors the way our own societies work. The best way of showing respect for these more awesome and remote gods was to build special buildings for them, buildings closer to the sky than ordinary dwellings, where humans could pay their respects by offering sacrifices and gifts. Where monumental architecture appears, we can be certain that there exist powerful leaders or managers, for someone has to coordinate the labor of hundreds, even thousands, of people. In this way, secular and religious power often went hand in hand. Leaders hoped to inspire awe by building such structures—awe at the power of the gods, and also at the majesty of the priests and rulers who dealt directly with such powerful gods and who supervised the building of their residences. Monumental architecture is both a sign of power and an instrument of power.

In Mesopotamia, the earliest monumental structure is probably the temple of Eridu, which dates to about 5000 BCE. The ziggurats of the late fourth millennium BCE were magnificent stepped structures, built with immense amounts of labor and displaying elaborate architectural detailing. They provided an awesome setting for religious and political ceremonies. In Mesoamerica, the earliest pyramids were built by the Olmec, early in the second millennium BCE. From as early as 2000 BCE, spectacular burial mounds appeared even in less densely settled regions, including the Eurasian steppes, where there were few towns and where most people were mobile pastoralists. The huge tomb of Arzhan in Tuva, which dates from the eighth century BCE, shows how much wealth and labor powerful steppe leaders could mobilize, often by exploiting the resources of neighboring sedentary communities. The Arzhan tomb included seventy chambers arranged like the spokes of a wheel; it contained about 160 saddle horses under a mound 120 meters wide.[20] At the center were buried a man and a woman, who wore furs and elaborate decorations. They clearly ruled a large and powerful tribal confederation, for subordinate princes or nobles were buried to their south, west, and north, and some may have been sacrificed as part of the burial rites. Monumental architecture of astonishing scale also appeared in one of the most remote communities of all in the era of agrarian civilizations: on Rapa Nui (Easter Island). Here, in a population of only a few thousand people, local chiefs competed with each other in the building of huge statues.

In densely settled regions, new communities began to arrange themselves in networks whose topology had less to do with natural features of the land than with the existence and distribution of other settlements. This is a familiar pattern even today in densely inhabited regions. Small villages tend to arrange themselves in roughly symmetrical patterns around larger villages that act as centers of gravity for local networks of exchange. In this way there may emerge hierarchical networks of small villages surrounding larger villages, which are grouped around small towns, which may be grouped around a major city. Even smaller towns will often contain institutions not present in the villages, such as temples, warehouses, and perhaps a priest's or a chief's residence. As a rule, then, the larger settlements, which act as local centers of gravity, will show greater internal differentiation than the surrounding villages. In Mesopotamia, there is clear evidence for the appearance of two-tiered systems by the period of Eridu, in the fifth millennium BCE. The large towns often have populations of 1,000 to 3,000 people, and many have ceremonial platforms of some kind as well as distinct storage areas, so they may have acted as marketplaces and religious centers.

Even more striking evidence of inequality is the appearance of large-scale conflicts or wars. The crucial markers here are fortifications and burials with weapons. In the Tripolye culture of Ukraine, which began as a typically egalitarian early agrarian era region, villages expanded after about 4000 BCE, and they appeared more often on sites that can be easily fortified. In the Eurasian steppes, warfare reflected growing conflict between sedentary communities of farmers and emerging communities of nomadic pastoralists. The rich pastoralist burials that appear from late in the third millennium show that by this time pastoralists sometimes extorted significant wealth from less well armed farming communities.

At the bottom of these emerging hierarchies were slaves and other dependents. These men and women were treated by their masters as stores of energy, as living batteries, as human cattle. In mechanical terms, humans are quite efficient converters of food into energy, so human slaves were often more valuable than animal slaves, if one could afford them.[21] The importance of human beings as a source of energy helps explain why forced labor was so ubiquitous in the premodern world, just as the existence of fossil fuels helps explain why human slavery has largely vanished today. Forced labor and human slavery existed in many forms in agrarian civilizations; and slaves or dependents occasionally rose to positions of power and wealth. But most were used as sources of stored energy for their owners: where human labor power was as important a source of energy as oil is today, controlling energy meant controlling people. To make slaves more amenable to control, they were often separated at birth from their families. And, like domestic animals, many were deliberately kept in a state of infantile dependence that inflicted a sort of psychic amputation on them—they remained like children, and their helplessness made them easier to control. Both animal and human slaves could be controlled best if kept economically and psychically dependent on their owners.

As emerging hierarchies also altered definitions of the social roles of men and women, hierarchies began to be ordered along lines of gender as well as of class and occupation. For the most part, elite men ended up dominating hierarchies. Why has hierarchy usually meant patriarchy? The simplest assumption—men were less vital than women within the household, the basic cell of human society—may provide the best explanation. New forms of power emerged above the level of the household as part of an increasingly elaborate division of labor. Power brokers were specialists in power, management, information collection, combat, or religion. But specialist roles in general were more available to those whose role within the household (the most fundamental unit of all societies) was least vital.[22] In societies without contraception or bottle-feeding, this meant the men (or the aristocratic women, some of whose functions could be fulfilled by other women). Thus, while weaving and spinning were regarded as one aspect of women's work in many societies, whether the products were consumed within the family or sold at market, specialist or *full-time* weavers were more likely to be men. As the division of labor became more elaborate, specialized roles, whether in warfare, in religious activity, or in government, were normally (not always) more open to men than to women, for men usually found it easier to place themselves at the hub of local exchange networks. And in this way there emerged in many larger agrarian communities a distinction between the household world, often dominated by women, and the public domain, often dominated by men.

Patriarchy is the way the emerging gradient of wealth and power was expressed in gender relations, because many of these specialized roles gave men access to new forms of wealth and power. Increased power, in turn, gave elite men more influence over public definitions of gender roles. The fact that written histories were constructed first within an emerging public domain, and were written mainly by men, helps explain why the written sources on which so much modern historical writing depends focused on the public domain and the doings of men. And it may be that works written by men have also made patriarchy seem simpler than it was, hiding from modern researchers the complex negotiations that went on in all households and the many ways in which both women and men evaded or softened social conventions that they found constricting.

New Forms of Power and Control: Power Based on Consent

How can we explain this steepening of the gradient of wealth and power in large agrarian communities? Anthropologists have shown that in small nomadic communities, individuals will normally resist attempts by individuals to assume power over them. How did hierarchies arise despite this resistance?

Modern studies of village communities, combined with archaeological evidence, suggest some of the stages through which particular groups or individuals may have begun controlling the labor and resources of others.[23] In many human communities, power and resources are surrendered willingly to trusted leaders. This we can call *consent-based power*, or power from below. In larger communities, however, leaders could use the increasing resources placed under their control to create new forms of power that enabled them to coerce at least some of the people they ruled. This is *coercive power*, or power from above.[24] The distinction corresponds to the distinction made earlier in this chapter between top-down and bottom-up theories of state formation. In practice, all states rest on both types of power, and the two are always intertwined. Nevertheless, there is a clear historical and logical sequence leading from power based on consent to power based also on coercion.[25]

In the absence of state structures, the resort to violence is available to everyone, so it is an unreliable way of controlling people or resources. But there are many reasons why village communities may willingly surrender *some* control over their resources and labor to trusted leaders. The logic is the same as that of a termite mound. As communities grow, new problems appear for which collective solutions have to be found. Agricultural, economic, and religious activities have to be coordinated more carefully; internal conflicts have to be defused; and conflicts with neighboring communities have to be managed. Handling these problems efficiently is often a matter of life and death, as failure can mean famine, sickness, or defeat in war. But they cannot be solved separately by each household, and thus households acquire an interest in delegating authority. In short, the majority of people in a community may willingly take part in building the simple social dams that concentrate surplus resources in reservoirs controlled by tribal or religious leaders. It may be appropriate to think of these early power structures as analogous to the earliest irrigation channels. These, as we have seen, were simple structures consisting of channels and small dykes, and they could be constructed and maintained with the more or less willing cooperation of entire communities.

Once the decision has been made to delegate authority, it is important to choose good leaders. Several factors may decide how leaders are chosen and what powers they are granted. That many leadership roles involved specialized tasks and skills explains why men occupied them more often than women, for men were less necessary within the household and had more opportunity to take up specialized tasks. Where ranked lineages existed, senior members of senior clans were likely to be chosen as representatives or managers, unless they were plainly incompetent. In internal conflicts, individuals known for their closeness to the gods, their diplomatic skills, or their wisdom were more likely to be chosen; in conflicts with neighbors, those with military skills. Where crises required the help of the gods, those thought to have privileged access to the gods, such as shamans or priests, made likely leaders. Using this authority, religious leaders often collected substantial resources to offer as sacrifices or gifts to the gods.

Sometimes, though, authority was granted in return for past favors, in a modification of the basic rules of reciprocity. This explains an institution whose rules can otherwise seem bizarre to moderns: the "big man." The label is appropriate, for the role was highly specialized and seems to have been occupied mainly by males. In some form, big men have appeared in many communities in recent times, and the role probably existed in many prehistoric communities as well. The classic studies were conducted in Melanesia in the early twentieth century by the Polish-born British anthropologist Bronislaw Malinowski. On Bougainville, the big men were known as *mumi.* The *mumi* would work hard to accumulate goods for a feast. He would harass his relatives and labor strenuously himself to produce extra goods that had high prestige, such as yams or pigs. Once he had accumulated enough goods, he would give them away at a huge feast. Here is an example, as described by Marvin Harris, from a study of big men on Bougainville: "At a great feast attended by 1,100 people on January 10, 1939, the host *mumi*, whose name was Soni, gave away thirty-two pigs plus a large quantity of sago-almond puddings. Soni and his closest followers, however, went hungry. 'We shall eat Soni's renown,' the followers said."[26] In commercial terms, such activity makes no sense at all. But in social terms, it does make sense, for gift-giving creates obligations. Gift-giving was to the world of kinship what investment is to the commercial world: the laying out of resources in the hope (which is never certain) of a greater return in the future. Though the holding of such feasts may impoverish the *mumi*, it also gives him the right to call on the services of those he has put under an obligation.

Anthropologists have observed these feasts or "giveaways" in many societies. One of the best-known examples is the potlatch of American Indian communities of the Pacific Northwest, such as the Kwakiutl. Among the Kwakiutl, chiefs accumulated blankets and other goods, and then gave them away in huge potlatch parties. Sometimes, the services owed to a big man could be turned directly into more significant forms of power—for example, if he asked those now obligated to him to join a raid on a neighboring tribe. The raid, in turn, might take goods that could be used in a new form of redistribution.

Anthropologists recognize an even more significant form of power in pre-state societies: the *chiefdom*. Definitions of chiefdoms are somewhat arbitrary, and none can capture the nuances of the real world, but anthropologists generally use the term to describe the heads of powerful aristocratic lineages who have authority over many lesser villages, groups, and clans, with populations of many thousands. Their authority is usually based on their status within a system of ranked lineages, which may enable them to mobilize considerable resources. In the Trobriand Islands, studied by Malinowski, chiefs could rule many different villages and have thousands of subjects. They often led raids on other islands, and subjects treated them with great deference. Malinowski once saw all the inhabitants of a village suddenly fall flat, as if "mowed down by a hurricane," at the appearance of their chief.[27] Villages supplied their chiefs with yams in fulfillment of kinship obligations. In this way, the chief, through rules of kinship, ended up controlling far more resources than did others. Often these yams were redistributed at feasts, which created new obligations, or they were used to pay for specialists, including warriors and canoe builders. Chiefdoms are not yet states, as they can easily split into separate tribal or clan segments. However, the resources concentrated in the hands of chiefs give them immense power, and sometimes chiefs can use this power to coerce individuals or groups reluctant to accept their authority.

Such forms of power are still limited and precarious. Rulers must conform to the demands of the kin-ordered world in which they live, because they are, in large measure, servants of those they rule. If they don't fulfill their obligations as leaders, they can quickly lose their influence and their following may fall apart. Anthropologists refer to such structures as *segmentary*, because they can easily fragment into the segments from which they are assembled.

Despite these limitations, power based on consent may grant leaders control over substantial material and human resources; this feature makes power based on consent the necessary foundation for the construction of larger and more durable power structures. What made possible the transition to more durable and coercive forms of power was the appearance of larger and more concentrated population centers—in particular, the appearance of the first cities.

The First Cities

Cities (at the lower end of level 5 in table 9.1) are more than large villages. In the first towns and cities, wholly humanized environments emerged for the first time. Here, large numbers of people depended entirely on other people to survive, and new forms of complexity and hierarchy appeared. The fundamental precondition for the existence of cities is that productivity levels reach a level such that rural populations can support themselves *and* a small surplus population of nonfarmers (see figure 9.3). The existence of cities presupposes a complex division of labor, both horizontal and vertical.

The first cities appeared in Mesopotamia. The process in that region has been most closely studied by archaeologists, so I will describe what happened here before asking how typical such processes were.[28] In the delta of the Tigris and Euphrates, populations grew rapidly in the fourth millennium BCE. Growth may have been stimulated by climatic changes, for climates became cooler and drier around 3500 BCE, and it was after this that the Sahara, long a region of steppe and savanna, was transformed into arid desert. In some parts of Mesopotamia, the change may have led to a decline in farming; but the south was a region of swamps, with scattered villages occupying small islands. Drier climates made more land available for settlement, and swamplands turned into rich farmland, producing several harvests a year

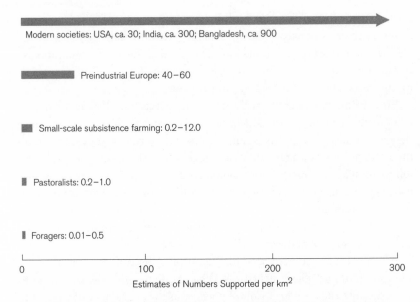

Figure 9.3.   Productivity thresholds in human history: population densities
under different lifeways. Data from Massimo Livi-Bacci, *A Concise History of
World Population*, trans. Carl Ipsen (Oxford: Blackwell, 1992), p. 27, and Allen W.
Johnson and Timothy Earle, *The Evolution of Human Societies: From Foraging
Group to Agrarian State*, 2nd ed. (Stanford: Stanford University Press, 2000),
p. 125.

with only the simplest forms of irrigation. The most important crops were
wheat, barley, and dates, as well as vegetables of various kinds. Livestock
was important and so were fish along the great rivers. Here was a later version of the "gardens of Eden" that lured foragers to settle down in the early
Holocene.

Another factor that may help explain population growth in southern
Mesopotamia was the changing topology of regional networks of exchange.
Andrew Sherratt has argued that

> in early 'Ubaid times, lowland Mesopotamia was, well, a backwater:
> just an area of mud. People did live there, in grass huts and using clay
> sickles, but it was not the most lively spot on earth. Where were the
> exciting things happening? Two nearby places: the northern arc of
> the fertile crescent, busily circulating all kinds of stones, metals, and
> painted pottery . . . ; and the coasts of the Persian Gulf, about which
> we know less because a lot of it is under the Mesopotamian mud that

accumulated at the river-mouths, but which certainly involved a lively maritime trade down to the present-day Gulf states. The bit in between these was the least important—*until the two got together.*[29]

The stream of traders moving along the great rivers was, Sherratt suggests, the opportunity that lowland Mesopotamia needed. As exchanges flourished, what had been a backwater suddenly became a hub region in far-reaching networks of exchanges of obsidian, metals, pottery, and semitropical goods from the south. There took place a sort of "sparking across the gap" between the two resource-rich areas of the Fertile Crescent and the Persian Gulf. Southern Mesopotamia just happened to be in the gap.[30] Its rising populations reflected not so much local conditions as the changing topology of networks of exchange reaching across much of Southwest Asia.

Perhaps these two explanations belong together. Increasing aridity forced populations into more concentrated regions of settlement; but it also created narrower corridors for the transmission of long-distance exchanges. Something like this happened in Egypt as well, where populations and, presumably, exchange networks became denser as the Sahara dried out, forcing more and more people to settle along the Nile.[31] Whatever the reasons, southern Mesopotamia attracted new settlers, some of whom may have come from lands that were now too arid to farm. Between 3500 and 3200 BCE, the region later known as Sumer became the most densely populated farming region in the world. New settlements were soon organized in hierarchies with three, perhaps even four levels. And at the head of these hierarchies was a group of large regional centers, including Uruk and Nippur.

In the final centuries of the fourth millennium, several of these towns expanded rapidly and turned into true cities—the earliest that we know of. Unlike the villages and towns of the early agrarian era, most of which consisted of similar, self-sufficient households, these had a complex internal division of labor and imported much of their food from elsewhere. Early in the fourth millennium Uruk was a regional center, with perhaps 10,000 inhabitants and several temples. By 3000 BCE it was a city of 50,000 people with well-fortified walls. It consisted of whitewashed mud-brick houses, of a type that can still be found today, with narrow streets running between them. While most were one-story high, wealthier houses often had two stories. In the center, on a ziggurat 12 meters high, stood the "White Temple" (see figure 9.4).

By the early dynastic period (ca. 2900–800 BCE), there were hardly any remaining small settlements in southern Mesopotamia. Almost the entire population of the region now lived in cities. Such dense concentrations of

Figure 9.4.   Early monumental ar-
chitecture: the "White Temple" at
Uruk in southern Mesopotamia, late
fourth millennium BCE. From A. Ber-
nard Knapp, *The History and Culture
of Ancient Western Asia and Egypt*
(Chicago: Dorsey, 1988), p. 44: from
Helen and Richard Leacroft, *The
Buildings of Ancient Mesopotamia*
(Leicester: Brockhampton Press;
Reading, Mass.: Young Scott Books,
1974).

humans had never existed before. Clearly, it was the rich and now well-
watered soils of the delta region that made it possible to support such dense
populations. But why did so many villagers move into the towns? There was
increasing warfare between the region's expanding towns and cities; villagers
may have taken refuge in the towns where they were safer, traveling by day
to farm nearby lands. But increasing aridity may also have driven nearby
villagers to the towns.

Cities, like stars, warped the social space-time of surrounding regions,
pulling in the goods, people, and skills of nearby villages and towns. So they
automatically become important foci for exchange. Regional networks of

exchange acquired more complex and more hierarchical structures, with greater activity, wealth, and knowledge concentrated in the cities. Outlying regions increasingly found that their own futures depended on finding a niche within these new networks of power and wealth.

Cities required new forms of social organization. Hans Nissen has argued that as southern Mesopotamia dried out, planned and carefully managed irrigation became more vital to support the region's soaring populations.[32] Archaeologists have mapped the rapid appearance of dense, planned networks of irrigation channels, particularly around the main population centers. Dependence on large irrigation systems, as well as the need for protection, forced villagers to cooperate more with each other and with the towns, which had the resources and the power to control and maintain the irrigation systems on which they depended. Town leaders could provide levies of labor for digging and clearing channels. They could also handle the complex disputes about water use that are inevitable in communities dependent on large-scale irrigation systems.

The First States: Power Based on Coercion

The "solution" to the many problems created in such dense communities was to establish the first states. But why? We have seen that rudimentary power structures are like simple dams that can create small reservoirs of surplus resources. Cities, however, required larger and more robust social dams. To manage their enormous reservoirs of wealth, they needed structures more like the huge irrigation systems of Sumerian cities. The politics of consensus could no longer handle social engineering on such a scale.

The city was crucial to these changes because it was, by its very nature, a concentrator of power.[33] On the one hand, it brought together in one place forms of authority and labor power that had previously been diffused over large areas and between many distinct communities. On the other hand, the creation of such large and dense communities *required* new forms of power; for as the size of communities grew, the organizational problems they faced became more acute. Cities needed special mechanisms to resolve disputes, to organize exchanges between farmers and specialists, to build warehouses in case of famine, to supply water and remove refuse, to build fortifications and irrigation canals, and to manage war and defense. Fortunately, the economic and demographic processes that created these needs also placed greater resources in the hands of leaders. As the need for central regulation increased, so did the resources available to central authorities. These two factors, in combination, explain why, with the emergence of really large concentrations of people, either in cities (such as southern Mesopotamia) or in regions of dense village settlements and small towns (which was the Egyptian pattern), something like the state was likely to emerge. Where most people lived in cities, as in Mesopotamia, the earliest states were normally city-states, but territorial states emerged in regions such as Egypt where populations were less concentrated and resources had to be mobilized from larger regions.[34]

The state (at level 5 in table 9.1) differs from the tribe (at the upper edge of level 4) primarily in its ability to coerce systematically and on a large scale.[35] States, like chiefs, will often claim to represent "senior lineages," though these tend to have less and less connection with real lineages. But where traditional forms of loyalty fail, states, unlike chiefs, have at their disposal methods of coercion, paid for out of the huge resources now at their disposal.

The simplest way to imagine state power in its most rudimentary forms is to think of a chief with sufficient resources to pay for an army or retinue. Marvin Harris gives as an example of such power the Bunyoro people of Uganda, who were governed in the nineteenth century by a hereditary ruler known as the *mukama*.[36] He ruled about 100,000 people, who lived mainly by growing millet and bananas. Formally, the *mukama* was merely the head of a series of chieftains. Like any traditional chief, he was seen as a "great provider" as well as a receiver of tribute. But in practice, his power was based on more than kinship obligations, for he used the large tributes he received to form a palace guard, as well as a retinue of servants, witch doctors, musicians, and so on. His armed attendants gave him the power to deprive individual chiefs or villages of their land. Like King Lear, he and his entire retinue toured the land, demanding that local chiefs and villages support them during their visits.

This is the pattern of many early states that have not yet evolved more bureaucratic forms of taxation. It matches what we know of the earliest Chinese state, that of the Shang.[37] The same logic is also apparent in the following passage from the chronicles of medieval Rus'. It concerns the tenth-century CE Grand Prince Vladimir:

> On one occasion, . . . after the guests were drunk, [his retinue] began to grumble against the prince, complaining that they were mistreated because he allowed them to eat with wooden spoons, instead of silver ones. When Vladimir heard of this complaint, he ordered that silver spoons should be moulded for his retinue to eat with, remarking that with silver and gold he could not secure a retinue, but that with a retinue he was in a position to win these treasures, even as his grandfather and his father had sought riches with their followers.[38]

Vladimir's remark captures only one half of this crude dialectic of power; in practice, as he knew perfectly well, silver and gold were necessary to buy the soldiers who could help him acquire more silver and gold. In the writings of the Byzantine emperor Constantine Porphyrogenitus (r. 913–59 CE), we have a superb description of exactly how Vladimir's "grandfather and his father" had secured tributes with the help of their armed retinues or 'druzhiny:

> When the month of November begins, their chiefs together with all the Russians at once leave Kiev and go off on "poliudie," which means "rounds," that is, to the Slavonic regions of the Vervians and Drugovichians and Krivichians and Severians and the rest of the Slavs who are tributaries of the Russians. There they are maintained throughout the winter, but then once more, starting from the month of April, when the ice of the Dnieper river melts, they come back to Kiev.[39]

Though states like that of the Bunyoro *mukama* or Kievan Rus' in the tenth century CE consist of little more than rulers, using the resources they control to pay a retinue of soldiers, they have clearly crossed the divide from consent-based power to power backed up systematically by coercion. Nevertheless, they are so rudimentary that many political scientists would hardly regard them as states at all, preferring to reserve that term for the more elaborate structures that emerge when rulers can create special bureaucracies and organized armies. At this stage, such structures begin to meet Charles Tilly's definition of states as "coercion-wielding organizations that are distinct from households and kinship groups and exercise clear priority in some respects over all other organizations within substantial territories. The term therefore includes city-states, empires, theocracies, and many other forms of government, but excludes tribes, lineages, firms, and churches as such."[40]

But we should not exaggerate the power of even these larger structures. Though they could exercise violence, sometimes in horrifying and spectacular ways, their actual control over the day-to-day activities of most of their populations, particularly in rural areas, was minuscule by comparison with that of modern states. In part this was a matter of the limited energy at their disposal: as John McNeill has pointed out, the energy they controlled consisted mainly of human muscle power, and in practice this means that "the Ming emperors and Egyptian pharaohs had no more power available to them than does a single modern bulldozer operator or tank captain."[41] In part the weakness of preindustrial states reflected their limited bureaucratic reach. In fact, the ready resort of early states to violence and their widespread use of the army to administer it were signs of weakness, not of strength. Traditional states often made up in brutality for what they lacked in administrative reach.[42] Anthony Giddens notes, "The ruler may have command over the lives of his subjects in the sense that if they do not obey, or actively rebel, he can put them to the sword. But the 'power of life and death' in this sense is not the same as the capability of controlling the day-to-day lives of the mass of the population, which the ruler is not able to do."[43] Traditional states rarely had total control over even the formal military organizations on their lands, and few knew exactly where their own authority ended and that of regional potentates began. Outside the cities, they usually had little authority over the more localized forms of violence used to collect taxes, prosecute offenders or deal with banditry, or right local injustices. These powers were exercised by local elites or kinship groups. For most individuals, the righting of wrongs remained the duty of the household or kin group, which might seek the support of local patrons or officials. And violence was, of course, pervasive even within households, where it was used to maintain the authority of males and seniors.[44]

But despite these limitations, and despite the absence of a real state monopoly on violence, early states were much more formidable structures than chiefdoms. And everywhere they appear, they are associated with the same cluster of features. These include new forms of specialization and an extensive division of labor, bureaucracies, systems of accounting and writing, armies, and fiscal systems.

Division of Labor   In southern Mesopotamia at the end of the fourth millennium BCE, the self-sufficient and relatively egalitarian villages of the early agrarian era were already a thing of the distant past. For at least two thousand years, agriculture had been productive enough to support non-farming populations of priests, potters, and other specialists. Growing specialization can be shown in the appearance of full-time potters beginning in the fifth millennium. Their presence is suggested by the excavation of workshops containing specialized equipment, including potters' wheels. From the late fourth millennium, there survives an extensive list of different professions, the so-called Standard Professions List.[45] This includes priests, officials, and many different kinds of artisans, such as silversmiths, stoneworkers, potters, scribes, and even snake charmers. Many of the professions seem to have been specially organized in guilds of some sort. There existed a complex class structure with god-kings, aristocrats, merchants, artisans, farmers, scribes, and, finally, slaves (most of whom were impoverished farmers or nomads or war captives). The wealth of the rulers can be illustrated from the astonishing tombs of Ur, dated to the end of the fourth millennium, which were excavated by Leonard Woolley. Here, rulers were buried with immense wealth, and humans were apparently sacrificed to serve their rulers after death. Merchants were a vital part of the urban division of labor, for cities like Uruk required more goods than could be supplied by nearby farmers. They also needed stone, wood, and luxury goods, which were traded in fleets of ships along the Tigris and Euphrates—some organized by the rulers, others by merchants. At the other end of the spectrum is the evidence for an impoverished class of people consisting of slaves, vagrants, war captives, and failed farmers. The existence of such groups is suggested by the appearance from the early Uruk period of crudely made but apparently mass-produced slant-sided bowls, which were probably used to feed levies of laborers. Lending support to this interpretation is the later symbol for eating, which seems to show a person pouring food into their mouth from one of these bowls.[46] It was workers such as these that probably made up the labor armies used to build fortifications and walls and to maintain irrigation channels.

By 3200 BCE, Sumerian society had attained a scale that could no longer be handled within traditional kinship ways of thinking. Society was simply too large and too complex to fit everyone into ever more elaborate models of kinship. Instead, new categories—by occupation, by city of origin, by what modern sociologists would call *class* or *estate*—begin to appear. However, kinship thinking remained the basis for relationships at lower levels of society, and this may be why symbolic forms of kinship thinking survived in the religious thought of early states. Rulers frequently portrayed themselves as the "parents" of their subjects, and the more powerful gods, too, were often treated as fathers or mothers of particular peoples.

Bureaucracy, Accounting, and Writing   Managing the huge resources concentrated in early states was a complex administrative and accounting task. Thus all early states supported officials who kept lists of the things they managed. The need to keep track of the large stores of foodstuffs and other resources stockpiled by the state explains why writing systems appeared in quite separate parts of the world, including Mesopotamia, Egypt, northern India, China, and Mesoamerica, as part of the process of state formation. Writing emerged first as a form of accounting and power, not as a way of recording speech.[47] (China may be a partial exception, as the earliest forms of writing there seem to be more concerned with religious activities than with accounting.)[48] However it evolved, writing constituted a new way of storing, and therefore controlling, information. Because it did not use the ambiguous symbolism of images, writing made it possible to store knowledge with the precision of spoken language. So writing stabilized and even rigidified empirical knowledge, shielding it from the variability that necessarily accompanies oral transmission. But the skills it demanded were for many millennia confined to elite groups, and primarily to males within those groups. Elites and males therefore benefited most from the capacity to hoard information in writing. Writing provided a powerful way of concentrating in the hands of a few the knowledge accumulated by millions.

In southern Mesopotamia, clay objects, which represented different types of goods, were used as marks of ownerships from as early as the eighth millennium. By the fourth millennium, it was common to bind them together in clay balls, or *bullae*. From the late fourth millennium period, as the first cities began to emerge, owners started using so-called cylinder seals, which could be rolled across bullae to list their contents. This procedure made the bullae redundant, and soon seals were used to mark flat tablets. Then, instead of using seals, officials began to mark the tablets using reed styli. These worked like pens, carving uniform wedge-shaped (i.e., cuneiform) symbols in the clay. Originally simply pictures of what they represented, these symbols soon became quite stylized (see figure 9.5). At first, even cuneiform writing could do little more than list objects, but this was sufficient for it to function quite effectively as a method of accounting. Most of the writing that survives from Uruk consists of lists of goods received and distributed.

Early in the third millennium, what had begun as a way of keeping records turned into a true writing system, as symbols for things and actions were slowly adapted to more abstract roles, describing emotions and even grammatical functions or separate syllables. Only at this point did writing become more than a system of accounting. The key to these changes was the rebus principle: that is, using an existing symbol for a particular object to represent another word that sounded similar to the first word. Thus, the Sumerian word for "arrow" was pronounced *ti*. Arrows can easily be drawn. But the word for "life," a more abstract notion, was also pronounced *ti*, so the symbol for an arrow could also be used to mean "life." Slowly, the system of symbols was simplified, though even in 1900 BCE it still had some 600 or 700 elements, making it closer in form to modern Chinese characters than to modern syllabic alphabets.

In Egypt, hieroglyphic writing was used at least from the time of Menes in ca. 3100 BCE. In the Indus valley, writing was used from ca. 2500 BCE. In China, writing systems were in existence by at least 1200 BCE, using signs many of which can still be read today. The first alphabetic systems were developed in the trading cities of Phoenicia in the eastern Mediterranean during the second millennium BCE. They were based on signs for consonants

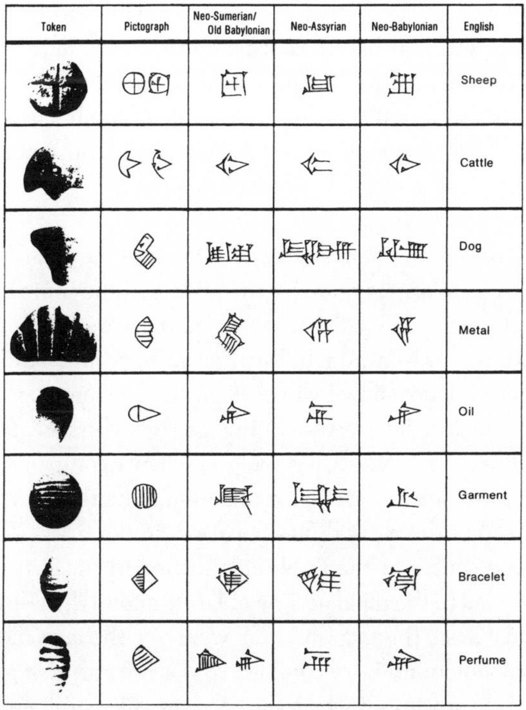

Token	Pictograph	Neo-Sumerian/ Old Babylonian	Neo-Assyrian	Neo-Babylonian	English
					Sheep
					Cattle
					Dog
					Metal
					Oil
					Garment
					Bracelet
					Perfume

Figure 9.5.   The evolution of cuneiform writing in Mesopotamia. From A. Bernard Knapp, *The History and Culture of Ancient Western Asia and Egypt* (Chicago: Dorsey, 1988), p. 55: courtesy of *Archaeology Magazine,* Archaeological Institute of America.

borrowed from Egyptian hieroglyphics. Letters for vowels were not used until the time of the classical Greeks. Creating alphabets with only a small number of letters simplified writing and reading and made literacy available for the first time outside the closed world of trained and highly specialized scribes. But despite this partial democratization of literacy, the power it generated remained a monopoly of elite groups until very recently.

In Mesoamerica, the first writing systems had appeared by ca. 600 BCE in southern Mexico. That the primary function of most early writing systems was to keep accounts is suggested by the counterexample of the Incas, who ruled the only major agrarian civilization without a writing system; they nevertheless had a large bureaucracy that used a system of accounting based on knotted strings, or *quipu*. It should come as no surprise that all agrarian civilizations have constructed elaborate systems of mathematics as well as writing. They also developed calendars, another vital tool for any complex society that had to coordinate the activities of thousands or millions of people to ensure they paid their taxes on time. Early calendars used the rich astronomical lore that accumulated in all early agrarian era societies and is evident even in remote Britain, in the construction of Stonehenge from the third or second millennium BCE.

Armies and Taxation   States can coerce because they can mobilize large retinues or groups of armed men. By the middle of the fourth millennium, most of the settlements in southern Mesopotamia were fortified, which suggests that warfare was common. In the third millennium, both archaeological and written evidence show a world of almost constant warfare. Conflicts were exacerbated by the steady drying of the rivers beginning in the mid–fourth millennium, which, particularly when accompanied by artificial interference in the flow of rivers, led to periodic changes in water courses. At about the end of the early dynastic period, in the first half of the third millennium, the course of the Euphrates shifted to the east of Uruk. The loss of the river led to the rapid decline of Uruk, and to the rise of cities such as Umma and Girsu (in Lagash) that were on the new channel. Such changes caused violent military conflicts, so it is not surprising that the first literatures and chronicles, which appeared in the third millennium, are concerned largely with warfare.

Armies enabled states to mediate in internal conflicts and to tax more effectively. In early states, taxes consisted, overwhelmingly, of foodstuffs collected from peasants and used to feed nobles or government officials, or of labor used to work on noble estates or on government projects.[49] Taxation differs from the methods used to collect resources in pre-state societies because of this element of coercion. Indeed, the anthropologist Eric Wolf has argued that this is perhaps the most critical distinction between state and pre-state societies.[50]

"Tribute-Taking" Societies

In what Wolf calls "kin-ordered" societies, resources are collected largely with the consent of those who contribute them. Once states have appeared, there is always an element of coercion, as resources are collected in the form of taxes, or what Wolf calls "tributes." This is justification for regarding societies with states as an entirely new type of social structure. Wolf treats the emergence of what he calls "tribute-taking" societies as a major transformation in the lifeways and the organization of human societies. Table 9.3 suggests how his classification of the major "modes of production" fits in

TABLE 9.3. TYPOLOGY OF MAJOR TECHNOLOGIES AND LIFEWAYS

Technology/ Lifeway	Mode of Production	Description	Era in Which Dominant
Foraging societies	Kin-ordered	Dominant technologies of the Paleolithic era used stone tools and were based on foraging lifeways; small-scale (organized up to level 3,* but hints of level 4 in regional cultures and intertribal systems); low population levels, but slow growth, particularly after ca. 50,000 BP	Paleolithic: before ca. 10,000 BP in world as a whole; up to present day in some areas
Agrarian societies (early agrarian societies)	Kin-ordered	Dominant technologies of agrarian era were based on domestication of plants and animals; supported both small-scale, pre-state, societies (up to level 4); and . . .	Agrarian: ca. 10,000–200 BP Early agrarian era: before ca. 5000 BP; in some areas starts later; in some lasts to now
Agrarian societies (agrarian civilizations)	Tribute-taking	Larger-scale societies organized in cities and states (up to level 6); more rapid population growth, but periodic demographic crises; pastoralism a distinctive Neolithic lifeway, based mainly on exploitation of domesticated animals, and often nomadic	Later agrarian era: ca. 5,000–200 BP; starts much later in some areas
Modern societies	Capitalist	Dominant technologies of the modern era are based on modern scientific technologies; capable of supporting a global system (level 7); rapid, unprecedented population growth	Modern: from ca. 1750 CE

* See Table 9.1 and Figure 9.1 for levels of social organization.

with some other familiar social typologies. The social theorist Anthony Giddens makes a similar point in slightly different terminology: "In class-divided societies [Wolf's "tributary societies"] the extraction of surplus-production is normally backed in a direct way by the threat or the use of force."[51]

In some combination, the ingredients described in this chapter were present in all areas of early state formation: in Afro-Eurasia, the Americas, and even the larger Pacific islands such as Tonga and Hawai'i. They include the emergence of dense populations, which generated a complex division of labor that posed new organizational problems, led to increased need for conflict resolution and to more frequent warfare, and encouraged the building of large monumental buildings as well as the creation of some form of writing. Here, there is room to give just one more example—this time from Mesoamerica.

In Mesoamerica, the earliest clear evidence for the existence of sedentary farming communities dates from ca. 2000 BCE. In the Andes, signs of such communities appear slightly earlier, from ca. 2500 BCE.[52] After this time, evidence of increasing social complexity, including monumental architecture and two- or three-tiered structures of settlement, appears quite rapidly, until the first state structures can be identified during the first millennium BCE. As in the Old World, it is tempting to see intensification and population growth as the primary motors of change. Settlements with large mounds or pyramids existed in the Andes and Mesoamerica early in the second millennium BCE. These may have been the ceremonial and perhaps market centers for many dependent villages. Their appearance suggests the existence of early forms of chiefdoms.

In the middle of the second millennium, in the lowlands of the gulf region of modern Mexico, there appeared the Olmec civilization. Its populations were supported mainly by swidden agriculture but also, in some regions, by the farming of rich alluvial soils. Like mid-fourth-millennium Mesopotamia, Olmec civilization consisted of large numbers of ranked towns. It also had monumental architecture and craftsmen of great expertise. At sites such as La Venta and San Lorenzo, huge ceremonial centers were constructed, some with pyramids up to 33 meters high. These were originally tombs, for most contain elaborate burials that provide clear proof of steep social and political hierarchies, as well as providing apt symbols of these hierarchies. Constructing the main pyramid at La Venta required at least 800,000 man-days, and some 18,000 people lived in the surrounding town.[53] The Olmec made huge and to modern eyes very beautiful statues or monumental heads from chunks of basalt imported from up to 80 kilometers away, presumably by the labor of hundreds of people. The brutality with which some Olmec sites were destroyed shows that there was also organized war. There are hints of early forms of writing, and it may be that the Olmec pioneered the writing systems that evolved in Mesoamerica, whose later versions have only recently been deciphered. Remains of a late Olmec carving that seems to use a dating system similar to that of the Maya suggest that the Olmec may also have invented the dating systems that spread throughout Central America.[54] Finally, there is evidence of extensive tributary or trade networks, as obsidian was imported in large amounts from the central Mexican highlands.

In Mesoamerica as in Mesopotamia, while the earliest civilizations developed in well-watered swamplands, civilization shifted gradually into regions of rainfall agriculture. In the Oaxaca valley, about 500 kilometers south of modern Mexico City, in a region of small villages ca. 1300 BCE there began to appear larger settlements, some with larger, apparently public, buildings. After ca. 1000 BCE, the size of these buildings increased rapidly. Populations multiplied, and agricultural production intensified with the building of a large system of canals. There are signs of increased specialization, particularly in crafts such as pottery, and systems of exchange and marketing expand. There are also signs that look like early forms of writing. Then, after 600 BCE, there appears clear evidence of a state-level polity with its capital at Monte Alban. By ca. 400 BCE, there were at least seven small city-states in the Oaxaca valley, so that the region began to look a bit like Sumer late in the fourth millennium. By 200 CE, the population of the entire valley may have reached almost 120,000. At its height, between 200 CE and 700 CE, the capital city of Monte Alban probably had a population of 17,000.[55]

Though agrarian civilizations appeared some two millennia later in the Americas than in Mesopotamia, the similarities in the history of these two regions suggest once again that state formation was a social explosion whose fuse was lit early in the agrarian era. The demographic dynamism introduced into human history by agriculture ensured that sooner or later, humans, like termites, would face the novel challenge of living in dense communities of their own species. For all the local differences, the solutions humans found in different parts of the world turned out to be remarkably similar to each other—and also strikingly similar to those found by termites and other social insects.

SUMMARY

The technological momentum of the early Holocene generated new technologies that raised output and supported larger and denser settlements. These technologies included shifting agriculture, the secondary products revolution, and irrigation. As communities grew in size, so did the managerial difficulties they faced, and humans found themselves confronting problems much like those of other social animals, such as the social insects. To solve these problems, communities found that they had to grant managerial powers to elite groups. At first, rulers governed with the active consent of their subjects. But over time, they acquired control over large amounts of resources; and in the largest communities, these resources enabled rulers to create more coercive forms of power. So it is no accident that the appearance of the first cities, late in the fourth millennium BCE, also coincides with the appearance of the first states. States mark the birth of a new type of community, which Eric Wolf calls the "tribute-taking" society. In such communities, elite groups used force or the threat of force to control surplus resources. Tribute-taking societies have been the most powerful and the most visible communities for most of recorded human history.

FURTHER READING

The account in this chapter makes much use of Hans Nissen's account of the rise of states in Sumer in *The Early History of the Ancient Near East* (1988); it has also borrowed some ideas from Marvin Harris's classic essay, "The Origin of Pristine States" (1978). Other general surveys can be found in Göran Burenhult, ed., *The Illustrated History of Humankind*, vols. 3 and 4 (1994); Michael Coe, *Mexico* (4th ed., 1994); Robert Wenke, *Patterns in Prehistory* (3rd ed., 1990); Charles Maisels, *The Emergence of Civilization* (1990); and Bruce Trigger, *Early Civilizations* (1993). There is a large literature on state formation; Elman Service gives a taste of some of its main ideas in *Primitive Social Organization* (2nd ed., 1971), as do Robert Cohen and Elman Service, eds., in *Origins of the State* (1978). Allen Johnson and Timothy Earle, *The Evolution of Human Societies* (2nd ed., 2000), is the most recent general survey; it adopts an evolutionary approach that many anthropologists would resist. Andrew Sherratt's essay "The Secondary Products Revolution" (1983) is the classic account of this major technological revolution, while Margeret Ehrenberg's *Women in Prehistory* (1989) discusses some possible impacts of these changes on the gender division of labor. Anatoly Khazanov, *Nomads and the Outside World* (2nd ed., 1994), and Thomas Barfield, *The Nomadic Alternative* (1993), offer good accounts of pastoralism; Peter Golden, "Nomads and Sedentary Societies in Eurasia" (2001), is a good short introduction. D. T. Potts, *Mesopotamian Civilization* (1997), is a recent study that stresses ecological issues.

10

LONG TRENDS IN THE ERA
OF AGRARIAN "CIVILIZATIONS"

The era of agrarian civilizations has dominated conventional accounts of human history, partly because agrarian civilizations were the first human communities to generate the written records on which most modern historical research has been based. So we know this era in great detail. However, on the scale of big history, recounting a detailed description of this era is not appropriate. Besides, many fine histories already exist. Instead, this chapter will examine some of the large structures and trends that shaped the era of agrarian civilizations. Traditional approaches, which focus on particular civilizations or cultures, can easily hide these large trends. As Robert Wright has put it, ancient world history can often seem a blur of civilizations and peoples, rising and falling. But "if we relax our vision, and let these details go fuzzy, then a larger picture comes into focus: As the centuries fly by, civilizations may come and go, but civilization flourishes, growing in scope and complexity."[1]

This chapter, which surveys the 4,000 and more years during which agrarian civilizations were the most powerful communities on earth, will concentrate first on the large-scale structures. Second, it will discuss some of the more important long-term trends of this era, focusing particularly on changes in the collective human capacity to manipulate the natural environment. These show up in population growth and in more productive technologies. Central questions posed in the chapter will be, what processes shaped long-term patterns of collective learning and innovation in the era of agrarian civilizations? and how did these processes play out in different parts of the world?

LARGE STRUCTURES

Two structural features stand out in this era. First, with the appearance of cities and states, human societies became more diverse than ever before. And diversity itself was a powerful motor of collective learning, for it increased the ecological, technological, and organizational possibilities available to different communities, as well as the potential synergies of combining these technologies in new ways. But states also increased the *scale* of human interactions. Because they were so much larger than all earlier human communities, their powerful gravitational fields sucked in resources, people, and ideas from great distances. By doing so, agrarian civilizations created vast new networks of exchange. These count as the era's second main structural feature. Networks of exchange that were more extensive, more varied, and more dynamic than those of any earlier era increased both the scale and variety of exchanges and the potential synergies of collective learning.

New Forms of Diversity

At the risk of seeming overly schematic, we can think of four main types of societies in this era: three—foragers, independent farmers, and pastoralists—lack states; one—agrarian civilizations—has states.

Foragers survived throughout the era of agrarian civilizations, living in small, usually nomadic communities and mostly depending on technologies without metals. Despite some intensification, Australia was occupied exclusively by foragers until just over two hundred years ago. Similar communities also lived, until several centuries ago, in most of North and South America, in Siberia, in many parts of southern and Southeast Asia, and in parts of Africa.

In many regions, large populations of farmers or horticulturalists lived much as they had in the early agrarian era, without large-scale power structures. Such communities made up most of the population of Papua New Guinea until recent decades; often, they had contacts of trade and sometimes of war with neighboring farmers or foragers, and sometimes with traders from Indonesia. Communities of stateless farmers could be found in much of Africa, and in areas of North and South America. They could also be found along the borders of great tributary empires, from Manchuria to northern Germany.

Where productivity increased and populations grew, farming communities and technologies spread into regions that had been only thinly populated before, thereby laying the foundations for new regions of agrarian civilization. In eastern Europe, for example, from the middle of the first millennium CE, large numbers of farmers, many of them speaking ancient Slavic languages, settled in what is now Russia, where they laid the demographic foundations for the first Russian states. Such changes have often been interpreted too simply as the result of migrations of whole peoples who brought with them more productive technologies. For example, the spread of Indo-European languages from somewhere north of the Black Sea to the Mediterranean, Iran, central Asia, and North India has been linked to the spread of agriculture or pastoralism. Similarly, the spread of Bantu languages from the Cameroon region to much of central and southern Africa has often been explained as a result of migrations of peoples who displaced indigenous communities already existing there because they used more productive agricultural technologies and practiced iron metallurgy. Modern interpretations of the spread of whole language groups are more complex, seeing them as the product of many different processes, including the diffusion of languages to local populations through trade or through political or cultural domination, as well as demographic expansion, technological change, and migration. Nevertheless, the expansion of entire language groups clearly does suggest the slow spread of more productive technologies of various kinds, from improved crops, such as rye in eastern Europe, to improved implements, including iron hoes and plows.[2]

One of the most astonishing of these expansionary movements, the creation of an entirely new "world zone" in the Pacific Ocean, clearly does reflect migrations of peoples. But "Remote Oceania," including the outer islands of Micronesia and Polynesia, was not settled until the appearance of specialized seafaring cultures with advanced boatbuilding and navigational technologies, perhaps 3,500 years ago. These peoples probably came from South China or Taiwan, which appear to be the homelands of the Austronesian languages that all these groups shared. In the Pacific, we can trace their migrations through the spread of a distinctive type of pottery, known as Lapita ware. The furthest limits of these migrations were Easter Island (Rapa Nui), first settled about 300 CE; Hawai'i and (far to the west) Madagascar, both first settled ca. 500 CE; and New Zealand (Aotearoa), first settled ca. 800 CE or possibly as late as 1000–1200 CE.[3] Jared Diamond has shown how the evolution of these Pacific island societies demonstrates the impact of ecological factors on social development: within one or two thousand years, there had appeared in the Pacific a vast array of different types of societies, ranging from technologically simple foraging societies to powerful proto-states with strict class systems and populations of 30,000 to 40,000 in Hawai'i and Tonga.[4]

The third type of community was confined to Afro-Eurasia, as it depended primarily on the exploitation of domesticated livestock. In many of the more arid regions of Afro-Eurasia, and in parts of northern Siberia, there lived nomadic or seminomadic pastoralists who herded cattle, sheep, horses, or reindeer. Like most independent farmers, pastoralists normally had contacts with neighboring agrarian civilizations through warfare, commerce, and the exchange of religious and technical ideas. Particularly in Eurasia, horse-riding pastoralists could pose a serious military threat to their neighbors, because of their mobility and the virtuosity with which they used horses and camels in warfare. From late in the first millennium BCE, some pastoralist communities created powerful empires in the Eurasian steppes by exacting resources from their richer sedentary neighbors. The greatest and most influential of these empires was founded by Genghis Khan in the thirteenth century CE: it was the first political empire to reach from the Pacific to the Mediterranean.

Communities without states played an immensely important role in the era of agrarian civilizations, though they generated few written records and have consequently been neglected by historians. Because they lived in the lands between the great agrarian civilizations, they could often link their powerful neighbors into larger networks of exchange, particularly in Afro-Eurasia. While the Silk Roads provide the clearest illustration of this mechanism,[5] stateless communities also linked the emerging civilizations of Mesoamerica and Peru. Agrarian civilizations tended to be localized, but the stateless communities that lay beyond their control had more diffuse borders; the contacts between these different types of societies created the largest exchange networks of all in the premodern world.

Communities *with* states were the real dynamos of change in this period, however. The distinctive features of agrarian civilizations were their size, the density of human settlement within them, and their social complexity. No previous communities had approached them in scale or complexity. Only the largest communities of the early agrarian era contained more than 500 people, and most had fewer than 50. In contrast, even one of the earliest cities, Uruk, contained at its height perhaps 50,000 people. And this huge mass of people depended on nearby rural communities for much of their food and labor, had close links with about thirteen other city-states in southern Mesopotamia, and traded through the Persian Gulf and the Mediterranean, even as far as northern India and central Asia. The total population of the region of interrelated city-states in southern Mesopotamia probably reached several hundred thousand people. This combination of dense populations and hierarchical networks of exchange that embraced hundreds of thousands or millions of people living in very different types of communities, and that extended well beyond the frontiers of particular polities, is one of the most important structural features of the era of agrarian civilizations.

Agrarian civilizations always contain several (at least three) tiers of management and exploitation. At their base were primary producers, mostly small farmers or horticulturalists living in villages. These lived in communities similar to those of the early agrarian era, except that now they had standing over them a hierarchy of rulers and tribute takers. Village communities produced foodstuffs, fibers, and fuels such as wood. They also supplied human and animal labor for large-scale endeavors such as irrigation projects, major building projects, and the waging of war. But the village world was still shaped mainly by the needs of the farming household, so here, more than elsewhere, men and women were partners. Above this sphere, as specialized roles became more important, men began to assume separate, usually dominant, roles, and patriarchy became more institutionalized.

Above the villages stood local elites and power brokers—chiefs, nobles, officials, or priests. Local power brokers extracted resources from primary producers, but they usually preferred not to interfere directly in the life of those below them. As a result, agrarian civilizations were normally characterized by a clear gap in status, wealth, lifeways, and habits of thought between the mass of primary producers and the tribute takers who stood above them.[6] Above local power brokers there was always at least one more level of cities and rulers, who supported themselves from the resources passed on by regional power brokers. Sometimes, there appeared even higher levels of rulers who ruled over rulers—"Shahs of Shahs," to use the Persian royal title.

So, even in the simplest of agrarian civilizations, many different types of community were caught up in the networks of political, economic, and ideological power through which elites mobilized the resources they needed. The way in which resources were mobilized shaped the lives of both elites and primary producers. And these methods reached even into the households that were society's productive foundation. Here, despite the rough equality within the peasant household, males often claimed (with varying degrees of success) a reflected authority modeled on the prominent role of males beyond the basic units of the household and village. Religious, cultural, and legal structures often supported these patriarchal claims.

The most important way of shifting resources from households to elites was through demands backed up by a combination of religious, legal, and physical threats. For this reason, Eric Wolf has described agrarian civilizations as "tribute-taking" societies.[7] Unlike gift-giving, which is the characteristic way of exchanging goods in kin-ordered societies (and analogous to mutualism in the biological world), tribute-taking is, by definition, an unbalanced form of exchange. It is closer to parasitism, a relationship in which one side gains more than the other and can usually impose its will on the other.

But as we have seen, there remains an element of reciprocity or mutualism even in tribute-taking societies. Power based on coercion and power based on consent can and do coexist in all tribute-taking societies. Primary producers often depended on tribute takers for protection and other services. In time of war, villagers hid behind castle or city walls. In peacetime, urban markets offered exotic goods and alternative forms of employment, while urban temples offered grander and more potent ways of reaching the gods. Besides, it was in the interests of tributary elites to make sure their peasants had enough land to feed themselves and produce a surplus. In this general sense, tributary rulers and overlords had to protect the rights of their peasant majorities to land. The result was that in practice (though not always in theory), productive resources were spread much more evenly in agrarian societies than they are in the modern world. Rather than demonstrating pure exploitation, agrarian civilizations embodied a complex, though unbalanced, form of symbiosis—analogous, in some ways, to domestication. William McNeill's analogy with parasitism, quoted in chapter 9, captures the nuances of this imbalance well, for parasites, if they are to survive, have to protect their hosts, just as humans have to protect their domestic animals and feed their slaves. In a more recent essay, McNeill has described this relationship between cities and villages, which lies at the heart of all agrarian civilizations, as "the civilized compromise."[8]

Relations of tribute-taking existed not only within but also between neighboring states, and some of these were extremely significant. Agrarian empires can be thought of as tribute-taking systems in which powerful states exact tributes from less powerful states. But occasionally the relationship was reversed. In the biological realm, after all, parasites can be as large as robins and as tiny as bacteria. Like the terrifying species of cichlids, which swoop down on other fish and cut out pieces of their flesh, small states sometimes organized dangerous military forces that could so harass giant neighbors that the latter were forced to pay tributes or protection rents. The best-analyzed examples of these relationships are between pastoral nomads of the Eurasian states and the great states of China, Persia, and the eastern Mediterranean.[9]

To summarize, certain structures are present, in varying degrees, in all agrarian civilizations. They include:

- *Agrarian communities* that provide most resources. These are largely separate from elite groups, but they contain most of the population and produce most of society's people, food, energy, and raw materials.
- *Gender hierarchies* that support male claims to domination at most levels of the social hierarchy.
- *Cities and towns.*
- *A complex division of labor* within cities and towns, and between cities and their rural hinterlands.
- *Hierarchies* of officials, judges, and rulers headed by kings.
- *Armies,* controlled by rulers, that provide protection from other tribute takers and also enable the rulers to exact tributes by coercion from their own subjects or from neighboring regions.
- *Literate bureaucracies* that keep track of and manage resources.
- *Networks of exchange,* through which states and cities procure resources that cannot be secured through naked force.
- *Systems of religion and ideology,* often managed by the state, that legitimate state structures and often give rise to monumental architecture and high levels of artistic achievement.
- *Wider hinterlands,* which are not directly under their control, whose resources are nevertheless vital for their successful functioning. These hinterlands may lie in other regions of agrarian civilization, or may be settled by independent farmers or pastoralists or foragers.

Networks of Exchange

In the era of agrarian civilizations, exchanges linked different types of communities more effectively and over larger areas than ever before. These complex networks of exchange count as the second major structural novelty of the era of agrarian civilizations.

World historians have become increasingly sensitive to the importance of large systems of interaction, and have often analyzed them using the notion of *world-systems*. Immanuel Wallerstein, the originator of such theories, argued that particularly in the modern era, it was necessary to analyze not just particular nations or civilizations, but rather the larger networks of power and commerce in which they were entangled, because these networks explained features that could not be explained solely from the internal history of particular regions. Wallerstein called these networks "world-systems," even though they did not literally embrace the entire world, on the grounds that in many regards they functioned as separate worlds. World-systems are multilayered, multiregional structures that incorporate different types of communities; within them, some regions are more influential than others.

Wallerstein focused on the early modern, capitalist world-system, which was dominated by European states. Indeed, he argued that this was the first time that a true world-system had ever appeared. To understand Europe's increasing power in the early modern world, he insisted that historians had to understand how Europe was entangled in, and benefited from, networks of exchange and power that embraced large parts of the world. Following Wallerstein's introduction of this notion, other writers have identified similar systems in earlier periods of world history. Janet Abu-Lughod has claimed that there was a Eurasia-wide world-system as early as the thirteenth century, and Andre Gunder Frank, Barry Gills, and others have argued that regional "world systems" (in a looser sense, without a hyphen) may have existed from as early as the third millennium BCE.[10] Christopher Chase-Dunn and Thomas D. Hall have gone even further, arguing that in all the world zones, even in regions without states, there existed networks of exchange with at least some of the features of world systems.[11]

These large networks mark the outer limits within which human communities can share information, technologies, and adaptations. They thus shape processes of collective learning on the largest scales and determine the pace and the geography of innovation over long periods. One of the most important insights of modified world systems theories is that there are different kinds of networks, working in different ways and at different scales. Michael Mann has argued that even states, which appear so neatly bounded, actually wield several different types of power that work in different ways, rather like different force fields.[12] He has identified four distinct "networks" of power and influence: ideological, economic, military, and political. Political power is usually limited by recognized frontiers. Military power, in contrast, can be projected beyond those frontiers in ways that are checked only by existing logistical and military technologies. Thus, Chinese generals of the Han period had a reasonably clear idea of how large an army they could send into the Mongolian steppes, and how long they could keep it in the field without excessive cost. Ideological power is more diffuse, because the cultural borders of a region such as China are hard to define; and economic power is even harder to pin down. So economic and informational networks tended to be larger and more diffuse than those controlled directly by force.

Building on this insight, Chase-Dunn and Hall have suggested that there are several distinct types of networks of exchange, each with its own typical range and characteristics. The main types that they identify are bulk-goods networks, prestige-goods networks, political/military networks, and information networks.[13] Portability determines the differing ranges of these networks. Until recently, bulk goods, such as grain, were difficult and expensive to transport, so they normally traveled short distances. Armies could usually travel farther, but with their immense baggage trains, they also moved slowly. However, prestige goods such as silk were more portable and traveled over greater distances, while information traveled even more easily. This is why exchanges of information and prestige goods generated the largest and oldest networks of all. (Indeed, prestige goods can often travel farther than information. Think of ornaments that get exchanged so many times that their original meaning is lost.) And this is why I have focused here on the largest exchange networks of all, the informational networks that embraced entire world zones.

Large networks of exchange have distinctive regional "topologies." It may help to return to the analogy of a social law of gravity. Under this imaginary law, human communities exert an attractive force on other communities and on the goods, the ideas, and the people within them. As human communities grew, this law began to operate in more powerful ways. Roughly speaking (in a surprisingly close analogy to Newton's law), the magnitude of the gravitational pull between communities is directly proportional to the size of the communities and inversely proportional to the distance between them.

In the Paleolithic era, exchanges were limited and small-scale, because no groups were large enough to exert a significant gravitational pull on other groups. But as larger communities emerged, some exchanged goods and information more briskly and over greater distances than others, because large communities could attract resources and people over large areas. Exchanges of information, of goods, and of people were most dynamic where there were many large communities. In such regions, more information and goods were pooled than anywhere else, so we can refer to them as *centers of gravity*. They sucked in people, ideas, and produce from huge hinterlands. But they also exerted a powerful gravitational pull over the less densely settled regions that lay between them. To understand how this influence worked, we need to imagine a modern, Einsteinian form of gravity, in which large bodies deform the space-time surrounding them, tilting and twisting it so as to alter the behavior and motion of smaller objects within their gravitational field. Large cities and states transformed the social topology of the regions between them, and sometimes by doing so they created what we can describe as *hub regions*. Hub regions were situated between regional centers

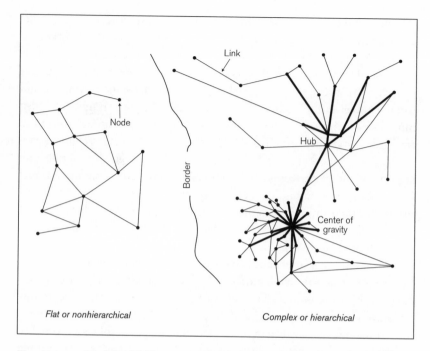

Figure 10.1.   Models of different types of exchange networks. As far as we know, all exchange networks of the Paleolithic era were "flat" or "nonhierarchical." That is, there was little difference in density from one region to another, and little variety in the pace or intensity of exchanges. With the development of dense agricultural settlement, networks of exchange became more complex and more hierarchical, and there appeared regions in which exchanges of information were so intensive that the pace of "collective learning" began to accelerate noticeably. As a result, the pace of innovation in the Holocene era was significantly faster than in the Paleolithic era.

of gravity. In these "gravitational corridors," or regions at the intersection of several different gravitational fields, they felt the pull of several different centers. Whether or not they were densely settled, hub regions carried a lot of traffic (see figure 10.1).

A glance at a world map suggests immediately that Mesoamerica and the corridor joining Mesopotamia and Egypt were likely to emerge as hub regions, because they linked large and diverse zones. Some regions, such as nineteenth-century CE Europe or Abbasid Mesopotamia, can perhaps be regarded both as hubs and as centers of gravity. They attracted information and goods because of their position and because they contained dense and rich populations. Other regions, such as late-nineteenth-century CE China, count as centers of gravity but not as hubs; conversely, regions such as fifth-century BCE Athens, central Asia 4,000 years ago, or Mongolia in the thirteenth century CE, count as hubs, though they were not populous enough to act as centers of gravity. Both centers of gravity and hub regions powerfully shaped change, because the volume of exchanges passing through them made them clearinghouses for information accumulated over large areas. Nevertheless, the differences between these two types of centers mattered. Centers of gravity gave structure and shape to large networks of exchange, while hub regions were more lightweight and were more easily transformed by the exchanges that swept through them. So it was often in hub regions that significant innovations first became important because here was where they could have the greatest impact, while the mass and momentum of centers of gravity ensured that those regions normally changed more slowly.

The increasing scale, diversity, and complexity of exchange networks energized processes of collective learning over huge areas, and they help account for the peculiar technological, political, and cultural dynamism of the era of agrarian civilizations.

LONG-TERM TRENDS

The Increasing Range and Power of Agrarian Civilizations

In 3000 BCE, the agrarian civilizations that had appeared in southern Mesopotamia and along the Nile were unique; despite their large populations, they included only a tiny proportion of the humans alive at the time. Most people still lived in communities without states. Four thousand years later, in 1000 CE, agrarian civilizations still controlled less than one-fifth of the earth's surface, but in most other respects they were the dominant communities on earth. They could be found in many parts of Afro-Eurasia, and in parts of the Americas. Small proto-states even existed in the Pacific (see table 10.1).

Why and how did agrarian civilizations become so dominant? Agrarian civilizations could not appear where the dense agrarian populations necessary to sustain them did not already exist. So the spread of agrarian civilizations was intimately connected with the spread of agriculture, and that, as we have seen, depended on technological innovations that enabled agriculturalists to farm an increasing variety of environments. The trends of innovation described in the second half of this chapter are the key to understanding the changes described in the first part. This section will examine the main stages in the spread of agrarian civilizations over 4,000 years.

TABLE 10.1. CHRONOLOGY OF EARLY AGRARIAN CIVILIZATIONS

Date	Event
ca. 3200 BCE	1st states in Sumer
ca. 3000 BCE	1st states in Egypt
ca. 2500 BCE	1st states in N. India/Pakistan (vanish 2nd mill.)
ca. 2200 BCE	1st territorial states/empires in Mesopotamia
ca. 2000 BCE	1st states in N. China (Yellow River)
ca. 1000 BCE	Revival of states in N. India/Ganges
ca. 500 BCE	1st states in Southeast Asia
ca. 500 BCE	1st "secondary empire" in Persia
ca. 500 BCE	1st states in Mesoamerica
ca. 500 BCE	1st territorial states/empires in Meosamerica
ca. 600 CE	1st states in sub-Saharan Africa
ca. 1400 CE	1st secondary empires in Mesoamerica/S. America

In 3000 BCE, agrarian civilization existed only in Mesopotamia and Egypt. By 2000 BCE, city-states had appeared in Sudan, to the south of Egypt (the powerful city-state of Yam, or Kerma), and had spread even more widely in Mesopotamia. During the reign of Sargon of Akkad (who ruled from ca. 2350 BCE for ca. 50 years), we have the first evidence for a new stage in state formation: the appearance of a state controlling several different city-states and their hinterlands.[14] Sargon claimed to feed 5,400 men every day, a figure that may indicate the size of his retinue.[15] Using what may have been the world's first standing army, he defeated rival city-states. Then, instead of merely exacting tributes from them, he incorporated them into his own empire by demolishing their walls and appointing his own sons as *ensis*, or governors. He also supported trade networks that reached throughout Mesopotamia and as far as central Asia and the Indus valley, as well as through Egypt and into sub-Saharan Africa. Mesopotamia acted as a main hub for these networks, but the density of settlement and the scale of political power under Akkadian rule probably also make it the first ever center of gravity of a regional network of exchange.

What it meant to be at the center of these widespread networks of exchange, where wealth and information were pooled in huge quantities, is suggested by this description of the Akkadian capital, Agade, from early in the second millennium:

In those days the dwellings of Agade were filled with gold,
its bright-shining houses were filled with silver,
into its granaries were brought copper, tin, slabs of
lapis lazuli, its silos bulged (?) at the sides . . .
its quay where the boats docked were all bustle . . .
its walls reached skyward like a mountain . . .
the gates—like the Tigris emptying its water into the sea,
holy Inanna opened its gates.[16]

By 2000 BCE, agrarian civilizations also existed in Crete and in the Hittite civilization of Anatolia. In the northwest of the Indian subcontinent, along the Indus River, a distinctive agrarian civilization appeared late in the third millennium. The Harappan civilization, like that of Sumer, was founded on the wealth and power of a number of large cities supported by irrigation agriculture in an arid alluvial plain. It had trading and cultural contacts with central Asia and Sumer, but its writing systems and artistic styles, like those of Egypt, seem quite distinctive. So it may be reasonable to regard the Harappan civilization as one of several regional hubs in a world system that also included the agrarian civilizations of the eastern Mediterranean.[17] The Harappan civilization declined in the first half of the second millennium. Its fall may have been the result of invasions from the north, or ecological problems linked to overirrigation, or shifts in the river systems on which it was founded.

In the second millennium, the center of gravity of Mesopotamian civilizations shifted north to Babylonia and, eventually, to Assyria. Babylon's gravitational pull made it one of the largest of all early cities, with a population perhaps exceeding 200,000 people.[18] Here, Hammurabi established a new imperial state ca. 1792 BCE. His law code, with its 282 laws carved onto forty-nine basalt columns, provides the earliest detailed written evidence about legal and bureaucratic structures (see figure 10.2). Meanwhile, expanding trade networks in the Mediterranean spread the technologies and styles of Mesopotamian and Egyptian civilization right around the Mediterranean shores. This expanding zone included the Aegean world of the Homeric epics. Egyptian trade networks also extended southward into Sudan and sub-Saharan Africa. In this way, there emerged a single region of exchanges embracing Mesopotamia, much of the Mediterranean shore, parts of sub-Saharan Africa, central Asia, and parts of the Indian subcontinent.

This system of exchanges, with its center of gravity in Mesopotamia but with links to hub regions in Egypt and Sudan, central Asia, and northern India, was the largest of several networks in the Afro-Eurasian world zone; that zone, in turn, was the largest interconnected region on Earth. So, on

Figure 10.2.   Hammurabi's law code (eighteenth century BCE). Hammurabi ruled
Babylonia from ca. 1792 to 1750 BCE. He
was the first ruler whose laws have come
down to us in some detail. The 2-meter-high
basalt pillar on which he engraved his laws
was rediscovered by French archaeologists in
1901; it is now in the Louvre, in Paris. This
part of the pillar shows the sun god investing
Hammurabi with the staff and ring of office.
© Erich Lessing/Art Resource, NY.

general principles alone, we should expect Afro-Eurasia to be the world zone
in which processes of collective learning were most intense and innovation
most rapid. Mesopotamia's position at the hub of Afro-Eurasia's exchange
networks may explain the central role played by this region in Afro-
Eurasian history and world history in general, from the earliest period of
state formation until the fundamental changes of the past half millennium
displaced it from its central position.

But Mesopotamia was never the only center of gravity, even within the

Afro-Eurasian world zone. During the second millennium BCE, agrarian civilizations emerged along the Yellow River in northern China. There is both archaeological and literary evidence that by 1600 BCE, a system of warring regional city-states covered much of northern and western China, reaching as far south as the Yangtze River. Many had wealthy and powerful rulers and some had literate bureaucracies. In the fourteenth century BCE, An-yang became a major ritual center for the semilegendary Shang dynasty, which claimed authority over many subordinate city-states. Modern research suggests that historians may have exaggerated the authority of Shang rulers over other regions, simply because their records are the only ones to survive. Nevertheless, Shang kings led armies of up to 13,000 men, which were supplied with mass-produced weapons and clothing from state factories. They also built huge and elaborate tombs, which often contained human sacrifices. The Zhou dynasty, from ca. 1050 to 221 BCE, presided over an early period of loose unity, followed by many centuries during which North China was controlled by more than a hundred independent kingdoms of various sizes; these were dominated by a core group of seven great "central states," the *zhongguo*, all near the central Yellow River. During the first millennium BCE, North China emerged as a second major center of gravity within the Afro-Eurasian world zone. The demographic, technological, and administrative foundations were laid for the Qin and former Han Empires (221–207 BCE and 207–8 BCE), and the technological, artistic, and intellectual foundations of traditional Chinese civilization were established.

Was the Chinese world system totally separate from those of northern India and Mesopotamia? The appearance, beginning ca. 4000 BCE, of mobile, pastoralist cultures that engaged in systems of exchange reaching right across the Inner Eurasian steppes means that there were at least indirect contacts between all regions of Eurasia throughout the era of agrarian civilizations.[19] We know that languages, technologies (such as the wheel and chariot), lifeways (including the basic technologies of pastoralism itself), and perhaps also methods of working bronze, as well as crops and staples such as wheat and barley (going east) and chicken and millet (going west), spread through the steppes during the third and second millennia. The appearance ca. 2000 BCE of a new hub region in the "Oxus" civilization—a cluster of trading city-states in central Asia, with links to Sumer and China as well as northern India—suggests that such exchanges were already significant 4,000 years ago, for central Asia is the natural hub for trans-Eurasian exchanges. Whether trans-Eurasian exchanges were significant enough to justify the claim that there existed a single Afro-Eurasian system by 2000 BCE remains a matter of debate.[20] We can be certain, though, that by this time no region of agrarian civilization anywhere in Eurasia was *totally* isolated from other regions.

In the first millennium BCE, the power and reach of agrarian empires in Afro-Eurasia increased decisively. The Assyrian Empire, based in northern Mesopotamia, dominated that region between the tenth and seventh centuries. The Achaemenid Empire, created in the sixth century by Cyrus the Great, was much larger than any earlier agrarian empire. Its position in Persia—at the center of networks of exchange reaching from Africa through Mesopotamia and eastward to India, central Asia, and China—presumably explains the enduring importance of Persia and Mesopotamia in Afro-Eurasian history. But the aridity of large parts of Persia also explains why this region always played the role of hub better than that of the center of gravity.

In the shadow of these great empires, agrarian civilizations spread around the Mediterranean and through Egypt into modern Sudan and Ethiopia. These new regions of agrarian civilization laid the foundations for the Greek, Carthaginian, Roman, and Sudanic Empires. At first, the newer regions of agrarian civilization consisted of small, competing states, many of which engaged in trade as well as conquest. But over time, some of these local hub regions turned into centers of gravity as well. The astonishing conquests of Alexander the Great, between 334 and 323 BCE, created a vast, if ephemeral, empire that included Greece, all of the Persian Empire, much of central Asia, and much of northern India. Alexander's empire embraced the entire hub region of Afro-Eurasia's extensive exchange networks. As it collapsed, regional dynasties, all touched by Hellenic culture, emerged in Persia, Egypt, and central Asia, and also farther west, in Italy and North Africa. The spread of agrarian civilizations around the Mediterranean laid the foundations for a new imperial system, under Rome. Roman expansion beyond Italy began with the conquest of Sicily in 241 BCE and with Rome's century-long duel with a second regional hub, Carthage, during the Punic Wars (264–146 BCE). At its height, before it split at the end of the fourth century CE, the Roman Empire controlled most of the Mediterranean, as well as huge colonies in the agrarian regions of Europe.

In the first half of the first millennium BCE, stimulated in part by new contacts with the Mediterranean world, agrarian civilizations also reappeared in the north of the Indian subcontinent, particularly in the rice-growing lands along the Ganges River. Here, significant regional hubs appeared, and eventually a regional center of gravity. The greatest Indian empire of the first millennium BCE, the Mauryan Empire (ca. 320–185 BCE), controlled most of the subcontinent; not for many centuries after this was a single ruler to control as much territory as the Emperor Ashoka (r. 268–233 BCE). Nevertheless, the emergence of densely populated civilizations in India created a new center of gravity that stimulated the appearance of new networks of exchange through the southern seas, beginning late in the first millennium BCE. As Lynda Shaffer has suggested, Indian exports of cotton and crystallized sugar, Indian control of the trade in Indonesian gold and Moluccan spices, and Indian developments in religion (particularly Buddhism) and in mathematics had an influence that reached around the huge arc from East Africa to South China. This process Shaffer has described as "Southernization," by analogy with the more familiar term *Westernization*.[21]

Late in the first millennium BCE, the agrarian civilizations of eastern, southern, and western Eurasia became more closely linked than ever before. Two developments bound several Eurasian centers of gravity more closely into a Eurasia-wide system of exchange. The first was a sharp increase in traffic along the Silk Roads after the Achaemenid rulers of Persia extended their influence into central Asia in the sixth century, and the Chinese government conquered Sinkiang early in the first century and began to actively promote trade with India, Persia, and the Mediterranean. The second development was the expansion of sea trade between Southwest Asia, India, and Southeast Asia, as sailors learned how to exploit the monsoon winds. These changes led to increased exchanges of trade goods, religious and technological ideas, and even diseases right across the Afro-Eurasian landmass. To the south of Egypt, the emergence of a significant state in Kush (in modern Sudan), which was briefly powerful enough to conquer most of Egypt (712–664 BCE), marks a significant stage in the incorporation of parts of sub-Saharan Africa into these larger networks.

In the first millennium CE, Afro-Eurasian networks were dominated by agrarian civilizations in the Mediterranean (with capitals at Rome and then Byzantium), Mesopotamia or Persia (the Parthian, Sassanid, and Abbasid Empires), India, and China (Han, Tang, and Song dynasties). Of these, perhaps the most influential were those that held the hub region of Mesopotamia and Persia—especially in the Islamic era, when Mesopotamia became once again a clearinghouse for goods, for technological ideas from the lateen sail to papermaking to the symbol for zero to new crops, and for new religious ideas incorporating elements from several different regions of Afro-Eurasia. But the Indian subcontinent may have played a greater part in these exchanges than is usually recognized, particularly by mediating the increasingly important seaborne trades from East Africa to the Mediterranean, and through Southeast Asia to China. As Shaffer has suggested, the Islamic world inherited the intellectual and technological traditions of the

Mediterranean world and those of the Indian subcontinent, while many significant developments in the religious, commercial, and technological history of Tang and Song China, from the import of Buddhism to the use of zero in mathematics to the introduction of Champa rice, may reflect Indian influences.[22] During this period, agrarian civilization spread to four new areas of the Afro-Eurasian landmass: South China, Southeast Asia, sub-Saharan Africa, and Europe. In all these regions, dense agrarian populations provided the demographic foundations for new cities and states or for the establishment of colonial empires by established civilizations.

South of Egypt, the Sudanese state of Kush was supplanted in the third century CE by the Ethiopian state of Axum, near the Red Sea, which controlled many of the trade routes linking Arabia with sub-Saharan Africa, and India with the Mediterranean.[23] In the sixth century, Axum converted to Christianity. In West Africa, the arid lands of the Sahara began to link the Mediterranean with sub-Saharan Africa, much as the steppes of Eurasia connected the Mediterranean world with China. Camels appeared in the Sahara early in the first millennium CE, and from the third century on, camel-riding pastoralists and traders such as the ancestors of the Tuareg linked sub-Saharan Africa into Mediterranean trade networks by carrying the gold and copper of West Africa (and sometimes its slaves) northward. The wealth of these trade networks triggered the appearance of cities and states within regions already settled by farmers relying mainly on sorghum, millet, and sometimes rice. The empire of Wagadu, led by a ruler called the Ghana, formed as a regional hub well before the ninth century CE on the borders of modern Mali and Mauretania. The trading empire of Kanem was founded north of Lake Chad in the mid–ninth century. Its ruling dynasty, the Sayf, survived for 1,000 years.

Usually, where new regions of agrarian civilization appear, it is easy to detect the influence of nearby centers of gravity—northern Chinese in South China and Vietnam, Indian in Southeast Asia, Mediterranean and Mesopotamian (later Islamic) in sub-Saharan Africa, Roman and Byzantine in western and eastern Europe, respectively. Such influences are clearest in South China, where populations expanded in a region long controlled by dynasties from North China. Whereas the north of China included more than three-quarters of the empire's population in 1 CE, by 1300 it included less than a quarter. At the western end of the Eurasian landmass a similar shift occurred, but this time northward into Europe.

Many regions continued to resist the spread of agrarian civilizations. Where the technological and ecological preconditions for dense settlement did not exist, traditional communities survived much longer, as is demonstrated most clearly in those areas (such as the Eurasian steppes) where agrarian civilization did not spread because they lacked dense farming populations.[24] In the lands that became Rus' and were later incorporated in the empire of Muscovy, farming had been practiced in only a few regions, including parts of modern Ukraine, since the early agrarian era. The region's harsh climates and the presence of warlike pastoralists prevented the formation of farming populations dense enough to support cities or states. Instead, farming communities remained isolated and weak, which made them easy prey for tribute takers such as the Scythians, described so well by Herodotus. Then, from the middle of the first millennium CE onward, new crops (including rye) and the use of metal plows, as well as overpopulation in eastern Europe, led to widespread immigration into the lands between Europe and the Urals. As in West Africa, dense settlement attracted outside traders. These came from the Eurasian steppes or from the Baltic and traded with central Asia and Byzantium. They created a number of regional states, the earliest of which was the Khazar Empire, whose capital lay north of the Caspian Sea. Along the routes from the Baltic to Byzantium, a number of petty city-states were linked under a single dynasty in the tenth century, creating the formidable power of Kievan Rus'. Given its early trade links with central Asia and Baghdad, Kievan Rus' could easily have converted to Islam and become an integral part of the Islamic world. But in 988 its grand prince, Vladimir (whom we met in chapter 9), converted to Orthodox Christianity; from then on, Rus' and its successor states belonged, culturally at least, to the world of Christendom.

Except for the unsuccessful Viking attempt to settle in Newfoundland ca. 1000 CE, there were no significant contacts between Eurasia and the Americas before the sixteenth century CE, which is why it is reasonable to treat the Americas as a distinct world zone.[25] Nevertheless, in the Americas, too, agrarian civilizations spread, came into contact with each other, and eventually created embryonic world systems. In Mesoamerica, as we have seen, the first agrarian civilizations appeared among the Olmec in the middle of the second millennium BCE, though some scholars would argue that the Olmec did not create true states. Nevertheless, they left a legacy in the cultural traditions of all later civilizations of Mesoamerica. True states had certainly appeared by the middle of the first millennium BCE, but it was not until the middle of the first millennium CE that there emerged imperial states based farther north, in Mexico. The history of Teotihuacán suggests how rapidly large imperial structures could appear once the appropriate foundations existed. It also reminds us how fragile early states could be. Teotihuacán, about 50 kilometers north of modern Mexico City, consisted of no more than a few small villages in 500 BCE. From 150 BCE it grew swiftly. Three centuries later, it had a population of ca. 60,000 to 80,000. Its growth—like that of the Anatolian city of Çatal Hüyük, which had flourished 6,000 years earlier—may have been based on the trade in obsidian, the premetallic equivalent of steel. At its height, ca. 500 CE, Teotihuacán had a population of 100,000 to 200,000 people, and its monumental architecture was as grand as anything that could be found in Afro-Eurasia (see figure 10.3).[26] Teotihuacán was supported by a network of nearby villages and towns, which grew crops using irrigation agriculture and *chinampa* systems (described below). But it also depended on foodstuffs imported along extensive trade networks embracing a larger, Mesoamerican world system. So it clearly counts as a regional hub, and perhaps even as the first American center of gravity. Then, between 600 and 700 CE, Teotihuacán collapsed. Overexploitation of the land as a result of rapid population growth may have ruined the ecology of the region, while rival towns may have cut off the trade networks that supplied it or even invaded and sacked the city. Within fifty years of its collapse, there was nothing left but a few villages. A source from the colonial period describes how the leaders of the city fled, taking "the writings, the books, the paintings; they carried away all the crafts, the castings of metals."[27]

In the contemporary cultures of the Maya, in the lowlands of the Yucatán Peninsula to the south, there appeared a number of regional centers linked into the same networks of exchange as Teotihuacán. These collapsed at about the same time, probably as a result of overpopulation, may be linked to climatic changes that undermined the fertility of regional farmlands. From late in the first millennium, urbanization and state building intensified in central Mexico. These processes culminated, eventually, in the creation of the Aztec Empire in the fifteenth century. In 1519 its capital, Tenochtitlán, contained between 200,000 and 300,000 people, and several other cities in the Valley of Mexico were almost as large. Here is how Cortés's lieutenant, Bernal Díaz del Castillo, described his first sight of Tenochtitlán in 1519:

> Next morning, we came to a broad causeway and continued our march towards Iztapalapa. And when we saw all those cities and villages built in the water, and other great towns on dry land, and that straight and level causeway leading to Mexico [Tenochtitlán], we were astounded. These great towns and *cues* and buildings rising from the water, all made of stone, seemed like an enchanted vision from the tale of Amadis. Indeed, some of our soldiers asked whether it was not all a dream. It was all so wonderful that I do not know how to describe this first glimpse of things never heard of, seen or dreamed of before.[28]

Figure 10.3.   Teotihuacán. The great city-state of Teotihuacán, 40 km outside of Mexico City, flourished from ca. 200 BCE to ca. 650 CE. At its height, its population may have reached 200,000, making it one of the largest cities in the world. It was certainly the largest and most powerful city in the Americas. It had contacts with many other parts of Mesoamerica, and its political traditions influenced later Mesoamerican states, including that of the Aztecs. From Brian M. Fagan, *People of the Earth: An Introduction to World Prehistory,* 7th ed. (New York: HarperCollins, 1992), p. 574: from *Urbanization at Teotihuacán, Mexico,* ed. René Millon, part 1, vol. 1 (Austin: University of Texas Press, 1973), © René Millon.

Up to 2 million people lived in and around Tenochtitlán in 1500. They were supported by raised-field farming, known as the *chinampa* system. Early settlers on the swampy lands around Tenochtitlán built up mounds of river vegetation and mud, held together with "fences" of willows. They cleared canals between the mounds and fertilized them with mud from the canal beds, with rotting vegetation, and with human refuse; by farming them carefully it was possible to raise up to seven crops a year. The diets of these early settlers were supplemented with fish and waterfowl.[29]

In South America, the first agrarian civilizations appeared in the first millennium CE, in a period of rapid population growth and urbanization. The first great empire to emerge here was that of the Incas, which was assembled in the fifteenth century CE. The agrarian civilizations of South America had significant contacts with Mesoamerica, but whether these were sufficient to create a single world system remains a matter of dispute. In North America, after the widespread adoption of maize cultivation late in the first millennium CE, populations grew rapidly; all the now-familiar signs of an agrarian civilization in embryo began to appear in the so-called Mississippian cultures. At its center were large towns with elevated ceremonial centers sometimes as high as 30 meters. Centers such as Cahokia, which may have had 30,000 to 40,000 people living nearby in 1200 CE, were similar to the ceremonial centers of the Eridu period in Sumer. Mississippian communities were probably organized in large-scale chiefdoms; but because most of their people still lived in small agricultural communities, it does not count as a fully developed agrarian civilization. Maize sustained population growth in this region, so it may be appropriate to regard the Mississippian region as a regional hub within a wider American network of exchanges whose centers of gravity lay in Mesoamerica and South America.

If it is viewed as such a hub, we can say that by the second millennium CE, the various regions of agrarian civilization throughout the world had become linked, through expanding networks of exchange, into two major world systems: Afro-Eurasia and the Americas. Of these, the Afro-Eurasian system was older, larger, more densely populated, and more powerful. The extent of its power became abundantly clear in the sixteenth century when the two regions finally came into contact. In neither of the other two world zones did agrarian civilizations exist, even though in some farming regions, including Papua New Guinea and islands such as Tonga and Hawai'i, there did emerge powerful chiefdoms, some on the verge of statehood.

Very loosely, we can quantify the expansion of agrarian civilizations over 4,000 years. Rein Taagepera has tried to measure the areas ruled by "imperial systems" of Afro-Eurasia at different dates. By *imperial systems*, he means large political entities that include several agrarian states. Though this definition would exclude some regions of agrarian civilization, it may still provide a rough index of the political and military expansion of agrarian civilizations in Afro-Eurasia. For each period, Taagepera estimates the total area controlled by state systems, and compares these estimates with the areas controlled by state systems today. Table 10.2 summarizes his data.

Three eras stand out. The first extends from the third millennium to the middle of the first millennium BCE. In this period, agrarian civilizations could be found only in the Afro-Eurasian zone, and they directly controlled just 2 percent of the areas ruled by state systems today. The second era begins

TABLE 10.2. AREAS OF AFRO-EURASIA WITHIN AGRARIAN CIVILIZATIONS

Era	Date	Area Controlled in Megameters (1 megameter = 100,000 km²)	Area as % of Modern Area Controlled by States
Late agrarian 1	early 3rd mill. BCE	0.15 (all in Southwest Asia)	0.2
	2nd mill.–mid 1st mill. BCE	1–2.5	0.75–2.0
Late agrarian 2	6th c. BCE	8	6.0
	1 BCE	16	
	1000 CE	16	13.0
Late agrarian 3	13th c. CE	33 (mainly Mongol Empire)	25.0
	17th c. CE	44 (now incl. Americas)	33.0
Modern	20th c. CE	ca. 130	100.0

SOURCES: William Eckhardt, "A Dialectical Evolutionary Theory of Civilizations, Empires, and Wars," in *Civilizations and World Systems: Studying World-Historical Change*, ed. Stephen K. Sanderson (Walnut Creek, Calif.: Altamira Press, 1995), pp. 79–82, relying heavily on Rein Taagepera, "Size and Duration of Empires: Systematics of Size," *Social Science Research* 7 (1978): 108–27.

in the middle of the first millennium BCE, with the appearance of the Achaemenid Empire, and extends to 1000 CE. By its end, agrarian civilizations controlled between 6 percent and 13 percent of the area controlled by modern states. During this period, agrarian civilizations appeared also in the Americas, but the areas they controlled were far smaller than in Afro-Eurasia. There is a further sudden increase in the area ruled by large empires after 1000 CE, with the rise of the Mongol Empire and the European empires of the past 500 years. American empires also expanded after 1000 CE, but they contributed much less to this rapid expansion. In 1500, the Inca Empire ruled ca. 2 megameters and the Aztec Empire only 0.22 megameters. Inclusion of regions of incipient state formation in the Pacific, such as Hawai'i and Tonga, would make no real difference to these calculations.[30]

Despite this long history of expansion, it is important to remember that even in the seventeenth century, just 300 years ago, state systems controlled no more than one-third of the lands incorporated within states in the twenty-first century. Even if they had come to dominate networks of exchange throughout the world and include most of the world's population, they never *controlled* the world in the way of modern capitalist states.

Accumulation, Innovation, and Collective Learning

The spread of agrarian civilization was made possible by a continuation of the processes of intensification that had begun early in the Holocene. Rates of innovation were thus a crucial determinant of the pace and nature of change in this era. What factors governed rates of innovation? Where was innovation most intensive, and how rapid was innovation in the era of agrarian civilizations?

Scale itself was a source of innovation, as the increasing size of exchange networks generated new intellectual and commercial synergies. But more specifically, three other factors shaped the pace and nature of innovation in this period: population growth, the expanding activity of states, and increasing commercialization and urbanization. I will describe these three sources of innovation separately, even though in practice they were intertwined. Though each contributed to innovation and growth in the long run, in the medium- and short-term each could also undermine growth. To contemporaries, these cyclical patterns over shorter time spans were usually the most apparent, which is why premodern historians have characteristically thought in terms of cycles rather than long-term trends. As we will see, the impact of these three sources of innovation was ambiguous and uncertain, characteristics that help explain why innovation was so much slower in the era of agrarian civilizations than it is in the modern era.

Scale as a Source of Innovation   At the most general level, the size and variety of information networks, as well as the intensity of exchanges within them, shaped average rates of innovation over long periods. The greater the volume and diversity of information being exchanged, the more likely it was that such exchanges would result in innovations, both large and small. In the period considered here, it is clear that in the Afro-Eurasian world zone, and to a lesser extent in the Americas, too, information networks expanded in both scale and variety. They also linked societies of many different types, so that innovations introduced by barbarian farmers in northern Europe could spread to the Mediterranean, while technologies of horse riding first developed in the Eurasian steppes spread to China or Mesopotamia, and metalworking technologies and crops spread throughout Afro-Eurasia, in areas of agrarian civilization and beyond their borders.

Improvements in transportation and communication technologies enabled innovations to spread more rapidly and more widely (see tables 10.3 and 10.4). The secondary products revolution was of fundamental importance in increasing the intensity and speed of exchanges within the Afro-

TABLE 10.3. TRANSPORTATION REVOLUTIONS IN HUMAN HISTORY

Era	Approximate Date	Ways of Moving Peoples and Goods
Paleolithic	From ca. 700,000 BP	First hominine migrations from Africa
	From ca. 100,000 BP	Modern humans in southern Eurasia; first migrations of modern humans out of Africa
	From ca. 60,000 BP	First migrations by sea to Australasia; earliest seaborne boats
Agrarian	From ca. 4000 BCE	Animal-powered transportation
	From ca. 3500 BCE	Wheeled transportation
	From ca. 1500 BCE	Long-distance ships in Polynesia
	1st mill. BCE	State-built roads and canals; coinage
Modern	1st mill. CE	Improvements in shipbuilding, navigation
	From early 19th c.	Railways and steamships
	From late 19th c.	Internal combustion engines
	From early 20th c.	Air travel
	From mid 20th c.	Space travel

Eurasian zone, as it made available new forms of transportation perhaps as early as 4000 BCE. The harnessing of oxen, asses, and horses may in turn have encouraged the evolution of wheeled transport. At sea, while no European mariners achieved the navigational sophistication of Polynesian navigators, the speed, reliability, and precision of their navigation undoubtedly increased, particularly in the first millennium CE. The building of major roads also stimulated transportation from China to Persia to Rome. The main innovations in forms of communication were associated with the evolution of scripts and methods of writing. But some empires, including the Achaemenids and the Han, organized long-distance courier systems (as did the Incas much later in Peru). Many societies also constructed rudimentary early-warning systems based on the lighting of signal fires, so that information could sometimes travel large distances with great speed.

The size and diversity of Afro-Eurasia and its relatively advanced systems of communication help explain why rates of innovation were faster here than in the other world zones of the Holocene era. Here the pool of information being exchanged was greater and more varied than anywhere else, so that the accumulation and exchange of new technologies proceeded more

TABLE 10.4. INFORMATION REVOLUTIONS IN HUMAN HISTORY

Era	Approximate Date	Ways of Moving Information
Paleolithic	Paleolithic, beginnings of human history	Modern forms of language; information sharing between different groups
	Upper Paleolithic	Cave paintings
	From Upper Paleolithic?	Communication at a distance using drums, beacons, smoke signals
Agrarian	From ca. 3000 BCE	Writing as congealed information
	From ca. 2000 BCE	Syllabic writing
	Era of agrarian civilizations	Government-sponsored or military courier system
	From 8th c. CE	Printing using wood blocks
Modern	16th c. CE	Global world system; worldwide systems of communication and transport
	18th and 19th c.	Print used for mass communication: newspapers, postal services
	From 1830s	Telegraph
	Late 1880s	Telephone
	20th c.	Electronic mass media: radio, film, TV
	Late 20th c.	Internet; instantaneous global communication of information

rapidly than ever before. But in the Americas, too, the emergence of large zones of agrarian civilization and extensive trade networks ensured that similar processes were at work. Information networks were large; they exchanged knowledge between regions with different lifeways, crops, technologies, and ecologies; and in their turn, ecological innovations accelerated the intensity and speed of exchanges. Literacy was part of the process here, as was the creation of impressive roadways; but what was lacking was the entire complex of innovations associated in the Afro-Eurasian zone with the secondary products revolution.

Population Growth   In the early agrarian era, population growth and technological change reinforced each other. In the era of agrarian civilizations, this relationship remained a major source of innovation and accumulation—particularly in regions of independent pastoralists or peasants, whose communities generated many important innovations, especially in agriculture and the use of livestock.

Between 3000 BCE and 1 CE, world populations rose from ca. 50 to ca. 250 million (see tables 6.2 and 6.3). This expansion marks a gentle acceleration in the demographic dynamism of the early agrarian era, suggesting that the creation of agrarian civilizations had a significant, but not revolutionary, impact on population growth. The long demographic trends create the illusion of steady growth. But on the scale of lifetimes or even centuries, the pattern that stands out is cyclical—a pattern of rise and fall. Historians have become increasingly aware of these large cycles of expansion and decline, though they differ on the questions of their periodicity and their causes. In a famous study first published in 1966, *The Peasants of Languedoc*, the French historian Emmanuel Le Roy Ladurie traced centuries-long cycles of boom and bust in the Languedoc region and the early modern French economy as a whole. These cycles affected all aspects of life: Robert Lopez described them as an "alternation of crest, trough and crest" that "can be observed not only in the economic field, but in almost every aspect of life: literature and art, philosophy and thought, politics and law also were affected, though not all to the same extent."[31] Le Roy Ladurie described these cycles, memorably, as "the respiration of a social structure."[32] One reason for their immense influence was the overwhelming importance of the agrarian sector. Where most forms of production relied on organic materials and energy sources, agricultural output set limits to the production not just of foodstuffs but also of clothing, housing, energy, productive implements, and even parchment and paper.[33] Because agriculture was the main motor of economic growth in the agrarian era, rates of innovation in agriculture dominated medium-term economic, political, and even cultural cycles. As populations rose, so did production, so did demand, and so did the supply of labor. Rising populations created the buoyancy needed to sustain expanding systems of trade, larger states, the building of monumental architecture, and the patronage of artists and artisans, which in turn stimulated cultural change. Agrarian civilizations flourished in such periods economically, politically, and artistically. Periods of declining populations undermined all these activities. As a result, the movements of economic expansion and contraction, of urbanization, of trade, and of political power are all linked to the same underlying rhythms. Long-term accumulation took place underneath these cycles in ways that were often invisible to contemporaries. Only in the long view of world history is it evident that each cycle normally rose higher than its predecessor.

What was true of early modern France was true of agrarian civilizations

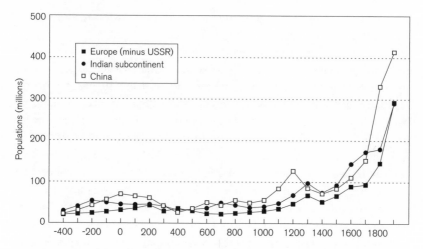

Figure 10.4.   Malthusian cycles in China, India, and Europe, 400 BCE–1900 CE.
This graph shows the Malthusian pattern of rising populations, punctuated by
sudden downturns, that was typical of the history of the agrarian era. From
J. R. Biraben, "Essai sur l'évolution du nombre des hommes," *Population* 34
(1979): 16.

in general. To get some feeling for the cycles described by Le Roy Ladurie,
it may help to look at population growth in several distinct regions of agrarian civilization. Figure 10.4 graphs the rough population figures calculated
by J. R. Biraben for China, the Indian subcontinent, and Europe between 400
BCE and 1900 CE.[34] It is immediately obvious that in each region there are
periods of rising population followed by periods of decline—sometimes
drastic decline. How should we explain these medium-term rhythms, which
seem to have shaped the histories of all regions of agrarian civilization?
Though they blend several overlapping trends in population growth, harvest fluctuations, warfare, and commerce and state policy, one set of factors
dominated them: a negative feedback cycle linking innovation (particularly
in agriculture), population growth, ecological degradation, declining health,
and increasing conflict, leading to population decline (see figure 10.5).

   The British pioneer of population studies, Thomas Malthus, was one of
the first to analyze the relations between population growth and available
resources. At the end of the eighteenth century, he argued that any species,
regarded purely mathematically, can multiply at a geometric rate, on the
upward-curving trend familiar from compound interest. Yet the resources
available to feed each species normally increase only at an arithmetic rate,

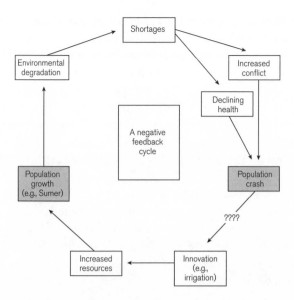

Figure 10.5.   A negative feedback cycle: population, agriculture, and the environment.

on a straight-line trend. This means, as we have seen in the final section of chapter 5, that available resources set the real limits to population growth. In the natural world, available resources are determined by the niches available to a particular species. But humans are different because they keep innovating: they explore, modify, improve, and even create new niches. Thus limits to human population growth are set only by the number and productivity of the niches that innovation has made available in any particular epoch. Every time there are significant innovations, the ceiling to population growth is lifted. When significant innovations occur, populations can climb until they overshoot the new ceiling. Then there will be a crash. The land will give out; famine will take the hungry; disease will take the malnourished; and wars, often launched by governments competing for scarce resources, will take soldiers and the people through whose cities or villages they move. Eventually human populations will settle down to a new level. Innovation ensured that each cycle normally reached a higher level than its predecessor, but innovation was normally too slow to prevent an eventual collapse within each cycle, as populations outstripped available resources.

There is an important ecological component to these rhythms, for population crashes were often triggered by overexploitation of fragile environments, particularly in areas where population growth depended on the irrigation of arid lands. This is the same rhythm we see in the evolution of excessively virulent parasites. In the era of agrarian civilizations, it is most apparent where it led to the collapse of entire civilizations. At the end of the third millennium BCE, the drying of southern Mesopotamia, joined with salinization caused by excessive irrigation, undermined the ecological foundations of Sumer. One sign in the archaeological record of increased salinity is the inhabitants' increased use of barley, which tolerates salt better than wheat does. Eventually, though, populations collapsed, falling from ca. 630,000 in 1900 BCE to ca. 270,000 by 1600 BCE, not to rise again until a millennium later under the Achaemenids.[35] Sadly, the same pattern was to be repeated again in the later history of Mesopotamia (see figure 10.6). A similar fate probably accounts for the collapse of the Mayan civilization at the end of the eighth century CE (see map 10.1). Michael D. Coe observes:

> The Classic Maya population of the southern lowlands had probably increased beyond the carrying capacity of the land, no matter what system of agriculture was in use. There is mounting evidence for massive deforestation and erosion throughout the Central Area, only alleviated in a few favorable zones by dry slope terracing. In short, overpopulation and environmental degradation had advanced to a degree only matched by what is happening in many of the poorest tropical countries today. The Maya apocalypse, for such it was, surely had ecological roots.[36]

In all these cases, new technologies or opportunities stimulated population growth, but neither the technological nor the managerial know-how was sufficient to support growth indefinitely. Innovation was sufficient in all these cases to initiate growth, but not to sustain it or avoid overexploitation and ecological collapse. This characteristic pattern of slow innovation ("technological drift," in Eric Jones's term) lagging behind potential rates of population growth is the primary explanation for the steplike cycles that can be observed throughout the agrarian era.[37] I will refer to these as *Malthusian cycles.*

Disease is as much a part of these cycles as environmental degradation. The evolving relations between humans and diseases is easiest to see in Afro-Eurasia—perhaps, as Jared Diamond has argued, because only there did humans live intimately enough with domestic animals to swap disease pathogens with them.[38] The figures in table 6.3 suggest that world populations

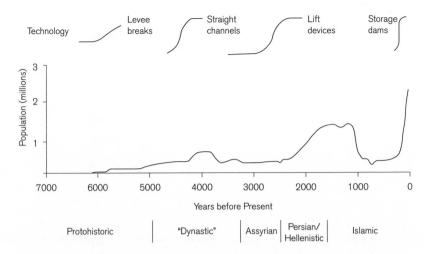

Figure 10.6.   Population and technological change: Malthusian cycles and irrigation technologies in lowland Mesopotamia. Adapted from Neil Roberts, *The Holocene: An Environmental History,* 2nd ed. (Oxford: Blackwell, 1998), p. 175: based on M. J. Bowden et al., "The Effect of Climate Fluctuations on Human Populations: Two Hypotheses," in *Climate and History: Studies in Past Climates and Their Impact on Man,* ed. T. M. L. Wigley, M. J. Ingram, and G. Farmer (Cambridge: Cambridge University Press, 1981), pp. 479–513.

grew exceptionally fast between 1000 and 1 BCE. (These figures are dominated by those for Eurasia, so any conclusions we draw from them apply mainly to Eurasia.) In the two thousand years from 3000 to 1000 BCE, the doubling time for world populations dropped from ca. 1,630 years in the early agrarian era to about 1,580; but in the period from 1000 to 1 BCE, it dropped to a mere 945 years. These calculations reinforce the impression revealed by many other trends: the first millennium BCE was one of exceptionally rapid growth, at least in much of the Afro-Eurasian world zone, and then growth slowed. Why?

William McNeill has suggested the most elegant explanation for this acceleration in population growth in Eurasia. It had to do with changing human relations with parasites, both large and small. "Macro-parasites" (tribute-taking states) learned to take tribute in less violent and more predictable ways; while population growth and epidemiological exchanges enabled each region to establish more stable relations with local diseases:

> During the first millennium B.C. in three important centers of human population [China, India, and the Mediterranean], the balances between

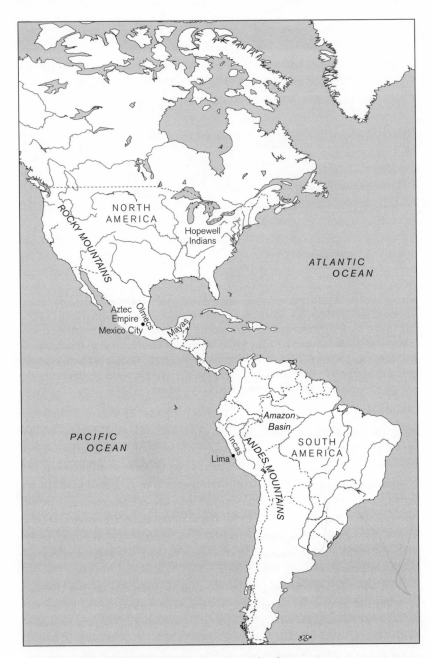

Map 10.1.   The American world zone before Columbus.

macro- and micro-parasitism adjusted themselves in such a fashion as to allow persistent population growth and territorial expansion of civilized types of society. As a result, by the beginning of the Christian era, the civilizations of China, India, and the Mediterranean had attained a size and mass comparable to that of the more anciently civilized Middle East.[39]

Political systems were getting better at judging appropriate levels of tribute exaction, and people (and their immune systems) were getting better at coping with infectious diseases.

Sadly, this argument has a flip side. When regions previously isolated from each other came into regular contact, they swapped diseases. And this exchange could prove devastating in regions that lacked the necessary immunities. For a time, plagues and epidemics could reverse or slow population growth on both sides of the old epidemiological frontier. In the first millennium of the modern era, up to 1000 CE, world populations did not rise at all. This demographic downturn is of fundamental importance, yet it has been largely ignored by historians. There may have been earlier periods of similarly slow growth, though the available evidence makes it hard to be sure. In the Mediterranean world system we find hints of significant demographic and political collapses late in the third millennium and toward the end of the second millennium BCE. Whatever the causes of these earlier declines, McNeill has suggested that the stagnation of the first millennium CE was caused by increasing traffic along the major exchange networks of Eurasia, such as the Silk Roads and the sea routes linking the Mediterranean and South and East Asia. Disease bacteria traveled these routes as well as people, goods, and techniques, causing massive and recurring plagues as each region faced new diseases for which its populations lacked biological or cultural antibodies. McNeill refers to this process as the "closing of the Eurasian ecumene."[40]

New diseases had their greatest impact at the extremes of the Eurasian world system, in the Mediterranean and China, where earlier contacts had been most restricted. They had less impact in Mesopotamia and India, regions that lay closer to the hub of the Eurasian network of exchanges and were therefore more disease-hardened. McNeill argues that more stable relations with regional disease bacteria had probably evolved within each of the regions of denser settlement by the first millennium BCE, which may explain the more rapid population growth of that millennium; the swapping of disease bacteria between the major civilizations of Eurasia may help explain the slow growth of the first millennium CE.

The relative immunity of the hub regions may lie behind the increasing significance in the first millennium CE of the Sassanid and then Islamic Empires, both with their heartlands in Persia and Mesopotamia, and of the Gupta Empire (320–535 CE) in northern India. But people suffered in the far east and the far west. As McNeill argues, "In the first Christian centuries, . . . Europe and China, the two least disease-experienced civilizations of the Old World, were in an epidemiological position analogous to that of Amerindians in the later age: vulnerable to socially disruptive attack by new infectious diseases."[41] It seems likely that in the Mediterranean world, diseases such as smallpox, measles, and rubella did not exist before the first millennium CE.[42] In Rome, large-scale epidemics struck first in 165 CE. They probably took the form of smallpox, but measles and bubonic plague appeared in succeeding centuries. The "plague of Justinian," which struck Byzantium in 542–43, was almost certainly bubonic plague, as we have a detailed account of it from the historian Procopius.[43] The plague recurred for at least the next two centuries.

These devastating bacteriological exchanges did more than alter demographic patterns. They also affected state structures and even religious and intellectual history. For example, population losses must have contributed to the decline of the Roman Empire. In China the picture is less clear, but there is evidence that severe infections, perhaps including smallpox and measles, broke out beginning in the middle of the second century CE. We also know that Chinese populations declined sharply, and that imperial political and ideological structures declined with them, during the period between the fall of the Han (220 CE) and the rise of the Tang dynasty (618 CE).[44] Meanwhile, the populations of Mesopotamia, Iran, and probably northern India held up much better, and thus these regions flourished late in the first millennium CE. The Black Death of the fourteenth century marks a new phase in the interchange of disease vectors, and on that occasion the dominant disease was bubonic plague.

Other killers also slowed demographic accumulation in the era of agrarian civilizations. Most important of all was the linked trio of famine, warfare, and urbanization. We will examine these in later sections.

States as Sources of Accumulation   Within the regions they controlled, states and cities were powerful concentrators of wealth and powerful sources of accumulation and innovation, for the strength of their rulers depended on their capacity to mobilize human and economic resources. Moreover, cities were, by their nature, important hubs for the exchange of ideas as well as goods. Yet cities and states could dampen innovation as well.

Though individual states also rose and fell, the long-term trends all in-

TABLE 10.5. AREAS RULED BY PARTICULAR STATES AND EMPIRES

Era	Date	State or Empire	Area Controlled (megameters)
Late agrarian 1	late 3rd mill. BCE	Sargon of Akkad, S. Mesopotamia	0.6
	late 3rd mill. BCE	Egypt	0.4
	mid 2nd mill. BCE	Thutmosis III of Egypt (1490–1463)	1.0
	late 2nd mill. BCE	Shang dynasty, China	1.0
Late agrarian 2	mid-1st mill. BCE	Achaemenid Empire, Iran (and successors)	5.5
	late 1st mill. BCE	Mauryan Empire, India	3.0
	late 1st mill. BCE	Han dynasty, China	6.0
	early 1st mill. CE	Roman Empire, Mediterranean	4.0
	mid–late 1st mill. CE	Early Islamic empires	10.0
	mid–late 1st mill. CE	1st Türk Empire, Inner Eurasia	7.5
Late agrarian 3	early–mid 2nd mill. CE	Mongol Empire, Inner Eurasia	25.0
Modern	ca. 1500 CE	Inca and Aztec Empires, Americas	2.2

SOURCE: Figures from Rein Taagepera, "Size and Duration of Empires: Systematics of Size," *Social Science Research* 7 (1978): 108–27.

dicate a sustained increase in the reach and power of the largest and most powerful states. Coupled with this was an increase in the number of smaller and more rudimentary state systems with limited bureaucracies and fragmented sovereignty, a type of political system often referred to as *feudal,* or simply as *early states.*[45] No state in the era of agrarian civilizations had anything like the capacity of modern states to regulate the daily lives of their subjects. Most ruled through chains of intermediaries, with little knowledge of or interest in the lives of the majority of those they ruled. Yet undoubtedly states slowly got better at what they did, and managed their power with great skill and efficiency as their methods of predation became less virulent and more restrained.

An indirect measure of the increasing power of states is the area ruled by the largest states. This trend has been roughly calculated by Rein Taagepera.[46] The figures in table 10.5 suggest that there were three distinct eras.

First, from ca. 3000 to 600 BCE, even the largest state systems controlled no more than a megameter (1 megameter = 100,000 km²) of land. The earliest imperial system, created by Sargon of Akkad, covered ca. 0.6 megameters, while the Egyptian dynasties of the third millennium, its nearest rivals, controlled at their peak about 0.4 megameters. Sargon's empire set a threshold that was not crossed until the middle of the second millennium when Pharaoh Thutmosis III of Egypt created a short-lived empire in Egypt and eastern Mesopotamia, which may have covered almost 1 megameter. In the thirteenth and twelfth centuries, the Shang dynasty in China may have ruled as large a territory.

Second, in the sixth century of the first millennium BCE, the Achaemenid Empire set a new benchmark. At its height, it ruled ca. 5.5 megameters of territory. For the next 2,000 years, under the Achaemenids, Seleucids, Parthians, Sassanids, and Abbasids, Persia was to be the heart of vast land empires that controlled similar areas. These set the standard for imperial reach throughout this period. In India, in the third century BCE, the Mauryan Empire briefly ruled over more than 3 megameters. No later Indian empire reached this level again until the creation of the Mogul Empire in the sixteenth century CE. By the first century BCE, the Han dynasty in China had begun to rule an even larger territory than Persia (more than 6.0 megameters). The empire of Alexander the Great was vaster than that of the Persians, but more ephemeral. By the first century CE, the Roman Republic controlled an empire larger than 4 megameters. In the seventh and eighth centuries CE, the Islamic conquests created empires, based in Mesopotamia and Persia, that controlled about 10 megameters of Afro-Eurasia's main hub region before they fragmented.

Third, with the remarkable exception of the Mongol Empire of the thirteenth century, which, at its peak controlled 25 megameters, and the much later example of the early modern European empires, which also controlled about 25 megameters of territory in the seventeenth century, 5 to 10 megameters remained the limit of most traditional empires. Not until the modern era did improved communications and transportation technologies, combined with modern military and bureaucratic techniques, make it possible to create even larger empires.

In the Americas, the growth of state systems followed parallel trends, but with a time lag of approximately two millennia. Agrarian civilizations similar in scale to those of third millennium Sumer or Egypt appeared late in the second or early in the first millennium BCE. The political threshold first crossed by the Achaemenid Empire was not breached in the Americas before the arrival of the Europeans. In 1500 CE, the Incas ruled an area of ca. 2 megameters, while the Aztec Empire was much smaller, controlling an area of only ca. 0.22 megameters.[47]

Changes in religious thinking mirrored the increasing power and reach of state structures, for religions could underpin state power by mobilizing loyalty and justifying tributary exchanges, particularly where there were institutionalized churches. Religions in early agrarian civilizations, like those of the Paleolithic world, tended to be local or regional in their claims and influence.[48] Their gods, like family members, were expected to protect particular tribes or cities and to smite their enemies. With the creation of the first empires, regional deities were incorporated into larger, more imperial pantheons; but religion remained a regional affair, tied closely to the fortunes of particular dynasties, cities, and empires. This connection can be seen in the religious art of Naram-Sin (ca. 2250–2220 BCE), the grandson of Sargon of Akkad, who is portrayed as a deity ruling over many other deities.

Not until the first millennium BCE do the first universal religions appear. Though associated in practice with particular dynasties or empires, they proclaimed universal truths and worshiped all-powerful gods. It is no accident that universal religions appeared when both empires and exchange networks reached to the edge of the known universe, controlling populations with diverse belief systems and lifeways. Nor is it an accident that one of the earliest religions of this type, Zoroastrianism, appeared in the largest empire of the mid–first millennium BCE, that of the Achaemenids, and at the hub of trade routes that were weaving Afro-Eurasia into a single world system. Indeed, most of the universal religions appeared in the hub region between Mesopotamia and northern India. They included Zoroastrianism and Manichaeism in Persia, Buddhism in India, Confucianism in China, and Judaism, Christianity, and Islam in the Mediterranean world. Their appearance persuaded the German existentialist philosopher Karl T. Jaspers, in a world history first published in 1949, to name this period the "axial age."[49] A powerful indication of the growing links between different parts of Afro-Eurasia was the way these religions traveled along trade routes, with Buddhism traveling to China, along the Silk Roads, as well as Manichaeism and Nestorian Christianity. Islam benefited from its control of the Mesopotamian hub region, and it eventually spread even more widely: westward to Spain, south to East Africa, east to central Asia and North China, and eventually east and south to India and large areas of Southeast Asia. Christianity, though successful in the Mediterranean region for a time, had to retreat before Islam for many centuries. Its time came late in the second millennium CE.

Tribute-taking states, like population growth, had a significant but contradictory impact on accumulation. On the positive side, they had good reason to encourage innovation and accumulation, in order to increase their own power and efficiency. Like viruses, they could take from their prey more or less efficiently, and more or less brutally. The most stable states and the wisest rulers protected the productive base of their societies by taxing lightly, maintaining basic infrastructure, upholding law and order, and encouraging growth in rural populations and agricultural output. Gentle taxation and stable rule could promote increased agricultural and artisanal production. But it was also important to stimulate growth in other ways by maintaining infrastructures such as roads and irrigation systems. The importance of such methods is a theme that crops up over and over again in the manuals of statecraft that appear in all Eurasian agrarian civilizations. Many ancient writers were concerned to describe and encourage less predatory and more sustainable forms of taxation. Thus, an eleventh-century CE Muslim prince from Tabaristan wrote in a book for his son, "Make it your constant endeavor to improve cultivation and to govern well; for understand this truth: the kingdom can be held by the army, and the army by gold; and gold is acquired through agricultural development and agricultural development through justice and equity. Therefore be just and equitable."[50] In a similar spirit, Chinese governments of the Song era ordered their officials to promote the use of more productive strains of rice in the south of the country. Earlier Chinese dynasties had invested huge amounts in the buildings of canals and improved roadways to ease the transportation of grain and other goods to the major cities. All governments whose populations depended on large irrigation works had to concern themselves with the maintenance of those systems.

States could stimulate accumulation in many other ways. Most tribute-taking states saw war as their central concern, because the conquest of neighboring societies was one of the quickest ways of acquiring new resources. So tribute-taking states always took great interest in military innovations. Sumerian governments traded in copper and tin because they needed bronze weaponry. Roman technology is remarkable in areas such as the building of bridges, aqueducts, and fortifications; the use of concrete; and the construction of machines of war such as catapults or siege engines, which were built using sophisticated systems of ratchets, pulleys, and gears. Han technology (and bureaucratic skill) was particularly impressive in such areas as fortification (as in China's Great Wall), the mass production of weapons and armor, the mobilization of resources for war, and the building of canals to transport foodstuffs.

Rulers often tried to enhance their prestige by supporting large building projects. The technologies used to maintain and beautify the capital city of Rome were particularly impressive: "The Rome of 100 A.D.," it has been suggested, "had better paved streets, sewage disposal, water supply, and fire protection than the capitals of civilized Europe in 1800."[51] Like the creation of large armies, these were projects that stimulated accumulation by encouraging trade and generating demand. Powerful states spent freely on large prestige projects, including cities such as the Achaemenid capital, Persepolis. Such projects were designed to overawe subjects and rivals, but they also provided employment and attracted merchants and artisans. In the pursuit of managerial efficiency, states also did much to promote improved forms of literacy, though usually only among their own officials. Changes that may have improved the effectiveness of bureaucracies include the introduction of alphabetic scripts in the cities of Phoenicia ca. 1000 BCE, as well as improvements in mathematics and astronomy, which gave states greater control over calendars and accounting. Larger and more efficient bureaucracies were also necessary to handle the large standing armies that began to be used at least from the time of the Assyrian Empire. Finally, stable government coupled with fiscal moderation encouraged peasants to produce larger surpluses and encouraged merchants to trade more extensively.

But while tribute-taking states often stimulated accumulation, they could also undermine it, sometimes severely. Indeed, the basic structures of agrarian civilizations ensured that this would be so. Tribute-taking elites could not exist unless primary producers had access to the land, for that was where most surplus resources were produced. So, in most agrarian civilizations, most people had access to land in some form. This broad allocation of productive resources limited the steepness of gradients of wealth and inhibited the concentration of resources in the hands of elite groups. It meant that though *surplus* wealth could be concentrated in the hands of governments and elite groups, the land, the fundamental productive resource in all agrarian societies, could not. Whatever symbolic claims elites might make to the land, they had to leave most of it in the hands of the peasant farmers who normally worked it. This requirement limited their ability to manage and supervise agricultural production. It also explains why tributary states could survive with such rudimentary bureaucracies: they left the most basic production tasks almost entirely to the skills and labor of rural households.

As Marx pointed out, these relationships explain why tribute-taking elites had to extract resources in ways that often inhibited innovation and depressed productivity.[52] If peasants have enough land to support themselves, then they have little incentive to surrender the huge sums often demanded by elites. For precisely that reason, elites normally had to use the threat of force to extract surpluses. In the short or medium term, such threats, whether used to exact regular taxes or to exact new flows of wealth through conquest, were simply the most effective way of getting resources, because *real* growth in output usually occurred too slowly to interest rulers. This is why Moses Finley has argued, with only slight exaggeration, that "what passed for economic growth in antiquity was always achieved only by external expansion."[53] In such an environment, it took an unusually far-sighted or confident ruler to invest large amounts in projects that required several decades to raise productivity levels. Faced with immediate crises, even the most able rulers became brutal and destructive predators. Rulers who were less able, or who were more desperate, used destructive fiscal methods as a matter of course, even when they or their advisers knew that they were undermining the bases of their own power. In Muscovite history, the reign of Ivan the Terrible offers a horrifying example of the dangers of excessive predation. After his death, the powerful Muscovite empire, which had been constructed over several centuries, came close to collapse in a period of civil war, famine, invasion, and depopulation known as the "Time of Troubles." The collapse was caused largely by Ivan's overpredatory policies, which drove peasants, the productive foundation of all agrarian civilizations, either to ruin or to flight.

Important consequences follow from these basic structural features of agrarian civilizations. First, elites in tribute-taking societies had to be specialists in coercion and management rather than in production. By and large, tribute-taking elites despised productive work and those who engaged in it, an attitude that left most of them ignorant of the productive technologies on which their wealth was based. The official and the warrior (managers and coercers) provided the models of elite lifeways, rather than the artisan, the peasant, or the merchant. Tributary elites were content, for the most part, to skim off what they needed, and to focus on the military and fiscal skills necessary to keep on skimming. Normally, they had to be wealth takers rather than wealth generators, and thus statecraft took precedence over economic calculation.[54] Machiavelli's descriptions of the strategic and tactical rules of this world are valuable despite an element of caricature:

> A Prince, therefore, should have no other object or thought, nor acquire skill in anything, except war, its organization, and its discipline. The art of war is all that is expected of a ruler; and it is so useful that besides enabling hereditary princes to maintain their rule it frequently enables ordinary citizens to become rulers. . . . The first way to lose your state is

to neglect the art of war; the first way to win a state is to be skilled in the art of war.

In such a world, it made sense for elite males to train themselves mainly in the exercise of coercion, rather than in intellectual or commercial activity. Thus, time spent hunting or jousting was more useful than time spent in the countinghouse.

> [The Prince] should never let his thoughts stray from military exercises, which he should pursue more vigorously in peace than in war. These exercises can be both physical and mental. As for the first, besides keeping his men well organized and trained he should always be out hunting, so accustoming his body to hardships and also learning some practical geography: how the mountains slope, how the valleys open, how the plains spread out.[55]

Such attitudes guaranteed that among tributary elites, violence was deployed with an unashamed exuberance rare in the industrialized world today, for they recognized it as the main instrument of rule. Nizam al-Mulk, the vizier of the Seljuk sultans, quoted the Abbasid caliph Ma'mun as follows: "I have two commanders of the guard who are occupied from morning till night in cutting off people's heads, hanging people, chopping off hands and feet, giving the bastinado and putting men in prison." And a twelfth-century French writer describes the joy of battle: "I tell you that I never eat or sleep or drink so well as when I hear the cry, 'Up and at 'em!' from both sides, and when I hear the neighing of riderless horses in the brush and hear shouts of 'Help! Help!' and see men fall . . . and the dead pierced in the side by gaily-pennoned spears."[56]

In some circumstances, elites operated at one remove from the exercise of violence, specializing instead in the management of coercion. In imperial China—where, from the time of the Qin dynasty (221–207 BCE), which founded the first unified imperial state, a large bureaucracy supervised armies and tax collectors—administrative and legal forms of coercion often won more prestige than physical coercion, and the ambitious spent more time studying than hunting. But what they studied was control, not farming or commerce.

At the same time, peasants (the primary producers) generally had little incentive to raise their productivity as long as they could survive, for increases in output could all too easily be skimmed off by their overlords. Stable and long-lived polities such as that of China thrived in part because they were rich enough and durable enough to maintain predictable and relatively light levels of taxation, which gave peasants a greater stake in productivity-raising innovations.[57] But even the peasants of less predatory states had limited incentives to innovate. Normally, they lacked the financial resources, the ability to take risks, and the training necessary to experiment with new technologies.

All in all, as Joel Mokyr has argued, technological innovation is unlikely to happen quickly where those who work lack wealth, education, and prestige, and those who are wealthy, educated, and have prestige know nothing about productive work. In agrarian civilizations, tributary elites had significant control over networks of informational exchange, and their hostility to technological thought must have significantly slowed the circulation of innovations in productive techniques.[58] To complete this vicious circle, slow growth rates themselves inhibited investment, for they meant that returns on investments could be expected only in the remote future, and few traditional rulers operated within such time frames. In a world of sluggish growth (by modern standards), investing in growth was far too slow a method of increasing revenues: conquest was normally a more promising gambit. In all these ways, the social and economic structures on which tributary states built their power slowed innovation in productive technologies.

Exchange, Commerce, and Urbanization   Yet another motor of innovation in the era of agrarian civilizations was commercial exchange. Those who specialized in such commerce had to be expert in the manipulation of consensual systems of exchange even if they were usually willing to use force when they could get away with it. But force normally played a lesser role, either because commercial exchanges occurred beyond the reach of coercive forms of power or because they concerned goods that did not interest those with the power to coerce. Because efficiency as well as consent is normally more important than force in commercial exchanges, it has generally been assumed that commerce is more likely than tribute-taking to generate efficiency-raising innovations.

Though tribute-taking dominated exchanges within most states before the modern era, rulers had less power to control resources beyond their borders. So, except where backed up by invading armies, international exchanges were often more consensual. As a result, consensual commercial exchanges could normally reach further than exchanges controlled by tributary rulers. The growth of populations, the spread of agrarian civilizations, and improvements in means of communication all tended, over long periods, to increase the volume and extent of long-range commercial exchanges. These, in turn, hastened the spread of new productive military and managerial techniques and of new products within expanding world systems, because merchants usually had good reason to seek out innovations that might give them a commercial advantage. (It was pressures such as these that led states and merchants in western Eurasia to successfully pursue the secrets of silk making in the first millennium CE.) But commercial exchanges also developed a synergy of their own, for innovations from different regions often formed new and even more fruitful combinations in their new home, thereby enhancing the scale and extent of their impact.[59] One spectacular example is the military synergy created by the introduction of horse riding from the steppes into agrarian civilizations, a process that revolutionized warfare. On the other hand, we will see that interregional exchanges could also retard processes of accumulation in the era of agrarian civilizations, mainly through their indirect impact on disease patterns; and at the same time, interregional contacts were often stifled by the greed of excessively predatory states. So commerce, too, could stimulate change, but not as powerfully as it does in the modern world.

Most tributary elites disdained the noncoercive exchanges of commerce and those who engaged in them. Their disdain is apparent in the official values of most imperial systems: in the Confucian system of values, in the Indian caste system, in Roman attitudes toward merchants, and, in general, in the low status accorded to merchants in most agrarian states. Nevertheless, in the very long run, systems of exchange expanded throughout the era of agrarian civilization; as they did, so did the volume of wealth handled by merchants and entrepreneurs, and so, eventually, did the influence of these groups.

As networks of exchange reached further, and long-distance exchanges became more frequent, hub regions gained increasing strategic importance as more information and wealth flowed through them. Because cities depend so much on commerce, urbanization provides a good indirect measure of these trends. The history of urbanization in Eurasia matches patterns we have already observed (see table 10.6). Here, too, a critical threshold seems to have been crossed in the first millennium BCE.[60] In the third millennium BCE, there were perhaps eight cities with at least 30,000 inhabitants. All were in the Afro-Eurasian hub region of Mesopotamia or Egypt, and their combined population was about 240,000. By 1200 BCE, there may have been sixteen cities of this size, with a combined population of half a million, but these were now scattered throughout the eastern Mediterranean, northern India, and China. In 650 BCE, there were still only twenty cities this size, with a combined population still under a million. But by 430 BCE their numbers exceeded fifty, by 100 CE over seventy, and their total populations were, respectively, 2.9 million and 5.2 million. The demographic downturn of the

TABLE 10.6. LONG-TERM TRENDS IN URBANIZATION IN AFRO-EURASIA

Date	No. of Largest Cities	Size of Largest Cities	Total Population in Largest Cities
2250 BCE	8	ca. 30,000	240,000
1600	13	24,000–100,000	459,000
1200	16	24,000–50,000	499,000
650	20	30,000–120,000	894,000
430	51	30,000–200,000	2,877,000
100 CE	75	30,000–450,000	5,181,000
500	47	40,000–400,000	3,892,000
800	56	40,000–700,000	5,237,000
1000	70	40,000–450,000	5,629,000
1300	75	40,000–432,000	6,224,000
1500	75	45,000–672,000	7,454,000

SOURCES: Stephen K. Sanderson, "Expanding World Commercialization: The Link between World Systems and Civilizations," in *Civilizations and World Systems: Studying World-Historical Change*, ed. Stephen K. Sanderson (Walnut Creek, Calif.: Altamira Press, 1995), p. 267; based on Tertius Chandler, *Four Thousand Years of Urban Growth: An Historical Census* (Lewiston, N.Y.: St. David's University Press, 1987), pp. 460–78.

first millennium CE meant that this was the high-water mark for urbanization before the second millennium. In 1000 CE, there were no more people and no more towns than in 1 CE.

Directly or indirectly, urbanization and state activity encouraged trade at all ranges in the era of agrarian civilizations. The earliest governments in Mesopotamia, Egypt, and China took an active part in organizing and managing exchanges both in essential and in luxury goods. By the middle of the third millennium, governments and temples traded extensively, used gold and silver to keep accounts, and even acted sometimes as lenders (at interest) or bankers. In general, market exchanges flourished where tributary methods didn't work effectively.[61] Thus, states had to trade where highly valued or strategic goods were beyond the reach of their armies. In those cases, William McNeill points out, "Rulers and men of power had to learn to deal with possessors of such commodities more or less as equals, substituting the manners and methods of diplomacy for those of command."[62] But military expansion, for all its brutality, could also provide a particularly powerful stimulus for commercial and intellectual exchanges. For example, the conquests of the Achaemenid and Hellenistic dynasties encouraged commercial and intellectual exchanges reaching from central Asia to India to the western Mediterranean. In the East, the expansion under the Han and Tang dynasties had similar catalytic impacts within China. The intellectual residue left by these exchanges shaped the cultural traditions of the Persian, Indian, Chinese, and Mediterranean worlds.[63]

In local trade networks, too small to interest tributary overlords, peddlers or market traders or peasants were usually better than government bureaucracies at dealing with petty exchanges. Thus competitive markets existed even in the earliest agrarian civilizations. And so did merchants, even though, in the earliest states, they often operated in close association with governments and had something of the position and status of high government officials.[64] By the early second millennium BCE, records from cities such as Ebla and Mari show the existence of independent trading firms in Mesopotamia, though their trade was probably supervised or licensed by governments.[65]

By the second millennium, commerce was vigorous enough in some regions to provide a more important economic basis for small city-states than did tribute-taking. An early example of this kind of state may have been Ebla in northern Syria, which flourished in the era of Sargon. Ebla's remarkable cuneiform texts, which contain detailed accounts of trade and of state involvement in trade, were found only in 1974. The records of the trading city-state of Kanesh (modern Kül-tepe) in central Anatolia also offer detailed evidence on markets, prices, and credit systems early in the second millennium BCE.[66] A more familiar example from the first half of the second millennium is the Minoan state, based on the island of Crete, which dominated trade networks linking agrarian regions throughout the Mediterranean. In central Asia, a flourishing system of trading cities emerged early in the second millennium BCE in the so-called Oxus civilization. Later examples of such trade-based polities include the Mycenaean Greeks who inherited Minoan trade networks from ca. 1400 BCE; the Phoenician cities, such as Tyre and Sidon in modern Lebanon; and Greeks of the archaic period, early in the first millennium. They also include the many trading city-states of East Africa, the coast of India, and Southeast Asia. The trading city-states of the Mediterranean world established networks of colonies, the Greeks mainly along the northern shores of the inland sea and the Phoenicians mainly along the southern shores. The most important of the Phoenician colonies was Carthage (in modern-day Tunisia), founded by Tyre in 814 BCE.

Afro-Eurasian trade networks expanded rapidly after 1000 BCE. In western Eurasia, commodities such as silver had functioned loosely as forms of money since the third millennium, while in China, cowrie shells and cloth had played a similar role from at least the middle of the second millennium.[67] But proper coinage, with its value stamped on it, first appeared in the middle of the first millennium BCE. Coins were circulating in Anatolia in the seventh century, and in northern China perhaps equally early; by the fourth century BCE, coins were in use in all the major regions of agrarian civilization in Eurasia. This innovation greatly simplified commercial exchanges. Equally decisive was the appearance, also from the middle of the first millennium, of vigorous commercial exchanges between eastern, southern, and western Eurasia, both by land and by sea.[68] Initially, states played a vital role in stimulating trade along these routes and in protecting their own merchants. Such action is particularly clear in the case of Han China, whose government expanded toward central Asia, at great expense, under the emperor Han Wudi late in the second century BCE. However, the actual task of carrying goods was normally left to merchants, who transferred goods—often with the help and protection of pastoral nomads, or local naval powers—in relays from one end of Afro-Eurasia to the other.

As any elementary economics textbook will explain, commerce, conducted in competitive markets by actors free to profit from their transactions, can provide a powerful stimulus to innovation. In a competitive environment, where naked coercion is ruled out, cost cutting is often the most effective way of competing with rivals; thus merchants generally have good reasons to keep prices low by operating at maximum efficiency. And they often have the knowledge to lower prices, because their wide contacts can alert them to new and more efficient ways of doing things. It follows from these general rules that commercial activity is most likely to stimulate cost-cutting innovation and raise productivity levels in those areas where markets really are competitive and where commercial activity is relatively free from the control of tributary elites, who were often more interested in taking than in generating wealth.

Within agrarian civilizations, two types of regions stand out in this respect. In peripheral regions or regions of agrarian civilization where fiscal pressure was lightest, peasants stood to benefit from productivity-raising innovations because they could often keep the surpluses they generated. In the "barbarian" lands of northern Europe in the classical era, rural producers had greater independence than within the Roman Empire and often found it worthwhile to experiment. Indeed, many new techniques appeared first within these communities. For example, Roman writers credited the Celts with "the invention of enameling, the spoked wheel, soap, improved agricultural implements, and advanced ironworking techniques."[69] In eastern Europe, the introduction of rye allowed the eventual settlement of the lands between eastern Europe and the Urals, which were poorly suited to the traditional crops of the Mediterranean or western Europe. Communities of farmers, such as the Goths, sometimes found it profitable to engage in a combination of pillage and trade, particularly along the borders of a decaying Roman empire. But even peasants within agrarian states were likely to take more interest in raising productivity when they had secure access to land and were not taxed too heavily. The surprisingly high yields of peasant farming in China in the centuries before the modern era almost certainly had something to do with the fact that tax levels were usually modest (because Chinese governments did not usually spend as much on warfare as did contemporary European states) and that the proportion of peasants who owned their land was high.[70]

The second area in which the productivity-raising potential of commerce tended to increase consisted of regions of small tributary states near the hubs of regional trade systems. Because they were small, such states had access to limited tributary revenues. But if the states were situated near imperial systems with flourishing networks of trade, their rulers could collaborate with local merchants to secure additional revenues from commerce. In such regions, markets were more likely to be genuinely competitive, because smaller states had less opportunities than the great tributary juggernauts to support themselves purely from tributary revenues. Indeed, in the era of agrarian civilizations, the real dynamos of innovation were often regions of petty states or city-states lying near the hubs of regional networks of exchange. If, in addition, such states existed in regions of intense interstate rivalry, the pressures to seek a mixture of commercial and tributary revenues were particularly strong. And those that succeeded commercially sometimes managed to tap into huge flows of wealth and information. By doing so, small states, such as classical Athens or early modern Genoa or Venice, sometimes became major powers despite their limited internal resources.

As Anthony Giddens has pointed out, in the era of agrarian civilizations, small, commercialized states often appeared not as isolated entities but in entire systems, which were usually highly competitive, and within which merchants generally had higher status than in larger tributary polities.[71] Such regions were particularly likely to pioneer innovations, especially in commercial methods, in transportation, and in warfare. It is no wonder that the earliest alphabetical scripts are Phoenician, or that modern mathematics and the classical military phalanx owe so much to the Greek city-states, or that the trading cities of Islamic central Asia did so much to preserve the technical and scientific knowledge of the classical world, or that modern commercial techniques are so indebted to the city-states of Renaissance Italy.[72]

Urbanization and commercialization encouraged accumulation of many different kinds: accumulation of wealth, of ideas, of new technologies and methods of doing things. But increasing commercial activity, like the state, could also undercut growth, and it did so primarily by affecting patterns of disease. Most precapitalist cities were unhealthy places to live in; their filth and congestion provided such benign environments for disease bacteria that city dwellers normally had lower life expectancies than village dwellers. Until the twentieth century, cities were the social equivalent of galactic black holes, sucking in and destroying surplus populations from their hinterlands. Thus urbanization itself dampened population growth, and it did so most decisively when cities grew fastest. As we have seen, expanding trade networks could have a similar effect, but on a much larger scale, by encouraging the exchange of diseases. For this reason, we cannot take the spread of cities and of trade as unambiguous measures of growth, though they are often indicators of increasing innovation.

Finally, the chances of commercial activity stimulating innovation were stifled throughout the era of agrarian civilizations by the fiscal methods of powerful tributary states. Though tributary states normally tolerated and sometimes encouraged commerce, their predatory methods and willingness to resort to force were ever-present threats to the freedoms needed for trade to flourish. There was therefore a fundamental long-term conflict between the methods of tribute takers and those of merchants; and as long as tributary elites dominated political systems, this conflict limited the productivity-raising potential of commercial activity.

Rates of Innovation

We have seen that population growth, increasing state power, and increasing commercialization all stimulated innovations and growth in the era of agrarian civilizations. But each of these factors could also inhibit accumulation. This contradictory pattern may help explain some important general features of the era of agrarian civilizations. First, despite the existence of new sources of innovation, long-term rates of population growth were not strikingly different from those of the early agrarian era. The negative impact of overpredatory tributary states and of new disease patterns counterbalanced the more positive influences of population growth, increasing state power, and commercial expansion. Second, throughout this era the rate of innovation was sluggish. There were innovations in many areas, of course, from bureaucratic management, to literacy, to warfare, to communications and metallurgy. Furthermore, increasing commerce ensured that technologies such as bronze- or ironworking or horse riding or chariot warfare would spread widely throughout Afro-Eurasia. Nevertheless, what is striking over the entire 4,000 years is how limited innovations were, particularly in *productive* technologies—in methods of farming and manufacturing. Finally, it is precisely this pattern of sluggish growth that explains the basic Malthusian rhythm of rise and fall that appears to have been characteristic of all agrarian civilizations. On the whole, in this era of human history, fundamentally new technologies contributed less to accumulation than did the gradual spread of small improvements to already-existing technologies, such as those of the secondary products revolution, which had been pioneered during the early agrarian era.

SUMMARY

Agrarian civilizations have dominated the stories told in modern historiography. From the moment of their appearance, about 5,000 years ago, they slowly expanded and became more powerful. Though they were never alone—they shared the world with many other types of communities not organized in states—they eventually became the most populous and most powerful social organizations in the world. Their power grew as new states appeared, as managerial techniques improved, and as the area controlled by states increased. Their size, together with the extent and vigor of exchanges within and between agrarian civilizations, ensured that innovation continued throughout this period. The main motors of innovation were population growth, state activity, and increasing commercial activity and urbanization. But each of these mechanisms could also slow innovation. As regional populations came into contact with each other, they swapped diseases in exchanges that sometimes led to catastrophic epidemics that undermined state power and led to regional declines. Tribute-taking states depended primarily on the coercive extraction of resources, and they were often ambivalent or hostile in their treatment of entrepreneurial activity. Yet entrepreneurial activity was itself one of the most important motors of innovation. Cities, too, were vital clearinghouses for informational and commercial exchanges, but their unhealthy environments slowed population growth and spread disease. In all these ways, the activities of agrarian civilizations stimulated but also slowed innovation. These contradictory influences had a contradictory outcome: though the history of the era of agrarian civilizations is characterized by innovations of many kinds, nowhere was innovation sufficient to keep up with the pace of population growth. And this is why the rhythms of historical change in this period were dominated by Malthusian cycles—long periods of demographic, commercial, and economic growth followed by periods of decline before growth began once again.

FURTHER READING

The era of agrarian civilizations has been the subject of a vast amount of scholarship, but surprisingly little of that scholarship has focused on the large trends. Works that do, include the remarkable articles of Rein Taagepera and the collections within the world systems tradition by Stephen K. Sanderson, Andre Gunder Frank and Barry Gills, and Christopher Chase-Dunn and Thomas Hall listed in the bibliography. They also include the many fine texts on world history available today, many of which focus on the period from ca. 3000 BCE to 1500 CE. Some of the best are Jerry Bentley and Herbert Ziegler, *Traditions and Encounters* (2 vols.; 2nd ed., 2003); Richard Bulliet et al., *The Earth and Its Peoples* (1997); and Howard Spodek, *The World's History* (2nd ed., 2001). There are also good surveys in William McNeill's classic study, *The Rise of the West* (1963), and in volumes 3 and 4 of the series of histories edited by Göran Burenhult, *The Illustrated History of Humankind* (1994). Michael Mann's *The Sources of Social Power* (1986) surveys the history of state power, while Tertius Chandler's *Four Thousand Years of Urban Growth* (1987) surveys urbanization. Finally, historical sociologists, such as Anthony Giddens (see especially *A Contemporary Critique of Historical Materialism* [2nd ed., 1995] and its second volume, *The Nation-State and Violence* [1985]), and Michael Mann, have taken up some of the themes discussed in this chapter.

PART V

THE MODERN ERA
One World

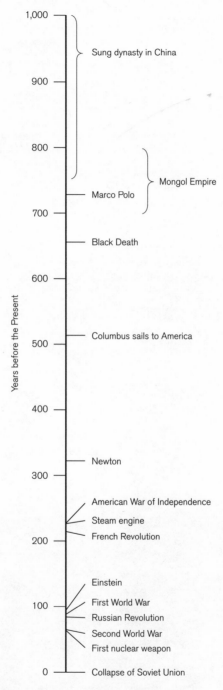

Timeline 11.1. The scale of modernity: 1,000 years.

11

APPROACHING MODERNITY

In the past thousand years, and particularly in the past two or three hundred years, a transformation more rapid and more fundamental than any other in human history has taken place. A new threshold was crossed, leading to a fundamentally new type of society. Anthony Giddens writes, "Over a period of, at most, no more than three hundred years, the rapidity, drama and reach of change have been incomparably greater than any previous historical transitions. The social order . . . initiated by the advent of modernity is not just an accentuation of previous trends of development. In a number of specifiable and quite fundamental respects, it is something new."[1] The change is not just important for humans; it is an event of planetary significance because the impact of humans on the biosphere has now taken on entirely new dimensions.[2]

Because we live in the middle of this transformation, it is hard to see its features clearly and objectively. So, in describing it, I will settle for a deliberately vague label: "the Modern Revolution."

THE WORLD ON THE EVE OF MODERNITY

In order to grasp the scale and significance of the Modern Revolution, it may help to begin with an imaginary tour of the world on the eve of modernity, in the early centuries of the second millennium CE.

In *Europe and the People without History* (1982), Eric Wolf takes his reader on a tour of the world in 1400.[3] This overview reminds us how much of the world was *not* yet incorporated in regions of agrarian civilizations even at that late date. Though they had steadily encroached on the lands of independent farmers, pastoralists, and even foragers, in 1000 CE agrarian civilizations still controlled less than 15 percent of the lands ruled by modern states. We therefore must not project back into the agrarian era the devastating impact that modern states have had on stateless communities in the past 500 years. In practice, stateless communities, including the peasant farmers of northern Europe or Manchuria, or the pastoralists of Mongolia or the Scythian steppes, could still pose severe military challenges to the mightiest of agrarian empires. At the same time, relations between different types of communities were shaped as much by exchanges as by conflict. Pastoral nomads exchanged horses and hides for city-produced silks or wines; Siberian foragers exchanged walrus tusks or furs for metal goods; and horticulturalists in the jungles of Central America or tropical Africa traded gold, feathers, jaguar skins, and slaves for city-produced goods of various kinds. Conversely, states from China to Rome needed the horses and the hired soldiers of the steppes; their merchants traded with and through the steppes, and with the forest zones beyond them. In the Americas, too, cities had to trade with or through regions controlled by communities without state structures, along trade routes that linked them with remote jungle communities.

Analytical categories encourage us to think of each lifeway as a world of its own, but as Wolf insists, this was never true: "Everywhere in this world of 1400, populations existed in interconnections. Groups that defined themselves as culturally distinct were linked by kinship or ceremonial allegiance; states expanded, incorporating other peoples into more encompassing political structures; elite groups succeeded one another, seizing control of agricultural populations and establishing new political and symbolic orders."[4]

The elites of agrarian civilizations normally regarded those who lived beyond their borders (and many who lived within them) as "barbarians." The barbarian communities included foragers, pastoralists, horticulturalists, and small-scale farmers, who often used seminomadic forms of swidden agriculture and still hunted and gathered some of their produce. Working within the networks linking these worlds were traders of various kinds—some ruthless and predatory, others more consensual in their methods. Most people still lived in small communities. Here, kinship was more important than state power. This was true even of the villagers who made up most of the populations and produced most of the resources of agrarian civilizations. Of course, villagers could not ignore the oppressive burden of landlords or tax gatherers, or the passage of armies that led so often to death, disease, or enslavement. But for most households, most of the time, the local communities of family, kin, and neighborhood were what counted.

In a vast frontierland reaching away from the regions of agrarian civilizations, there lived communities of farmers organized in villages, often under kin-based leaders. Some of these communities were on the verge of statehood. Much of the Amazon basin was populated by small communities of horticulturalists who also engaged in hunting and gathering. In North America, along the Mississippi River, farmers lived in dense communities structured almost like states. Some sites of the Mississippian culture, such as Cahokia, near St. Louis, may have contained more than 30,000 people. Cahokia was a huge political and ceremonial center, with some 100 earthen mounds. Elements of the Mississippian culture survived until the sixteenth century, though the major sites such as Cahokia had declined long before; and surviving communities were decimated by Eurasian diseases when first contacted by Europeans. But we have one recorded eyewitness account from a French explorer, Le Page du Pratz, who lived briefly among the Natchez tribes of the Mississippi valley. As Brian Fagan summarizes, "He found himself in a rigidly stratified society—divided into nobles and commoners and headed by a chieftain known as the Great Sun—whose members lived in a village of nine houses and a temple built on the summit of an earthen mound. Pratz witnessed the funeral of the Great Sun. His wives, relatives, and servants were drugged, then clubbed to accompany him in death."[5]

Large communities of farmers could be found in much of western and central Africa as well. In some regions, such as those occupied by modern Zimbabwe, or the lands north of modern Ghana, high population densities and extensive trade networks supported state systems from the middle of the first millennium CE, or perhaps even earlier. The states of West Africa depended largely on their control of trade networks specializing in gold that crossed the Sahara and either reached the Mediterranean shores near modern Morocco or reached Egypt and the Islamic world. The states that emerged in central and East Africa traded with coastal cities whose Muslim merchants carried their goods (above all, gold and slaves) to the Islamic world and to South and Southeast Asia. In the fourteenth century, the Chinese fleets that sailed under the Muslim eunuch Zheng He reached the eastern shores of Africa. But even these expeditions were novel only insofar as they cut out the middlemen in ancient trading networks. There were African slaves in China from at least the seventh century CE, and, Wolf reports, "by 1119 most of the wealthy people of Canton were said to have possessed Black slaves."[6]

Northern Europe also supplied slaves to neighboring agrarian civilizations, and until late in the first millennium CE, much of Europe remained a world of stateless farmers. Such regions, though lacking the large standing armies of agrarian empires, could prove dangerous to their "civilized" neighbors. This was particularly true where the wealth of neighboring agrarian civilizations prompted emulation and attempts at conquest. Gothic raiders founded a series of dynasties in the remains of the Roman Empire in the fifth and sixth centuries CE, while Manchurian dynasties created several states in North China from as early as the fourth century CE—including the last dynasty of premodern China, the Qing or "Manchu" (r. 1644–1911). Such conflicts often prompted the spread of state structures beyond the established frontiers of agrarian civilizations. In the middle of the first millennium CE, states began to appear throughout northern Europe. In eastern Europe, agricultural populations expanded rapidly and migrated into what is today Ukraine and Russia; thus, by the end of the millennium, states had also appeared in much of eastern Europe.

In the New World, too, agrarian civilizations were often threatened by neighboring "barbarians." In Mesoamerica, several great cities, including Teotihuacán and Tula, suffered devastating raids from communities farther north, with whom they already had links of culture and trade. The career of the Aztecs parallels that of the Goths. Originally known as the Mexica, the ancestors of the Aztecs came from horticulturalist or foraging communities north of the Valley of Mexico whose worlds were influenced in many ways by the cultural heritage of central Mexico. From their homelands, the Aztecs moved into the Valley of Mexico, where they survived in marginal lands between the region's major city-states. In the fourteenth century, they began to hire themselves as mercenary soldiers, until in 1428 they overthrew their masters and created a dynasty of their own.[7] Large regions of stateless farming communities also flourished in much of Southeast Asia and on the borders of an expanding China. The most isolated communities of this kind could be found in the islands of Melanesia and Polynesia.

In Afro-Eurasia, there was another important type of frontier: between farming regions and regions of pastoralism. Pastoralists survived in lands too arid to support dense farming populations. These lands stretched from Mongolia through the steppes of Inner Eurasia and Iran, through to Mesopotamia and the Sahara, and south into East Africa.[8] Pastoralism based mainly on horses, goats, sheep, and camels was the most widespread lifeway throughout the arid steppelands and deserts of Eurasia. Camel pastoralism was particularly important in the heartland of Arabia and in the Sahara Desert. Much of central and eastern Africa was populated by large communities of cattle-herding pastoralists. Pastoralist communities normally consisted of kin-based groups organized in clans, tribes, and occasionally (particularly in times of widespread conflict) larger intertribal alliances. In times of peace, pastoralists traveled in small groups of a few households along established migration routes. They either erected tents at each new campsite or traveled in covered mobile homes. The Greek writer known as pseudo-Hippocrates described the wagons used by the Scythians north of the Black Sea more than 2,000 years ago: "The lighter wagons have four wheels but some have six, and they are fenced about with felt. They are built like houses, some with two divisions and some with three, and they are proof against rain, snow and wind. The wagons are drawn by two or three yokes of hornless oxen; hornless because of the cold. The women live in these wagons while the men ride on horseback, and they are followed by what herds they have, oxen and horses."[9]

Everywhere, pastoralists had a significant impact on neighboring communities because their limited productivity and great mobility encouraged them to trade with agrarian or forager neighbors, while their virtuosity as warriors meant that raiding was often a profitable alternative to trade. Their raids provoked similar strategies of counterinvasion and wall building from North China to central Asia to the Balkans.[10] The horse-riding pastoralists of the Inner Eurasian steppes created powerful military alliances from perhaps as early as the second millennium BCE. Because the steppes supported small populations, such alliances could turn into more durable structures only if they managed to extract large amounts of wealth from neighboring agrarian civilizations; the most powerful pastoralist armies thus appeared along trade routes or at the borders with agrarian neighbors. Some of these structures deserve the name of *states*, despite their differences from the states of the agrarian world. They were not products of pastoralism, agriculture, or trade but of a complex intertwining of these different lifeways.[11] The best-known pastoralist empire is that of Genghis Khan. Created in the thirteenth century, in campaigns of conquest more spectacular and more durable than those of Alexander the Great, the Mongol Empire eventually controlled all of the Inner Eurasian steppelands, most of Iran, and all of China. It was the first political system to touch all major regions of Eurasia.

The frontiers between agrarian civilizations and pastoralism may have been the most active and complex of all frontier zones. Here, perhaps more than anywhere else in the world, we can see the powerful intellectual synergies that could be generated when communities with different technologies and lifeways regularly exchanged ideas, goods, and people. And such exchanges made these frontiers a powerful motor of innovation throughout the Afro-Eurasian world zone. Through them were transferred new technologies, including those of horse riding, metallurgy, and warfare, as well as religious ideas from shamanism to Buddhism, Islam, and Christianity. They also transmitted diseases, genes, and languages. Indo-European languages expanded, probably from somewhere in modern Russia, to the borders of China, to India, to Mesopotamia, and to Europe, carried by pastoralist migrants. The armies of agrarian civilizations often contained contingents of cavalry from the steppes. And sometimes pastoralist leaders, from the Parthians to the Seljuks and Moguls, set up successful dynasties in the borderlands and then moved into the cited heartlands.

Smaller and less powerful kin-based communities of foragers could be found in much of Siberia, along the shores of the Arctic, in parts of Africa, in much of North America, in much of the southern half of South America and the Amazon basin, and throughout Australia. Their lifeways varied greatly, and no generalizations can do justice to that variety. A verbal photograph of a single group will have to do here.

The Khanty and Mansi lived in western Siberia, east of the Urals Mountains. They spoke languages distantly related to modern Finnish and Hungarian. In the seventeenth century, as Muscovite traders and soldiers entered their territory, they probably numbered about 16,000 people. (The population of Muscovy at this time was about 10 million, a powerful reminder of the huge demographic difference between foraging and agrarian communities.) According to the accounts of Muscovite travelers, the Khanty and Mansi lived mainly by hunting and fishing. But they also borrowed techniques from neighbors. Some southern clans cultivated barley and herded cattle and horses, while some northern clans engaged in reindeer herding, like their neighbors the Samoyed. Their outer garments were made from reindeer and elk skins or furs, though some clans also used feathers or fish skins. In the south, some even wove cloth from vegetable fibers. Most Khanty and Mansi lived in semipermanent winter camps; in the summer they moved to hunting and fishing grounds, where they lived in birch-bark tents. They traveled the extensive rivers of their homelands using birch-bark canoes in warm weather and skis in winter. Though small in numbers, the Muscovites found them formidable military opponents, for they used metal armor, longbows, and iron spears.

The following account of their lifeways was recorded by a Muscovite ambassador to China in 1675. As with all accounts by literate travelers from agrarian civilizations, we learn from it almost as much about the attitudes of the writer as about the society being described:

> All the Ostyaks [Khanty] catch great quantities of fish. Some eat it raw, others dry it and boil it, but they know neither salt nor bread, nothing but fish and a white root *susak*, of which they collect a supply in summer, dry it, and eat it in winter. Bread they cannot eat; or if any do eat their fill of it, they die. Their dwellings are yourts [i.e., yurts]; and they catch fish not merely for the sake of food, but to make themselves cloth-

ing out of the skins—also boots and hats, sewing them with sinews of the fish. They make use of the lightest possible boats, built out of wood, holding five or six men, and even more. They always carry with them bows and arrows, to be ready to fight at any moment. Wives they have in plenty—as many as they wish, so many do they keep.[12]

Like the Khanty and Mansi, many foraging communities had significant contacts with larger communities, with whom they exchanged technologies and goods of various kinds. Some systems of exchange were thousands of years old. These include the trades in arctic goods such as walrus tusks and precious furs that linked Siberian communities of foragers with communities of farmers or pastoralists to their west or south, and, indirectly, with the cities even farther south. In South America, the large agrarian populations of the western slopes of the Andes traded with stateless communities of the eastern slopes for prestige goods such as feathers, coca, and jaguar skins, or sought access to the gold of the Amazon basin through indirect systems of exchange that passed in relay through many different hands. Even some of the domesticated crops of western South America, such as sweet potato and peanuts, may have come from the tropical forests of the Amazon basin.[13] Such trades sometimes enabled local chiefs to build up more powerful political systems than would have been possible otherwise. The military confederations that formed in the northeast of North America and in southeastern Canada during the eighteenth century were built on the weapons and the liquor traded inward from Europe in return for furs. But though at first such exchanges might have appeared quite equal, they were dangerous for indigenous communities in the long run. Fur lured the Russian state deep into Siberia, and French and British traders deep into North America and Canada, with immense and tragic consequences for the many communities of foragers and horticulturalists with whom they traded.

Even the remotest communities often had some contact with agrarian communities, or practiced minimal forms of domestication. Along the northwestern shores of Australia, in recent centuries, communities were visited periodically by traders from Sulawesi whose crews brought trade goods including glass, pottery, tobacco, and metalwares in return for the prized trepang, or sea cucumber. These they sold in Southeast Asia and China as a gastronomic delicacy and an aphrodisiac.

In these and many other ways, communities of farmers, pastoralists, and foragers within and beyond the borders of agrarian civilizations helped shape each others' histories. But for most of the agrarian era, the balance of power between agrarian civilizations and other communities was much less uneven than it has been in the modern era. The ecological and cultural heterogeneity that could be found throughout the inhabited world in 1000 CE was one of the main casualties of the Modern Revolution.

THE MODERN REVOLUTION

Many features of the world described above had existed for millennia—yet most had vanished by 2000 CE. The world of the early twenty-first century is utterly different from the world of seven or eight hundred years ago. Indeed, the transformations caused by the Modern Revolution have been so all-embracing that it is hard to think of areas of life that they did *not* transform. What follows is little more than a checklist of some of the more important changes.

Population Growth

Population growth has accelerated sharply, as can be seen from a glance at figure 11.1 or table 11.1. In 1960, an attempt to calculate the mathematical tendency of global population in the past 2,000 years concluded that human populations would reach infinity on Friday, 13 November 2026.[14] This calculation (which came to be known as the "doomsday equation") is a reminder that such rates of growth cannot be sustained forever. In 1000 CE, the world's population stood at about 250 million. At the end of the twentieth century, it had multiplied twenty-four times, to reach 6 billion. Most of this growth occurred in the second half of the second millennium. In 1500, world populations stood at about 460 million; in 1800, they stood at 950 million, or just under 1 billion; and by 1900, they had reached just over 1.6 billion. In the 800 years before 1800, populations increased by about four times, while in the two centuries after 1800, they multiplied by six times. As a result, the doubling time for world populations declined sharply, particularly in the past two centuries (see table 6.3). As table 11.1 shows, in the past two centuries, populations grew throughout the world.

The bloom of human populations revealed in table 11.1 will probably peak in the twenty-first century. Even so, it is a phenomenon of planetary significance, for it has affected the entire biosphere. Humans have become, as Lynn Margulis and Dorion Sagan put it, "a sort of mammalian weed."[15] Carlo Cipolla comments: "A biologist, looking at the diagram showing the recent growth of world population in a long-range perspective, said that he had the impression of being in the presence of the growth curve of a microbe population in a body suddenly struck by some infectious disease. The 'bacillus' man is taking over the world."[16] One large species, our own, has

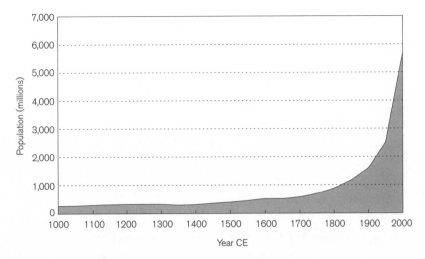

Figure 11.1.   Human populations, 1000 CE to now. Based on table 6.2.

acquired a quite unprecedented capacity to divert the planet's resources to its own uses. As we have seen, humans are currently co-opting at least a quarter of the energy that enters the biosphere through sunlight and photosynthesis (see p. 140). No wonder human population growth has been accompanied by a sharp decline in the number of other species.

Technological Virtuosity

Sustained population growth presupposes an increase in the resources available to feed, clothe, and sustain human populations. But growth this rapid requires more than just an increase in the available land; it also requires greater productivity, which implies increased rates of ecological and technological innovation. Thus rapid population growth has been accompanied by (indeed, made possible by) a dazzling display of technological virtuosity. In the past two centuries, innovation has ceased to be sporadic and occasional; it has become general and pervasive. And there is no sign that this burst of innovation is ending. On the contrary, in the late twentieth century the rate of innovation was faster than ever.

New technologies have affected demographic trends directly, by improving the quality of medical knowledge and medical care and thereby enabling babies and adults to live longer. But their indirect impact has been even greater, for they sharply raised the productivity of both agriculture and industry. Productivity in agriculture has crossed the decisive threshold

TABLE 11.1. WORLD POPULATIONS BY REGION, 400 BCE–2000 CE

Year	Region (population in millions)												
	China	India Subcontinent	S.W. Asia	Japan	Rest of Asia	Europe	USSR	N. Africa	Sub-Saharan Africa	N. America	C. & S. America	Oceania	World
400 BCE	19	30	42	1	3	19	13	10	7	1	7	1	153
300	30	42	47	1	3	22	13	12	8	1	7	1	187
200	40	55	52	1	4	25	14	14	9	2	8	1	225
100	55	50	50	1	4	28	13	14	10	2	9	1	237
0 CE	70	46	47	2	5	31	12	14	12	2	10	1	252
100	65	45	46	2	5	37	12	15	13	2	9	1	252
200	60	45	46	2	5	44	13	16	14	2	9	1	257
300	42	40	45	3	6	30	13	14	16	2	10	1	222
400	25	32	45	4	7	36	12	13	18	2	11	1	206
500	32	33	41	5	8	30	11	11	20	2	13	1	207
600	49	37	32	5	11	22	11	7	17	2	14	1	208
700	44	50	25	4	12	22	10	6	15	2	15	1	206

800	56	43	29	4	14	25	10	9	16	2	15	1	224
900	48	38	33	4	16	28	11	8	20	2	13	1	222
1000	56	40	33	4	19	30	13	9	30	2	16	1	253
1100	83	48	28	5	24	35	15	8	30	2	19	2	299
1200	124	69	27	7	31	49	17	8	40	3	23	2	400
1300	83	100	21	10	29	70	16	8	60	3	29	2	431
1400	70	74	19	9	29	52	13	8	60	3	36	2	375
1500	84	95	23	10	33	67	17	9	78	3	39	3	461
1600	110	145	30	11	42	89	22	9	104	3	10	3	578
1700	150	175	30	25	53	95	30	10	97	2	10	3	680
1800	330	180	28	25	68	146	49	10	92	5	19	2	954
1900	415	290	38	45	115	295	127	43	95	90	75	6	1,634
2000	1,262	1,327	181	127	680	514	290	151	659	313	516	30	6,057

SOURCES: J. R. Biraben, "Essai sur l'évolution du nombre des hommes," *Population* 34 (1979): 16; figures for 2000, using roughly comparable areas, are based on *World Development Indicators* (Washington, D.C.: World Bank, 2002), table 1.1, "Size of the Economy," pp. 18–20.

beyond which a minority on the land can support a majority off the land (see figure 9.3). In industrial production, changes have been even more spectacular. As David Landes puts it in an influential modern history of the Industrial Revolution:

> Improvements in productivity of the order of several thousand to one have been achieved in certain sectors—prime movers and spinning for example [e.g., compare horses with jumbo jets]. In other areas, gains have been less impressive only by comparison: of the order of hundreds to one in weaving, or iron smelting, or shoe-making. Some areas, to be sure, have seen relatively little change: it still takes about as much time to shave a man as it did in the eighteenth century.[17]

In textiles, which was perhaps the second most important sector of consumer goods production in the premodern world, traditional Indian hand spinners took about 50,000 hours to spin 100 pounds of cotton; the machinery invented in the eighteenth century in Britain lowered that figure to 300 hours by the 1790s, and it took a mere 135 hours by the 1830s.[18] New technologies have also transformed methods of communication and information exchange, making it possible for modern networks of exchange to work more rapidly, more efficiently, and over greater areas than ever before. Whereas eighteenth-century messages traveled, at best, at the speed of horse-riding couriers or sailing ships, today the telephone and the Internet allow instantaneous communication by millions of people to any part of the world (see tables 10.3 and 10.4).

Perhaps most important of all, new technologies have enabled humans, as a species, to cross a fundamental ecological threshold by granting them access to vast and previously untapped sources of energy, far greater than those supplied by plants, animals, and other humans. No longer do human societies have to rely mainly on human or animal muscles or firewood, wind, and water to supply their energy needs. Instead of depending on these sources of recently captured solar energy, humans have begun to exploit the vast stores of ancient sunlight accumulated in coal, oil, and natural gas, which is why it makes sense to talk of a "fossil fuel revolution." Learning how to use coal and oil to generate steam power or electricity has been the equivalent of finding several whole new continents for human exploitation. As Anthony Wrigley has argued, in Britain alone the energy extracted from coal ca. 1820 was equivalent to the amount of energy that could have been extracted using traditional technologies from an area of forest greater than all of Britain's pasture and arable land taken together.[19] Roughly speaking, the amounts of energy used by human societies increased by about five times in the nineteenth century, and then by another sixteen times in the twentieth century. Even per person, the amounts of energy used increased by some four to five times in the twentieth century.[20] John McNeill suggests that "we have probably deployed more energy since 1900 than in all of human history before 1900"[21] (see table 6.1). All in all, the fossil fuels revolution offered an astonishing bonanza, multiplying by perhaps 100 times the total energy available to humans and making possible projects such as the transportation of grains halfway around the world—projects that were unthinkable earlier, because the needed technologies were unavailable and the energy costs would have been prohibitive. For a time, at least in the more industrialized countries, energy seemed more or less free. In this sense, the Modern Revolution was similar to other episodes in human history when a new resource has become available in such abundance that for a while it appears infinite. Just as land, game, and other resources undoubtedly seemed limitless when humans first entered the Americas or Australia or New Zealand, as did water when humans first began using irrigation on a large scale, or land and other resources when Europeans reentered the Americas and Australasia from the sixteenth century on, so too it was tempting in the era of steam, coal, and oil to think of fossil fuels as limitless and, effectively, free. In later as in earlier periods, the discovery of vast new resources often encouraged dangerously short-sighted methods of exploitation.

Increased Political and Military Power

Linked to these demographic and technological changes have been profound changes in social, political, and military structures. The sheer volume of resources produced by modern economies, together with the degree to which they are concentrated in the hands of minorities, means that while modern states dispose of much greater resources than premodern states, they also have to defend steeper gradients of wealth and face vastly more complex organizational challenges. Like dams, their size, strength, and complexity have to be commensurate to the volume of resources backed up behind them. Since the French Revolution, states throughout the world have acquired the ability to regulate the daily lives of their subjects in ways that were unthinkable in all earlier epochs. Indeed, the ability of modern states to keep their subjects within a tight mesh of legal and administrative rules helps explain why they resort less often to the terroristic methods of rule normal in the era of agrarian civilizations. But in addition to these new powers, modern states can inflict violence on a scale that is also unprecedented, for the productivity of weaponry has also increased rapidly—so rapidly that humans now have the power, if they choose, to destroy themselves and much of the biosphere within a few hours.

Transformed Lifeways

Personal lives have been transformed. In the late agrarian era, most house-holds lived in the countryside and engaged in small-scale farming. Today, small farming has vanished in many regions and is declining where it still survives. The small number of surviving foragers now live at the sufferance of states, often on marginal land; sooner or later they are all incorporated into modern economic and legal networks that undermine traditional cul-tural and economic structures. Pastoralism, too, has become marginal. Within just a few centuries, the Modern Revolution has destroyed or pushed to the margins lifeways that had flourished for thousands of years.

Instead of living on the land and producing their own food, which is what *work* meant for most humans throughout most of history, typical modern households live in urban environments where they earn incomes through some form of wage work and buy food produced by others. In the more in-dustrialized economies, ca. 65 percent of populations lived in towns in 1980, and globally, ca. 38 percent; it is probable that even global levels of urban-ization will cross the symbolic threshold of 50 percent early in the twenty-first century.[22] In towns, the household remains a basic unit of consump-tion, but it has ceased to be a fundamental unit of production and the basic structure within which people are socialized. The mesh of kinship has been supplanted by the mesh of state regulation. In addition, new forms of con-traception, new methods of child rearing, and new forms of education and public welfare have provoked a fundamental renegotiation of gender roles.

The meaning and texture of life have changed. In prosperous regions, bet-ter medical care has postponed death. Average life expectancies in the more affluent societies of the late twentieth century were perhaps double those typical in the more prosperous agrarian societies, and perhaps triple those of Stone Age societies. In the year 2000, a child born in Burkina Faso could expect to live for 44 to 45 years, a child born in India for 62 to 63 years, and a child born in the United States for 74 to 80 years (see table 14.4). In more affluent societies, moderns have access to a level of material wealth unimag-inable in all earlier societies. On the other hand, by many standards mod-erns work harder than the peasants and foragers of earlier societies. And with the rise of a modern sense of clock time, they work increasingly to rhythms that are not their own.[23] Furthermore, it is less clear what they are working for. Whereas those living in self-sufficient farming households or foraging bands knew perfectly well what the "meaning" of work was, be-cause it was directly concerned with subsistence, the link is less direct for the highly specialized workers of modern businesses and corporations. For better or worse, the decline in kinship networks and traditional social roles has deprived people of the sharply defined identities that gave them a sense of purpose and place in many traditional societies. Huge movements of people, whether through the slave trade, mass migration, or forced displacement, have also deprived many of the sense of community known by their parents and grandparents.

In the most industrialized countries, personal relations are, on the whole, less violent today. In England, for example, modern murder rates are about $\frac{1}{10}$ those of 800 years ago, and $\frac{1}{2}$ those of 300 years ago. This drop occurred because most modern states have disarmed their populations and assumed a monopoly on the use of violence. Charles Tilly notes, "Disarmament of the civilian population took place in many small steps: general seizures of weapons at the ends of rebellions, prohibitions of duels, controls over the production of weapons, introduction of licensing for private arms, restrictions on public displays of armed force."[24] But though they are on the whole less violent, personal relations in modern urban communities also lack the intimacy and continuity of those in most traditional societies. Increasingly, they are casual, anonymous, and fleeting. These changes may help explain the loss of a clear sense of values and meaning in modern lives, a subtle and disorienting alteration in the quality of modern life that the French sociologist Émile Durkheim referred to in the late nineteenth century as "anomie."

The German sociologist Norbert Elias has argued that these changes have reached deep within our psyches, as modern forms of work and time discipline, enforced through the market, have shaped behavior in interpersonal relations, table manners, and attitudes toward sexuality. He has shown how the "emotional economy" typical in the modern world arises out of a relaxation of external restraints combined with an intensification of internal restraints: "The compulsions arising directly from the threat of weapons and physical force gradually diminish, and . . . those forms of dependency which lead to the regulation of the affects [feelings or emotions] in the form of self-control, gradually increase."[25] The internalization of new forms of discipline seems to be closely linked to new perceptions of time. As populations have grown, and as an ever greater proportion of people live in towns, the scheduling of daily activities has been geared increasingly to the activities of other humans rather than to the natural schedules of our own bodies, the seasons, and day and night. The growing influence of modern calendars and clocks and the emergence of conventions such as the international date line and local time zones based on Greenwich mean time (established in 1884) are the best index of these changes, for calendars and clocks provide a precise measure of social rather than ecological or psychic time. They therefore measure the extent to which humans had to adapt their behaviors and attitudes to a social rather than a natural ecology—to an ecology whose main elements were created by other human beings. The Modern Revolution has also given consumers access to a much broader range of mind-altering substances, in what David Courtwright has called the "Psychoactive Revolution."[26] These substances, ranging from opiates to coffee, tea, and sugar, have helped millions to cope with the sometimes harsh pressures and disciplines of modern life.

New Modes of Thought

The scientific modes of thought characteristic of modern society have generated both confidence and widespread alienation. Modern science has given humans unprecedented power over the natural world. But its universe is dominated by inanimate forces, a very different place from the rich world of spirits in which most people lived before the modern era. The ancient gods have been expelled, and the world of modern science is controlled by impersonal scientific laws. Gravity and the second law of thermodynamics now rule in the place of gods and demons. Scientific knowledge also lacks the specificity and sense of place of most premodern systems of knowledge, as it attempts to construct generalizations that work in all societies and all eras.[27] Such a system of knowledge cannot offer the consolations or the moral guidance of traditional religions, even if it is much better at helping us to manipulate our material environment. But in a world with so many people, the trade-off is unavoidable. A system of knowledge that is good at manipulating the material world is exactly what we need. Without such knowledge, we could not possibly support a human population of 6 billion.

Acceleration

The speed of these transitions is itself a distinctive feature, for the pace of change has also increased. Indeed, this change is so decisive that it forces us to approach the Modern Revolution differently than all earlier revolutions. Unlike the transition to agriculture, which occurred, region by region, over several millennia, the Modern Revolution has been virtually instantaneous, lasting no more than two or three centuries. And it took place in a globally connected world, in which innovations spread so fast that there was little room for independent invention. At such high speed, the decisive thresholds could be crossed only once. This singularity gave a huge advantage to the regions that modernized first and ensured that most other communities would experience the transition to modernity as a violent imposition of new norms from outside, a brutal social tornado over which they had little control. The rapid transmission of change explains why the forms taken by the Modern Revolution were influenced so much by the cultures of one part of the world, Europe. Yet if Europe had not been first, we can be sure that other parts of the world would soon have crossed the same threshold.

THEORIES OF MODERNITY

How can we explain these astonishing transformations? There is as yet no consensus about the nature of the Modern Revolution, or about its causes. A century of detailed historical research has generated a colossal fund of information about modern history, particularly in Europe and North America, but no single theory of the rise of modernity has achieved general acceptance. The difficulties caused by lack of consensus and the sheer volume of information and ideas are compounded by the fact that we are still living through the Modern Revolution. We do not know its overall shape; perhaps in a few centuries' time it will be apparent that the transformations had barely begun by 2000 CE. Even our most general definitions of the Modern Revolution may turn out to have been grossly misleading.

A book such as this cannot "solve" the problem of modernity. But we do have to try to see what this revolution looks like on the scale of big history, and from the point of view of the early twenty-first century. If there is anything distinctive about the argument that follows, it is that it views the Modern Revolution in the large context of human and even planetary history, rather than just as a problem about recent centuries and particular regions of the world. Its perspective is therefore global—a feature that sets it apart from many standard accounts. All too often, accounts of the Modern Revolution start from the (usually unstated) assumption that modernity was created in and by European societies; they thus imply that explaining modernity means looking at European history. Unfortunately, the assumption of European "exceptionalism" has discouraged the careful comparative analysis needed to check whether these arguments really work.[28] If modernity is, as I will argue, a global phenomenon, a Eurocentric approach is bound to mislead us. More recently, historians interested in world history have tried to see modernity as a global problem that requires a global explanation.[29] The account that follows does not ignore the distinctive role of Europe and the Atlantic world in the Modern Revolution, but it is constructed within the parameters of world history and focuses on the global aspects of the problem.

Population Growth and Rates of Innovation

To clarify some of the problems faced in trying to explain the Modern Revolution, I will take the methodological gamble of beginning with population growth. I will argue that if we can explain the astonishing bloom of human populations in the last two or three centuries, we should also be able to explain many other aspects of the Modern Revolution. But an explanation of population growth leads us rapidly to the problem of innovation. Rapid and sustained population growth *must* imply an acceleration in rates of innovation. Thus changing rates of innovation must be at the heart of any explanation of the Modern Revolution. As Joel Mokyr argues, "Technological change . . . accounted for *sustained* growth. It was not caused by economic growth, it caused it."[30]

The problem, then, is to explain a sharp and global acceleration in innovation, which is the key to the Modern Revolution. We have seen that accelerating innovation is in some sense implicit in the notion of collective learning, so the Modern Revolution really represents a gear shift in the pace of collective learning in the last two centuries. As Daniel Headrick writes: "Knowledge is both cause and effect of economic growth, and the information industry has been the primary cause of the acceleration of technological change in the past 200 years."[31] We have seen some of the mechanisms that accelerated or retarded rates of innovation in previous eras and in different parts of the world, among them the size and variety of exchange networks and the intensity of exchanges within those networks. They also include population growth itself, which not only increases the size of exchange networks but also exerts a more or less gentle pressure to raise productivity in regions of high population density. In the era of agrarian civilizations, states and commercial exchanges acted as new sources of innovation. But they could also inhibit growth, as could population pressure when it led to overpopulation or the spread of disease. As a result, these pressures, even in combination, could never generate rates of innovation rapid enough to match potential rates of population growth. Hence there were periodic famines and Malthusian cycles that shaped the basic rhythms of human history in the era of agrarian civilizations.

The most striking feature of innovation in the past two centuries is that for a time at least, it has been so rapid and so sustained that levels of productivity have kept pace with and in some respects outpaced population growth. In fact, as we will see later, the large-scale rhythms of modern history are shaped less by Malthusian cycles, which were a result of insufficient productivity, than by business cycles, which are generated by *over*production. Of course, there have been many, sometimes devastating, regional famines; but on a global scale, food production has more than kept pace with population growth, which is precisely why populations have risen so fast. And what is true of food production is also true of production in other areas, from clothing and housing to consumer goods to energy and weaponry. It is this sharp global *acceleration* in the pace of collective learning, in rates of innovation, and in levels of productivity that we need to explain.

Some Possible Prime Movers

We can clarify the options by listing some of the prime movers that have been offered in existing attempts to explain the Modern Revolution. The rich tradition of scholarly debate on the Modern Revolution has yielded several promising candidates.[32] Usually, these have been used to lever Europe into the modern world. But in principle, they should work equally well on a global scale.

Demographic Theories   Demographic theories (often associated with the work of Ester Boserup) rely largely on population pressure to explain increased rates of innovation.[33] We have seen that population growth exerted pressure to innovate throughout the agrarian era. And it is true that when combined with increasing commercialization, population growth sometimes acted as a stimulus by increasing the supply of labor and by increasing demand. For example, in eighteenth-century Britain, increasing demand for timber for fuel, building, and manufacturing led to deforestation, which in turn created pressure to find improved ways of using alternative fuels. And some of the key inventions of the British Industrial Revolution, including the invention of the coal-fired steam engine and methods of manufacturing iron using charcoal rather than wood, can be seen as responses to this pressure.

Nevertheless, population pressure on its own can explain only a small part of the sudden acceleration in growth rates characteristic of the modern era. The trouble is that all too often, population pressure has *not* generated the necessary innovations, and people have consequently starved or done without. After all, Great Britain was not the only country short of wood, and the problem may have been even worse elsewhere—in China, for example.[34] Necessity is *not* always the mother of invention.

Geographical Theories   Geographical theories rely heavily on particular geographical features to explain rising rates of innovation. For example, during the Industrial Revolution, Britain was able to replace timber with coal only because it had abundant and accessible reserves of that fuel. In the hands of E. A. Wrigley, such observations have been used to support an argument that focuses on "contingent" geographical factors to explain the distinctive role of Europe in the Modern Revolution.[35] Several regions of the world, such theorists point out, had large populations and high levels of productivity and commercialization; so it may have been geographical accidents such as the location of coal or the relative closeness of the Americas that best explain the different trajectories of Europe and, say, China, in the nineteenth and twentieth centuries.

Geographical features of this kind are undoubtedly important, and they will play an important role in the account offered below, but on their own they cannot explain much, simply because they were always there. Opportunities for change do not guarantee that change will occur. Indeed, in Britain iron makers had tried to use coal for almost two centuries before Abraham Darby showed them how to use coke in the early eighteenth century. As Mokyr argues, geographical factors of this kind may *shape* change, but they are not a fundamental *cause* of change.[36] What we must explain is why geographical factors such as the presence of coal suddenly began to be exploited more effectively, an effort that encourages us to look for distinctive features in the intellectual, economic, or social history of modern industrial societies.

Idealist Theories   A third group of theories can be described as *idealist*. They argue that rates of innovation are affected by different ways of thinking. The simplest theories of this type explain the Modern Revolution as the outcome of a sustained wave of new inventions. T. S. Ashton caricatured this approach in his summary of the typical school essay on the subject: "About 1760 a wave of gadgets swept over England."[37] Of course, in a simple sense, such theories are correct. The number of innovations increased, and each particular innovation did help raise the general level of productivity. But the difficulty even with sophisticated versions of this approach, including Ashton's own account,[38] is that they cannot explain *why* the rate of innovation should have increased as and when it did. Why so many innovations? Why such interest in more productive or efficient technologies and material techniques? Why then, and why there?

Subtler idealist theories have suggested that deeper changes in attitudes and methods of thinking stimulated new commercial and technological methods. The most famous example of this approach (one that its author later retracted, at least in part) is Max Weber's thesis about the connection between Protestantism and capitalism, first published in 1904–05. He argued that Protestantism, unlike Catholicism, embodied a new ethic of hard work, saving, and rationality that encouraged entrepreneurs to save and innovate in new ways.[39] But such theories are difficult to handle. Religions are not monoliths: like all systems of thought, they are complex, multifaceted, and malleable enough to adapt to many different environments. At different periods of their history, Buddhism, Islam, Confucianism, and even Catholicism have all encouraged at least some of the qualities that Weber associated with Protestantism and capitalism. "Freedom" (particularly for entrepreneurs) has often been included as a significant prime mover of innovation; so has the "rise of science." But for these arguments, too, the problem is to explain why and how those particular factors suddenly acquired such salience.[40] Even the most subtle idealist theories have difficulty explaining why attitudes should have changed so decisively at a particular period in human history. If Protestantism led to science or rationality or modernity, what led to Protestantism? Changing attitudes are certainly an important part of the explanation for rising rates of innovation, but they are symptoms of something deeper, not independent motors of change.

Commercial Theories   A fourth cluster of theories focuses on the role of commercial exchanges. Economic historians, working in a tradition that goes back at least to the writings of Adam Smith, have highlighted the role of expanding networks of commercial exchange. Smith argued that rates of innovation were directly related to levels of commercialization. He begins chapter 1 of *The Wealth of Nations* (1776): "The greatest improvement in the productive power of labour, and the greater part of the skill, dexterity, and judgment with which it is any where directed, or applied, seem to have been the effects of the division of labour." In other words, increased specialization raised productivity. But Smith explained increasing specialization as itself a consequence of the rise of the market. Chapter 2 begins, "This division of labour, from which so many advantages are derived, is not originally the effect of any human wisdom, which foresees and intends that general opulence to which it gives occasion. It is the necessary, though very slow and gradual, consequence of a certain propensity in human nature which has in view no such extensive utility; the propensity to truck, barter, and exchange one thing for another."[41] As networks of exchange expand, cheap imports will undercut more expensive local producers and force them either to become more specialized, so that they in turn can produce more efficiently, or to focus on other products that they can produce more efficiently. In this way, wide networks of exchange ensure that the most productive methods soon become the best practice. Besides, where markets are extensive, people can afford to specialize more narrowly, for they will have

Figure 11.2.   An eighteenth-century pin factory. Adam Smith used the pin fac-
tory as an example of the advantages of the division of labor. From Joel Mokyr,
The Lever of Riches: Technological Creativity and Economic Progress (Oxford:
Oxford University Press, 1992), p. 78: from René-Antoine Ferchault de Réaumur,
L'art de l'épinglier (1762).

enough customers to support them entirely from their specialized crafts (see
figure 11.2). Chapter 3 of *The Wealth of Nations* explains the link between
markets and the division of labor in its title: "That the Division of Labour
Is Limited by the Extent of the Market." In other words, expanding networks of exchange encouraged specialization, which stimulated innovation
in productive techniques—a type of growth we can refer to as *Smithian*.[42]

As the previous chapter has argued, there clearly is a deep link between
the expansion of trade networks, increased specialization, and increased rates
of innovation. By and large, commercial activity (that is, the generation of
revenues through relatively consensual exchanges in which coercion is not
the paramount factor) tends to encourage innovation more than does tribute-
taking (the generation of revenues through exchanges dominated by the
threat of coercion), because those generating commercial incomes have to
make up in efficiency for what they lack in coercive power. But we have seen
that there are many exceptions to this rule; tribute takers are sometimes interested in efficiency-raising innovations, while merchants have never been
averse to using force when they can get away with it. Moreover, the very
nature of most premodern states suggests that as a general rule, in agrarian civilizations tribute-taking generated more wealth and certainly more
power than commercial exchanges. This differential helps us understand
what might at first appear puzzling: though commercial networks are as old

as agrarian civilization, their impact on rates of innovation has been limited until the past two or three centuries. Why, then, did commercial exchanges suddenly become so much more significant in the modern era? Did they reach some critical threshold? If so, can we describe it? Or did some other factor suddenly increase their significance? To explain modernity, we have to explain how and why the role and significance of markets have changed in recent centuries.

One common approach (often associated with idealist theories of modernity) has been to argue that Europe was unusually commercial, and European markets were unusually vigorous. The trouble with arguments of this kind is that recent research has shown that as late as the end of the eighteenth century, general levels of commercialization and even overall productivity were probably as high in China, Japan, and northern India as in Europe, yet only in the Atlantic world did rates of innovation begin to rise rapidly in the nineteenth century. Andre Gunder Frank has recently argued that Asian economies had the largest populations, and the largest and most productive economies perhaps as late as 1750 or even 1800. Indeed, he maintains that per capita incomes in China may have been higher than those of Europe as late as 1800.[43]

Social Structure Theories   However, downplaying European exceptionalism makes it extremely difficult to explain the different trajectories of these regions in the nineteenth century. An answer that has played a significant role in these debates since at least the time of Karl Marx is that even if western Europe does not stand out in 1800 from a Smithian perspective, it is remarkable from an institutional and social perspective. This idea is typical of a fifth possible approach to explaining the surge in rates of innovation. Social structure theories argue that different social structures affect rates of innovation in different ways. In general, they attempt to explain how the productivity-raising capacity of commerce may be transformed as powerful social groups become dependent on commercial rather than tributary exchanges of various kinds. In earlier chapters, I have already made some use of arguments of this kind to suggest why rates of innovation tended to be low in kin-ordered societies, and why the structures of tributary states did encourage innovation, but ambiguously and hesitantly. A social structure account of modernity must demonstrate that new social structures emerged that gave a much more powerful stimulus to innovation. Such theories owe much to Marx, who called the social structure characteristic of modernity "capitalism." In his highly formal argument, Marx contended in *Capital* that the generalization of exchanges that is typical of capitalism encourages a new and especially powerful technological synergy, whose properties he analyzed in great detail. The simplified account that follows draws on Eric Wolf's modified version of Marx's scheme of "modes of production."[44]

Marx's thought is currently unfashionable; indeed, some have claimed it has been "refuted" by the collapse of socialism in the 1980s, and certainly much of it is dated today. Nevertheless, like Anthony Giddens, I believe that Marx's analysis of capitalism "remains the necessary core of any attempt to come to terms with the massive transformations that have swept through the world since the eighteenth century."[45] In the writings of Marx, each "mode of production" characterizes a type of society in which particular life-ways and technologies are associated with particular social structures. We have already used Eric Wolf's model of the kin-ordered and tribute-taking modes of production. Here, we must look more closely at the capitalist mode of production. As an ideal type, it has three main elements: (a) a dominant class of entrepreneurs or "capitalists" who own productive resources (i.e., capital) and use them to generate commercial profits that sustain their elite lifeways; (b) a class of people who, unlike peasants, have *no access to productive property* and can therefore support themselves only by selling their own labor power, thereby becoming wage earners or "proletarians"; and (c) competitive markets that link these two groups through commercial exchanges governed by market forces rather than by legal or physical coercion. In an idealized capitalist world, elite groups consist mainly of capitalists, the rest of the population consists mainly of proletarians, and most exchanges pass through markets.

By definition, wealth in such a world is distributed more unevenly than in the tributary world, for most proletarians do not have direct access to productive resources such as land. Speaking generally, it is the steepness of this gradient of wealth that accounts for capitalism's remarkable dynamism, just as the large temperature gradient between the Sun and the space surrounding it drives complex processes on Earth. The large inequalities of capitalism help explain why resources no longer have to be moved primarily by the crude use (or threat) of physical violence as in tributary societies. Instead, states exert force mainly to maintain the structures of law and ownership that protect the concentration of wealth. It is the steepness of the gradient that drives wealth so efficiently through capitalist societies and that helps explain why, paradoxically, modern states have to be so much larger and more complex than the states of the tributary world.

Why should such structures encourage innovation? The point of the argument is that both main classes of society find themselves in environments that force them to innovate, constantly and endlessly. Just as ecological changes force species to evolve rapidly in periods of rapid environmental change such as the ice ages, so the new and constantly changing social ecology of capitalism forces humans of all classes to adapt by continually seeking more productive ways of working. In this way, the structures of capitalism led to the evolution of new behaviors that mobilized the innovative capacity of human beings in revolutionary ways.

At this point, Marx's argument differs little from that of orthodox economists. In a world of entrepreneurs, competitive markets, and wage earners, both entrepreneurs and wage earners have to pursue innovation as a condition of survival. Entrepreneurs have to do so because in competitive markets the most successful long-term strategy will always be to cut the costs of production, and therefore of sale, and implementing such a strategy requires introducing cost-cutting innovations in production, transportation, and management. Like evolution in the nonhuman world, the process is endless because competitors will rapidly copy successful innovators, making entrepreneurial innovation general, constant, and accelerating.

Wage earners also have to actively seek ways of improving productivity. As sellers of labor power, they compete with other wage earners. To find purchasers for their labor power, wage earners have to offer labor that is more productive than that of potential rivals but costs less. Here, too, the ratchet of competition ensures that the productivity of labor will steadily increase. These rules explain the odd paradox that what Leon Trotsky called the "economic lash" of capitalism—the threat of unemployment—is a far more effective tool for increasing labor productivity than the lash of slavery or serfdom. Owners cannot afford to starve their slaves or serfs, but they have no incentive to give them a high standard of living. Such a system cannot stimulate the creativity of the worker. However, capitalist employers do not own their workers and have no need to protect them from starvation or poverty. In fact, they will generally regard the threat of unemployment or poverty as a healthy stimulus to harder work. So the onus is on the workers to ensure that their labor is productive enough to find a buyer. In this way, the economic lash can stimulate genuine, even creative, self-discipline, whereas the overseer's whip can generate no more than grudging conformity. Capitalism generates a discipline that touches the intellect, the psyche, and the bodies of wageworkers with a power unattainable through the more direct and brutal methods typical of tributary societies. It is as if the structures of capitalism forced people to load new types of software into their brains. Or, to adopt a less rigid metaphor, it is as if the structures of capitalism filled people's heads with quite new types of motivations and meanings (or "memes," in the language of Richard Dawkins).[46]

This is a model of a society in which innovation is never-ending, because both main classes of society find themselves on a relentless treadmill of constantly rising productivity. Social structure theories of modernity imply that if we can explain how and why modern societies began to conform to this ideal type, we will have gone far toward explaining the Modern Revolution.

But here, too, there is a difficulty. Recent research has suggested that it is not as easy as it once seemed to distinguish between a capitalist Europe and a noncapitalist China or India. In much of East Asia, wage earning was widespread, and so was capitalist production. Indeed, two thorough comparative studies, by Kenneth Pomeranz and R. Bin Wong, have shown that levels of capitalist development in China and western Europe were so similar that it is no longer possible to explain the Industrial Revolution simply by referring to Europe's higher levels of capitalism.[47] In fact, the likenesses are so close that both authors leave us with the impression that the move to accelerated growth, which was to prove critical in the history of the modern era, turned on a few contingent differences, such as the distribution of coal.

In the next two chapters, I will attempt an explanation of modern rates of innovation that introduces many of these different prime movers but adds one more.

The Scale and Synergy of Exchange Networks   In chapter 7, I argued that on large scales, rates of innovation were shaped by the size and heterogeneity of information networks. In other words, the sheer scale and variety of interactions may have been a powerful determinant of changing rates of innovation. In chapters 12 and 13, I will argue that the sudden increase in the scale and, perhaps even more important, in the variety of information exchanges in the early modern period may have given a sharp stimulus to processes of collective learning, particularly in hub regions where these exchanges were most concentrated and most diverse. But I will integrate this suggestion into an argument that also makes use of many of the other prime movers familiar in the literature on the rise of modernity. First, I will describe some of the factors that led to accelerating rates of innovation in general. Second, I will explain why the acceleration first became apparent in Europe. It might help to preview this argument somewhat schematically.

- A global explanation of rising rates of innovation
  Accumulation. Particularly in the Afro-Eurasian zone, processes

of accumulation over several thousand years had created several regions in which innovation had proceeded about as far as it could have gone within the traditional tributary framework of the era of agrarian civilizations. By the eighteenth century, these regions included China, Japan, parts of India, and parts of western Europe.[48]

Expanding networks of exchange. The creation of a global system of exchanges from the sixteenth century on gave a sudden and decisive boost to global processes of collective learning and commercialization. Expanded networks of information exchange opened new possibilities for innovation that helped break through the technological ceiling reached in the most densely populated regions of the world. As a result of this change, the amount and variety of information being exchanged increased sharply, and so did the speed with which it circulated, leading to a marked expansion in the pool of knowledge that could be drawn on by societies throughout the world. Increased commercial exchanges also boosted commercial activity, thereby accelerating the processes of innovation familiar from both Smithian and Marxist accounts of modernity.

An explanation of Europe's distinctive role in the Modern Revolution

A new topology of exchange. A small number of societies were well placed geographically to benefit from the sudden acceleration in global processes of collective learning. The emergence of a global system of information exchanges transformed the topology of large-scale exchange networks. The Atlantic coasts of the Eurasian landmass, which had been at the periphery of Afro-Eurasian exchange networks, suddenly found themselves at the hub of new, global networks of exchanges. Europe, and then the Atlantic shores of North America, emerged as the first hub of a new world system, even if the center of gravity of that system remained for a long time in India and China. The volume of exchanges remained greatest in East Asia until well into the nineteenth century, but a greater variety of ideas, goods, wealth, and technologies began to flow through Europe and the Atlantic zone.[49] This rearranged topology granted western Europe both a commercial and an intellectual windfall. At the same time, Mesopotamia, which had been the hub of Eurasian exchange networks for millennia, suddenly found itself less central within the new system of global exchanges. These rapid changes in the topology of global exchange networks gave Europe significant advantages.[50] Seen in this light, modernity is not something that began in Europe and spread to other parts of the world; rather, it is a product of global processes, which cast the lands bordering on the Atlantic in a quite novel role.

European preadaptations. But why was Europe so well able to exploit these unexpected advantages? Because Europe alone enjoyed both a central position within the newly emerging world system and a high level of commercialization. Europe's advantages were not just a matter of geographical good fortune. On the contrary, western European societies were, in an important sense, preadapted to exploit the opportunities created within new, global exchange networks. The social, political, and economic structures of many regions of western Europe helped Europe take advantage of the new systems of exchange that appeared with the emergence of a global network of exchanges, and it is at this point that I will return to arguments of a more familiar type about some of the distinctive features of European history. As Wong has put it, in his important studying comparing China and western Europe in the early modern era: "European political economy did not create industrialization, nor was the European political economy deliberately designed to promote industrialization. Instead, European political economy created a set of institutions able to promote industrialization once it appeared."[51]

SUMMARY

The world has been transformed in the past two or three centuries. Explaining that transition will be the task of the next two chapters, using the strategy described in this chapter. I focus on population growth, in the hope that a successful explanation of the astonishing population growth of modern times may also help make clear many other aspects of modernity. Such an account will have to explain why and how humans have learned to extract from their environment the huge resources needed to support populations of several billions. This means explaining the astonishing increases in innovation and productivity typical of the modern world.

There have been many attempts to explain the revolutionary increases in innovation that are the key to the Modern Revolution. Each focuses on a different explanatory prime mover—population pressure, geographical factors, changing ideas, the expansion of markets and networks of exchange, changes in social structures, and so on. The account of the Modern Revolution offered in the next few chapters will make use of several of these elements, though it will focus mainly on the changing topology of networks of exchange and on changes in social structures. I will argue that the emergence of a global network of exchanges hugely stimulated commercial activity and ecological innovation throughout the world. The enhanced scale of information exchanges in a global information network boosted rates of ecological innovation, while increased commercial exchanges accelerated the sorts of innovation identified in Smithian and Marxist models of modernity. Within the global system, Europe appeared, quite suddenly, as a new hub region, so it was particularly well placed to exploit the huge commercial opportunities created in the new global system. But I will also argue that the social and economic institutions of Europe helped it take advantage of its fortunate location within the new global network of exchanges.

FURTHER READING

J. L. Anderson's *Explaining Long-Term Economic Change* (1991) is a useful introduction to the theoretical literature. Among the more important recent studies are Anthony Giddens, *A Contemporary Critique of Historical Materialism* (2nd ed., 1995); Joel Mokyr, *The Lever of Riches* (1990); and E. A. Wrigley, *Continuity, Chance, and Change* (1988) and *People, Cities, and Wealth* (1987). E. L. Jones's *The European Miracle* (1987) and *Growth Recurring* (1988) are classics that have stimulated much debate on the rise of the modern world. Andre Gunder Frank, *ReOrient* (1998); Kenneth Pomeranz, *The Great Divergence* (2000); and R. Bin Wong, *China Transformed* (1997), have reminded us powerfully of the relative backwardness and weakness of Europe before 1800, thereby undercutting the once-popular theories that traced modernity to medieval Europe. These works demonstrate the extent to which the Modern Revolution was the product of global processes. Margaret Jacob, *The Cultural Meaning of the Scientific Revolution* (1998), has been immensely influential in suggesting the importance of the scientific revolution for explaining the rise of modernity in Europe; Charles Tilly, *Coercion, Capital, and European States, AD 990–1992* (rev. ed., 1992), is the most thorough general account of the modernization of the European state system. Eric Wolf, *Europe and the People without History* (1982), reminds us of the crucial role played in modern history by peoples without state structures. Beyond these books, there is a vast literature on particular aspects of "the rise of the modern world," some of which will be listed at the end of the next two chapters.

GLOBALIZATION, COMMERCIALIZATION, AND INNOVATION

> [T]he Aborigines of Arnhem Land called the first Europeans
> they saw *Balanda,* a Bahasa Indonesia term for Europeans which
> is derived from "Hollander," as the Dutch were once known.

This chapter will survey world history in the period from 1000 CE to ca. 1700 CE, setting forth some of the changes that prepared the way for the Modern Revolution. It will concentrate first on global processes, showing how expansion in the size of exchange networks, slow before the sixteenth century and then much faster, created new possibilities both for the exchange of information and goods and for innovation. It will argue that the creation of a truly global exchange network in the sixteenth century decisively increased the scale, significance, and variety of informational and commercial exchanges. The coming together of the different world zones of the Holocene era marks a revolutionary moment in the history of humanity.

Second, this chapter will describe the changing topology of global exchanges. As the geography of exchange networks was transformed, flows of information and wealth entered new channels. These effects were particularly significant in western Europe, which had previously been at the margin of exchanges within the Afro-Eurasian world zone but now suddenly found itself at the hub of humanity's first global system of exchanges. These changes in the scale and geography of exchange networks laid the intellectual and commercial foundations for the Modern Revolution, and determined its geography.

It may be helpful to think of three distinct scales of explanation. First, in one sense the Modern Revolution was and is a global process; it cannot be properly understood without appreciating this feature. Its intellectual, material, and commercial raw materials came from all parts of the world. And the new level of creative synergy generated by linking the two largest world zones—Afro-Eurasia and the Americas—was and remains perhaps the most powerful single lever of change in the modern world. The Modern Revolution was also global in its effects, both creative and destructive. In some form, its impact was felt very soon in all parts of the world.

But in different world zones modernity was experienced in different ways, and the need to understand the diversity of its impacts requires a second level of explanation. The coming together of the different world zones proved a brutal and destructive process for indigenous populations (both human and nonhuman) in all three of the smaller world zones: the Americas, Australia, and the Pacific. The advantages accumulated disproportionately within parts of the Afro-Eurasian zone—and later in the "neo-Europes" of the Americas, Australia, and the Pacific, the new societies created by peoples from Afro-Eurasia after migrating (willingly or unwillingly) to the other three world zones. In some sense, the history of the Afro-Eurasian zone ensured that when its peoples encountered societies from the other world zones, the Afro-Eurasian societies prevailed.

We have already seen some of the reasons for this dominance. Some have to do with the existence of domestic livestock in Afro-Eurasia. Used for transportation and haulage, domestic livestock magnified the advantages of scale by extending and quickening processes of exchange within what was already the largest and most varied of the world zones. Extensive and vigorous exchange networks help explain some of the technological advantages enjoyed by Afro-Eurasian societies. But animal domesticates also swapped diseases with their human owners; thus cohabitation with domesticates, combined with the efficient systems of communication they provided, ensured that the populations of Afro-Eurasia were more disease-hardened than those of the other world zones.[1] And the diseases of Afro-Eurasians may have been more useful to them in their attempts at conquest than their advanced naval and military technologies. For example, smallpox, as Alfred Crosby writes, "played as essential a role in the advance of white imperialism overseas as gunpowder—perhaps a more important role, because the indigenes did turn the musket and then rifle against the intruders, but smallpox very rarely fought on the side of the indigenes."[2]

But even within the huge Afro-Eurasian world zone, the advantages of the Modern Revolution accumulated erratically and lopsidedly, an observation that brings us to the third or regional scale. If we think of the Modern Revolution as a product of the new intellectual and commercial synergies of the first global system, it initially seems natural that the intellectual and commercial raw materials for modernity should have accumulated preferentially within established hubs of exchange and centers of gravity, perhaps in the Mediterranean world, or Mesopotamia, or northern India, or

China. And indeed perhaps something like this did happen. Rates of growth, and even of innovation, were high and sustained in all of these regions throughout the period covered by this chapter.[3] But although all the old core regions were shaped by the emerging global network of exchange, the full power and significance of the Modern Revolution emerged elsewhere. The sharp rise in innovation that signifies modernity first became apparent at the western edge of the Afro-Eurasian world zone, in a region that had not been incorporated within the expanding zone of agrarian civilizations until the first millennium CE and had seemed of secondary importance before the middle of the second millennium. That the adaptive significance of the Modern Revolution would first become apparent here was not obvious even as late as 1776, when Adam Smith remarked that "China is a much richer country than any part of Europe."[4]

An adequate explanation of the Modern Revolution must attempt to explain its origins at all these different scales. As the Islamicist Marshall Hodgson put it in an essay first published in 1967:

> just as civilization on an agrarianate level had appeared in one or at most a very few spots and spread from there to the greater part of the globe, so the new modern type of life did not appear everywhere among all cited peoples at the same moment, but first in one restricted area, Western Europe, from which it has spread everywhere else. It was not that the new ways resulted from conditions that were limited entirely to the Occident. Just as the first urban, literate life would have been impossible without the accumulation among a great many peoples of innumerable social habits and inventions, major and minor, so the great modern cultural mutation presupposed the contributions of all the several cited peoples of the eastern hemisphere. Not only were the numerous inventions and discoveries of many peoples necessary—for most of the earlier basic ones were not made in Europe. It was also necessary that there exist large areas of relatively dense, urban-dominated populations, tied together in a great interregional commercial network, to form the vast world market which had gradually come into being in the eastern hemisphere, and in which European fortunes could be made and European imaginations exercised.[5]

Today, more than thirty-five years after Hodgson wrote, it is even easier to see the extent to which the Modern Revolution was a product of global processes, even if its full significance first became apparent at the western edge of the Afro-Eurasian world zone.

We have seen in earlier chapters that at large scales, the size, variety, and intensity of exchange networks could be important determinants of rates of innovation, while at slightly smaller scales, population growth, state activity, and commercial expansion were also significant. All of these factors were influenced considerably by the Malthusian cycles that characterized the history of most agrarian civilizations. Networks of commercial, political, and information exchanges expanded most vigorously during eras of demographic expansion; they often contracted in periods of demographic decline. And during the phases of expansion, the increased scope of exchanges, population growth, state activity, and commercial activity all tended to generate innovations. In the millennium preceding the Industrial Revolution, two large Malthusian cycles were crucial in shaping the history of the entire Afro-Eurasian world zone and, indirectly, that of other zones as well (see figure 10.4). The first cycle began with a demographic revival in the second half of the first millennium and ended abruptly with the Black Death in the middle of the fourteenth century. The second, which began after the Black Death, ended in a less drastic slowdown during the seventeenth century.

THE POSTCLASSICAL MALTHUSIAN CYCLE: BEFORE THE FOURTEENTH CENTURY

The Expansion Phase

Malthusian cycles are easiest to see in the rhythms of population growth (see table 11.1 and figure 10.4). In all Malthusian cycles, it is possible to identify some significant innovations that enabled populations to grow to a new level. The postclassical cycle was linked partly to developments in agricultural technologies, such as the introduction of heavier, horse-drawn plows in Europe, or the introduction of new crops such as rye or new strains of rice (encouraged by government activity, though the strains of rice were improved by peasant farmers) and better-managed systems of irrigation. In China, northern Europe, and the Islamic world, agricultural methods were revolutionized between the eighth and the twelfth centuries. Everywhere, population growth stimulated colonization. Indeed, growth was most rapid in those lands, such as central Asia, northern and eastern Europe, and South China, that had been frontier zones in the classical era. In China, 60 percent of the population lived in the northern lands, dominated by the Yellow River; 250 years later, only 40 percent lived there, and South China had become the demographic heartland of the Chinese Empire.[6]

Far to the west, in the borderlands we now call Europe, internal colonization shifted the demographic center of gravity northward as lands once regarded as wastelands began to be farmed. In England, moorlands, woodlands, and marshlands were brought into cultivation in the twelfth and thirteenth centuries. Asa Briggs notes, "The 'wastes' of Dartmoor, for example, were cultivated; terraced hillsides . . . were farmed at Mere in Wiltshire and in Dorset; the monks of Battle Abbey in Sussex constructed successive sea-walls to reclaim the marshes. By the late thirteenth century a bigger area was cultivated than at any period before the wars of the twentieth century."[7] Along the northwestern shores of Europe, colonists and their landlords from the Rhine to the Loire reclaimed land from coastal swamps and marshes, beginning a process that evolved in the Netherlands into a great national art. In eastern Europe, a massive and largely unrecorded peasant migration from the sixth century onward created the demographic foundations for the first great Russian states.

Population growth stimulated urbanization. In Europe and Russia, the number of cities with populations exceeding 20,000 rose from 43 to 103 between 1000 and 1300.[8] Cities flourished particularly in the Islamic world. In the ninth century, the Abbasid capital, Baghdad, may have had a population of half a million. But even at the edge of the Islamic world, in Khorezm on the Aral Sea, towns prospered at the hub of trade routes linking the woodlands of Siberia, the steppes, and the urbanized lands of the south. Khorezm displays the mix of high culture and squalor characteristic of most premodern cities. The Arabic geographer al-Muqaddasi wrote that its capital, Kath, had a superb mosque and a royal palace, and its muezzins were renowned throughout the Abbasid domains for "beauty of voice, expressiveness in recitation, deportment, and learning." Yet "the town is constantly flooded by the river, and the inhabitants are moving (farther and farther) away from the bank. The town . . . contains many refuse drains, which everywhere overflow the high road. The inhabitants use the streets as latrines, and collect the filth in pits, whence it is subsequently carried out to the fields in sacks. On account of the enormous quantity of filth strangers can walk about the town only by daylight."[9]

Cities also flourished in China, particularly in the more commercialized south. By the twelfth century, China may have been "the most urbanized society in the world," with levels of urbanization perhaps as high as 10 percent.[10] Hangzhou (Marco Polo's "Kinsai," the capital of the southern Song) was then possibly the world's largest city, with at least a million inhabitants. It contained many different neighborhoods: working-class suburbs with crowded multistory houses; foreign quarters with Christians, Jews, and Turks; a large Muslim quarter with many foreign traders; and a wealthy southern region dominated by government officials and rich merchants.[11] Some idea of the variety of trades conducted in the town is conveyed by the historian Jacques Gernet's list of the guilds of Hangzhou. These included, in the words of Janet Abu-Lughod, "jewellers, gilders, gluemakers, art and antique dealers, sellers of crabs, olives, honey, or ginger, doctors, soothsay-ers, scavengers, bootmakers, bath keepers, and . . . money-changers[.]"[12] In this period, the largest cities in the world could be found in China.[13]

Urbanization stimulated commerce, both local and international. A whole hierarchy of markets appeared. At the lowest levels, markets were still dom-inated by barter, as a twelfth-century Chinese description suggests:

> The small market—
> People with their bundles of tea or salt,
> Chickens cackling, dogs barking,
> Firewood being exchanged for rice
> Fishes being bartered for wine.
> Here and there—
> Green tavern flags
> Where elderly gentlemen sit propped,
> Drowsy with drink.[14]

But regional and international markets flourished as well. In northwestern Europe, in 1000 CE, most people were still self-sufficient peasants; farther south, too, most production was rural, even in old urban regions such as northern Italy. But early in the second millennium, as populations and cities grew, so did networks of trade and commerce. The famous fairs of the Cham-pagne region linked Flanders to the ancient trade networks of Italy and the Mediterranean. In Europe, the expansion of trade and of cities was so spec-tacular that one historian, Robert Lopez, has called the "commercial revo-lution of the Middle Ages" a fundamental turning point in modern world history. For another historian, Carlo Cipolla, "the rise of the cities in Eu-rope in the tenth and twelfth centuries marked a turning point in the his-tory of the West—and, for that matter, of the whole world."[15] Such com-ments convey the pace of change in Europe, though they underestimate the extent and significance of changes elsewhere in Afro-Eurasia.

That commercialization was significant throughout the Afro-Eurasian zone is shown by the consolidation and intensification of a thriving inter-regional trading system. The thirteenth-century world system, described so well in an influential study by Janet Abu-Lughod, linked China, South-east Asia, the Indian subcontinent, the Islamic world, central Asia, parts of sub-Saharan Africa, the Mediterranean, and Europe into a single commer-cial network that carried much more traffic than the networks of the clas-sical era.[16] As Thomas Allsen has shown, significant amounts of political, cultural, and technological information flowed through these networks, as well as trade goods and diseases.[17] Pastoralists played crucial roles in this system as protectors, as guides, and sometimes as traders. The extent of these

Islamic-dominated networks of commerce and culture is vividly illustrated in the memoirs of Ibn Battuta, a Moroccan scholar who visited most parts of them as he traveled from Morocco to Mecca, to the Eurasian steppes, to India, to China, and across the Sahara between 1325 and 1355.[18] Under the Mongols, trans-European trade networks were even more vital, for the Mongols actively protected trade in the lands they ruled. While these land networks stimulated exchanges of many different kinds throughout Eurasian trading networks, sea routes may have been still more important— particularly those linking China, India, and the Islamic world. It was an early sign of the commercial precocity of Europe that its traders played an active role in many of these systems. By the tenth century, Viking traders and settlers could be found from Greenland (even briefly from Newfoundland) to Baghdad and central Asia. Early in the fourteenth century, Italian merchants (following in the footsteps of Marco Polo) traveled so regularly between the Mediterranean and China that guidebooks were published to help them on their way. But they were not alone. Armenian and Jewish merchants played crucial roles in trans-Eurasian exchanges.[19] Religions, including Christianity, Zoroastrianism, Buddhism, Manichaeism, and Islam, also moved surprisingly freely along the major Afro-Eurasian trade networks. And so did diseases. Eventually, traveling from east to west, came the bubonic plague. Its spread indicated the scale and intensity of Afro-Eurasian exchange networks, even though it was to end the postclassical cycle of expansion.

The hub of these networks remained in the Islamic world, so it is not surprising that Islam expanded throughout this period. In the centuries before 1000 CE, the importance of the Mesopotamian/Persian hub region was apparent in the critical role of the Sassanid and Islamic Empires in Afro-Eurasian exchange networks. In the first thousand years of their history, the Islamic civilizations that controlled this zone encouraged exchanges of ideas, goods, and technologies between many different parts of Afro-Eurasian networks, thereby stimulating population growth and increasing the synergy of commercial and information networks. As Andrew Watson has shown, the expansion of Islam was sustained in part by the openness of early Islamic states to innovation, particularly in agriculture.[20] In the course of several centuries, agriculturalists in the Islamic world imported and learned to use a wide range of new crops—including fruit trees, vegetables, and cereals, as well as fiber crops, condiments, and narcotics—in what we might call the *Abbasid exchange,* by analogy with the later Columbian exchange. Many new crops came from India, Africa, or Southeast Asia. And because information was pooled in the Islamic world as well as crops and technologies, it became the center of Eurasian science as well as commerce.

It was here, not in Europe, that the greatest achievements of classical Mediterranean philosophy and science were preserved for the future. In 1000 CE, there could be little doubt that the hub of the Afro-Eurasian ecumene lay in the Islamic world, and the expansion of Islam continued throughout the postclassical Malthusian cycle. By 1500, Islamic states included the Ottoman Empire, the most powerful empire in the Mediterranean world; the Safavid Empire in Persia; and a series of states reaching from the Philippines through Southeast and South Asia to sub-Saharan Africa.

But though the hub of Afro-Eurasian exchange networks was in Southwest Asia, their center of gravity lay in India and China. While the exchanges that passed through the eastern Mediterranean may have been more diverse and drawn from a larger area, the greatest *volume* of exchanges could be found in East Asia. European merchants were drawn to Asia, and particularly to China, because that was where the largest markets could be found, sustained by the largest populations and the most dynamic economies in the world. East Asian economic history has not been studied as intensely as that of Europe; and ever since the eighteenth century, models of Asian economic history have been shaped too much by images of a fundamentally static "Asiatic" type of economy and society. The reality was different.[21] Not only were Asian economies the largest in the world, they may also have had the highest levels of commercialization, at all levels of society, and the highest levels of productivity, both in the countryside and the towns.

Indeed, as noted in chapter 10, Lynda Shaffer has argued that the main geographical feature of this era of world history was "Southernization."[22] Akin to the more recent phenomenon of Westernization, Southernization began, she suggests, with technological and commercial innovations in textile production, metallurgy, astronomy, medicine, and navigation, all pioneered in the Indian subcontinent and in Southeast Asia. In the ninth century CE, a Muslim writer, al Jahiz, wrote:

> As regards the Indians, they are among the leaders in astronomy, mathematics . . . and medicine; they alone possess the secrets of the latter, and use them to practice some remarkable forms of treatment. They have the art of carving statues and painted figures. They possess the game of chess, which is the noblest of games and requires more judgment and intelligence than any other. They make Kedah swords, and excel in their use. They have splendid music. . . . They possess a script capable of expressing the sounds of all languages, as well as many numerals. They have a great deal of poetry, many long treatises, and a deep understanding of philosophy and letters. . . . Their sound judgment and sensible habits led them to invent pins, cork, toothpicks, the drape of clothes and the dyeing of hair. . . . They were the originators

of the science of *firk*, by which a poison can be counteracted after it has been used, and of astronomical reckoning, subsequently adopted by the rest of the world. When Adam descended from Paradise, it was to their land that he made his way.[23]

Innovations pioneered or preserved in the Indian subcontinent spread to Southeast Asia and China and then to the Islamic world, providing much of the driving force for the postclassical Malthusian cycle. Shaffer notes, "By 1200 the process of southernization had created a prosperous South from China to the Muslim Mediterranean."[24]

Commercialization and Its Impact

The expansion of Afro-Eurasian markets during the postclassical Malthusian cycle ensured that commerce and those engaged in it acquired a cultural, economic, and political importance they had never enjoyed before. We have seen that merchants had played a significant role in all agrarian civilizations, but usually they constituted a subordinate and sometimes despised layer of the upper classes. However, as trade networks expanded within agrarian civilizations over several millennia, so did the volume of wealth passing through mercantile hands, and so did the number and importance of those who managed or depended on commercial wealth. By the end of the postclassical Malthusian cycle, merchants formed an important, wealthy, and distinctive social category in most states in the Mediterranean and Islamic worlds, the Indian subcontinent, and China. In some regions and countries, such as the city-states of Italy or the Netherlands or Southeast Asia, merchants dominated small states.

In such states, increased reliance on commercial revenues led to fundamental changes in attitudes, in state structures, and in policy. We have already seen that those polities near major trading systems often had to depend more on commercial than on tributary revenues. In Europe, small states multiplied during the postclassical Malthusian cycle, because here (unlike in the eastern Mediterranean, northern India, and China) no large tributary empire emerged to succeed the imperial juggernauts of the classical era. Thus Europe, like parts of South and Southeast Asia, developed as a region of many small and highly competitive states. Their size limited the volume of tributary revenues that could be exacted; intense competition raised the cost of survival; and the proximity of major trade routes provided opportunities to siphon off commercial revenues. In such environments, commercial sources of revenue ceased to be an embarrassing expedient: they not only provided fiscal salvation for many small states but also shaped their economic and political structures, and even their values and social composition.

Clusters of aggressively commercial city-states appeared in Italy and also in northwestern Europe—particularly in Flanders and in the many cities of the Hanseatic League, which traded in the furs and fish of the Atlantic and the Baltic. Because these states were so dependent on trade, their rulers were often closely allied to merchants; sometimes the rulers *were* merchants. Not surprisingly, such states backed mercantile activity with all the political and military force at their disposal, engaging in a confusing mix of tributary and commercial exchanges, using force where they could, but trading with commercial finesse where necessary. In Italy, Thomas Brady observes, "States ruled by merchants or by merchants and landowners arose . . . very shortly after A.D. 1000. Pisa, Genoa, and Venice led the pack, but all up and down Central Europe, from Tuscany to Flanders, from Brabant to Livonia, merchants not only supplied warriors—as they did all over Europe—they sat in governments that made war and, sometimes, buckled on armor and went into battle themselves."[25] On occasion, these trading polities proved powerful enough to inflict military defeats even on powerful tributary polities, as the Athenian city-states had 1,500 years before, when they defeated the Persian Empire at Marathon and Salamis (in 490 and 480 BCE, respectively). In 1176, at the battle of Legnano, a league of northern Italian communes defeated the German emperor, Frederick Barbarossa, and freed themselves from imperial control. An uncle of Barbarossa's noted the oddity of this phenomenon: "In the Italian communes they do not disdain to grant the girdle of knighthood or honorable positions to young people of inferior station, and even to workers of the vile mechanical arts, whom other peoples bar like the plague from the more respectable and honorable circles."[26]

Militarily powerful commercial states such as these reflect the long-term rule that as commercial networks expanded and the wealth passing through them increased, so too did the potential influence of mercantile elites, until they sometimes found that they could challenge neighboring tributary elites not just commercially but in war as well. One of the decisive markers of the Modern Revolution was to be the rising economic and military influence of states whose economies were based on commercial exchanges rather than on more traditional tributary activities such as the gathering of taxes from the land. But not until the nineteenth century did it become apparent that as more and more wealth circulated within international commercial networks, such states would eventually eclipse even the most powerful of tributary empires, and on their home ground—the use of military force.

An Aborted Industrial Revolution in Song China?

China offers an interesting example of the potential impact of commercialization even within powerful tributary empires. By the first millennium BCE, commercial activity was already widespread in much of China; even land could be bought and sold. And by the middle of that millennium, the appearance of a powerful and independent merchant class was noted in the literary classics of the late Zhou dynasty, including the writings of Confucius (the Latinized form of Kong Fuzi, or "Master Kong," who lived ca. 551–479 BCE). By the time of the early Han dynasty, there were wealthy merchants who catered to the needs of rulers and nobles, petty merchants who bought and sold in provincial centers, and peddlers who bought and sold in the villages, thereby bringing villagers, too, within networks of commerce. Chang-an (modern Xian), the Han capital, covered almost 34 square kilometers, a much larger area than contemporary Rome, which covered only 13 square kilometers.[27] In the large towns, according to the imperial historian Sima Qian (who wrote at the end of the second century BCE), one could buy "alcoholic drinks, prepared foodstuffs, silks, hemp cloth, dyes, hides, furs, lacquerware, copper and iron goods."[28] A description from the same period suggests the increased visibility of a distinctive and wealthy merchant class, and also conveys the air of disapproval with which merchants were normally treated by members of the traditional nobility:

> Well-to-do merchants accumulate goods and redouble their profits, while the less well-to-do sit in their shops and sell. They control the markets and daily enjoy their ease in the cities. They take advantage of the pressing needs of the government to sell at twice the normal price. Their sons do not plough or hoe. Their daughters do not raise silkworms or weave. They have fancy clothing and stuff themselves on millet and meat. They earn fortunes while suffering none of the hardships which the farmers suffer. Their wealth enables them to hobnob with princes and marquises, and to dispose of greater power than the officials.[29]

Increased commercial activity, by offering new forms of revenue to states, could eventually have subtle but important effects on state systems. But it was least likely to transform states that had easy access to traditional forms of revenue, such as land taxes—most notably, large, tributary empires such as Han China, which controlled huge areas of land. Nevertheless, where traditional fiscal methods proved ineffective, commercialization could transform even the most powerful tributary states. The shift can be seen clearly in China during the postclassical Malthusian cycle. After the long period of imperial breakdown that followed the collapse of the later Han dynasty early in the third century CE, China was reunited during the Sui (589–617) and Tang (618–906) dynasties. Under the Tang, strong central rule and relatively orderly government made possible rapid growth of urban populations and of commercial activity, particularly in the south. And the Tang proved exceptionally open to foreign influences, whether in religion (this was the great era of Chinese Buddhism) or in trade. But the Tang were not particularly supportive of private commercial activity. Their tax base lay in the land, and until the An Lushan rebellion (755–63), they administered land taxes with an efficiency never again matched. They thus had little need for or interest in commercial revenues. Accordingly, the Tang for the most part maintained a traditional disdain for commerce and commercial activities, both at home and abroad. For example, merchants were not allowed to take the civil service exams.

However, the rulers of the Song dynasty (960–1276) found themselves in a much weaker position. After the Tang dynasty collapsed in the tenth century, much of northern China was lost to the Khitan (Liao) dynasty. In 1125, the Song dynasty lost its remaining control of the north to the Manchurian Jurchen (Jin) dynasty. Forced to relocate to the more commercially minded south, the Song moved their capital from Kaifeng to Hangzhou. Facing constant military challenges in the north, without the huge tributary revenues available within a unified China, and in the more entrepreneurial environment of southern China, the rulers of the later Song dynasty began to look more benignly on commercial activity and those who engaged in it. In the twelfth century, they even allowed successful merchants to buy official rank; and Marco Polo was told that the Song emperor had invited wealthy merchants to his palace, which would have been unthinkable under the Tang.[30] This shift in attitudes was driven by hard fiscal realities. By the mid–thirteenth century, 20 percent of Song revenues came from tolls on foreign trade, compared to only ca. 2 percent 200 years before.[31] Not surprisingly, the southern Song began to actively promote commercial activity and technological innovation. Whereas under the Tang, Guangzhou (Canton) was the only port allowed to take part in foreign trade, under the Song, seven more ports were opened. Aiding such trade were the very advanced junks built under the southern Song. They used compasses and sternpost rudders and had watertight bulkheads and special buoyancy chambers.[32] Internal commerce also flourished, particularly in the south, where populations were booming and trade networks with Southeast Asia and Japan were developing rapidly. To support increased monetization, the Song produced huge amounts of coinage; by 1080, they were minting about 6 million strings of coins a year (or ca. 200 coins per person), whereas the Tang had normally issued no more than 100,000 to 200,000 a year (ca. 10 coins per person).[33]

We have seen already that commercial exchanges are more likely to generate efficiency-raising innovations than are tributary exchanges, in which coercion can take the place of efficiency. And where states look benignly on commercial activity and create a supportive political and legal environment, it is reasonable to expect signs of increasing openness to innovation. This theoretical prediction certainly seems to be borne out by the history of the Song dynasty, for despite their political weakness, the Song presided over a period of astonishing growth and innovation.

By the middle of the eleventh century, China was divided between three large powers: the Song, the Khitan in the north and northeast, and the Tangut or Xia kingdom in the northwest. This period of divided rule was the prelude to an era of extraordinary technological innovation, which provided a kind of climax to the long process of Southernization. First, the agricultural foundations of the Song economy were revolutionized. According to Mark Elvin,

> the agricultural revolution . . . had four aspects. (1) Farmers learned to prepare their soil more effectively as the result of new knowledge, improved or new tools, and the more extensive use of manure, river mud and lime as fertilizers. (2) Strains of seed were introduced which either gave heavier yields, or resisted drought better, or else by ripening more rapidly made it possible to grow two crops a year on the same land. (3) A new level of proficiency was reached in hydraulic techniques, and irrigation networks of unprecedented intricacy constructed. (4) Commerce made possible more specialization in crops other than the basic foodgrains, and so a more efficient exploitation of varying resource endowments.[34]

Indeed, he concludes that by the thirteenth century, China likely had the most productive agricultural sector in the world, with the possible exception of India.

Supportive governments also encouraged innovation in other areas of the economy. The widespread use by governments and officials of woodblock printing to disseminate technical knowledge ensured that new techniques and new methods of inquiry in metallurgy, armaments, farming, medicine, and engineering spread widely. Coal and perhaps coke were used in the manufacture of iron; and official statistics show that by 1078, iron production had reached 113,000 metric tons a year, the equivalent of ca. 1.4 kilograms per person. This level of production was six times what had normally been produced under the Tang, and it was not matched in Europe be-

Figure 12.1.   Commercial activity in China under the Song dynasty. Courtesy Palace Museum, Beijing.

fore the eighteenth century.[35] At about the same time, two government arsenals were producing as many as 32,000 suits of armor a year, in three different sizes. Copper production rose so sharply that Greenland glaciers today show a sudden increase in atmospheric copper pollution in this era.[36] The technology of gunpowder was also pioneered under the Song, though its explosive properties were first deployed in war by their northern rivals, the Jurchen (Jin), in 1221. By the end of the thirteenth century, the first guns were in use in northern China.[37] By the eleventh century, a silk-reeling machine had been invented—the earliest known attempt to mechanize textile production.[38] There was significant innovation also in commercial methods (see figure 12.1). Early in the eleventh century, the government even started backing the issue of paper money.[39]

The innovations of this period were not purely Chinese. Rather, they reflected an increasing willingness on the part of governments and elites to exploit new productive and commercial ideas, whatever their source. Much Chinese innovation was based on a pooling of knowledge that had accumulated in other parts of the Afro-Eurasian system. For example, the new strains of rice that sustained the population boom in the south were imported from Vietnam. Many other techniques were imported from India or the Islamic world. Hydraulic techniques were particularly well developed in the Islamic world, where irrigation had a history of several thousand years; methods of textile manufacture were highly developed in India. The research of Joseph Needham into Chinese technology has highlighted the technological virtuosity of China, but it may also have unwittingly obscured the innovative technologies of other regions of the Afro-Eurasian world system.[40]

Nevertheless, rates of innovation under the Song were exceptional. Indeed, so astonishing is the extent of commercialization and innovation under the Song dynasty that it is tempting to think that medieval China was on the verge of an industrial revolution of its own. But if there was a revolution, it was not sustained and thus could not revolutionize the world. There are three main reasons that it failed to spark widespread change. First, the factors that encouraged the Song rulers to back commerce and entrepreneurial activity proved ephemeral; second, China's location at the edge, not at the hub, of Afro-Eurasian exchange networks slowed the spread of its innovations to other regions; and third, the world system as a whole was not yet large or integrated enough to ensure that innovation in China would rapidly infect other regions.

A system of competing states proved to be an unstable configuration in China. Well-established traditions of political and cultural unity, and well-integrated communications systems, made it likely that sooner or later, China would be reunified and the commercial and technological wealth of the Song period would be devoted once again to supporting a powerful united dynasty. In fact, this process was completed by 1279 after the conquest of South China by the Mongols under Kublai Khan. After reunification, two of the three conditions encouraging states to support commercialization (small size and intense rivalries) vanished, and the third (easy access to rich trading systems) lasted only slightly longer. China ceased to be a region of vulnerable and competing states keen to take revenues from any available source. Under the Yuan and Ming dynasties, government revenues shifted back to more traditional tributary sources, such as taxes from the peasantry.[41] The sheer size of a united China meant that commercial revenues would find it hard to compete with more traditional sources of revenues. And over the centuries that followed, the colossal inertia of this huge system made the shift from traditional revenues more complex and difficult than it might have been in a region of smaller, competing states.

In the fifteenth century, Chinese governments disengaged almost entirely from world trade networks, even if many of their subjects continued to trade despite the obstacles they now faced. The naval traditions of the Song survived until the fifteenth century. Indeed, between 1405 and 1433, under the command of a Muslim eunuch, Zheng He, seven separate fleets containing up to sixty ships and 40,000 soldiers were sent on journeys to the west (see figure 12.2).[42] They reached Ceylon, Mecca, and East Africa and may even

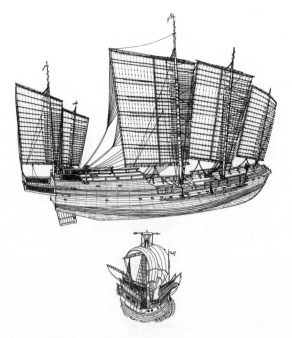

Figure 12.2.   Boatbuilding in China and Europe in the fifteenth century. The larger boat is a reconstruction of one of the ships used by the Chinese admiral Zheng He. Between 1405 and 1433, Zheng He led seven huge Chinese fleets with up to 60 ships and 40,000 soldiers in voyages to India, Southwest Asia, and East Africa, navigating with surprisingly accurate maps of all the lands he visited. His largest ships were at least five times the length of Columbus's ship, the *Santa Maria*, and they had watertight internal compartments. Columbus sailed fifty years after Zheng's last voyage. His fleet lacked many of the technological refinements of the Chinese fleets, and he was far more ignorant of where he was going. Nevertheless, Columbus's ships were more maneuverable, and probably better suited to the exploration of unfamiliar seas. Courtesy Relics Publishing House, Beijing.

have touched land in northern Australia. But these were not primarily trading missions, and the government that backed them was not looking for commercial revenues so much as symbolic submission to China. Not surprisingly, it had a limited stake in their continuance, particularly as they cost the state dearly. Eventually, deciding that the money was better spent defending its vulnerable northern borders, the Ming government lost interest in these expensive expeditions. Within a few decades the government had banned all Chinese shipping, though determined Chinese traders could usually find ways around these restrictions.

The second factor that lessened the impact of the economic revolution under the Song was China's geographical position, at the edge of the Afro-Eurasian network of exchanges. Though the volume of exchanges in China was vast, Chinese exchange networks neither reached as far nor carried as diverse information and goods as the exchange networks of hub regions such as the Islamic heartland in Mesopotamia typically did; thus, their influence on other parts of the Afro-Eurasian world was limited. Chinese innovations certainly had consequences elsewhere: many inventions, including the use of movable type for printing, the use of paper money (and the technology to make paper), and the use of gunpowder, reached the West, where their eventual impact was revolutionary. In addition, China's vast commercial momentum drew traders eastward by both land and sea. But these development made little *immediate* impression outside of China itself.

Third, and related to this last point, was the loose integration of the Afro-Eurasian networks, and their isolation from the networks of other world zones. The sluggishness with which Chinese innovations were taken up elsewhere suggests that the preconditions for a worldwide industrial revolution existed neither in China nor in the rest of the world. Exchanges of goods, ideas, and wealth were still restricted by technologies of communications that had changed little since the first millennium BCE. One sign of the limits of information exchanges is the immense ignorance about China that prevailed in medieval Europe, an ignorance matched only by Chinese unfamiliarity with western Afro-Eurasia.

In short, during the postclassical Malthusian cycle, Afro-Eurasian exchange networks, though not as connected as in modern times, were more integrated than ever before, and commercial activity flourished in all major agrarian civilizations. Innovation was more rapid than in the classical era, particularly during the astonishing era of growth under the Song. And much innovation, in this as in previous eras, came from states in which rulers were closely allied to mercantile elites and linked in widespread networks of commercial and information exchange.

THE EARLY MODERN MALTHUSIAN CYCLE: THE FOURTEENTH CENTURY TO THE SEVENTEENTH CENTURY

The First Global Network of Exchanges

In the fifteenth century, after the long slump associated with the Black Death, populations rose again throughout Afro-Eurasia. Once again, population growth stimulated commerce and urbanization. The commercial networks of the preceding cycle, which had decayed for much of the late fourteenth and fifteenth centuries, revived in the early sixteenth century—but now they reached even further. European traders played an important role in creating these links, now operating mainly by sea. And the activities of European merchants and sailors, usually with government backing, eventually led to one of the most significant breakthroughs of this period: the appearance of the first exchange networks to circle the globe. The bridging of the Atlantic early in the sixteenth century was an event of truly world-historical significance, and it is no accident that most historians of modernity, particularly within the Marxist tradition, have counted it as one of the defining events of the last millennium. As Marx himself put it: "World trade and the world market date from the sixteenth century, and from then on the modern history of Capital starts to unfold."[43]

The system of exchanges that first appeared in the sixteenth century linked the markets of Afro-Eurasia, the Americas, sub-Saharan Africa, and eventually even Melanesia, Australia, and Polynesia into the first truly global world system.[44] The new system was almost twice the size of any that had earlier existed, and it contained a much greater variety of goods and resources. The size of the new system and the scale of the exchanges that took place within it meant that more wealth was in circulation than ever before in world history. The sheer volume of wealth now flowing through international systems of exchange made the gradient between the largest and the smallest reservoirs of wealth in the world much steeper, and it increased the influence of the merchants and financiers who handled these exchanges. The widening gulf between the rich and the poor energized commercial flows of many kinds, and the economic "voltage" accumulating within the new global system drove a commercial motor of unprecedented power. Silver looted from the Americas by Spain energized European and world commerce, as it worked its way through Europe or moved via the Philippines to India and then to China. Chinese demand for silver (driven by the devaluation of paper and copper currencies, widespread commercialization in the countryside, and the monetization of taxation) fueled the global trade in silver.[45]

But other exchanges were as important. The transmission both ways of crops, of technologies, of peoples, and even of disease bacteria made possible by the linking of the Afro-Eurasian and American world systems is described in Alfred Crosby's *The Columbian Exchange* (1972). The swapping of diseases ensured that global integration was a destructive process for all the smaller world zones. By 1500 CE, exchanges of diseases within the more densely settled parts of Afro-Eurasia had increased overall immunities throughout Afro-Eurasia. But no such toughening had occurred in the Americas or the even more isolated communities of the Australasian and

Pacific world zones. Thus, when Europeans reached the Americas in the six-teenth century, bringing their diseases with them, it was overwhelmingly the Americans who died of Eurasian diseases rather than the other way around.[46]

Our figures are only educated guesses, but in the more densely settled regions of Mesoamerica and Peru, the fall in populations during the sixteenth century was truly catastrophic: populations may have fallen by as much as 70 percent, while populations in the Americas as a whole may have fallen anywhere from 50 to 70 percent.[47] Contemporaries, on both sides of the di-vide, were aware of imbalance in the exchange of diseases. As a native Amer-ican from Yucatán put it, before the Europeans came "there was then no sickness; they had no aching bones; they had then no high fever; they had then no smallpox [see figure 12.3]; they had then no burning chest; they had then no abdominal pain; they had then no consumption; they had then no headache. At that time the course of humanity was orderly. The foreigners made it otherwise when they arrived here."[48] English colonists from the Roanoke Island settlement of 1585 made the same observation, but from the other side of the epidemiological frontier. Thomas Hariot, the colony's surveyor, wrote that after they had visited native towns or villages,

> within a few dayes after our departure from everies such townes, that people began to die very fast, and many in short space; in some townes about twenties, in some fourtie, in some sixtie, and in one sixe score, which in truth was very manie in respect to their numbers. . . . The disease also was also so strange that they neither knew what it was, nor how to cure it; the like by report of the oldest men in the countrey never happened before, time out of mind.[49]

Other populations suffered just as severely when Europeans migrated to Australasia and the Pacific. Sub-Saharan Africans were usually spared, be-cause they had always been part of the wider Afro-Eurasian networks; in any case, they inhabited an even more dangerous bacteriological environ-ment than most Eurasians. Elsewhere, Eurasian diseases removed native pop-ulations, making settlement easier for the Eurasian migrants that eventu-ally turned large areas of the smaller world zones into Eurasian colonies, with Eurasian crops, domesticates, pests, and diseases.[50]

While the introduction of Eurasian domesticates transformed the econ-omies, the social structures, and the exchange networks of the Americas, the introduction of American domesticates had almost as great an impact in Afro-Eurasia. From the Americas came maize, beans, peanuts, many vari-eties of potato, sweet potato, manioc (cassava, tapioca), squashes, pumpkin, papaya, guava, avocado, pineapple, tomato, chili pepper, and cocoa.[51] Man-

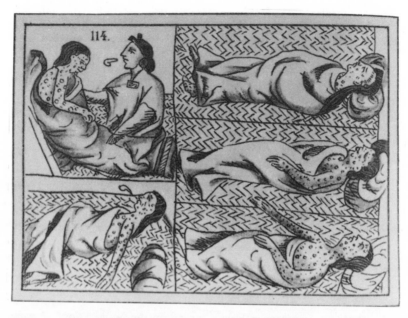

Figure 12.3.   Aztec victims of smallpox in the sixteenth century, from a sixteenth-century Spanish history of "New Spain." From Alfred Crosby, *Ecological Imperialism: The Biological Expansion of Europe, 900–1900* (Cambridge: Cambridge University Press, 1986), plate 9: from *Historia de Las Casas de Nueva Espana*, Volume 4, Book 12, Lam. cliii, plate 114. Used with permission of the Peabody Museum of Archaeology and Ethnology, Harvard University.

ioc has become a staple in many tropical regions of the Afro-Eurasian zone; maize and potatoes are staples in many temperate zones. In China, American crops were adopted more rapidly than anywhere else in Afro-Eurasia, after being introduced by the Portuguese in the sixteenth century.[52] Sweet potatoes were being grown as early as the 1560s, and more than a third of all crops today grown in China are of American origin.[53] Because such crops flourished where more familiar staples grew less well, American crops effectively increased the area under cultivation and thereby made possible population growth in many parts of Afro-Eurasia from the sixteenth century onward.

Patterns of Growth and Innovation

The vastly expanded flows of wealth and information generated in this period had a profound impact on states and societies throughout the world. In the Americas, their initial impact was rapid and destructive. Global integration caused the death of millions of individuals and the end of traditional empires, states, cultures, and religions. And the pattern was to repeated in each new world entered by Europeans, from Mauritius to Hawai'i.

In the Afro-Eurasian zone, the effects were subtler and slower to manifest themselves. But in much of Afro-Eurasia, and certainly in the most densely populated core zones, new and expanded exchange networks, population growth, state activity, and commercialization encouraged growth and innovation. As Joel Mokyr puts it:

> The age of discoveries was . . . the age of exposure effects, in which technological change primarily took the form of observing alien technologies and crops and transplanting them elsewhere. The aggressive Europeans adopted crops from America in exchange for the livestock, wheat, and grapes they transplanted into the New World. Furthermore, they also transplanted non-European flora from America into Africa and Asia and back in a massive act of what could be called ecological arbitrage. Thus, they introduced bananas, sugar, and rice into the New World, and cassava (also known as manioc) into Africa, where it eventually became the staple crop in many areas.[54]

Population increases were partly caused by the "innovation" of using American crops in Europe, China, and Africa. Cultivation of these new crops required a whole series of minor agricultural innovations, including different types of crop rotation, plowing, and irrigation. In China, the new crops were particularly important because they could be sown in regions unsuitable for rice; they also brought great change in Africa.[55] But there were significant developments as well in seafaring and in warships (which provided the technological resources for the creation of a unified world system in the sixteenth century), in mining technologies, in warfare, and in commercial methods.

However, rates of innovation were in some ways unimpressive; nowhere do they rise to the level found during the Industrial Revolution. Even in Europe, where the appearance of a global world system had the greatest immediate commercial impact, technological innovation in the middle centuries of the millennium—in areas outside warfare, shipping, instrument building, and metallurgy—was surprisingly sluggish.[56] As Peter Stearns observes:

> Western technology [in 1700] and production methods remained firmly anchored in the basic traditions of agricultural societies, particularly in terms of reliance on human and animal power. Agriculture itself had scarcely changed in method since the fourteenth century. Manufacturing, despite some important new techniques, continued to entail combining skill with hand tools and was usually carried out in very small

shops. The most important Western response to new manufacturing opportunities involved a great expansion of rural (domestic) production, particularly in textiles but also in small metal goods.[57]

In Afro-Eurasia, the large-scale effects of the new global networks of exchange were subtler and less direct. In all the core regions, population grew and commercial activity expanded. In China, between 1400 and 1700, population rose from ca. 70 to ca. 150 million. In India, in the same period, population rose from 74 to 175 million, and in Europe, from 52 to 95 million (see table 11.1). According to one recent estimate, Asian populations continued to grow faster than those of Europe well into the eighteenth century, at which time Asia contained ca. 66 percent of the world's population and produced almost 80 percent of the total value of the world's goods and services.[58] Historians used to assume that population growth in East Asia must have led to greater poverty in this era, but that assumption was wrong. On the contrary, as Andre Gunder Frank argues, it seems that

> Asians were preponderant in the world economy and system not only in population and production, but also in productivity, competitiveness, trade, in a word, capital formation until 1750 or 1800. Moreover, contrary to latter-day European mythology, Asians had the technology and developed the economic and financial institutions to match. Thus, the "locus" of accumulation and power in the modern world system did not really change much during those centuries. China, Japan, and India in particular ranked first overall, with Southeast Asia and West Asia not far behind.[59]

Indeed, as noted above, the dominance of Asian economies was understood by European observers such as Adam Smith even in the late eighteenth century. Nor was Europe yet dominant technologically. Philip Curtin writes that in the seventeenth century,

> the "European Age" in world history had not yet dawned. The Indian economy was still more productive than that of Europe. Even per capita productivity of seventeenth-century India or China was probably greater than that of Europe—though very low by recent standards. Europe's clear technological lead was still limited to select fields like maritime transportation, where design of sailing ships advanced enormously through the sixteenth and seventeenth centuries. Otherwise, Europe imported Asian manufactures, not the reverse.[60]

That surpluses of silver gravitated toward Asia throughout this period also suggests the centrality of Asia in the emerging world system of trade. And these changes were not just skin-deep: commercial activity affected all levels of society. In China, the government began collecting most of its taxes in cash rather than in kind in the sixteenth century, a clear sign of the extent of commercial exchanges even in the countryside. As Kenneth Pomeranz has shown, measures such as consumption of sugar or textiles or other nonessentials, as well as statistics on life expectancy, suggest that in the eighteenth century, standards of living were as high in China as in Europe.[61]

By all these measures, the expansion phase of the early modern Malthusian cycle—despite the vastly increased scale of exchange networks—was fairly typical in stimulating a moderate degree of innovation, not the high levels characteristic of the modern era. We might thus expect that much of the world was headed, sooner or later, for a Malthusian decline of some kind. There was a slowdown in growth in much of Afro-Eurasia in the seventeenth century, though not nearly as sharp a decline as that seen at the end of the previous cycle. Soon afterward, growth resumed in many different parts of the world, even in areas such as India and China that were to lag significantly in the nineteenth century. As late as 1800, observers such as Adam Smith and Thomas Malthus had good reason to assume that the pattern of Malthusian cycles that we have seen operating in the era of agrarian civilizations was a permanent feature of economic life.[62] And some modern researchers have argued that, absent one or two largely chance circumstances such as the existence of large coal reserves in Britain, they may have been close to the mark.[63]

However, there were other changes during the early modern Malthusian cycle that prepared the way for the decisive developments of the nineteenth century.

The Impact of Commercialization in Tributary Societies

Social structure models of innovation suggest that we should expect innovation to be most rapid where all sectors of society are tightly integrated into commercial networks, so that all sectors of society are influenced by the rules of efficiency and productivity that ensure success in a competitive commercial environment. The simplified version of Marxist models described in the previous chapter points to the importance of focusing on the impact of increased commercialization in two main areas: first, the increasing influence and power of mercantile elites; and second, the increasing involvement of the rural population (the majority of the population in most agrarian civilizations) in commercial activities of various kinds, until finally, when debts or expropriation sever them entirely from the land, they become wage earners whose lives are totally dominated by commercial networks.

Much research within the Marxist tradition has explored these issues, demonstrating that in most core areas of the Afro-Eurasian zone such processes had made great headway. Merchants and markets were of vital importance to the functioning even of the most traditional agrarian societies. Less commercialized states such as Poland or Muscovy actively supported commerce and, where possible, colonial expansion, particularly into such potentially profitable regions as the fur-rich lands of Siberia. In these ways, large parts of the world occupied by kin-ordered societies of various kinds were drawn into commercial exchange networks, often with profound effects on their traditional ways of life.[64]

Such changes could transform states, as increased reliance on commercial sources of revenue reduced the relative significance of traditional tributary revenues from feudal dues or land taxes, forcing even large tributary states to take more interest in commercial activities. Like many traditional states, Muscovite governments established monopolies over the most lucrative trades, including those in precious metals or furs. But in the seventeenth century, they began to explore ways of taxing domestic trades as well by levying sales taxes on salt and particularly on vodka. These items were strategic, for in a country where most peasants were largely self-sufficient, they were the only staples that could not be produced in the household, and therefore had to be purchased. Salt was necessary to preserve foods, while vodka soon became a vital ingredient of religious and social rituals in the villages. In 1724, taxes on liquor sales already accounted for 11 percent of government revenues; by the early nineteenth century, vodka taxes constituted the single largest source of revenue, providing between 30 percent and 40 percent of the government's total income.[65] As commercial revenues became increasingly important, the Russian government found that despite its professed hostility toward merchants, it had to dicker with them. At several points in the 1850s, the government feared that failure to offer sufficiently attractive terms to the powerful merchants who ran the liquor tax farms might lead to its bankruptcy. This example of a shift in fiscal practices is particularly striking because the Russian Empire in so many respects remained a paradigmatic tributary society well into the nineteenth century.

Commercialization affected rural areas as much as it did towns and states. In fact, by the middle of the second millennium CE, there were few rural areas within the great civilizations of Afro-Eurasia where peasants were not caught up in commercial activities of some kind. And in all these civilizations, large and growing numbers of people were dependent solely on wage labor. The Chinese countryside was commercialized early. Mark Elvin notes that in Song China, as early as 1000 CE, increased contact with the market made the Chinese peasantry into a class of adaptable, rational, profit-oriented, petty entrepreneurs. A wide range of new occupations opened up in the countryside. In the hills, timber was grown for the booming boatbuilding industry and for the construction of houses in the expanding cities. Vegetables and fruit were produced for urban consumption. All sorts of oils were pressed for cooking, lighting, waterproofing, and to go in haircreams and medicines. Sugar was refined, crystallized, and used as a preservative. Fish were raised in ponds and reservoirs to the point where the rearing of newly-hatched young fish for stock became a major business. . . . Growing mulberry leaves became in itself a profitable undertaking, and there were special markets for mulberry saplings. Peasants also made lacquer goods and iron tools.[66]

The more commercialized the regions the peasants lived in, the more varied their possibilities for generating commercial revenues. They could sell their surpluses of crops or specialize in commercial crops such as litchis or tangerines; they could make and sell fibers or engage in other part-time crafts; they could send individual members of the family to town in search of wage labor. As more of their income came from commercial activities, peasants throughout Afro-Eurasia had to abide by the rules of commerce rather than the rules of tribute. It was no longer enough to pay the tribute demanded by landlords or states; they also had to conform to the standards of productivity or quality demanded by their customers or employers, whether the wool merchants in Europe or the timber or wood merchants in nearby towns. And in this way—sometimes rapidly, sometimes almost imperceptibly—peasants found themselves turned into petty entrepreneurs or wage earners.

But in China as in much of the rest of Afro-Eurasia, such processes had their limits. Though deeply involved in commercial activities of many kinds, peasants everywhere resisted the final step of severing contacts with the land, and governments used to raising traditional land taxes often supported their resistance. In China in the eighteenth century, R. Bin Wong points out, "Many peasants owned at least some of their own property, and many also rented some land. Virtually all land was worked at the household level of production; landlords who expanded their bases of direct production in response to market opportunities were few."[67] Traditional peasants often retained a deep ethical commitment to the ancient principle that they had a right to the land; they believed that the land was not something to be bought and sold, like so many sacks of grain. Such attitudes survived well into the twentieth century in many countries. In Russia as late as 1906, in a petition sent to pro-peasant deputies in the newly established parliament or duma, mutinous peasant soldiers insisted:

> In our view, the land is God's, the land should be free, no one should have the right to buy, sell or mortgage it; the right to buy is fine for the rich, but for the poor it is a very, very bad right. . . . We soldiers are poor, we have no money to buy land when we return home from service, and every peasant needs land desperately. . . . The land is God's, the land is no one's, the land is free—and on this, God's free land, should toil God's free workers, not hired laborers for the gentry and *kulaks* [rich peasants].[68]

Though the Chinese countryside was highly commercialized in the eighteenth century, structures of ownership and control of the land limited the extent to which the majority of the population could be involved in commercial networks. And, according to traditional Marxist models, such limits were bound to restrict long-term rates of innovation.

Commercial attitudes and practices had entered deeply into rural life and had even affected the practice of governments in some of the most traditional tributary empires, but they had not yet undermined the structures of power and production typical of traditional agrarian societies.

A NEW GLOBAL TOPOLOGY: THE CHANGING ROLE OF EUROPE

In western Europe, the commercialization of social, political, and economic structures went further than anywhere else in the Afro-Eurasian zone. European societies were younger and more malleable than those of the older core regions; their states were smaller and more susceptible to international commercial pressures; they were also, for reasons that will be discussed later, more open to commercial activities; and perhaps most important of all, the changing topology of global exchange networks ensured that during the early modern Malthusian cycle, the relative volume, variety, and intensity of information and commercial exchanges passing through Europe were greater than anywhere else.

The Changing Topology of Global Exchanges

The creation of a global network of exchanges affected Europe decisively because it was accompanied by a rearrangement in the topology of global exchanges. At the largest scale, the structures of exchange systems in the Afro-Eurasian system had been relatively stable for several millennia, with hub regions at the eastern end of the Mediterranean, in northern India, and in central Asia; since the first millennium BCE, the center of gravity had shifted eastward toward the more densely settled regions of northern India and China. But with the linking of the Afro-Eurasian and American world zones, western Europe and the entire Atlantic coast suddenly emerged as a new hub region, through which flowed most of the exchanges connecting the world zones of Afro-Eurasia and the Americas. What had been a peripheral region in the Afro-Eurasian zone abruptly became the most important hub in the largest network of exchanges that had ever existed. Even if the center of gravity of global exchange systems remained in the Far East as late as 1800 CE, the greatest diversity of exchanges could be found in the new hub region of western Europe.

This was a fact of immense consequence, in particular for the future of Europe. It was in a sense contingent. Europe was in the right place to take advantage of the vastly increased exchanges of the emerging global exchange network. Having been on the margins of the Afro-Eurasian system for millennia, Europe in the sixteenth century had the good fortune to suddenly find itself at the hub of the largest and most varied network of exchange in history. Its repositioning in the center of the new global network revolutionized the life of the entire region. The exchanges now passing through Europe were much larger than any previous flows of this kind. The transfers of silver from the Americas to Europe to the Islamic world and on to the Far East between the sixteenth century and the nineteenth century provide just one instance of Europe's crucial role as middleman.[69] Clearly, we do not need to rely on European exceptionalism to explain Europe's distinctive role in the modern world, just as we do not need to treat the emergence of urban civilization in Sumer as a sign of that region's exceptionalism. As Andrew Sherratt has argued,

> western Europe only stumbled on a new role with the discovery of the New World and the growth of Atlantic links. *There is thus no predetermined relationship between social or economic sophistication and the way a region develops; from a local stand-point, change is often arbitrary and unpredictable.* Enlargement of the world-system, and alterations of its shape and connections, thrusts areas into new roles for which at the time they often seem unfitted.[70]

As in Sumer, more than 4,000 years before, a sudden increase in the scale of exchanges and a sudden rearrangement of exchange networks encouraged and made possible a quite new scale of investment in what had previously been a backwater.[71]

But we should not make too much of contingencies, for Europe's strategic position was not entirely accidental. Other regions within the Afro-Eurasian world zone might have built and financed merchant fleets capable of traveling around the entire globe, perhaps fleets similar to those commanded by the Ming admiral Zheng He early in the fifteenth century. If they had done so, they, rather than the Atlantic rim, might have become the hubs of the new global system. Indeed, a world in which hub and center of gravity coincided in China might have generated an even more rapid and chaotic modern revolution than the one we know, whose hub and center of gravity long lay in different parts of the world. The topology of the new system was not determined purely by geography; Europe became the hub of the new global system of exchanges in part because it was preadapted for the role.

In two ways, western European societies were exceptionally well prepared to survive in the new, global commercial system that emerged in the sixteenth century. First, they were young and flexible. States emerged in northwestern Europe within the past 1,500 years. By then, powerful and successful states had already existed for more than 3,000 years in Mesopotamia and at least 2,000 years in China. The success of these large, tributary states is a measure of the extent to which their political and military structures, their class alliances, and their values were adapted to the social and political ecology of the agrarian era. In contrast, the younger polities of Europe evolved in a more commercialized world. Their structures and traditions of government, their characteristic class alliances and attitudes, and their traditions of warfare were adapted to this very different social and political environment. Of course, there were striking differences between different European states, differences that are described superbly in Charles Tilly's *Coercion, Capital, and European States, AD 990–1992* (rev. ed., 1992). Nevertheless, the general rule stands: the state systems of Europe north of the Mediterranean (and the new colonial states of the Americas to an even greater degree) evolved their basic structures and attitudes in a more commercialized world than that of the classical era.

Second, the European state system was characterized by a cluster of features that we have seen before (in chapter 10), whose combined effect was to encourage elites to look more benignly on commercial activity. In western Europe, unlike in Mesopotamia or China, no new tributary empires emerged after the collapse of those that had ruled the region in the classical era. The Holy Roman Empire aspired to but failed to achieve this role. As a result, western Europe emerged during the postclassical Malthusian cycle as a region of many small states, in constant competition and close to the major trading routes of the Mediterranean world. This is a familiar setup.[72] In periods of limited commercialization, such as the era in which the city-states of classical Greece flourished, this cluster of factors created commercially and militarily adventurous polities that could prove surprisingly powerful. Their traders traveled much of the known world; and, as noted above, their armies could sometimes challenge even the juggernauts of the tributary world, as the Greek city-states did at the battles of Marathon and Salamis that drove back the Persians. But they could not hope to replace the great empires permanently. In the much more commercialized world of the eighteenth century, analogous differences between states and regions could prove more decisive.

These two factors explain why European societies were already well adapted to the economic, political, and military realities of a highly commercialized world. More specifically, they help explain the extremely competitive and often brutal commercialism typical of European trading systems from the fifteenth century onward. In the expansion phase after the Black Death, European states engaged in a life-and-death struggle for a share of the commercial wealth available in expanding Eurasian trade networks. Even the most traditional of states, such as the militaristic polities that expelled Islamic rulers from Spain, or the powerful French state ruled by Louis XVI, understood the importance of commercial revenues. The Spanish monarchy, in its heyday, relied overwhelmingly on commercial revenues and loans, while seventeenth-century governments in France depended on a wide range of new consumption and commercial taxes.[73] Increasing commercial activity and government interest and support helped drive European improvements in ship design and navigation, in textile working (textiles were the second-largest economic sector in most premodern economies), in canal locks, and even, perhaps, in printing. Indirectly, they were factors in the Iberian conquest of Atlantic trade networks and the subsequent conquest of the agrarian civilizations of the Americas.[74] The huge wealth extracted from the Americas—and the immense commercial, political, and military power that could be built on these revenues, as the examples of Spain and Portugal made clear—encouraged an intensification of this aggressive commercialism, which became the hallmark of European states in the early modern period. This complex of state power dependent on commercial revenues also explains why European ships could be found in all parts of the world by the sixteenth century.

So it was not entirely by accident that Europe found itself at the hub of the new, global system of exchanges. The appearance of a highly competitive world of expansionist and commercialized states on the shores of the Atlantic guaranteed that eventually the Atlantic would be bridged. Indeed, a fragile and all-too-temporary bridge had already been built by Viking navigators in the previous Malthusian cycle, prefiguring the aggressive expansionism of European states in later centuries.

The Impact of Global Exchange Networks in Europe

Europe's strategic position made it certain that Europe would be affected by changes in this new global system more decisively than any other part of the world. Information exchanges have too often been ignored in accounts of modern world history. Yet, as I have argued in previous chapters, at large scales, changes in the amount and variety of information exchanged between different communities may be a crucial determinant of rates of innovation. Europe in the early modern period found itself swamped by new information. At the hub of the new global exchange system, it was the first to receive a mass of new knowledge about the New World, as well as about other regions of Afro-Eurasia. Europe became a sort of clearinghouse for new geographical and cultural lore. Thus it was here that the torrent of new information flowing through the first global exchange network had its earliest and greatest impact on intellectual life and activity.

The digestion of this new mass of information transformed European intellectual life. Margaret Jacob writes that "the cumulative effect" of the travel literature of the sixteenth and seventeenth centuries "had been to call into question the absolute validity of religious customs long regarded, especially by the clergy, as paramount."[75] As the arena of information exchange expanded, and as the printing press circulated ideas more quickly, traditional systems of knowledge faced ever harsher tests of their truth claims and had to shed many of their more parochial features. As Andrew Sherratt has recently written in an essay stressing the role of widespread exchanges in human history, " 'Intellectual Evolution' . . . consists principally in the emergence of modes of thinking appropriate for larger and larger human groupings. . . . This transferability has been manifested in the last five hundred years in the growth of science, with its striving for culture-free criteria of acceptance[.]"[76] This infusion of new information and knowledge is the best available explanation for the radical skepticism about traditional accounts of reality that lies at the heart of the modern scientific project and that first becomes apparent in Europe in the sixteenth century. From the seventeenth century onward, European "natural philosophers" knew that they were working with a vastly expanded body of information, much of which undermined the credibility of traditional maps of reality. Steven Shapin observes, "Philosophical schemes based on restricted knowledge were likely to be faulty for just that reason, and the expanded experience afforded, for example, by the voyages of discovery to the New World was an important support for currents of early modern skepticism about traditional philosophical systems."[77] Skepticism about the very foundations of knowledge, a search for increasingly universal conclusions (exemplified by Newton's laws of gravity), and more rigorous testing procedures (such as those used by Galileo) can thus be seen as important consequences of the expanded framework within which knowledge systems were tested in an emerging global system of information exchanges.

The impact of global networks of exchange on European social, political, and economic structures is more familiar but was equally significant. European merchants and the rulers who backed them were rewarded spectacularly and rapidly. Spanish soldiers conquered the agrarian heartlands of Mesoamerica and Peru, while Portuguese, French, Dutch, and British expeditions began to colonize other regions of the Americas previously occupied by stateless communities of farmers or foragers. The windfall of American silver sustained Spanish power in the sixteenth century. Indeed, Spain depended so heavily on American silver that when the supply ran out in the seventeenth century, its commercial and political influence declined. American silver also helped European traders, usually with the backing of their governments, fight or buy their way into the rich trade networks of Asia. As Andre Gunder Frank has suggested, the piratical way they gate-crashed the commercial networks of South and Southeast Asia during this period is analogous to how the Mongol armies had seized the trade routes of the Silk Roads three centuries before.[78] European traders now began to take on the linking role played by the Mongols in the world system of the thirteenth century, but they did so in the far larger arena of the new global trading system.

The rewards of such activities encouraged mercantile elites and states to build on the tentative alliances they had constructed earlier. Serious dependence on commercial revenues gave states a particular structure and distinctive policies. First, merchants often enjoyed unusually high status in such polities; in some, such as Venice or Holland, they *were* the state. Second, states dependent on commercial revenues had to support commercial activity, and they therefore protected the rights of merchants with an enthusiasm uncommon in the larger and more traditional agrarian empires. Such policies often drew states directly into commercial warfare. Finally, such an environment could have even subtler effects on the attitudes of ruling elites, by encouraging them increasingly to think not just about ways of capturing tributes but also about ways of generating new entrepreneurial wealth. The mercantilist policies of European states in the seventeenth century—

such as the Navigation Acts of the English commonwealth, which protected British commerce within British colonies—are good examples of new government attitudes toward commerce and the actions that these changes encouraged. Also illustrating this trend is the proliferation throughout Europe of patent laws, which were pioneered in Venice in the fifteenth century. Governments also began to promote innovation through the founding of scientific societies or the offering of prizes. (The most famous of these prizes belongs, strictly speaking, to the next chapter. In 1714, the British government offered an award for the construction of a clock robust and reliable enough to be carried on a ship so that sailors could determine longitude. The prize was not claimed successfully until 1762 by John Harrison.)[79]

Over time, commercialization transformed traditional tributary elites. Such transformation was most likely to occur when demands on elite incomes rose sharply in environments where commercial revenues were available for the taking. The English wool trade offers a classic example, for it tempted landowners to clear the land of tenants and replace them with sheep, particularly in the sixteenth century, when new land became available as a result of the dissolution of the monasteries. In England, a traditional tributary aristocracy became increasingly engaged in commerce, either by producing wool for the markets of Flanders, or by investing in foreign trade or piracy (such as the expeditions of Sir Francis Drake and Sir John Hawkins), or by marrying into the merchantry. Behind the complex rituals of aristocratic precedence that survived in the early modern period, we can see a slow change in the personnel and in the nature of the nobility. Throughout western Europe, nobles were being imperceptibly transformed from tribute takers into commercial and entrepreneurial landowners. Many nobles, like the French judicial scholar Charles Loyseau, doubtless continued to believe throughout this period that "it is gain, whether vile or sordid, that derogates from nobility, whose proper role is to live off rents."[80] But in practice, this idealized image of the nobility as a class of tribute extractors was becoming more and more anachronistic. Inspection of their account books would have shown that many were turning slowly into capitalists, even though they might have been horrified to be told so. At the same time, merchants "commercialized" the nobility by marrying into it, by buying titles (particularly in France), or by establishing partnerships with aristocrats keen to exploit their financial and commercial expertise. Where aristocrats refused to act more entrepreneurially or to ally with merchants who could help them do so, they eventually failed. In nineteenth-century Russian literature, the classic symbols of such failure include Stepan Oblonsky (Anna Karenina's brother) and Mme Ranevsky in Chekhov's The Cherry Orchard.

Alliances between merchants and governments could eventually turn into symbiosis. Many governments had worked closely with merchants before, and some had included merchants within their structures; but now such co-operation began to take place even in moderately large states with a global reach. In some cases, merchants became integral to government. At the one extreme was Holland, where the merchantry *was* the government; at the other was Spain or Russia, where traditional governments relied only reluctantly on merchants for loans or the conduct of important commercial operations. In between were states such as Britain and France, in which merchants and mercantile activities of various kinds were gradually incorporated within the structures of government.[81]

One of the most spectacular products of the emerging symbiosis between governments and merchants was a highly commercialized style of warfare, which eventually enabled mercantile states to compete successfully with tributary empires in war as well as in commerce. The violently competitive environment within Europe itself ensured that the commercialization of European states would commercialize warfare, too. This process, sustained by massive flows of American silver, led to a revolution in military technology that raised both the destructiveness and the cost of warfare to entirely new levels. Charles Tilly has argued that in Europe, states formed primarily in response to the needs of war.[82] As in early Sumer and so many other regional systems of competing small- or medium-sized states, warfare was endemic. Thus preparing for war and mobilizing the needed soldiers, weaponry, and supplies were central tasks of government. The military consequences of such systems are captured well in a conversation that a Jesuit, Giuldo Aldeni, had with a Chinese friend who asked, "If there are so many kings, how can you avoid wars?" Aldeni replied that intermarriages between rulers or the authority of popes was sufficient to maintain the peace, but his answer was disingenuous. In reality, his Chinese friend was perfectly right: Aldeni's conversation took place during the Thirty Years' War.[83] China itself provides an interesting contrast, for in the middle of the seventeenth century, the Manchu overthrow of the Ming dynasty led to a period of intense warfare. In these wars, cannon and musket, based on Ottoman or South Asian designs or crafted to Chinese specifications by Europeans in China, played a vital role. But once the Manchu (Qing) dynasty had established its superiority, military innovation slowed again and the gap in military technology between China and Europe began to widen rapidly, leaving China extremely vulnerable by the nineteenth century.[84]

Yet though the general pattern is an old one, the particular ways in which European states mobilized for war are distinctive. Tilly notes that before the fifteenth century, mobilization for war was handled through methods that we can recognize, broadly, as tributary: "tribes, feudal levies, urban militias, and similar customary forces played the major part in warfare, and monarchs generally extracted what capital they needed as tribute or rent from lands and populations that lay under their immediate control[.]"[85] However, from the fifteenth to the early eighteenth century, it became increasingly common for states to purchase or hire troops, often relying on loans from large capitalists to do so. In this way, military victories increasingly became a measure of success in commerce. As early as 1502, a French veteran of the Italian wars, Robert de Balsac, ended a study on warfare with the remark that "most important of all, success in war depends on having enough money to provide whatever the enterprise needs."[86] During the next hundred years, the influx of new forms of wealth sharply raised the stakes in this centuries-old European arms race.

The shift to more commercial methods of war making reflected in part the commercialized nature of the European state system. But equally important was the fundamental change in military techniques known as the *gunpowder revolution*.[87] Its technological roots reached throughout the Afro-Eurasian system. Chinese experiments with gunpowder, under the Song, were probably influenced by knowledge of Byzantine techniques for using oil in incendiary devices (which created Greek fire). This knowledge was transmitted to Southeast Asia, and then China, by Arab intermediaries. The explosive properties of gunpowder were first used in war by the Jin, the northern rivals of the Song, in 1221.[88] But it was in Europe that these technologies were developed most fully. Already in the fifteenth century, siege guns had begun to revolutionize warfare, as they required the building of more complex and expensive fortifications. Mobile siege guns spread these costs even further. The increased use of portable muskets in the sixteenth century transformed infantry warfare, making necessary an entirely new level and type of training and discipline. The placing of cannons on ships similarly transformed naval warfare. The rising cost of armies and navies favored those states with the largest treasuries and the sources of funding that could be most readily mobilized—that is, highly commercialized states, such as Holland. But even traditional states, such as Muscovite Russia, turned to new, more commercial, sources of income to pay for military reforms. It was Ivan the Terrible who, in the sixteenth century, began to create the Russian state's vodka monopoly; by the nineteenth century, it had become the most important of all sources of revenue for the Russian state, covering most of its defense spending.[89]

Scholars broadly agree that commercial activities profoundly affected European states in the early modern period. There is less agreement about their impact in rural areas of Europe. In traditional historiography, the western European countryside has been viewed as decisively capitalistic and thus utterly different from the countryside of, say, China or India. Recent research has forced us to qualify these conclusions, as we have realized how profoundly commercialized country regions might be, even in East Asia. Nevertheless, it still seems likely that at least in some parts of western Europe (particularly Britain), commercialization of the countryside had proceeded further than in much of East Asia, beginning to transform traditional patterns of ownership and control of the land and to break down traditional structures that protected peasant access to the land.

In Europe, as elsewhere, it was never difficult for commerce to get a toehold in rural areas. Exotic urban trinkets or more vital goods such as salt readily found rural markets, if only in a sort of barter trade. But such trades were unlikely to revolutionize rural lifeways. More important were pressures that forced peasants to seek wage labor as a supplement to farming. Many types of pressure drove European rural dwellers, like their counterparts in East Asia, to supplement their agricultural activities by entering the marketplace. Such pressures could be fiscal. As landlords and governments came to depend on commercial activity, they often demanded that traditional dues or taxes be paid in cash rather than in goods or services, a change that forced taxpayers to earn cash. Population pressure, by creating land shortages, could have the same effect. In many parts of Europe, the population growth of the postclassical Malthusian cycle meant that by the thirteenth century, perhaps half of all peasant households lacked sufficient land to support themselves without seeking wage work of some kind. In their study of preindustrial Europe, Catharina Lis and Hugo Soly note that

> in Picardy around 1300 . . . 12 per cent of the population consisted of landless paupers and beggars, dwelling in huts outside the village, living off wage labor; . . . 33 per cent cultivated a morsel of ground and were likewise obliged to market their labor to make ends meet; . . . 36 percent were poor, not owning a plough-team of oxen or horses, but generally succeeding in buying off labor services; . . . 16 percent had a holding sufficiently large to escape all difficulties; and . . . 3 percent dominated everyone else.[90]

Where the land could not generate enough to feed households and cover their obligations to states, landlords, and others (including churches), the peasants had several options. They could try to sell rural produce more advantageously on local markets, though here they often faced competition from larger producers. They could borrow from local moneylenders, which, in an era of expensive credit, was often the most dangerous way of entering the moneyed world. They could also engage in commercial activities within the household, such as spinning or weaving. Such processes, which have come to be known as *proto-industrialization*, could create entire regions in which rural incomes came primarily from domestic industrial activities. Maxine Berg's account of domestic industries in Staffordshire in the late seventeenth century gives some idea of their immense variety:

> There were wood turning, carpentry and tanning in the Needlewood Forest, coal in south Staffordshire, as well as iron and metal goods including locks, handles, buttons, saddlery and nails, coal and iron in Cannock Chase. Kinver Forest in the southwest had scythesmiths and makers of edge tools, and there were glassworkers on the Staffordshire-Worcestershire border at Stourbridge. Bursham in the northwest had a pottery industry, and there was ironstone mining in the northeast. Leather working and textile weaving in hemp, flax and wool were scattered throughout the country.[91]

She adds that in Essex in 1629, there were already 40,000 to 50,000 people so dependent on the manufacture of cloths that they "were not able to subsist unless they be continually set on work, and weekly paid," and a trade crisis could cause instant impoverishment of thousands.[92] Households could send some members of the family away to earn wages, either on the land or in nearby towns. Finally, at the bottom of this long and slippery slope, some workers found they had to abandon the land entirely and try to survive as wage laborers.

The strategies familiar in rural areas today could be found in all regions of agrarian civilization where peasants came under significant commercial, fiscal, or demographic pressure. Each maneuver increased the cash component of household budgets, drawing it further into the commercial world. Peasants thus found themselves reluctantly entering the world of capitalism. Here is a description of these processes taken from a social history of seventeenth-century France:

> Faced with a staggering and chronic imbalance between the grain they could call their own and the minimum needed for survival, most peasants resorted to improvisation. They rented a few extra acres to supplement their own. They hired out in the busy summer season to work on the larger farms. They cultivated their gardens intensively, selling vegetables and fruit at nearby markets. A single, skinny cow provided milk. Pigs were few, in the Beauvaisis, where

they tended to compete with human beings for nourishment. Four or five chickens in the barnyard, a few sheep out to pasture with the communal herd, this was as much as the ordinary peasant family could afford. Add the meager wages earned in the winter months from spinning and weaving cloth, and the deficit could almost be made up, in a good year. In bad times, peasants could not pay their taxes. The time came, inevitably, when they had to borrow grain. These debts resulted, sooner or later, in the loss of a portion of their remaining land. Land-hungry and indebted, peasants faced the risk of losing their prized status in the community, of sinking to the level of the landless poor.[93]

As peasants and their landlords entered networks of entrepreneurial activity, both groups found their relationship with the land changing. For elite groups whose incomes came increasingly from commercial sources, in environments where ever greater amounts of agricultural produce were marketed, it was no longer vital to supply peasants with land. Because landlords now had sources of income that did not depend on peasant farming, they could, as in the extreme case of sixteenth-century England, replace peasants with sheep and still survive. As a result of such changes, states, landlords, and even some wealthier peasants began to see the land itself as a source of commercial profits rather than just a source of produce. In some countries, such as England, governments encouraged the commercialization of the land by abolishing or buying up ancient rights to it and expropriating tenants whose claims were merely customary. There, in the course of a mere three centuries (from 1500 to 1800), the expropriation of traditional peasant rights to the land through enclosures destroyed the traditional peasantry. Elsewhere, peasants were squeezed from the land by slower and sometimes more agonizing pressures of taxation, indebtedness, crop failure, and land shortage. Sometimes, as in postrevolutionary France, their rights to the land were protected, but commercial pressures ensured that to survive, they had to become petty entrepreneurs. Everywhere, commercialization, as it penetrated the countryside, turned the land into a commodity and turned peasants into wage earners or petty entrepreneurs. In this way, capitalism began to pervade all corners of rural life.

Commercialization of the land steepened the gradient of wealth, for it began to undermine the basic rule of agrarian civilizations: rural producers *had* to be provided with land. The analogy Marx used to describe the change was electrical. What he called the "primitive accumulation" of capitalism was, unlike the simpler, cumulative forms described in the previous chapter, a sort of social "electrolysis," like the accumulation that occurs in a car battery. There, potential power is generated by the gravitation of one ion to the negative pole of a battery and of another ion to its positive pole.[94] During primitive accumulation, property and wealth gravitated toward a property-owning class, while absence of property characterized an emerging class of proletarians. Particularly in its earlier stages, this was a painful and predatory process; primitive capitalism, like any novice predator (and like the earliest and simplest forms of tribute-taking), was at first more concerned to consume than to protect its prey.[95] Yet, as Marx argued, the increased potential energy generated by this social electrolysis is what explains the dynamism of the capitalist system. Removing peasants from the land forced them to engage, decisively and permanently, in wage-earning activities. As wage earners, they found themselves competing with other wage earners, while as traditional peasants, their main goal had been sheer survival. As wage earners, the price they paid for inefficiency was dismissal and possible destitution; as peasants, it had merely been poverty because they still had some land to feed themselves from. So, as Marx claimed, driving peasants from the land was a crucial step in creating a world in which competition forced the bulk of the population to concern themselves, like merchants, with issues of efficiency and productivity. Like merchants, they had to buy and sell (because they could no longer produce their own food and clothing); and, like merchants, they had to work harder and harder just to survive in an increasingly competitive world. Marx used the notion of "absolute surplus value" to explain the increasing burden of work in the early history of capitalism. Recently, Jan de Vries has argued that in Europe at least, an "industrious revolution" preceded the more familiar "Industrial Revolution" of the eighteenth and nineteenth centuries.[96]

What remains uncertain is whether these processes had proceeded much further in western Europe than in many other parts of Afro-Eurasia. It is tempting to say that while most peasants were engaged in market activities of some kind by 1700, the region in which most had actually suffered land expropriation was western Europe, particularly Britain. Nevertheless, as recent research has shown, the differences are not clear enough to justify the claim that western Europe, or Britain, was now "capitalist," whereas, say, China was not.

A WORLD RIPE FOR TRANSFORMATION?

This is a frustrating conclusion. The sudden creation of a global network of exchanges had transformed economic and social systems in many parts of the world. Though devastating the indigenous populations of other world zones, it had magnified the wealth concentrated in the more commercial regions of Afro-Eurasia. Both Afro-Eurasia and the Americas were integrated into global systems of exchange so that by 1700 the world was significantly more commercialized than it had been a few centuries earlier. In some regions, social structures now approximated more closely than ever before the ideal type of a capitalist economy. Rural producers were deeply implicated in entrepreneurial or wage-earning activities of some kind, and commercial activity was breaking down the traditional isolation of the village world. Furthermore, many other parts of the world outside the core areas of agrarian civilization were also enmeshed in networks of entrepreneurial activity. These included settler regions in North and South America and Siberia, as well as significant parts of Africa and, by the end of the eighteenth century, much of the Pacific and Australasia. Moreover, as in earlier periods, expansion in exchange networks and growth in populations, commercial activity, and state activity had stimulated major developments in some sectors of the economy, including commerce, mining, and warfare, as well as many smaller but highly significant innovations in agriculture (e.g., the introduction of new crops). Finally, and perhaps most important of all, the sheer size of the modern world system magnified the possibilities for commercial and intellectual synergy by increasing the volume of trade and the extent to which new goods and new ideas from one region could stimulate economic activity in other parts of the world system. In this huge, global arena, commercialization was not only more extensive; it was also more dynamic in its social, political, and economic effects. It is tempting to think that the world had attained the threshold of capitalism as Marx defined it: "an accumulation of use values sufficiently large to furnish the objective conditions not only for the production of the products or values required to reproduce or maintain living labor capacity, but also for the absorption of surplus labor."[97]

Stimulated by the sudden emergence of a global exchange system, and a sudden increase in the volume, variety, and intensity of exchanges of many different kinds, the modern world system had approached the threshold of modernity—but had not yet crossed it. There are important respects in which the world in 1700 remained decisively premodern and precapitalist. Modernity is unthinkable without levels of agricultural productivity high enough to release a majority of producers from agricultural work. Yet nowhere in the world had such thresholds clearly been crossed by the early eighteenth century (though things looked rather different by the end of that century). England comes as close as any region to an exception, for there, by the late seventeenth century, a relatively entrepreneurial class of landlords controlled ca. 70 to 75 percent of all cultivable land, and 40 percent of the population was no longer in agricultural employment.[98] But these figures also demonstrate that more than half of the population continued in agricultural employment of some kind, and about three-quarters of the population still lived in hamlets or villages.[99] Even England remained primarily an agrarian country like all tributary societies for the previous 4,000 years, with ca. 50 percent of its population employed in agriculture as late as 1759.[100] "What remains true," observes Peter Mathias, "is that the greatest single flywheel of the economy was the land, the greatest source of wealth in rents, profits and wages, and the greatest single employer. Directly and indirectly much of industry depended upon the domestic harvest for its raw materials. The brewer, miller, leatherworker, chandler, weaver, even the blacksmith in village England were supporting and being supported by agriculture."[101] Elsewhere, the changes were considerably less noticeable; in France, for example, ca. 85 percent of the population may have been peasants, ca. 13 percent town dwellers, and ca. 1 percent nobles.[102]

The limits to social and economic changes before the eighteenth century explain the other striking aspect of the early modern era: the continued slowness, by modern standards, of innovation. As of 1700, it would be extremely hard for a visitor from another planet to have detected two of the most important features of the modern world: the dominant role of Europe and the accelerating rate of innovation.

SUMMARY

During two large Malthusian cycles, the first before the fourteenth century and the second between the fourteenth and seventeenth centuries, there was a sustained and accelerating increase in rates of accumulation in the major regions of agrarian civilization. Throughout these core regions commercialization also increased significantly, particularly after the emergence of a global network of exchanges in the sixteenth century. In some areas, such as Song China or Europe beginning in the sixteenth century, commercialization led to the emergence of polities committed more to commercial than to tributary forms of wealth. In short, in some regions there began to emerge what we can call *capitalist states*, and world markets as a whole were becoming much larger and more integrated.

Nevertheless, no revolutionary changes occurred in this period. In the eighteenth century, it would still have been appropriate to describe the dominant political structures of the emerging world system as tributary rather than capitalist. Despite the high levels of commercialization to be found in many areas, the most powerful governments remained traditional in their attitudes and their economic and social policies. Perhaps the clearest sign of this continuity with the past was that Asia remained the heartland of the world system—a fact that historians have only recently understood clearly.

Even in Europe, where commercialization had gone furthest in breaking down traditional political structures, it had a limited impact on how goods were produced in the countryside. Though capitalist structures dominated trading systems and shaped the policies of major states, they did not yet dominate production. As Charles Tilly has written, "Through most of history, indeed, capitalists have worked chiefly as merchants, entrepreneurs, and financiers, rather than as the direct organizers of production"[103]—an observation that remained true in 1700. Capitalism was transforming commerce, but it had not yet transformed methods of mass production. The basic unit of production remained the household: the peasant household on the farm or in domestic industries, and the artisan household in the towns. Though wages were increasingly important to them, these people were not yet wage laborers. Thus commercial methods and attitudes had not yet significantly affected production, which remained small-scale and traditional. Europe's social structures also remained traditional in many ways, as can be seen most clearly in the dominance of agriculture and of the peasantry.

So, in the eighteenth century, there existed a global world system, in which traditional tributary structures were still dominant. However, all regions of this system were now highly commercialized as a result of a long and accelerating process of accumulation of knowledge and of resources, particularly commercial resources. Furthermore, in some regions, particularly in Europe, capitalist structures were powerful enough to dominate state structures and governmental policies, and some of these neo-capitalist state structures were powerful enough to militarily challenge major tributary states. This combination—a world system that was already highly commercialized, and some regions in which political structures were also being transformed—provided the preconditions for the rapid creation of an entire world system driven by the dynamic imperatives of capitalism.

FURTHER READING

The literature on world history in the last 1,000 years is huge and rich, but there is no consensus on many of the key issues. Mark Elvin offers what is still one of the best accounts of economic growth under the Song in *The Pattern of the Chinese Past* (1973). Robert Lopez offers a traditional, Eurocentric account of expansion in medieval Europe and its significance in *The Commercial Revolution of the Middle Ages, 950–1350* (1971), which can be supplemented by Carlo Cipolla's *Before the Industrial Revolution* (2nd ed., 1981). Eric Jones stirred up a new round of debate on global processes leading to modernity with two immensely influential studies, *The European Miracle*, 2nd ed. (1987) and *Growth Recurring* (1988). These provoked many replies, the most recent of which have downplayed the role of Europe and highlighted the high levels of productivity and the high living standards of East Asia in the early modern period. Among the most recent works in this vein are studies by Janet Abu-Lughod (*Before European Hegemony* [1989]), Andre Gunder Frank (*ReOrient* [1998]), Kenneth Pomeranz (*The Great Divergence* [2000]), and R. Bin Wong (*China Transformed* [1997]). Alfred Crosby has done more than anyone to emphasize the importance of ecological exchanges between and within the Afro-Eurasian and American world zones, in *The Columbian Exchange* (1972) and *Ecological Imperialism* (1986). Works by William McNeill (*The Pursuit of Power* [1982]) and Geoffrey Parker (*The Military Revolution* [2nd ed., 1996]) have explored the military revolution of the early modern period, while Charles Tilly's *Coercion, Capital, and European States, AD 990–1992* (rev. ed., 1992) offers the best one-volume account of state formation in Europe in the last millennium.

BIRTH OF THE MODERN WORLD

In the past 250 years, the Modern Revolution has transformed the world. Tables 13.1 and 13.2 and figure 13.1 offer some comparisons of industrial output over most of this era. And the first thing they suggest is that global industrial output has increased by almost 100 times. The figures are, of course, very rough-and-ready: the raw statistics are unreliable, as are definitions of "industrial potential," and not all countries are included. Nevertheless, the general conclusions we can draw from these tables are very clear, and even significant adjustments to the details would not alter them.

On the scales of big history, the large changes shown in these tables may seem both universal and instantaneous. But to understand them properly, we must use a smaller lens and study the form and timing of the transformation in different regions of the world. On timescales of a century or two, the transformation has a clear sequence. And that sequence matters, for it affected the *form* and *impact* of the Modern Revolution decisively. Those regions that found themselves at the hub of the new global network of exchanges were the first to experience the high rates of innovation and the extraordinary energy flows characteristic of modernity. By the late nineteenth century, their industrial lead gave them a decisive economic, political, and military advantage, which enabled them to put their stamp on the nature and form of modernity throughout the world.

The transformation first became apparent in western Europe. Within a century, it had revolutionized European rates of growth and European social and political structures. These changes fundamentally altered Europe's role in the global world system. In 1750, the United Kingdom, Germany, France, and Italy accounted for about 11 percent of global industrial production; in

TABLE 13.1. TOTAL INDUSTRIAL POTENTIAL, 1750-1980 (UK IN 1900 = 100)

	1750	1800	1830	1860	1880	1900	1913	1928	1938	1953	1963	1973	1980
Developed	34	47	73	143	253	481	863	1,259	1,562	2,870	4,699	8,432	9,718
UK	2	6	18	45	73	100	127	135	181	258	330	462	441
Germany	4	5	7	11	27	71	138	158	214	180	330	550	590
France	5	6	10	18	25	37	57	82	74	98	194	328	362
Italy	3	4	4	6	8	14	23	37	46	71	150	258	319
Russia/USSR	6	8	10	16	25	48	77	72	152	328	760	1,345	1,630
USA		1	5	16	47	128	298	533	528	1,373	1,804	3,089	3,475
Japan	5	5	5	6	8	13	25	45	88	88	264	819	1,001
Third World	93	99	112	83	67	60	70	98	122	200	439	927	1,323
China	42	49	55	44	40	34	33	46	52	71	178	369	553
India/Pakistan	31	29	33	19	9	9	13	26	40	52	91	194	254
World	127	146	185	226	320	541	933	1,357	1,684	3,070	5,138	9,359	11,041

SOURCES: Daniel R. Headrick, "Technological Change," in *The Earth as Transformed by Human Action: Global and Regional Changes in the Biosphere over the Past 300 Years*, ed. B. L. Turner II et al. (Cambridge: Cambridge University Press, 1990), p. 58; based on Paul Bairoch, "International Industrialization Levels from 1705 to 1980," *Journal of European Economic History* 11 (1982): 292, 299.

NOTE: These figures include handicrafts as well as industrial manufacturing. Figures are rounded off, and are based on triennial annual averages except for 1913, 1928, and 1938. Because of rounding errors, figures under *World* do not match exactly the totals under *Developed* and *Third World*. Figures under these two categories were created using countries not listed, as well.

TABLE 13.2. TOTAL INDUSTRIAL POTENTIAL, 1750-1980, AS A PERCENTAGE OF GLOBAL TOTAL

	1750	1800	1830	1860	1880	1900	1913	1928	1938	1953	1963	1973	1980
Developed	26.8	32.0	39.7	63.3	79.1	88.9	92.5	92.8	92.8	93.5	91.5	90.1	88.0
UK	1.6	4.1	9.8	19.9	22.8	18.5	13.6	10.0	10.7	8.4	6.4	4.9	4.0
Germany	3.2	3.4	3.8	4.9	8.4	13.1	14.8	11.7	12.7	5.9	6.4	5.9	5.3
France	3.9	4.1	5.4	8.0	7.8	6.8	6.1	6.0	4.4	3.2	3.8	3.5	3.3
Italy	2.4	2.7	2.2	2.7	2.5	2.6	2.5	2.7	2.7	2.3	2.9	2.8	2.9
Russ./USSR	4.7	5.4	5.4	7.1	7.8	8.9	8.3	5.3	9.0	10.7	14.8	14.4	14.8
USA		0.7	2.7	7.1	14.7	23.7	31.9	39.3	31.4	44.7	35.1	33.0	31.5
Japan	3.9	3.4	2.7	2.7	2.5	2.4	2.7	3.3	5.2	2.9	5.1	8.8	9.1
Third World	73.2	67.3	60.9	36.7	20.9	11.1	7.5	7.2	7.2	6.5	8.5	9.9	12.0
China	33.1	33.3	29.9	19.5	12.5	6.3	3.5	3.4	3.1	2.3	3.5	3.9	5.0
India/Pakistan	24.4	19.7	17.9	8.4	2.8	1.7	1.4	1.9	2.4	1.7	1.8	2.1	2.3
World	100	100	100	100	100	100	100	100	100	100	100	100	100

SOURCE: Table 13.1.

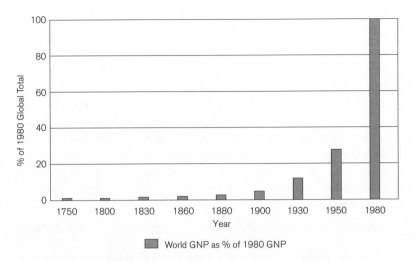

Figure 13.1.   Global industrial potential, 1750-1980. Based on table 13.2.

1880, they accounted for almost 42 percent. Today's "developed world," as a whole, accounted for about 27 percent of global production in 1750, for 63 percent in 1860, and for almost 94 percent in 1953. The United Kingdom clearly took the leading role during the first century of industrialization. In 1750, the United Kingdom accounted for less than 2 percent of global production; in 1880, more than 20 percent.

The changing balance of industrial power revolutionized the balance of military and political power. By 1800, European powers controlled some 35 percent of the earth's land area; by 1914, they controlled ca. 84 percent.[1] The demographic balance of power was also transformed, though less decisively. The figures in table 11.1 suggest that between 1000 and 1800, Europe's share of world populations fluctuated between 12 percent and 14 percent (with a temporary increase to 16 percent in the fourteenth century). Then in 1900 it rose to 18 percent before falling to about 9 percent at the end of the twentieth century. These figures underestimate Europe's demographic significance, as they ignore the millions who left Europe to settle in the neo-Europes of the Americas and Australasia.

For much of the nineteenth century, industrialization appeared to be a European phenomenon. In the twentieth century, however, it showed itself to be global, as production began to rise outside the hub zone of the Atlantic economies. As the populations, the economies, and the military powers of European and Atlantic societies grew, governments in other regions realized that they would have to try to imitate Europe's economic, political, and military successes. As a result of their efforts and of the increasing economic and cultural integration of the world, European patterns of modernity were imposed on the rest of the world. The speed and scale of these changes closed off the possibility of separate regional industrial revolutions, analogous to the separate regional transformations of the Neolithic era. Instead, European patterns of modernity provided global templates of industrialization, just as the technologies of pioneering agricultural regions provided templates that were copied within the regional exchange networks of the early agrarian era. It is no accident that today businesspeople throughout the world wear suits rather than caftans, or that English has become the universal language of business and diplomacy.

Why did the transformation first become apparent in Europe? Why did the European transformation not fizzle out, as the economic revolution of the Song era had done? In what ways was the trajectory of modernity set during its first century, when it was largely confined to Europe and North America? And what were the main features of these early transformations? These are the main questions tackled in this chapter.

Because of the significance of the first transitions to modernity, the rest of this chapter will concentrate on the hub regions of western Europe and the North Atlantic. For the sake of clarity it will distinguish three aspects of the modern revolution: economic change, political change, and cultural change. In reality, though, these were facets of a single, complex, and interrelated transformation that occurred with terrifying speed.

ECONOMIC REVOLUTION IN BRITAIN

As economic historians have focused on the details of economic change (historical "pointillism," as Patrick O'Brien calls it), many have questioned the notion of an "industrial revolution," just as archaeologists have questioned the idea of a "Neolithic revolution." Viewed in close-up, the details stand out, not the larger patterns. But from the wide-angle perspective of world history, it is hard to miss the revolutionary nature of the economic changes. In a recent survey, O'Brien writes:

> on all the indicators that have since been constructed and reconstructed for the measurement of rates of economic change, when we compare the first half of the nineteenth with the first half of the eighteenth century, the evidence for an intervening period of pronounced discontinuity still seems unmistakable. Nothing like that sustained degree of acceleration had ever occurred either in Britain (or elsewhere in Europe

and America). In short, between 1750 and 1850 the long-term rate of growth of the British economy became historically unique and internationally remarkable.[2]

In discussing the early stages of the Industrial Revolution, I will focus on Britain. This does not mean that Britain was typical: on the contrary, its priority ensured that it was atypical.[3] As O'Brien and Caglar Keyder have argued, the French path to modernity, though different from that of Britain, was not by any objective criterion "inferior." The French peasantry survived longer, and even consolidated its position after the French Revolution; as a result, French agriculture remained more traditional than British agriculture well into the nineteenth century, and its social structures were probably less unequal. Yet the differences between the two countries in long-term rates of growth of output from 1780 to 1914 are not significant.[4] Nor, really, are differences in the pace of innovation. Many strategic technological breakthroughs were "Western" rather than British. These include early developments in the design of steam engines; the invention in France of the Jacquard loom, which pioneered the use of digital coding as a form of mechanical control (1801); the invention of the cotton gin in the United States (1793); new bleaching processes, also pioneered in France (1784); the manufacture of porcelain (Meissen, 1708); new techniques in glassmaking and papermaking; and the beginnings of aviation when two papermakers, the Montgolfier brothers, launched the first ever controlled flight, at Antonnay, in southwestern France (1783). Nevertheless, Britain was the region where the economic transformation has been studied most intensively (see table 13.3). It was also the region where the revolutionary nature of these transformations first became apparent to contemporaries. As early as 1837, the French revolutionary Blanqui used the term *industrial revolution* to suggest that the economic transformations occurring in Britain were quite as revolutionary as the more obvious political and social changes of the French Revolution.[5] Thus Britain remains a good vantage point from which to observe the moment of takeoff and to see what it meant at a regional scale.

Unfortunately, Blanqui's term exaggerates the importance of industrial change. In Britain, changes in methods of industrial production were just one part of a threefold economic revolution. First, the social and political structures within which economic activity took place were transformed with the emergence of a characteristically capitalistic system of social classes and economic exchanges. Second, the agrarian sector was transformed as profit making displaced subsistence as the primary goal of agricultural production, and widespread innovation raised agricultural productivity. Though the tech-

TABLE 13.3. ESTIMATES OF ECONOMIC GROWTH RATES IN BRITAIN, 1700-1831

Years	Growth Rates of National Product		Growth Rates of National Product per Head	
	National Product (% per annum)	Implied Doubling Time (years)	National Product per Head (% per annum)	Implied Doubling Time (years)
1700–1760	0.69	100	0.31	223
1760–1780	0.70	99	0.01	6,931
1780–1801	1.32	53	0.35	198
1801–1831	1.97	36	0.52	134

SOURCE: Adapted from N. F. R. Crafts, *British Economic Growth during the Industrial Revolution* (Oxford: Clarendon, 1985), p. 45.

NOTE: "National product" is an estimate of the combined output of agriculture, industry, and services.

nical changes in agriculture were not as startling as those in industry, their real impact was greater, at least before the early nineteenth century. The calculations of N. F. R. Crafts suggest that for most of the eighteenth century, agricultural productivity rose at least as rapidly, and sometimes more rapidly, than industrial productivity.[6] Third, new methods of production, based on mechanization and the use of new sources of power (such as coal and steam), revolutionized the scale and productivity of many sectors of British manufacturing—cotton, coal, and iron production in particular. Most of this vast increase in productivity was made possible by technologies that tapped the colossal reserves of ancient solar energy locked up in fossil fuels.

The Social Context

Like many regions of Afro-Eurasia, eighteenth-century Britain was highly commercialized. But in two respects—the structures of government and those of rural society—it was particularly so. Supportive governments and elites help explain why, at least in the early stages of the Industrial Revolution, British entrepreneurs were so effective in exploiting new technologies, including those pioneered elsewhere.[7]

Britain's strategic position in the global exchange networks of the eighteenth century certainly had something to do with geography, which placed Britain at the epicenter of the new global world system. Its physical location guaranteed that British governments would take exceptional interest in commerce. But as we have already seen, British governments were already preadapted for such a transformation. Britain's high level of commercialization depended in large part on persistent and aggressive financial and military investments of successive British governments, backed by important sectors of the nobility and merchantry, in protecting British commercial interests overseas.[8] The government had good reason to support commercial activity both at home and abroad, because by the eighteenth century, most of its revenues came from customs and excises of various kinds. By creating the Bank of England and supporting expansion overseas, it was protecting its own interests as well as those of a large and influential commercial elite. The contrast with Ming China—whose government despised commerce, depended mainly on noncommercial revenues such as land taxes, and refused to back foreign trade—is striking. But so is the geographical contrast between the two societies: the one now at the center of global exchange networks, the other at the edge of a huge and ancient but subglobal network of exchange.

Commercial activity had also transformed British rural society. Even in Tudor and Stuart England, landless rural laborers may have constituted 25 to 30 percent of the population.[9] In the 1640s, an English writer insisted that "the fourth part of the inhabitants of the parishes of England are miserable people, and (harvest time excepted) without any subsistence." Recent research, based on the pioneering estimates of the English statistician Gregory King, suggests that in 1688, about 43 percent of the population consisted of "cottagers and paupers" or of "laboring people and out servants," who did not earn enough to fully support themselves.[10] Most of these people had no land at all, and those who had land did not have enough for subsistence, making them (in Marx's terms) proletarians. Many left for the towns, which grew rapidly. By 1700, 10 percent of the population of Great Britain lived in London. Here, living conditions were in many respects worse than in the villages (death rates were notoriously high—42 per 1,000, according to Gregory King), but at least there was a chance of finding work.[11]

Which were the most important sectors of the British economy in the early eighteenth century? Modern estimates suggest that 37 percent of national income came from agriculture, 20 percent from industry, 16 percent from commerce, and another 20 percent from rents and services, while the government's income accounted for the remaining 7 percent. In other words, over half of the incomes earned in Britain came from industry, commerce, or rents and services.[12] With perhaps half of its population depending mainly on wages rather than subsistence farming, and a national economy in which commercial activities generated well over 50 percent of national income, British society was beginning to conform more closely to the capitalist ideal type than to that of a traditional tributary society. Social structure models of growth predict that innovation should flourish in such an environment; and that is precisely what we see.

Agriculture

Most significant of all was the spread of commercial attitudes and methods in the agricultural sector, the most important sector of most premodern societies. During the seventeenth and eighteenth centuries, capitalistic methods began to transform British farming. This was a fact of fundamental importance, for agriculture remained the engine of the British economy, as it had been in all traditional agrarian civilizations. In the eighteenth century it was still Britain's largest productive sector, responsible for most of the country's food, clothing, and raw materials. In the seventeenth and eighteenth centuries, the changing social structure of landownership stimulated a technological transformation that, though slow by modern standards, was revolutionary on the scale of world history.

In most agrarian civilizations, the primary function of agriculture was to feed those who worked the land. In Britain, however, over some two centuries, more and more land had been consolidated in the hands of large owners, for whom land was a source of profit rather than subsistence. Meanwhile, increasing numbers of small peasants had been driven from the land or deprived of traditional rights to use pastures, meadows, and woods. Since the sixteenth century, governments had periodically encouraged these changes by allowing *enclosures*—procedures that allowed landlords to ignore traditional rights to the land—in order to create large, consolidated, and enclosed landholdings. Perhaps half of English land was enclosed before the mid–eighteenth century; in the late eighteenth century, the process was largely completed, mainly through acts of Parliament. The British peasantry vanished as a result, and Britain became the first large-scale society to flourish without a peasant class.

For most rural dwellers, these changes were catastrophic. No longer able to depend on what they produced themselves, rural families found themselves at the mercy of the erratic and unreliable employment market. W. G. Hoskins describes the change in one English village, Wigston Magna in Leicestershire, as agricultural "improvements" brought it money, but not wealth:

> The domestic economy of the whole village was radically altered. No longer could the peasant derive the necessaries of life from the materials, the soil, and the resources of his own countryside and his own

strong arms. The self-supporting peasant was transformed into a spender of money, for all the things he needed were now in the shops. Money which in the sixteenth century had played merely a marginal, though a necessary, part, now became the one thing necessary for the maintenance of life. Peasant thrift was replaced by commercial thrift. Every hour of work now had a money-value, unemployment became a disaster, for there was no piece of land the wage earner could turn to. His Elizabethan master had needed money intermittently, but *he* needs it nearly every day, certainly every week of the year.[13]

For Wigston Magna, the enclosure act of 1765 was a cataclysm. Small owner-occupiers disappeared as a group within about sixty years, becoming rural laborers or framework knitters or paupers.[14]

As peasant holdings declined, those of their former landlords rose, and so did the average size of farms in general. In the South Midlands, the proportion of farms larger than 100 acres rose from ca. 12 percent in the early seventeenth century to ca. 57 percent two centuries later.[15] These figures suggest how quickly the gradient of inequality could steepen during the Modern Revolution. In most agrarian civilizations, a majority of the population had access to farmable land; indeed, low rates of agricultural productivity guaranteed such access, for societies had to allocate most of their labor to the production of food. But now the land was concentrated in the hands of a minority. The changing pattern of ownership revolutionized the economics of agricultural production. Because those who farm on a large scale cannot possibly eat what they produce, they must farm for profit. The increasing size of landholdings therefore offers a good indirect measure of the commercialization of British agriculture.

Commercialization on this scale changed both attitudes toward and methods on the land. To generate profits from enclosed land, landlords had to either produce for the market or put in commercial "farmers"—that is, tenants who would produce for the market and pay rents out of their profits. Both approaches turned agriculture into a business rather than a means of survival. But the second method had the advantage of letting aristocratic landowners keep a polite distance from the crass business of moneymaking, even as they enjoyed their profits. Eric Hobsbawm concludes: "We have no reliable figures, but it is clear that by 1750 the characteristic structure of English landownership was already discernible: a few thousand landowners, leasing out their land to some tens of thousands of tenant farmers, who in turn operated it with the labor of some hundreds of thousands of farm-laborers, servants or dwarf landholders who hired themselves out for much of their time."[16]

Changes in the way the land was controlled revolutionized farming techniques. Commercial farmers had to produce for competitive markets, so they had to produce in volume and they had to produce efficiently. But they also had better access than peasant farmers to capital that they could invest in more efficient methods of production. Finally, after enclosure, they normally had access to large blocks of land that enabled them to exploit economies of scale, using modern farming methods that were beyond the means of small producers. To be sure, most of the techniques introduced in the late seventeenth and the eighteenth centuries were not new; at this stage, efficient implementation of existing techniques counted the most. Indeed, not until the nineteenth century did farm machinery and artificial fertilizers start transforming the technology of modern agriculture. Before then, most of the methods introduced by enterprising farmers had been familiar since the Middle Ages, and many were already in use in different parts of Europe. What was new in Britain was the numbers who embraced these techniques, who had the money to invest in them, and who used them effectively.

British farmers borrowed methods pioneered in the Low Countries since the Middle Ages, often known as "the new husbandry." These integrated crop and livestock farming in new ways to increase yields and reduce the amount of land left fallow. Many farmers began to plant fallow crops such as clover or turnips. Turnips provided cattle feed and increased livestock numbers, and more livestock provided more manure. Legumes, which are effective fixers of nitrogen, helped regenerate the soil. New crop rotations therefore increased the amount of crops and livestock that could be supported from a given area of land. But there were many other changes—including improved forms of irrigation, land reclamation, and more systematic livestock breeding—all driven by the need for a commercialized agriculture to produce goods in large volume and at low cost.

As these changes were taken up more widely, the productivity of British agriculture rose and the proportion of agricultural workers fell. While the share of employment in agriculture dropped, agriculture's contribution to national income remained at ca. 37 percent between 1700 and 1800.[17] The total output of British agriculture rose about 3.5 times between 1700 and 1850, while the percentage of the male labor force engaged in agriculture fell from 61 percent (in 1700) to ca. 29 percent (in 1840). It has been estimated that by 1840, every male agricultural worker in Britain produced about 17.5 million calories, in comparison with 11.5 in France and even lower figures for most other European countries.[18] Table 13.4 shows growth in output of particular crops.

The increasing productivity of British agriculture in the eighteenth cen-

TABLE 13.4. OUTPUT OF PRINCIPAL AGRICULTURAL COMMODITIES
IN BRITAIN, 1700-1850 (IN MILLIONS)

	1700	1750	1800	1850
Commodities				
Corn (bushels)	65	88	131	181
Meat (lbs.)	370	665	888	1356
Wool (lbs.)	40	60	90	120
Cheese (lbs.)	61	84	112	157
Volume in 1815 prices (£)				
Corn and potatoes	19	25	37	56
Animal produce	21	34	51	79
Total volume	40	59	88	135

SOURCES: Maxine Berg, *The Age of Manufactures, 1700–1820: Industry, Innovation, and Work in Britain*, 2nd ed. (London: Routledge, 1994), p. 81, citing R. C. Allen, "Agriculture and the Industrial Revolution, 1700–1850," in *The Economic History of Britain since 1700*, ed. Roderick Floud and Donald McCloskey, 2nd ed. (Cambridge: Cambridge University Press, 1994), 1:109.
NOTE: "Corn" includes wheat, rye, barley, oats, beans and peas, net of seed and oats consumed by livestock. "Animal produce" includes meat, wool, dairy products, cheese, hides, and hay sold off the farm.

tury was of profound importance. In the first place, it made possible rapid population growth. Calculations by Crafts suggest that in the eighteenth century, productivity rose just fast enough to support the rapid rates of population growth that Malthus observed; but in the nineteenth century, productivity rose even faster, thereby averting a serious Malthusian crisis of the kind that did strike in many other parts of the world, from Ireland to India and Pakistan and to China.[19] In Britain, rising populations enlarged markets for agricultural produce, encouraged further investment, and released more labor for nonagricultural sectors of the economy.

Why was so much commercial capital attracted to the land? One answer is that population growth and the decline of subsistence farming increased the internal market for rural produce. Those who had no land had to buy food, however poor they were. So farmers could normally count on expansion in the markets for their produce. These processes created an entirely new type of market—a large market for cheap consumption goods. Such markets could hardly exist to any significant degree in a society of subsistence farmers, a fact that had always limited the scope of and possibilities for commercial farming in the preindustrial world. Cities like Beijing or

Baghdad or imperial Rome needed enormous supplies of food; so, too, did many elite households, which demanded luxury as well as subsistence foods. But outside these huge cities, most people ate what they had grown themselves. The appearance of societies in which *most* people depended entirely on markets for their subsistence was a new phenomenon, and it gave a tremendous stimulus to commercial production of goods of mass consumption.

The change was particularly rapid because in Britain, as in several other European countries, external markets for rural produce were also growing rapidly in the eighteenth century. These were mainly colonial markets, protected (sometimes at great cost) by increasingly commercially minded governments. In Britain, colonial expansion and the Navigation Acts of 1651 and 1660 provided a large protected market for British producers. The West Indies were particularly important, as their cash crop economies (concentrating on sugar from the mid–seventeenth century onward) meant that they had to import almost all their food. This was one of the many ways in which Britain's position at the epicenter of global exchange networks gave a critical extra impetus to commercial activity.

Industry

Given the growing number of landless would-be wage earners, ruling elites increasingly dependent on commercial revenues, a highly commercial agricultural sector, and exceptionally good access to expanding world markets, the surprising thing is how long it took to transform industry as well as agriculture. One reason for the delay is that the levels of investment needed to set up a factory or to buy a steam engine were higher than those needed to "improve" agriculture or innovate in domestic industries. As a result, most industrial production remained traditional in late-eighteenth- and early-nineteenth-century Britain. Most production still took place in artisan workshops, operating on a scale little different from that of Sumer, 4,000 years before, or exploiting the labor of peasant families who spun or knitted or wove in their own homes. Indeed, for a time the Industrial Revolution gave a new stimulus to small-scale production. A second reason for the delay may be that in a world still dominated by the countryside, demand for industrial goods remained lower than demand for agricultural produce.

Eventually, however, the pursuit of profit began to transform industry as it had agriculture. It is hard to determine exactly when the trickle of innovations characteristic of the premodern world turned into a flood. In the seventeenth and early eighteenth centuries, there were innovations in industrial production throughout Europe. But it would be hard to prove that innovation was more rapid in Britain than elsewhere before, perhaps, the middle of the eighteenth century. In 1709, faced with the rising cost of timber (which increased about ten times between 1500 and 1760, while prices in general rose only about five times), Abraham Darby began to experiment with the use of coke in blast furnaces manufacturing iron at Coalbrookdale in Shropshire.[20] Such techniques had been used in China in the eleventh century, but there is no evidence that the techniques used by Darby borrowed directly or indirectly from Chinese practices.[21] Indeed, his methods were not particularly efficient and did not spread widely until improved in the 1760s. But they did cut costs and raise output, as did another change, the invention of the puddling process by Henry Cort, in 1784. All in all, British iron production rose ten times in the eighteenth century.[22]

Another technique whose full significance became apparent only later was the use of steam power to pump water out of mines. The idea that atmospheric pressure was a potential source of mechanical power had a history going back at least to the sixteenth century, and it may have been familiar in China as well as in Europe.[23] A French inventor, Denis Papin, who was well aware of the scientific theory behind the idea of atmospheric pressure, first demonstrated the potential use of steam as a source of mechanical power in 1691. A working steam pump was built by Thomas Savery in 1698; its engine used the vacuum created by condensing steam to suck up water. Thomas Newcomen built an improved version in 1712. Its use was limited because it was inefficient, depending on the repeated heating and cooling of a single cylinder. It also needed vast amounts of coal, so the earliest industrial steam engines were sited at large coal mines, where fuel was abundant and cheap. There they raised productivity, particularly in mines subject to regular flooding. In 1742, in Darby's ironworks at Coalbrookdale, a steam engine was used for the first time not to pump water but to power the bellows of a blast furnace. By the mid–eighteenth century, enterprises in many parts of Europe and the Americas were using Newcomen engines.

Textile producers also experimented with new techniques to meet increasing demand in what was the second-largest productive sector of most premodern economies. A factory with special Dutch machines for twisting silk, powered by a waterwheel, was set up in Derby as early as 1702. In 1718, a new owner, Thomas Lombe, in an early example of planned industrial espionage, stole techniques already in use in Italy to set up an improved factory. By the 1730s, producers of linen and cotton were trying to construct similar machines, as well as machines to mechanize weaving; these included the flying shuttle, invented in 1733. Government support in the form of bans on imported cotton textiles encouraged innovation from the 1730s on. In the 1770s and 1780s, three new machines began to transform cotton spinning: Richard Arkwright's water frame, James Hargreaves's spinning jenny, and Samuel Crompton's spinning mule, a modification of the jenny.[24] They all increased output significantly, but at first they were used mainly in domestic industry. In the two decades after 1780, these and subsequent innovations reduced the price of cotton textiles by 85 percent, making cotton, for the first time in Europe, a mass consumption good rather than an expensive import.[25]

Arkwright built his first water frame on a large scale and used it in a factory, where it was powered by a waterwheel. His machines did not require factory organization, but factories did give employers greater control over discipline and quality. This is a reminder that the key changes of this period were managerial as well as technological. In the preindustrial world, most nonagricultural production had been organized in households or small workshops. Productive enterprises consisted of small groups of people, sometimes linked by kinship, who worked together, often on similar tasks; and for a time such enterprises may have multiplied as a result of early inventions of the Industrial Revolution, such as the spinning jenny. The factory was a much larger and more anonymous production unit, more like an army than a family. And it normally required a more complex division of labor, skills, and authority. The eventual spread of the factory form had something to do with technological change: large prime movers could best be exploited by concentrating the workforce in one place. But the factory form also gave entrepreneurs the sort of power to guide work processes that was necessary to maximize efficiency and cut costs. After all, wage laborers recruited one by one could not be expected to show the solidarity of a family working together in its own household. So the spread of the factory had as much to do with the need to improve work discipline as it did with technology.[26] It was a way of controlling both workers and machines. The managerial technologies of the Industrial Revolution also had roots throughout the emerging global world system. The disciplined control of large groups of people had been pioneered in the armies of Europe since the sixteenth century,[27] as well as in the slave plantations of the Americas. But other techniques of control, such as using examinations to select personnel, came ultimately from China.

The changes described so far may suggest the strength of the impulse toward innovation in both technique and organization, at least in the crucial sectors of textiles, coal mining, and iron. But as yet, there was little that could not have been paralleled by developments elsewhere in the Afro-Eurasian world system—in China, in India and Pakistan, in the Islamic world, or in other parts of Europe. What revolutionized British industry was the bringing together of steam power, improved machinery, and factory organization.

James Watt made several improvements to the steam engine in the 1760s. First, he separated the condenser from the cylinder, which eliminated a major cause of heat loss and allowed his machine to run on far less fuel. Second, instead of exploiting the atmospheric pressure created by condensing steam to form a semi-vacuum (as did Newcomen's engine), Watt's machine used the expansive power of steam directly to drive a piston (see figure 13.2). These and other changes made the steam engine more economical, more powerful, and more adaptable. By the 1790s, spinning wheels were being driven by steam engines rather than people or waterwheels, and productivity soared. By 1800, a power-driven spinning mule could produce as much as 200 or 300 human cotton spinners. The improved steam engine marks the first significant increase in the power available to humans for many millennia. Not since about 6,000 years earlier, when humans had first learned to harness the traction power of other animals, or perhaps 5,000 years earlier, when they had first learned to systematically exploit other human beings on a large scale, had there been such a change in the availability of power sources to produce basic necessities. With the introduction of steam power, then of electricity and oil, human societies began at last to draw on the huge sources of energy locked up in the inorganic world. (The most important earlier example, gunpowder, was mainly used in destructive rather than productive technologies.) Each change opened a new range of ecological niches for human exploitation.

Improved steam engines quickly raised productivity in a few select industries. They also required changes to the way work was organized, for to justify their expense they had to drive a number of machines, which made them incompatible with domestic industry. They worked most effectively in factories, where supervision could be more or less continuous, and human workers became little more than minders of machines—fixing their broken threads, supplying them with raw material, and keeping them running smoothly. As they spread, steam engines became major consumers of coal and metal. Their production thus stimulated mining, iron production, and advances in engineering. Within a few decades, they had also revolutionized methods of land transportation. The idea of using steam to power locomotion had been around for several decades (indeed, a steam cart had been invented in France in the 1760s), but the earliest steam engines were too large. The first practical steam locomotive was built in 1802 at Coalbrookdale by Richard Trevithick, who had designed a smaller high-pressure

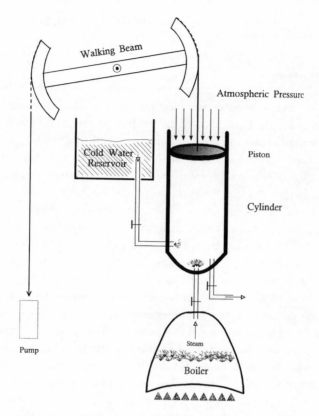

Walking Beam

Atmospheric Pressure

Cold Water
Reservoir

Piston

Cylinder

Pump

Steam

Boiler

Figure 13.2.   Evolution of the steam engine in eighteenth-
century Britain. **a.** In the Newcomen "atmospheric engine,"
which was first used in 1712, steam was pumped into a cylinder, a jet of cold water was sprayed in, and the steam con-
densed, creating a vacuum that sucked down a piston, which
worked the pump. By later standards this was highly inefficient, mainly because the cylinder was alternately heated
and cooled. It thus used huge amounts of coal, and was economical only in mines, where coal was plentiful and cheap.
b. James Watt patented an improved steam engine in 1769.
Among several improvements, he separated the condenser
from the cylinder, so that the temperature of the cylinder
remained more constant. He also began to use the pressure
from the steam, rather than the vacuum created by the
steam's condensation, to drive the piston. The greater fuel
efficiency of the Watt engine made it possible to use steam
engines away from coal mines. From James E. McClellan III
and Harold Dorn, *Science and Technology in World History: An Introduction* (Baltimore: Johns Hopkins University Press, 1999), p. 282, fig. 13.1; p. 284, fig. 13.2. © 1999
Johns Hopkins University. Reprinted with permission of
The Johns Hopkins University Press.

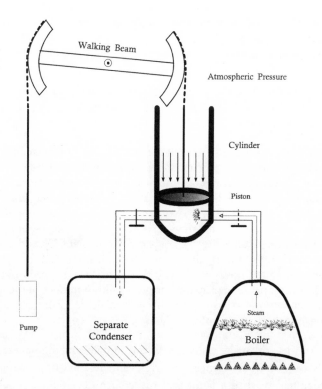

steam engine. The locomotive was initially used as a mechanical horse, to haul coal more quickly. Over the next thirty years, the quality both of the rails and of the engines was improved. The Stockton and Darlington Railway, the first designed to carry passengers as well as coal, opened in 1825.

The first thing to note as we analyze this series of innovations is that though their impact was fundamental, they were developed incrementally, built on developments and resources drawn from the entire world system. British inventors drew on traditional skills and on knowledge of techniques diffused in a complex network of ideas throughout the global world system. Thomas Lombe's "silk throwing" machine has a genealogy that may lead back, via Italy, to medieval China. Knowledge of the commercial potential of cotton reflected the importance of Indian textile imports since the seventeenth century, while dyeing techniques owed much to Indian, Persian, and Turkish methods.[28] In an essay titled "The Pre-Natal History of the Steam Engine," Joseph Needham, a historian of Chinese science, argued that its antecedents lay in China and Greece as well as in Europe, and concluded: "No single man was 'the father of the steam engine'; no single civilization either."[29] The technologies of the first Industrial Revolution were Afro-

Eurasian, even global, though their full productivity-raising potential was initially demonstrated in England.

Furthermore, in industry as in farming, the techniques needed in the early stages of the Industrial Revolution depended more on traditional artisans' skills than on fundamentally new methods or techniques. Many of the pioneers were practical workers rather than scientists or theoreticians. Peter Mathias points out that

> by and large, innovations were not the result of the formal application of applied science, nor a product of the formal educational system of the country. . . . Most innovations were the products of inspired amateurs, or brilliant artisans trained as clock-makers, millwrights, blacksmiths or in the Birmingham trades. . . . They were mainly local men, empirically trained, with local horizons, often very interested in things scientific, aware men, responding directly to a particular problem. Up to the mid–nineteenth century this tradition was still dominant in British manufacturing industry. It was no accident that the Crystal Palace in 1851, a miracle of cast iron and glass like the great railway stations of the nineteenth century, was the conception of the head gardener of the Duke of Devonshire. He knew about greenhouses.[30]

This is not to say that the task of inventing and developing new technologies was easy, or that science was irrelevant; but existing technological knowledge had reached a point that made such developments possible.[31]

The second explanation for this wave of innovation is commercial and social. Placed at a major crossroads of expanding commercial networks by the changing topology of global networks of exchange and the aggressive commercialization of its elites, and controlling huge, protected markets in India and Pakistan and in North America, British entrepreneurs could exploit raw materials such as cotton that were not available in Britain. They could also sell in large protected markets that were developing rapidly enough to absorb the vast increases in output made possible by new machinery. But the British internal market was also growing rapidly as Britain's class structure was revolutionized, with ever larger numbers of people leaving the subsistence economy of the village to become urban wage earners. Rapidly expanding markets within a global world system and a high level of commercial competition stimulated innovation, particularly in the production of goods for the mass market, such as textiles (see table 13.5). Acting on that stimulus were not just well-known inventors but thousands of tinkerers and investors and managers who made a commercial success of the major breakthroughs. The innovations that shaped the British industrial revolution represent the response of a highly commercialized society

TABLE 13.5. VALUE ADDED IN BRITISH INDUSTRY, 1770-1831 (£ MILLIONS)

Sector	Product	1770	1801	1831
Textiles	Cotton	0.6	9.2	25.3
	Wool	7.0	10.2	15.9
	Linen	1.9	2.6	5.0
	Silk	1.0	2.0	5.8
Coal and metals	Coal	0.9	2.7	7.9
	Iron	1.5	4.0	7.6
	Copper	0.2	0.9	0.8
Building	Buildings	2.4	9.3	26.5
Consumer goods	Beer	1.3	2.5	5.2
	Lather	5.1	8.4	9.8
	Soap	0.3	0.8	1.2
	Candles	0.5	1.0	1.2
	Paper	0.1	0.6	0.8
Total		22.8	54.2	113.0

SOURCE: Maxine Berg, *The Age of Manufactures, 1700–1820: Industry, Innovation, and Work in Britain,* 2nd ed. (London: Routledge, 1994), p. 38.

to new commercial challenges and opportunities. Eric Hobsbawm summarizes the role played by demand:

> Exports, backed by the systematic and aggressive help of government, provided the spark, and—with cotton textiles—the "leading sector" of industry. They also provided major improvements in sea transport. The home market provided the broad base for a generalized industrial economy and (through the process of urbanization) the incentive for major improvements in inland transport, a powerful base for the coal industry and for certain important technological innovations. Government provided systematic support for merchant and manufacturer, and some by no means negligible incentives for technical innovation and the development of capital goods industries.[32]

However, the fundamental reason for the increased pace of innovation in eighteenth-century Britain and Europe was that there was intense *pressure* to innovate in a world shaped by the competitive forces of an increasingly global capitalism. The importance of commercial pressures is apparent in the motivations of particular inventors. James Watt, for example, wrote in his autobiography that he was interested in making machines that were "*cheap* as well as *good*."[33] Even better evidence is provided by the huge number of innovations in eighteenth-century Europe. And as pressures to innovate increased elsewhere in the course of industrialization, rates of innovation accelerated in all industrializing regions. This suggests that there had emerged in western Europe a culture of innovation—an environment that encouraged entrepreneurs actively to seek out and make effective use of new techniques. Such arguments provide the strongest justification for explanations of the Industrial Revolution that look to both commerce and social structure.

POLITICAL REVOLUTION IN FRANCE

Alongside the economic revolution, there took place a political revolution. Gradually in the seventeenth and eighteenth centuries, and rapidly in the nineteenth century, the power and reach of states grew, and so did the resources available to them. As a result, their relationship to the populations they ruled was transformed. Today's political systems are to the great tributary empires of the past what those empires were to the chiefdoms and "big man" systems that they had displaced. Charles Tilly amplifies the point:

> Over the last thousand years, European states have undergone a peculiar evolution: from wasps to locomotives. Long they concentrated on war, leaving most activities to other organizations, just so long as those organizations yielded tribute at appropriate intervals. Tribute-taking states remained fierce but light in weight by comparison with their bulky successors; they stung, but they didn't suck dry. As time went on, states—even the capital-intensive varieties—took on activities, powers and commitments whose very support constrained them. These locomotives ran on the rails of sustenance from the civilian population and maintenance by a civilian staff. Off the rails, the warlike engines could not run at all.[34]

The power of European states had been growing for some centuries, partly as a result of the increased resources available to commercially aggressive states and partly as a response to the fiscal and organizational demands of the gunpowder revolution.[35] But these changes, which culminated in the "absolutism" of the seventeenth and eighteenth centuries, were merely playing catch-up. In comparison to the huge imperial states of China or the Islamic world, the European states of 1000 CE were small and fragile affairs. Fierce military competition, heightened by the advent of gunpowder, eventually squeezed out the smaller and less viable states. Those that survived lived through a torrid adolescence during which they learned many of the lessons and gained many of the skills acquired much earlier by the great agrarian empires. Yet even the power and reach of European absolutist states are not particularly striking if they are compared with the Ottoman or Chinese states.

What changed in the period after the French Revolution was the extent to which state power reached directly into the lives of a majority of its subjects. As Tilly points out,

> After 1750, in the eras of nationalization and specialization, states began moving aggressively from a nearly universal system of indirect rule to a new system of direct rule: unmediated intervention in the lives of local communities, households, and productive enterprises. As rulers shifted from the hiring of mercenaries to the recruitment of warriors from their own national populations, and as they increased taxation to support the great military forces of eighteenth-century warfare, they bargained out access to communities, households, and enterprises, sweeping away autonomous intermediaries in the process.[36]

The change can be seen most clearly in revolutionary France, largely because the revolution itself swept away so many of the intermediary authorities that had ruled during the ancien régime. But change was also driven by the need to assemble from scratch powerful new armies. In turn, the conquests of French armies spread the new methods of government (along with the decimal system) to other parts of Europe.

The management of warfare was crucial to these changes. Whereas states in early modern Europe had relied mainly on mercenary armies, from the French Revolution onward, states began to take part directly in recruiting, organizing, and funding national armies. As a result, the organizational and fiscal role of states expanded, and they found they had to start worrying about entirely new problems (such as the health and education of potential recruits).[37] All these pressures forced governments to collect more information on the demographic and economic resources they controlled. Later in the nineteenth century, states began to take an interest in public health and to support systems of public education. The political ideologies and the commitment to electoral politics of the French revolutionary governments also forced them to take responsibility for popular welfare and for law and order. The organization of citizens' armies turned a sense of nationhood into a crucial legitimizing device, encouraging states to become active supporters of nationalist thought and of the historians and writers who constructed nationalist ideologies.

Electoral politics forced states to court ever wider sections of the population, and they did so, at least in part, by presenting themselves as representatives of "the people." To the surprise of many traditionalists, democratic politics, when handled with care, turned out to strengthen states, not weaken them. Elections also made available to governments new sources of information about shifts in attitudes and demands within the populations they ruled, while they limited the extent to which officials and other intermediaries could filter the information being passed upward to rulers. Whatever their precise form, new methods of information gathering—or "surveillance," to use Anthony Giddens's term[38]—were crucial to the success of rulers in the complex new environments of modern politics.

Policing was a particularly important aspect of these changes, for it was part of the process by which modern states began to create a real monopoly over the means of coercion. In ancien régime France, the state took little interest in police matters, which were usually handled by local authorities; in extreme cases, the army was used. At the end of the 1790s, the French government created for the first time a bureaucratized police organization that assumed a preventive rather than merely a reactive role in dealing with crime and disorder. It was initially headed by the ex-Jacobin Joseph Fouché, now acting as minister of police. As Tilly concludes, "By the time of Fouché, France had become one of the world's most closely-policed countries."[39]

In all these ways, France pioneered what has become the typical modern state: a huge bureaucratic organization with a scale, a power, a wealth, and a reach that would have been inconceivable in the premodern world. This political revolution of modernity is both a cause and a consequence of economic revolution. It is a cause insofar as effective and commercially minded states were necessary if capitalism was to achieve its full dynamism. The steepness of the modern gradient of wealth put far more wealth in the hands of a minority than ever before, and preserving these vast reservoirs of affluence required larger and more elaborate dams than had existed in the agrarian era. States, in short, had to be powerful enough to protect the wealthy and the entrepreneurial. Giddens observes that

> private property, as Marx so consistently stressed, has as its other face the dispossession of masses of individuals from control of their means of production. . . . [T]he "freeing" of wage-labor was undeniably a major aspect of the early establishing of capitalist enterprise on the grand scale. Without the centralization of a coercive apparatus of law, it is doubtful either that this process could have been accomplished, or that the rights of private property as capital could have become firmly embedded.[40]

The work of defending the emerging gradient of wealth went on in many areas of life. In Britain it encompassed the passing of enclosure acts; the defense of the royal forests (as described vividly by E. P. Thompson); the imprisonment, deportation, or execution of petty thieves; and the protection of entrepreneurial rights against industrial violence (a subject also dealt with brilliantly by Thompson).[41] But it took place in many other areas as well. For example, the creation of a modern monetary system was unthinkable without the existence of a powerful state with significant fiscal and managerial resources, as well as effective control over the laws and courts.

On the other hand, the modern state is also a product of the economic transformations of modernity. Just as the first states emerged in part in response to the challenges of managing and organizing the huge concentrations of people and resources gathered into the earliest cities, so the modern state was, at least in part, a response to the entirely novel challenges and possibilities created by the abundant wealth generated within industrial economies. The sheer scale of the resources available to modern states would have demanded new managerial methods even without the state's new need to manage and fine-tune the commercial machinery that generated growth. But modern states also benefited from new technologies, especially in military matters. New forms of communication transformed the movement of troops and supplies, while new methods of manufacture transformed not just the production of weaponry but its nature. The American Civil War was the first truly industrialized war of modern times. At the same time, improved communications and greater literacy enhanced the capacity of states to handle the mass of information needed to rule effectively. And as they became increasingly dependent on the technologies and the huge revenues generated by modern economies, modern states also had to learn how best to encourage growth by adjusting the balance of their interference in and regulation of entrepreneurial activities. As Karl Polanyi argues in a classic study of modernity, the widespread belief that the modern state is less interventionist than premodern states is misleading. For the most part, modern states intervene more widely and more effectively than traditional agrarian states; but they are also more aware of those areas of economic activity where excessive intervention can be counterproductive.[42]

In the past two centuries there have been many exceptions to these generalizations. Many modern states have never managed to closely regulate the lives of their citizens; others have found it difficult to create the framework for a viable capitalist economy. But for citizens of the many states that *have* undergone these transformations, the results have been paradoxical. On the one hand, the modern state regulates the lives of its citizens in ways that would have been inconceivable, and might often have seemed inappropriate, in the era of tributary states. It requires that children be taken from parents for compulsory education; it demands detailed information on the lives of individuals, in areas ranging from their incomes to their religious beliefs; it regulates in detail how we may and may not behave. Moreover, these requirements are backed up by formidable police powers. The modern state has taken over many of the educational, economic, and policing functions once handled by households and local communities. In these ways, our lives are more regulated by the state than ever before. Like the nerve centers of multicelled organisms, modern states regulate the lives of individuals because communities of individuals so much larger and more interdependent than in premodern states cannot exist without some degree of central coordination.

On the other hand, the most modern states foster participation in creating and implementing policy through public debate and through elections in which ordinary citizens can stand for office. In these ways, modern states encourage their citizens to see themselves as active agents rather than mere subjects. Modern governments also set clear limits to their own power, because they know the extent to which the wealth they manage depends on avoiding overregulation of entrepreneurial activity. And though they have more force at their disposal than did any premodern states, they normally deploy it with more restraint. In addition, modern states enable many activities that would be impossible without them. They provide infrastructure, protection, and services of many kinds, from education to public health care, and they maintain the legal and administrative framework necessary if modern capitalist economies are to flourish.

While the regulatory powers of the modern state have led some critics to describe it as "totalitarian," its efforts to include and nurture its citizens explain why so many see it as an ally and a defender of liberty and freedom. Much of modern political life arises from the constant renegotiation of this balance between regulation and support in the activities of the modern state.

CULTURAL REVOLUTION

The movement of ex-peasants into the towns, an increasing concern with technological innovation, government involvement in education, and the spread of new forms of mass media are among the many changes that transformed cultural life.

The most important single change was perhaps the spread of mass education and literacy. Literacy, as we have seen, had emerged first as a way of coping with the large managerial tasks of the earliest states. But for most of the agrarian era, it had remained a privilege of elites, a form of power denied to most of the population. Modern states engaged with their citizens in entirely new ways, which required that the mass of the population itself become involved, albeit perhaps in minor ways, in the huge organizational tasks of modern society. The crucial precondition for popular involvement in productive and managerial tasks was that literacy become general. The effects of this cultural revolution were profound. For example, mass literacy began the process of "disenchanting" the world by undermining the authority of traditional, often semi-magical, forms of thinking. In this way mass education helped spread a different worldview—if not a rigorous understanding of modern science, then at least a certain skepticism about nonscientific maps of reality.

Such developments were accompanied and have been affected by a profound change in the nature of high culture and in attitudes toward knowledge. The usual modern attitude toward knowledge can be characterized as competitive, by analogy with the market. In agrarian civilizations, where most people relied on orally transmitted information, knowledge was largely shaped by the authority of particular teachers. Education consisted of the transmission of traditional skills and a traditional body of knowledge. Where literacy spread, knowledge became more abstract and less personal, and abstract knowledge began to acquire an authority quite independent from the prestige of particular teachers. Furthermore, where societies became more commercialized, habits of *testing* traditional knowledge became more common, as can be seen in classical Greece, in Abbasid Persia, in Song China, and in early modern Europe. In Europe, methods for the testing of knowledge had their precedents in the dialectical traditions of Socratic philosophy as transmitted through the Islamic world, with its madrassas in which important problems were solved by debate.[43] By the Renaissance, thinkers such as Leonardo da Vinci or Christopher Columbus found it natural to hawk their ideas from court to court like intellectual peddlers.[44]

The emerging market for ideas, in which ideas survived not because of the authority of particular teachers but because they found buyers who had tested their quality, was the proving ground of modern science. Though science's impact on methods of production was still limited, the scientific style of thinking was already present in a world increasingly dominated by market forces in ideas and politics as well as in trade. As Margaret Jacob has argued, "To an extraordinary extent scientific knowledge had penetrated the thinking of literate Englishmen by the late eighteenth and early nineteenth century, and . . . such knowledge contributed directly to the process of industrialization, to creating the world in which we now live."[45] But the market in ideas, like that in goods, was now global; new technologies, such as those of printing, ensured that new ideas would circulate more rapidly as well as more extensively. In the nineteenth century, beginning in Germany, science itself began to be incorporated into entrepreneurial activity as companies set up laboratories specifically to raise productivity and profits. By the late nineteenth century, scientific research was taking a leading role in processes of innovation that might have simply petered out if they had continued to rely on the technical and practical skills of individual entrepreneurs and artisans.

Science's grip on modern culture may reflect other, subtler changes as well. Wage earners, unlike traditional subsistence peasants, lived in a world in which the dominant forces were not particular overlords or rulers who could be identified, named, and complained about. The modern world is ruled by larger and more impersonal forces, from faceless bureaucracies to abstractions such as "inflation," or "the rule of law." Where abstract forces take over the work of coercion from the landlord, the executioner, and the overseer, it is not surprising that there should emerge cosmologies ruled by equally abstract forces. Perhaps the face of God was bound to vanish behind the neutral mask of gravity in a world shaped more by commerce than by coercion.

THE SECOND AND THIRD WAVES

Recent research has stressed the limits of the early Industrial Revolution. In Britain, productivity rose rapidly in agriculture, in cotton, in metallurgy, and in several other branches of manufacturing, but rates of growth for the British economy as a whole were not particularly fast before the 1830s. The first innovations that appeared in British industry affected particular sectors of the economy, but others were little changed before the middle of the nineteenth century (see table 13.5). Despite the rising productivity of British agriculture, the production of food lagged slightly behind population growth until the 1830s.[46] And the slowdown in growth in the British economy after the 1870s suggests that on its own, the British industrial revolution could have generated only a limited momentum. If it had occurred, as did the industrial revolution of Song China, at the margins of a regional world system, it would have had a more limited impact and might have fizzled out within a century.

But Britain, unlike Song China, was at the center of the largest and most vigorous exchange networks that had ever existed, and the world as a whole was more unified and commercialized. In addition, the Industrial Revolution proved self-energizing, as inventions in transportation and communications—including the railway, the steamship, the bicycle, the modern printing press, and the telegraph and telephone—accelerated the exchange of information in general and of new technologies in particular. "With improved mobility," Joel Mokyr notes, "technology itself traveled easier: the minds of emigrants, machinery sold to distant countries, and technical books and journals all embodied the technological information carried from country to country. More mobility also meant more international and interregional competition. Societies that had remained impervious to technological change, from Japan to Turkey, felt left behind and threatened as distance protected them less and less."[47] Improved communications ensured that innovations that cut costs and raised profits would soon be adopted elsewhere within the already commercialized regions of the North Atlantic world. The result was a chain reaction that eventually spread throughout the world, rather than a regional burst of innovation that slowed after a century or two.

Regional patterns of industrialization varied greatly. As Alexander Gerschenkron pointed out in the 1960s, the sequence of change was itself important.[48] By the early nineteenth century, many outside observers were becoming aware of the changes occurring in Britain. From that point on, industrialization was bound to be a more conscious process, with greater dependence on deliberate and more or less planned government intervention (a process that was to culminate in the command economies of the twentieth century). It was possible to borrow British technology, and increasingly governments began to promote development. By the end of the nineteenth century, governments and large banks were actively managing industrial change. But differences in existing endowments, social structure, government structures, and geography also counted for much. While industrial production was at the heart of early changes in Britain, Belgium, Germany, and Czechoslovakia, a large, modern industrial sector developed later in France, the Netherlands, and Sweden. Nevertheless, rates of economic growth in general were impressive in all these regions in the nineteenth century.

If we focus on the broad picture, we can identify a sequence, a series of "waves" of industrialization, each shaped by different technologies and with different centers of dynamism.[49] The first wave, in the late eighteenth century, had little effect outside of Britain. The full impact of steam technology in particular became apparent only in the middle of the nineteenth century, during a second main wave of innovation. Serious industrialization began in Belgium, Switzerland, France, Germany, and the United States between the 1820s and 1840s. By the 1870s, these areas were creating new industries such as chemical manufacture (in particular for making dyes and producing artificial fertilizers), electricity, and steelmaking, in what Daniel

Headrick treats as a third wave of innovation. The Industrial Revolution now spread rapidly throughout the Atlantic economies; indeed, many developments, such as the exploitation of electricity, depended on multiple innovations pioneered in many different parts of this hub region, including Italy, the Balkans, Germany, Scandinavia, France, Britain, and the United States.

German industrialists pioneered the systematic application of science to production, while the United States was at the forefront of the industrialization of agriculture, the mass production of interchangeable components for commodities such as rifles, and, during the Civil War, the industrialization of warfare. By 1900, the United States had overtaken the United Kingdom as a producer of manufactured goods, and Germany was close behind: the United States was responsible for almost 24 percent of world manufacturing output, the United Kingdom for almost 19 percent, and Germany for 13 percent (see table 13.2). Germany and the United States had also pioneered two new, multicellular forms of industrial organization: the national corporation, which vertically integrated tasks previously shared by many separate enterprises, from the production of raw materials to manufacture, wholesaling, and retailing; and the multidivisional corporation, which horizontally integrated what had previously been different sectors of production.[50] The second and third waves together created a long boom in production in the late nineteenth century, unmatched until the second half of the twentieth century.

In a great tsunami of change, the second and third waves of industrialization carried the Modern Revolution to the rest of the world, where its impact was largely destructive. Just as the first stage in globalization had destroyed the traditional societies of the Americas, so this new round of global integration ruined traditional political, social, and economic systems beyond the emerging industrial heartlands of the Atlantic seaboard. As productivity rose in the industrialized hub region and the prices of goods such as British machine-made textiles fell, producers in other regions found their livelihood undermined by European imports. In entering global markets, small producers found themselves competing with large corporations using the most up-to-date technologies, and in the long run there was no doubt as to who would lose that competition. Wherever they had the power to do so, as in India and Pakistan, European powers accelerated such processes by juggling with tariff barriers or by forcing weaker powers and colonies to accept European exports. In this project, the power of newly industrialized armies, with modern, mass-produced weapons, and of better transportation systems such as steamships and railways could prove decisive—so decisive that Europe was able to import Indian grain even during the subcontinent's horrifying famines of the late nineteenth century.[51] Even China's once self-sufficient economy buckled as the increasing gravitational pull of the Atlantic economies warped the topology of international trade. Britain forced China to accept European exports, beginning with opium, after the First Opium War of 1842, when British forces threatened to cut the canal routes that supplied the north with grain. Over the next sixty years, industrialized European powers began to take economic and political control of China, as Britain had already taken control of the huge economy of Mogul India. In the final two decades of the century, European states imposed direct imperial control over much of Africa in the last great wave of political imperialism. Europe's economic and political colonies saw nineteenth-century capitalism in its most predatory forms.

The transformations of the late nineteenth century created a world divided between those that did and those that did not have industrial economies. The same processes that enriched the societies of the Atlantic seaboard ruined much of the rest of the world; and the gradients of inequality within nations, which had widened so spectacularly with the decline of the traditional peasantry, now became gradients between regions and nations. As the balance of economic and military power shifted, China's share of world industrial production fell from 33 percent in 1800 to 6 percent in 1900 and 2 percent in 1950; that of India and Pakistan fell from 20 percent in 1800 to less than 2 percent in 1900. The twentieth-century term *the third world* could have made no sense in 1750, when today's third world countries accounted for almost 75 percent of global industrial production. By the late twentieth century, they counted for less than 15 percent. Third world industrial production crashed in the second half of the nineteenth century, with total production falling to 37 percent in 1860, to 21 percent in 1880, and to about 7 percent for much of the first half of the twentieth century (see table 13.2 and figure 13.3).

The gap between "first" and "third" worlds, which was so familiar a part of the international landscape in the twentieth century, first appeared in the late nineteenth century. As Mike Davis writes:

> When the Bastille was being stormed, the vertical class divisions inside the world's major societies were *not* recapitulated as dramatic income differences *between* societies. The differences in living standards, say, between a French *sans-culotte* and Deccan farmer were relatively insignificant compared to the gulf that separated both from their ruling classes. By the end of Victoria's reign, however, the inequality of na-

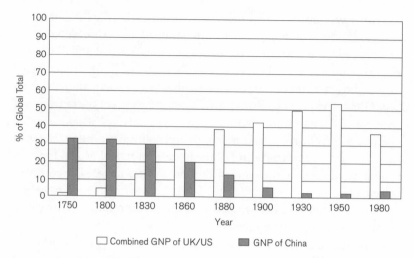

Figure 13.3.   The "rise of the West": the industrial potential of China and the United Kingdom/United States, 1750–1980, as a percentage of the global total. Based on table 13.2.

tions was as profound as the inequality of classes. Humanity had been irrevocably divided. And the famed "prisoners of starvation," whom the *Internationale* urges to arise, were as much modern inventions of the late Victorian world as electric lights, Maxim guns and "scientific" racism.[52]

The famines of the late 1870s, which touched equatorial and subequatorial regions right around the globe, were a watershed event in the history of the modern world, for the disruptive economic and social effects of European imperialism magnified the impact of a traditional, El Niño–related drought and thereby caused some of the worst famines since the fifteenth century.[53] Even worse was to come in the next twenty-five years, as the embryonic third world was integrated more tightly into global transportation networks, within which famines and epidemic diseases spread more widely and more rapidly than ever before. Many more people died in these crises than were to die during the First World War.

As traditional rulers outside the industrializing core became aware of their vulnerability, they began to wonder if they would have to industrialize the lands they ruled. But how? The conclusions reached in the previous chapter suggest that the problems they faced were as much political and cultural as economic. Matching the rates of innovation of the North Atlantic hub region meant changing political systems and cultural attitudes as well as economic structures in order to create well-integrated capitalist societies. This was bound to be a delicate and painful political maneuver—particularly for more traditional governments such as that of Tsarist Russia, which maintained many of the anticommercial attitudes of traditional tributary empires. Eventually, traditional governments would have to compromise with the new world of industry; but whatever form these compromises took, they were bound to threaten the established support base of these governments and undermine their stability. In the late nineteenth and early twentieth centuries, two highly traditional governments, whose societies were already moderately commercialized, presided over state-led industrialization drives. While the Meiji government in Japan rode the rapids of industrialization with considerable success, the Tsarist government did not; and it was left to the Communist government of Stalin to attempt the paradoxical project of an industrialization drive undertaken in the absence of entrepreneurs. Though the Stalinist industrialization drive had remarkable early successes, its eventual failure illustrates the difficulty of sustaining innovation without a competitive market environment.[54] Other once powerful regions—including the Islamic world, India and Pakistan, and China—underwent halfhearted attempts at reform, which left them in increasing economic and sometimes military dependence on Europe.

SUMMARY

The threshold to modernity was first crossed in the eighteenth and the early nineteenth centuries, in western Europe. The change had three interlinked aspects: economic, political, and cultural. The Industrial Revolution (the label applied to the economic aspects of the change) has been best studied in England, the country in which the changes first became apparent. England's social structures already conformed closely to the model of a capitalist society in the eighteenth century, with a rapidly growing class of wage earners and with governments closely allied to mercantile interests. The innovative potential of British capitalism first became apparent in agriculture, where commercially minded landowners raised agricultural productivity by introducing improvements on a large scale. Industrial breakthroughs came later; the crucial innovation was the use of steam power in large factories, which provided access to the energy bonanza of fossil fuels. Growing wealth, and the need to manage market economies and protect new forms of wealth, posed new challenges for governments, which had to begin mobilizing both resources and political support in new ways. These changes can be seen most clearly in the revolutionary changes that transformed government in France from the late eighteenth century on. For the first time, government began to reach into the daily lives of a majority of its subjects, concerning itself with their education, health, and attitudes. The most fundamental cultural change of the period was probably the increasing importance of scientific approaches to the world. Though scientific attitudes did not affect a wider population until their spread by mass education in the twentieth century, they played an important role in the technological innovations of the Industrial Revolution. The impact of science gained increasing significance during the nineteenth century's second and third waves of innovation. The Industrial Revolution then spread to western Europe and North America, as rates of innovation slowed in Britain. Beyond the industrializing core, the early stages of the Modern Revolution were largely destructive. In the late nineteenth century, differentials in wealth between different parts of the world became, for the first time, as great as the differentials within nations, and traditional structures that had worked, after a fashion, for millennia broke down, failing those who still depended on them for survival.

FURTHER READING

There is a huge literature on the Industrial Revolution. Classic studies that retain much value, though some of their details are now dated, include E. J. Hobsbawm, *Industry and Empire* (1969), and David Landes, *The Unbound Prometheus* (1969). More recent are Maxine Berg, *The Age of Manufactures, 1700–1820* (2nd ed., 1994); Pat Hudson, *The Industrial Revolution* (1992); and E. A. Wrigley, *Continuity, Chance, and Change* (1988). N. F. R. Crafts's *British Economic Growth* (1985) is an econometric survey. Margaret Jacob's *Scientific Culture and the Making of the Industrial West* (1997) is a classic study of the relationship between industrialization and the emergence of science. A good recent survey of industrialization on global scales is Peter Stearns, *The Industrial Revolution in World History* (1993). In *Economic Growth in Britain and France, 1780–1914* (1978), Patrick O'Brien and Caglar Keyder compare two distinctive paths to modernity. Kenneth Pomeranz, *The Great Divergence* (2000); R. Bin Wong, *China Transformed* (1997); and Andre Gunder Frank, *ReOrient* (1998), have put the argument that China was in many ways as close to industrialization as western Europe, even as late as the late eighteenth century. Joel Mokyr, *The Lever of Riches* (1990), and James McClellan III and Harold Dorn, *Science and Technology in World History* (1999), survey technological developments. Charles Tilly, *Coercion, Capital, and European States, AD 990–1992* (rev. ed., 1992), is good on some of the political changes associated with the Industrial Revolution. Peter Mathias and John Davis, eds.,

The First Industrial Revolutions (1989), is a collection of essays on European industrialization. Mike Davis's *Late Victorian Holocausts* (2001) is superb on the destructive impact of the Modern Revolution outside of the industrial core. *The Birth of the Modern World* (2003), by Chris Bayley, a fine global history of the "long" nineteenth century, stresses the link between warfare and state building.

14

THE GREAT ACCELERATION OF THE TWENTIETH CENTURY

If I had to sum up the twentieth century, I would say
that it raised the greatest hopes ever conceived by
humanity, and destroyed all illusions and ideals.

ACCELERATION

The twentieth century is so close to us that we may think we understand
it. But in some ways, it is harder to grasp than any other epoch discussed in
this book. Of all periods of human history, the twentieth century may be
the most difficult to see in the large perspective of big history. We cannot
know what will stand out a few centuries into the future. In Eric Hobsbawm's
superb history of the "short" twentieth century, *The Age of Extremes* (1994),
the things that loom large are the world wars of the first half of the century, the Great Depression, the Communist experiment, decolonization, and,
above all, the long boom after the Second World War. But on the scale of
big history, other aspects of the twentieth century stand out. Most striking
are the astonishing changes that have occurred in the relationship between
human beings and the biosphere. In a recent environmental history of the
twentieth century, John McNeill has argued that "the human race, without
intending anything of the sort, has undertaken a gigantic uncontrolled experiment on the earth. In time, I think, this will appear as the most important aspect of twentieth-century history, more so than World War II, the
communist enterprise, the rise of mass literacy, the spread of democracy, or
the growing emancipation of women."[1]

This chapter will focus on the sharp acceleration in the pace and scale of
change in the twentieth century. Not until the twentieth century did the
full significance of the Modern Revolution begin to reveal itself. Change accelerated so rapidly, and the ramifications of change were so universal, that
this period marks an utterly new stage in human history and in the history
of human relations with other species and with the earth. Indeed, it may be

no exaggeration to say that the twentieth century marks a decisive moment in the history of the entire biosphere.

On the cosmological scale, changes mostly occur at the stately pace of millions or even billions of years. In the biological realm, where natural selection sets the pace, significant alterations take place on scales ranging from thousands to millions of years. In human history, shaped increasingly by cultural change, the pace is more rapid. In the Paleolithic era, significant changes took many thousands of years. Agricultural societies, with their greater demographic dynamism, reduced that scale so that the entire history of agrarian societies could fit into a mere ten thousand years, while the history of agrarian civilizations was only half as long. The extraordinary dynamism of the Modern Revolution has accelerated the pace of global historical change once more. Time itself seems to have been compressed in the twentieth century.

Our sense of space has been revolutionized as well by modern forms of transportation and communication, from air travel to the Internet. It is not just that telescopes can reach nearly to the edge of the universe and the beginnings of time. Within the compass of human society, information and money can now be transferred more or less instantaneously across the globe, while people can travel only slightly less rapidly. Collective learning now embraces the world, but on the timescales of a private conversation. Robert Wright observes, "The fitful but relentless tendency of invisible social brains to hook up with each other, and eventually submerge themselves into a larger brain, is a central theme of history. The culmination of that process—the construction of a single, planetary brain—is what we are witnessing today, with all its disruptive yet ultimately integrative effects."[2] Space has contracted as rapidly as time. The epidemiologist D. J. Bradley has illustrated vividly what these changes mean for the life experience of individuals by plotting the "life-time travel tracks" of four generations of males in his own family. His great-grandfather's lifetime travel track could be contained within a square whose sides were only 40 kilometers long. In each of the next three generations, the sides of this square multiplied by about ten times. The lifetime travel track of his grandfather occupied a square with sides about 400 kilometers long, while that of his father had sides at least 4,000 kilometers long, and Bradley's own travels covered the entire globe.[3]

In 1940 the German cultural critic Walter Benjamin offered a haunting image of the hurricane of change that human societies have experienced in the twentieth century:

A Klee painting named "Angelus Novus" shows an angel looking as though he is about to move away from something he is fixedly contemplating. His eyes are staring, his mouth is open, his wings are spread. This is how one pictures the angel of history. His face is turned toward the past. Where we perceive a chain of events, he sees one single catastrophe which keeps piling wreckage upon wreckage and hurls it in front of his feet. The angel would like to stay, awaken the dead, and make whole what has been smashed. But a storm is blowing from Paradise; it has got caught in his wings with such violence that the angel can no longer close them. This storm irresistibly propels him into the future to which his back is turned, while the pile of debris before him grows skyward. This storm is what we call progress.[4]

As Eric Hobsbawm has argued, this hurricane of change has threatened to cut us loose from the past so decisively that it has transformed how we think about history itself.[5]

On many important scales, more change has occurred in the twentieth century than in all earlier periods of human history. The fact that this chapter covers only one century, while the equivalent chapter on the era of agrarian civilizations (chapter 10) covers four millennia, is merely one index of the transformations of scale engineered by modern society.

In describing these changes, it makes sense, once again, to begin with population growth, for whatever the impact of other factors, such as new technologies and new forms of social organization, every increase in human populations inevitably places new demands on the earth's resources (see table 14.1).[6] In 1900, world populations stood at about 1.6 billion. A century later, they had quadrupled to about 6 billion. It took 100,000 years for human populations to reach the first billion, and just over a century to add another five. In this century, the doubling time for human populations fell to eighty years in the first half of the century, and to a mere forty in the second half.

CHANGES WITHIN HUMAN SOCIETY

Waves of Innovation in the Twentieth Century

The acceleration of technological change was the primary agent of transformation. In the first place, technological changes are what made it possible to support such huge populations. A thoroughly commercialized agriculture had appeared in northwestern Europe by the eighteenth century, but the most significant increases in agricultural productivity have occurred in the twentieth century. Between 1900 and 2000, the productivity of the world's croplands has increased by three times, and the total grain harvest has multiplied five times, from 400 million to almost 2 billion tons.[7] Agricultural output rose faster than populations in the twentieth century.

TABLE 14.1 WORLD POPULATIONS, 1900-2000

Date	Population (billions)
1900	1.634
1910	1.746
1920	1.857
1930	2.036
1940	2.267
1950	2.515
1960	3.019
1970	3.698
1980	4.450
1990	5.292
2000	6.100

SOURCES: Massimo Livi-Bacci, *A Concise History of World Population,*
trans. Carl Ipsen (Oxford: Blackwell, 1992), p. 147; 1910 figure interpolated;
2000 figure from Lester R. Brown, *Eco-Economy: Building an Economy
for the Earth* (New York: W. W. Norton, 2001), p. 212.

Increased food production depended partly on the increased use of an old
technology, irrigation, and on the continued exchange of crops such as maize
and soybeans between different parts of the world. But new techniques were
also crucial. Particularly important were the use of artificial fertilizers and
the systematic breeding of new strains of crops, the most important of which
have been varieties of high-yield cereals and hybrid corns.

Outside agriculture, the most important technological changes of the
twentieth century came in waves whose impact and size dwarfed those of
the nineteenth century.[8] A fourth wave of innovation began at the end of the
nineteenth century and lasted for much of the first half of the twentieth
century. The internal combustion engine, whether installed in cars, trucks,
tanks, or planes, was the crucial new technology, and oil was the vital energy source, though other fossil fuels (including coal and natural gas) were
also important. In this phase, large, multidivisional corporations based in
the most industrialized countries began to break out of the national framework in which they had evolved, turning into multinational corporations,
with operations in several different countries.[9] The appearance of multinational corporations was one expression of the increasing dominance of

the most highly industrialized countries. The geographical spread of industrialization slowed in this period, and the productive capacity of those regions that had already begun to industrialize soared ahead of that of the rest of the world. Paul Bairoch's calculations (see tables 13.1 and 13.2) suggest that the absolute as well as the relative industrial output of regions beyond the emerging industrial heartland was falling for almost a century, from the mid–nineteenth century to the mid–twentieth century.

A fifth wave of innovation, after the Second World War, was dominated by atomic power and electronics. Electronics raised the efficiency of many other technologies. But because they also slashed the cost of using, acquiring, and processing information, they accelerated the pace and efficiency of collective learning and ensured that collective learning would now take place on global rather than local scales. This wave saw a sharp upturn in industrial production in many regions that had barely been touched by earlier waves, particularly in Latin America, East Asia, and Southwest Asia. It also witnessed an increase in the wealth and influence of multinational corporations. Particularly in the more industrialized regions—the dynamos of the world economy—the postwar boom seemed to slow in the late 1970s and 1980s.

Then growth accelerated once more in a sixth wave of innovation. This wave is still in motion, early in the twenty-first century. Its dominant technologies are electronic and genetic, while its most striking early effect has been to draw all parts of the world more tightly together than ever before. Manuel Castells has argued that the last two decades of the twentieth century mark the transition to an entirely new phase of capitalist history, which he labels the "Information Age."[10] In this phase, he maintains, flows of information are the key to profit making; the boundaries between individual enterprises are being erased, as production and services are organized in constantly shifting alliances or networks of enterprises, many of which subcontract much of their work to individuals or smaller companies. The control and movement of information have become perhaps the largest single sector of industry.[11] Global flows of information and wealth have become so rapid, and have such little respect for traditional boundaries, that they have blurred the borders between states as well as between enterprises. In 2000, many international corporations had as much market value as many major states, and most of these massive corporations dealt in communications (see table 14.2).

Taken together, the fifth and sixth waves of innovation sustained a boom in production far larger than that of the late nineteenth and early twentieth centuries. Between 1900 and 1950, the total output of the global economy rose from just over $2 trillion to just over $5 trillion. In the next fifty

TABLE 14.2. ECONOMIC ENTITIES RANKED BY MARKET VALUE, JANUARY 2000

Rank	Political Unit	Corporation	Value ($ billions)
1	USA		15,013
2	Japan		4,244
3	UK		2,775
4	France		1,304
5	Germany		1,229
6	Canada		695
7	Switzerland		662
8	Holland		618
9	Italy		610
10		Microsoft (US)	546
11	Hong Kong		536
12		General Electric (US)	498
13	Australia		424
14	Spain		390
15		Cisco Systems (US)	355
16	Taiwan		339
17	Sweden		318
18		Intel (US)	305
18		Exxon-Mobil (US)	295
20		Wal-Mart (US)	289
21	South Korea		285
22	Finland		276
23		Nippon JT (Japan)	274
24		AOL Time Warner (US)	289
25	South Africa		232
26		Nokia (Finland)	218
27	Greece		217
28		Deutsche Telekom (Ger.)	218
29		IBM (US)	213
30	Brazil		194

SOURCE: *Sydney Morning Herald,* 15 January 2000.

years it rose to ca. $39 trillion. These figures indicate that global production multiplied almost twenty times in the twentieth century. Growth in just the three years from 1995 to 1998 is estimated to have been greater than total growth in the 10,000 years before 1900.[12]

Creation: Consumer Capitalism and New Lifeways

The positive side of change is apparent in the staggering wealth of the most industrialized regions. Large populations in these regions have enjoyed high and rising levels of material affluence. In the nineteenth century, critics of capitalism saw its capacity to create poverty but underestimated its capacity to create material wealth. Some of those who did appreciate its productive potential (such as Rosa Luxemburg) argued that capitalism's extraordinary dynamism would prove its downfall. The more it produced, the greater the difficulty of finding buyers. Whereas for most earlier epochs of human history scarcity had been the fundamental problem faced by peoples and governments, now the main issue was to cope with abundance. (Marxists called this the "realization" problem—the problem of realizing profits through sales.) However, from late in the nineteenth century on, capitalist economies began to find a solution by treating their own workers not simply as factory fodder but also as potential markets for the goods they produced in such vast quantities. Just as viruses often evolve so as to protect their prey, so capitalism learned (in a move that Marx seems not to have anticipated) to protect, and even woo, its own proletariat into a new, and less unbalanced, form of symbiosis. This move is what generated the consumer capitalism of the twentieth century. Its distinctive feature has been the requirement that the mass of the population should consume commodities in ever-increasing amounts for the good of the entire system. To ensure the existence of a mass consumer market, wages had to be raised, consumer goods had to be marketed aggressively, and there had to be an end to the old-fashioned ethos of saving and conserving, the economic morality that had dominated most communities for most of human history. These changes began in the nineteenth century, but the shape of modern consumer capitalism first came into focus in the United States in the 1920s. Some of the earliest critiques of consumer capitalism—for example, Sinclair Lewis's 1922 novel, *Babbitt*—also appeared in the United States early in the twentieth century.

For governments, of course, disposing of surpluses is a more congenial problem than managing scarcity, the central task for most earlier states. The highly productive social and economic system of modern capitalism could defuse the hostility of subordinate classes by offering them living standards that would have gratified many a monarch in earlier historical epochs. In this way, consumer capitalism transformed traditional political problems, making it possible for modern elites to generate loyalty through gift-giving on a massive scale. It is this change, more than anything else, that explains the survival and resilience of liberal capitalist societies in the world's most industrialized regions.

Consumer capitalism has transformed the rhythms of historical change. The agrarian world was governed by Malthusian cycles, as population growth periodically outstripped productive capacity. During the "great depression" of the 1870s, it became apparent for the first time that economic growth could falter because of overproduction as well as underproduction. Manufacturers in sectors of rapidly growing productivity found that markets were too small to absorb what they could now produce. Over the following decades, it became clear that in a world of steadily increasing productivity, the problem of finding (or creating) markets would shape the rhythms of economic activity much as the problem of insufficient productivity had done in the agrarian era. As a result, the modern era is dominated by cycles of activity with a different (normally a shorter) periodicity, which we know as *business cycles.* Coping with those cycles has generated quite new types of behavior on the part of entrepreneurs, governments, and consumers in the most industrialized countries. At first, many governments and entrepreneurs reacted to rising levels of productivity by demanding protection for their own markets and creating protected markets in colonial regions. But this proved a self-defeating strategy, not only generating intolerable military conflict but also partitioning the huge world markets that had fueled much industrial growth in the nineteenth century. In the long run, John Maynard Keynes and others realized that avoiding cyclical downturns means maintaining and sustaining markets rather than monopolizing them. Thus a central concern of twentieth-century consumer capitalism was to create and expand markets. This change helps explain the ethical revolution that has made consumption a virtue as fundamental as abstinence had been in the precapitalist world. And it explains the emergence of a powerful new priesthood of advertisers, most visible on television in their ceaseless advocacy of consuming.

The beneficiaries of these changes enjoyed unparalleled levels of material prosperity and entirely new forms of freedom. In the richer countries, medical advances have improved health and eliminated many once unavoidable causes of physical misery. Indeed, lifestyles have changed so greatly that they may be exerting a significant evolutionary impact on human bodies. Studies in the United States suggest that people of the late twentieth century were not only taller than their predecessors of a century earlier but also had less skeletal toughness. Improved nutrition and medical care, coupled with lazier lifestyles, may be exerting much more evolutionary pressure on our species than we had realized.[13]

Personal relations have also been transformed. Though levels of interpersonal violence remain high, modern democratic societies frown on such behavior; most people are freer from the threat of violence than they would have been in traditional tributary societies, in which physical coercion was a more acceptable form of control. The political structures of democratic states, despite their many failings, also offer an unprecedented degree of legal protection to individuals. And the control over information that sustained the privileges of elites in the past has been lessened by the progress of mass education. Particularly striking is the slow breakdown of traditional gender roles that restricted the opportunities available to women. The spread of contraception and new forms of employment that depend less on physical strength have made it easier for women to take on many of the more specialized roles outside the household that men had monopolized in traditional societies. As a result, though women's wages and rates of promotion still lag well behind those of men in most sectors of most industrialized economies, the long-term trends in the more developed countries have seen a significant increase in their educational levels and employment opportunities. In 1990, in industrialized countries there were as many women as men in institutions of secondary and tertiary education, and there were almost 80 women in paid employment for every 100 men. By comparison, in the world as a whole there were only about 80 women in secondary education and 65 in tertiary education for every 100 men, and there were only about 60 women in paid employment for every 100 men.[14]

The immense gains enjoyed in the twentieth century by those living in the richer countries illustrate the astonishing creativity of the Modern Revolution. And that creativity holds out the tantalizing promise of a better future for human beings everywhere.

The Contradictions of Capitalism: Inequality and Poverty

Yet despite the remarkable positive changes seen in the twentieth century, in many ways and for many people the impact of the Modern Revolution was much less benign. In principle, the increased productivity of modern societies held out the possibility of building, for the first time, societies in which all sectors of society were free from the oppression of material poverty. This was the grand vision of socialism. But it was also clear to most socialists that although capitalism created the material preconditions for such a society, its basic structures were fundamentally inegalitarian. The productive dynamism that seemed to be capitalism's greatest virtue was driven by an unequal distribution of control over productive resources. Capitalism, it seemed, needed steep gradients in the distribution of wealth in order to survive and flourish. Marx argued that the system could not work without an appropriate mixture of owners and non-owners of productive resources. His conclusion seemed to mean that as long as capitalism existed, inequality would increase. For socialists, it followed that to build a society in which the benefits of high productivity would be available to all sections of the population, capitalism itself would have to be overthrown. But would a socialist society be able to match capitalism's high levels of productivity? Would more egalitarian societies be capable of matching the high productivity of capitalism, productivity on which socialist hopes for a world free of material poverty ultimately rested? The twentieth century was to suggest some answers to these agonizing questions.

Developments in the twentieth century vindicated much in the socialist critique of capitalism. The same forces that generated the extraordinary material abundance of the twentieth century also magnified global inequalities both within and between nations. Wealth was increasingly piled up in huge reservoirs that made the immense valleys of poverty between them seem all the more squalid. Capitalism has proved its capacity to generate abundant material wealth; but so far it has proved incapable of distributing global wealth in equitable, humane, and sustainable ways.

Though our attempts to measure these inequities are crude and approximate, they suggest some clear trends. Estimates of global income per head suggest that this gauge rose from $1,500 in 1900 to $6,600 in 1998. During the same period, global life expectancy, one of the most critical of all indicators of well-being, has risen from ca. 35 years to ca. 66 years.[15] These are significant gains, but they have been distributed unevenly, as tables 14.3 and 14.4 show. While the per capita gross national income of the United States was ca. $34,100 in 2000 (and that of the highest income countries averaged ca. $27,680), the per capita gross national income of Brazil was ca. $3,580, that of China (an economic superpower just two centuries ago) was ca. $840, and those of India (another former economic giant) and Burkina Faso stood respectively at $450 and $210. The ratios make these disparities even more striking (see table 14.3). These figures indicate that the per capita gross national income of Burkina Faso was less than 1 percent of the average for highest income countries, while the figures for India and the whole of sub-Saharan Africa were just over 1.5 percent of that average. The ratios for life expectancy statistics are not as extreme, of course, and modern medical knowledge has raised life expectancies throughout the world. Nevertheless, the statistics tell a clear story of lives shortened by relative poverty (see table 14.4).

TABLE 14.3. GROSS NATIONAL INCOME PER CAPITA, 2000

Country or Region	Income ($)
World	5,170
United States	34,100
Average of high-income countries	27,680
Burkina Faso	210
Sub-Saharan Africa	470
India	450
China	840
Brazil	3,580
Latin America and Caribbean	3,670

SOURCE: *World Development Indicators* (Washington, D.C.: World Bank, 2002), table 1.1, "Size of the Economy," pp. 18–20.

TABLE 14.4. LIFE EXPECTANCIES AT BIRTH, 2000

Country or Region	Life Expectancy at Birth (years)	
	Men	Women
World	65	69
United States	74	80
Average of high-income countries	75	81
Burkina Faso	44	45
Sub-Saharan Africa	46	47
India	62	63
China	69	72
Brazil	64	72
Latin America and Caribbean	67	74

SOURCE: *World Development Indicators* (Washington, D.C.: World Bank, 2002), table 1.5, "Women in Development," pp. 32–34.

In the last decades of the twentieth century, the wealth gap seems to have widened. In 1960, the wealthiest 20 percent of the world's population earned ca. 30 times as much income as the poorest 20 percent; in 1991 that multiple had soared to 61.[16] Conditions have especially deteriorated in South America and sub-Saharan Africa. In the early 1970s, Africa was self-sufficient in food production, and even exported surpluses. It is therefore shocking to realize that in the 1990s, if South Africa is excluded, the total gross domestic product of sub-Saharan Africa's population of 450 million was less than that of Belgium, with a population of only 11 million.[17]

These statistics are a reminder that for millions of people, modernity has led to worse living conditions. The number of adults suffering from acquired immunodeficiency syndrome (AIDS) has been kept below 1 percent in wealthier countries because they have the medical and educational resources to take the necessary preventive measures. In contrast, in Zimbabwe 26 percent of all adults were HIV-positive in the mid-1990s, and levels were almost as high in Botswana, Namibia, Swaziland, and Zambia.[18] Food shortages provide another shocking measure of inequality. Famine is only their most extreme form; more often, shortages mean lives ground down by the misery of chronic malnutrition. As Paul Harrison writes: "The everyday reality of malnutrition in the Third World is . . . adults scraping through, physically and mentally fatigued and vulnerable to illness. It is children—often dying, not so frequently of hunger alone, as of hunger working hand in hand with sickness; but more often surviving impaired for life."[19] In the late 1990s, more than 800 million people (ca. 14 percent of the world's population) were estimated to be undernourished, while 1.2 billion (ca. 20 percent) did not have access to clean and safe water.[20] Table 14.5 gives some summary demographic and economic statistics for 1994.

The Destruction of Traditional Lifeways

Figures such as those in the tables above reflect more than a falling behind in relation to the richer countries. They also tell us of the destruction of traditional lifeways—and of the safety nets built into them in local traditions of charity or specialized institutions such as emergency granaries. The decline in production of those countries that did not industrialize before the middle of the twentieth century is apparent from the figures in table 13.2, and declining production unraveled all the traditional safety nets. The fate of eighteenth-century English peasants facing enclosure is repeated today as population pressure or debt or taxation or war undermines established rural lifeways. Statistics on urbanization offer an indirect index of this change. In 1800, 97 percent of the world's

TABLE 14.5. SOME GLOBAL DEMOGRAPHIC AND ECONOMIC INDICATORS, 1994

Region	Population (millions)	Natural Increase (annual %)	Birth Rate per 1,000	Death Rate per 1,000	Life Expectancy (at birth)	GNP per Capita (1992 $)
World	5,607	1.6	25	9	65	4,340
More developed*	1,164	0.3	12	10	75	16,610
Less developed	4,443	1.9	28	9	63	950
Africa	700	2.9	42	13	55	650
Asia	3,392	1.7	25	8	64	1,820
Latin America and Caribbean	470	2.0	27	7	68	2,710
Europe	728	0.1	12	11	73	11,990
North America	290	0.7	16	9	76	22,840
Oceania	28	1.2	20	8	73	13,040

SOURCES: Allan Findlay, "Population Crises: The Malthusian Specter?" in *Geographies of Global Change: Remapping the World in the Late Twentieth Century,* ed. R. J. Johnston, Peter J. Taylor, and Michael J. Watts (Oxford: Blackwell, 1995), p. 156, based on 1994 *World Population Data Sheet,* compiled by the Population Reference Bureau, Washington, D.C.

*As determined by then-current UN convention: North America, Europe (including Russia), Australia, Japan, and New Zealand.

population lived in settlements of fewer than 20,000 people. By the middle of the twentieth century, the number had fallen to ca. 75 percent, and by 1980 to ca. 60 percent. In 2000, for the first time in human history, as many people lived in settlements containing more than 20,000 people as in smaller communities.[21] In 1800, Britain and Belgium were the only countries in the world in which less than 20 percent of the population found employment in agriculture or fisheries. Today, peasant farming remains the dominant lifeway in only three major regions—sub-Saharan Africa, South and Southeast Asia, and China—and in many of their communities, peasants are barely surviving. Eric Hobsbawm has argued that "the most dramatic and far-reaching social change of the second half of this century, and the one which cuts us off for ever from the world of the past, is the death of the peasantry."[22]

The statistics are a bloodless way of describing such changes; the following description conveys a sense of what they could mean to families and individuals. It comes from Paul Harrison's account of an interview he conducted in the 1980s with the head of a household in Burkina Faso, the African nation that lies north of Côte d'Ivoire (Ivory Coast), Ghana, and Togo. As in much of the Sahel, farming in Burkina Faso was based on shifting cultivation. Inhabitants prepared land that had not been farmed for a few decades by chopping down vegetation and setting it alight. Crops were then planted in the ashy soil: millet and sorghum for food, cotton or groundnuts for sale. Fertility was usually high for a year or two, then rapidly declined, so that communities had to move on and prepare a new patch of ground. Such methods can support only thin populations, for obvious reasons: at any given time, most of the land is fallow. But in recent years, population pressure had forced farmers to speed up the cycle and return to each patch before its fertility had been restored. Eventually, overuse threatened to destroy the soil itself, irrevocably.

Paul Harrison met and interviewed a 60-year-old farmer named Moumouni, who had lived through several stages in the developing crisis that has ruined much of the traditional croplands across the southern borders of the Sahara.

> Moumouni remembered that, when he was a child, only twelve people lived in his father's compound. Now there were thirty-four, with five young men working away from home on the Ivory Coast. Land in the village is allocated by the chief on the basis: to each according to his need. . . . Yet the village's traditional lands had not expanded at all. . . . The additional land needed had been taken out of the five sixths that

usually lay fallow. Fallow periods had been slowly whittled down over the decades, until they were now only four or five years, when at least twelve would have been needed to restore the exhausted fertility of the soil.

Moumouni showed Harrison his land.

Even close in to the compound, the soil looked poor enough, stony and dusty, without a trace of humus. And this was the only area they ever fertilized, with the droppings of a donkey and a couple of goats. Outside a circle of about fifty yards' diameter round the houses, the ground was a dark red, baked hard. It had been cultivated the year before but had yielded very little. Moumouni said he didn't think anything would grow there this year.[23]

The impact of such difficulties can be calculated on a national basis. A World Bank report estimates the cost of "crop, livestock, and fuelwood losses from land degradation" in Burkina Faso in 1988 as ca. 8.8 percent of the country's gross national product.[24]

The lives of traditional foragers have been subjected to an equally violent assault in the twentieth century. But the change they have undergone is curiously incomplete, despite the colossal mismatch in scale and resources between foraging communities and modern capitalist states. Indeed, it may be the mismatch that explains the remarkable capacity of many such communities to preserve something of their past. Where their lands were needed for settlement or mining, they were removed with brutality and little ceremony; otherwise, they were often left in peace. Their military encounters with modern societies often took the form of guerrilla wars or small-scale military confrontations. The conflicts were real enough, and sometimes states became directly involved, as they did in the Indian wars in the United States, or the many guerrilla campaigns fought with kin-ordered communities from Australia to Siberia. But when the fighting was over, kin-ordered communities could often find niches within the societies that had taken so much from them. So, in a sense, they survived, and continue to the present day, preserving far more of their past than was possible for the peasant communities of the agrarian world. And the modern world has much to learn from communities whose lifeways have lasted so much longer than those of industrial capitalism.

The Destruction of Traditional Tributary Empires

Modern capitalism has also destroyed the larger political structures of the era of agrarian civilizations. The great tributary empires that dominated the era of agrarian civilizations vanished with remarkable swiftness. In 1793, when George Macartney was sent as ambassador by George III to ask for equal diplomatic representation and trading rights in China, his request was refused by the Qing emperor, Qianlong, who referred to England as a "remote and inaccessible region, far across the spaces of ocean." However, the emperor congratulated George III for his "submissive loyalty" in sending "this tribute mission" and encouraged him to show obedience in the future "so that you may enjoy the blessings of perpetual peace."[25] These were the arrogant attitudes that Europeans themselves would display toward the rest of the world a century later. At the time, they seemed perfectly realistic; after all, Europe produced little that China could not produce better and more cheaply, so Europeans had to buy most Chinese goods in exchange for silver.

Soon, however, British traders found something else that Chinese consumers wanted: Indian-produced opium, a substance whose consumption had been banned in China. At first trading illegally, in the 1840s British traders, backed by gunboats, forced the Chinese government to allow this new, if destructive, trade during the so-called Opium Wars. In 1839, the local Chinese official at Guangzhou compelled British ships to surrender their opium, and proceeded to destroy it. To Queen Victoria a Chinese official, Lin Zexu, wrote: "We have heard that in your honourable country, the people are not permitted to inhale the drug. If it is so regarded as deleterious, how can seeking profit by exposing others to its malefic powers be reconciled with the decrees of Heaven?"[26] Claiming that the real issue was free trade rather than opium, the prime minister of England, Lord Palmerston, sent a fleet to blockade Guangzhou, and it clashed with Chinese naval vessels. During the next two years, British ships began to attack other ports. Eventually they took control of the Yangtze cities from which Beijing itself was supplied along the Grand Canal, forcing the Chinese to back down in 1842. Chinese military and naval technology, which had changed little since the time of Marco Polo, was no match for British equipment. The gulf in technologies and productivity levels that industrialization had opened up caused the collapse of the Chinese empire in the early twentieth century. By the end of the century, there existed no political or economic structures that fit Eric Wolf's model of "tributary states," even though such states had dominated the world a mere two centuries before.

While the speed with which the ancient tributary empires collapsed was one striking feature of the past two centuries, another has escaped general notice: many characteristics of the traditional tributary world survived in the great Communist empires of the twentieth century.[27] Communist governments appeared in Russia and then in China, led by modernizing revolutionary movements. But their ideologies were as much anticapitalist as antiautocratic. This feature helps explain their appeal to societies whose elites felt acutely the insult of capitalism's assault on their traditional prestige and culture. Stalin's radical rejection of capitalism during the collectivization drive of the early 1930s meant that the Soviet Union had to compete with the major industrial powers without enjoying the innovatory dynamism of capitalism. Central control of economic and intellectual exchanges stifled the commercial and intellectual transactions that are the lifeblood of capitalism, and censorship clogged the networks of collective learning that generated so many petty innovations in market economies. Maoist China followed a similar course after 1949. Where market forces were banned, there was little choice but to mobilize resources in more traditional ways, using techniques of taxation and social and economic organization similar to those of the great tributary empires—but with the addition of some twentieth-century technologies, from telephones to tanks. Like traditional tributary empires, the command economies of the Communist world were better at mobilizing resources than at raising productivity. Recent estimates suggest that improvements in efficiency levels can explain no more than 24 percent and perhaps as little as 2 percent of the increases in output in the Soviet Union during the first three Stalinist five-year plans. Most of the achievements of the Soviet industrialization drive rested on a massive, and highly coercive, mobilization of capital, raw materials, and labor.[28] Neither the labor of the Soviet population nor its resources were spared in the government's determination to match the industrial and military might of its capitalist rivals.

For a while—particularly in the 1930s, when the capitalist world itself was in crisis, and again in the 1950s—it seemed as if these new, state-managed structures might generate a dynamism to match that of capitalism. What they lacked in entrepreneurial flair they made up for in systematic commitment to high levels of education, in the introduction of modern technologies, and in the massive organizational capacity of powerful and ruthless states using modern technologies of communication. But in time their innovatory sluggishness, the same quality that had slowed innovation throughout the era of agrarian civilizations, ensured that they would fall behind their capitalist rivals in productivity levels, in innovation, and eventually in military capability. The wasteful habits of the construction phase proved hard to shake, and the Soviet command economy never managed to shift from resource-intensive to resource-economizing forms of growth; eventually, it ran out of resources. The collapse of the Soviet Union was, as Mikhail Gorbachev understood, a failure to compete economically and technologically. Mobilizational capacity could not, in the long run, compensate for sluggish innovation:

> At some stage—this became particularly clear in the latter half of the seventies—something happened that was at first sight inexplicable. The country began to lose momentum. . . . A kind of "braking mechanism" affecting social and economic development formed. And all this happened at a time when scientific and technological revolution opened up new prospects for economic and social progress. Something strange was taking place: the huge fly-wheel of a powerful machine was revolving, while either transmission from it to work places was skidding or drive belts were too loose.
>
> Analyzing the situation, we first discovered a slowing economic growth. In the last fifteen years the national income growth rates had declined by more than a half and by the beginning of the eighties had fallen to a level close to economic stagnation. A country that was once quickly closing on the world's advanced nations began to lose one position after another. Moreover, the gap in the efficiency of production, quality of products, scientific and technological development, the production of advanced technology, and the use of advanced techniques began to widen, and not to our advantage.[29]

Gorbachev's attempts to introduce a new dynamism by relaxing the grip of the planners on the economy and society ended with the collapse of the entire system. In the 1990s, Russia had to start rebuilding capitalism, almost from scratch.

China has faced similar challenges, but it has taken a different route. Under its surface, Communist China is becoming a capitalist society, using methods that were not open to the Soviet leadership because the structures and habits of capitalism were not eliminated as decisively in China as they had been in the Soviet Union. The experiences of the Communist era suggest that scrapping capitalism will not necessarily provide a solution to the many problems that capitalism creates. The Communist societies of the twentieth century could not match the productivity of their capitalist rivals; but neither were they strikingly egalitarian.

Conflict

Not surprisingly, a world of such instability, in which gradients of inequality are steadily rising, has been ridden with conflict. The past hundred years saw more violent conflict than any previous century in human history. The scale of the human and material damage caused by war reflects the increased "productivity" of armies and weapons in the modern era, as well as the increased size of the armies and the populations at war. William Eckhardt has

TABLE 14.6. WAR-RELATED DEATHS, 1500-1999

Years	War Deaths (millions)	Deaths per 1,000 People
1500–1599	1.6	3.2
1600–1699	6.1	11.2
1700–1799	7.0	9.7
1800–1899	19.4	16.2
1900–1999	109.7	44.4

SOURCES: Lester R. Brown et al., *State of the World, 1999: A Worldwatch Institute Report on Progress toward a Sustainable Society* (London: Earthscan Publications, 1999), p. 153; citing William Eckhardt, "War-Related Deaths Since 3000 BC," *Bulletin of Peace Proposals 22*, no. 4 (December 1991): 437–43, and Ruth Leger Sivard, *World Military and Social Expenditures 1996* (Washington, D.C.: World Priorities, 1996).

roughly calculated that 3.7 million people died in war in the 1,500 years up to 1500 CE. In the sixteenth century, he estimates that 1.6 million died in war; in the seventeenth and eighteenth centuries, 6.1 million and 7.0 million; and in the nineteenth century, 19.4 million. In the twentieth century, war deaths reached 109.7 million, or almost three times the sum of all deaths in the preceding 1,900 years (see table 14.6).[30] Deaths in the Second World War alone reached 53.5 million. Equally spectacular were the casualties that might have been suffered but were not, through the (fortunate?) avoidance of nuclear war. But nuclear wars were prepared for. In 1986, there were almost 70,000 nuclear warheads, with a total explosive power equivalent to 18 billion tons of TNT—3.6 tons for every individual on earth.[31] If they had been used, these weapons would have caused a catastrophe similar in its scale and many of its consequences to the extinction event at the end of the Cretaceous era, which destroyed most species of large dinosaurs.

Smaller wars have taken as great a toll as the world wars and the cold war. Between 1900 and the mid-1980s there were some 275 different wars.[32] Between 1945 and 2000, there were nine regional wars in which more than 1 million people died; and in these wars, civilian casualties exceeded military casualties. The Korean and Vietnam Wars caused the deaths, respectively, of 10 percent and 13 percent of national populations.[33] Since the end of the cold war, changes have occurred that may prove very significant in the long run. In the 1990s, global military expenditures declined by perhaps 40 percent, and stocks of weapons of all kinds fell. (The "war on terrorism" initiated after the attacks on New York and the Pentagon on September 11, 2001, may reverse this trend.) Warfare has become more localized and has increasingly taken place within states, or between states and guerrilla armies of various kinds, a shift that implies a reduction in scale (though no reduction in the horror for those involved).[34] These figures mark a shift in the nature of warfare rather than a real reduction in the amount of violent conflict. The tensions and dislocations of the hurricane of change affecting the entire globe will ensure that conflict remains endemic, and modern weaponry will ensure that local conflicts continue to cause great suffering.

CHANGES IN HUMAN RELATIONS WITH THE BIOSPHERE

The scale of human society and the extent of its productive (and destructive) capacity in the twentieth century have ensured that the Modern Revolution has had an impact on the world environment that is no longer just regional but is also global. This is why most significant indicators of the environmental impact of human populations "mimic the same exponential curve over the past three centuries."[35]

One rough measure of human environmental impacts arises out of attempts to measure the changing energy requirements of human societies (see table 6.1). These figures make clear that total human energy consumption multiplied many more times in the twentieth century than in all of previous human history. At the end of the twentieth century, the total amount of energy consumed by humans may have been 60,000 to 90,000 times that used by humans early in the Neolithic era. As a result of these changes, human societies became, in the twentieth century, a major force acting on the entire biosphere. As we have seen already, estimates of the distribution of global "net primary productivity" on the land suggest that up to 25 percent, and perhaps as much as 40 percent, is being co-opted by our own species (see p. 140).

Given that the resources of the biosphere are finite, human use of energy, resources, and space on this scale has inevitably reduced the resources available for other species. Declining biodiversity is one unavoidable consequence. Teamed up with domesticates and fellow travelers such as rabbits, goats, and weeds, humans have reduced biodiversity by destroying or appropriating the habitats of other species. In 1996, about 20 percent of all vertebrate species were in significant danger of extinction.[36] As Richard Leakey has argued, the scale of modern extinctions could prove similar in magnitude to the five other great extinction events known to paleontologists in which at least 65 percent of all marine species vanished.[37]

Will there even be enough resources to support our own species at an acceptable level? How easy will it be to feed 10 to 12 billion people in a century's time? It is possible that new technologies, perhaps depending on genetic engineering, will ensure that food production continues to rise at the rapid rates typical of the twentieth century. In the meantime, there is good reason to think we may be approaching some critical limits. We feed ourselves from croplands, grazing lands, and fisheries. The area of grazing lands can hardly be increased much more, and much of the land available is severely degraded. There is also general agreement that the harvest of fish cannot be increased significantly. Meanwhile, the output of croplands depends heavily on increased use of irrigation; since 1950, irrigated land has risen from 94 to 260 million hectares, and now accounts for 40 percent of all food production.[38] Yet in many areas, the introduction of modern diesel-powered water pumps has led to a fall in water tables, an outcome suggesting that here, too, there is little room for expansion. Ecologically speaking, what is happening is that underground stores of fresh water that were built up over millions of years are being emptied in just a few decades.

Overuse of resources measures one side of the human impact on the biosphere; disposal of wastes is the other. Perhaps the most powerful illustration of the significance of human-generated pollutants is the possibility that we may be fundamentally changing the earth's atmosphere. Lester Brown argues that "while the Agricultural Revolution transformed the earth's surface, the Industrial Revolution is transforming the earth's atmosphere."[39] The temperature at the surface of the earth depends on a precarious balance between the amount of sunlight captured within the earth's atmosphere and the amount released or reflected back into space. Mars, without a significant atmosphere, retains little of the Sun's energy, so it is too cold for life. Venus, with a greenhouse atmosphere dominated by carbon dioxide, is ca. 450°C, too hot for life. The crucial factor (though not the only one) in determining how much of the Sun's energy is retained at the surface of our planet is the amount of carbon dioxide in the atmosphere. During the last ice age, average temperatures were ca. 9° lower than today, and levels of atmospheric CO_2 were ca. 190–200 parts per million. By about 1800 CE, CO_2 levels had risen to ca. 280 parts per million. At that point, the Industrial Revolution began the massive exploitation of nonorganic fuels such as coal and oil, which greatly increased the amount of CO_2 pumped into the atmosphere. Now CO_2 levels have reached ca. 350 parts per million, or twice the levels during the ice ages. By 2150 they could nearly double again, to between 550 and 600, if present rates of emission continue. Carbon that was stored in trees and then buried in the ground during many tens of millions of years in the Carboniferous era has been flung back into the atmosphere in just a few decades. A part of the carbon cycle that normally runs on a timescale of many millions of years has been accelerated by several orders of magnitude. Natural processes simply cannot absorb carbon at these rates.

What this explosive release of carbon dioxide will mean in practice is not clear. There is general agreement that it will lead to warming, and global temperatures are already warmer than at the beginning of the twentieth century. Warming may increase ecological productivity in some areas, but it will certainly have a worldwide impact, whether benign or harmful. Average temperatures seem to have been rising for at least two decades, causing unusual periods of heat and dryness, as well as unusual weather patterns. A temperature rise of 2.5°C (a modest estimate) by 2050 would be the equivalent of the changes at the end of the last ice age. Sea levels will go up as the volume of water expands, and ice caps will melt. This will have tragic effects on low-lying areas: the island nations of the Pacific, the Netherlands, Bangladesh, and elsewhere. Warming will also affect existing species, including some that are vital for humans. Rice does not tolerate high temperatures, so its productivity may decline as climates warm.[40]

Perhaps the most worrying aspect of global warming is its unpredictability. Climatologists know that climatic systems, like many other chaotic systems, are subject to sudden, sharp changes. They may change slowly and predictably for a time, then become unstable before switching, quite abruptly, to a new state. The end of the last ice age may have marked one such sudden shift. If the scale of warming today is on a similar scale, we cannot rule out a sudden *qualitative* shift in global climates—possibly occurring on the scale of a human lifetime.

Declining biodiversity and increased carbon emissions are among the two most significant indicators of human impacts. Lester Brown, former project director of the annual survey *State of the World,* writes that at the end of the twentieth century, the most dangerous effects of human activity were apparent in six different areas: fresh water, rangelands, oceanic fisheries, forests, biological diversity, and global atmosphere.[41] While the impact on the last three is for most people indirect, and therefore easier to ignore, the first three areas are affected more obviously, and in ways that appear to set clear limits to our capacity to feed growing populations. Lack of access to fresh water threatens the health of millions and hinders the potential growth of irrigation agriculture. Moreover, the exploitation of fisheries and rangelands appears to have reached its maximum level.[42]

In a massive survey of the human impact on the environment published at the beginning of the 1990s, Robert W. Kates, B. L. Turner II, and William C. Clark make an interesting attempt to measure the extent of human environmental impacts on several different scales. They take ten fundamental

TABLE 14.7. HUMAN-INDUCED ENVIRONMENTAL CHANGE,
10,000 BCE TO THE MID-1980S CE

Form of Transformation	Dates of Quartiles (compared to 1985 levels)		
	25%	50%	75%
Deforested area	1700	1850	1915
Terrestrial vertebrate diversity	1790	1880	1910
Water withdrawals	1925	1955	1975
Population size	1850	1950	1970
Carbon releases	1815	1920	1960
Sulfur releases	1940	1960	1970
Phosphorus releases	1955	1975	1980
Nitrogen releases	1970	1975	1980
Lead releases	1920	1950	1965
Carbon tetrachloride production	1950	1960	1970

SOURCE: Robert W. Kates, B. L. Turner II, and William C. Clark, "The Great Transformation,"
in *The Earth as Transformed by Human Action: Global and Regional Changes in the Biosphere over the Past 300 Years,* ed. B. L. Turner II et al. (Cambridge: Cambridge University
Press, 1990), p. 7.

measures of human effects on the environment, estimate the total impact
between 10,000 years ago and 1985, then attempt to identify the dates when
each type of change attained 25 percent, then 50 percent, and finally 75 percent of its 1985 level; their figures are presented in table 14.7. The quickest
way of appreciating the significance of this table is to look at the dates at
which each type of impact reached 50 percent of its 1985 level. For seven of
these variables, more change occurred in the 40 years from 1945 to 1985
than had occurred in the previous 10,000 years.[43] As for the remaining three
variables—deforestation, rates of extinction of vertebrate species, and carbon release into the atmosphere—50 percent of all change has occurred since
the middle of the nineteenth century. Table 6.1 tells a similar story for human use of energy. Chronologically speaking, the twentieth century is a tiny
chunk of history, but the scale of the transformations it has witnessed dwarfs
all of previous human history.

In the course of the twentieth century, human beings caused changes so
decisive, so rapid, and so vast in their scale that they force us to see human
history, once again, as an integral part of the history of the biosphere. The

statistics collected in this chapter give some impression of the scale and speed of change. What they cannot do is give us any clear indications of its long-term implications, creating instead the impression of something very large moving at very high speed. And that, perhaps, is the most worrying aspect of this brief survey of twentieth-century history—the fear that it is like a traffic accident in slow motion. Can change continue to accelerate without dangerous consequences for human society and the biosphere as a whole? Or will the astonishing creativity that is the other side of the Modern Revolution lead us toward a more stable and more sustainable relationship with our natural environment? The next chapter, which considers possible futures at several different scales, will begin by considering these questions.

SUMMARY

The changes that occurred in the twentieth century are, by many measures, greater than the changes that took place in all previous eras of human history. As the Modern Revolution hit its stride, productive capacity soared; so, too, did living standards in the industrialized hub regions, as governments and business began to see the material satisfaction of their own populations as the key to a flourishing capitalist society. But outside the hub regions, much of the impact of the Modern Revolution was destructive. Here, traditional lifeways, and the safeguards built into them, were largely destroyed, as were the states that presided over them. The Communist states of the mid–twentieth century sought to match the economic and military successes of capitalist societies, while avoiding the inequalities inseparable from capitalism. But they succeeded neither in keeping up with their rivals nor in creating societies that offered attractive alternatives. As spectacular as the economic and technological changes of the twentieth century was the increasing human impact on the biosphere, which escalated more rapidly than in any other era. Early in the twenty-first century, human societies were beginning to have a major effect on the entire biosphere, and evidence grew stronger that humans were beginning to live beyond sustainable limits. The acceleration in the pace and scale of change is perhaps the most striking and (for contemporaries) the most frightening aspect of twentieth-century history. The scale of human impacts on the biosphere and on other humans is now so great that the changes of the twentieth century will stand out on the scale of planetary history.

FURTHER READING

J. R. McNeill, *Something New under the Sun* (2000), and E. J. Hobsbawm, *The Age of Extremes* (1994) offer contrasting introductions: the first focusing on ecological issues, the second on more conventional historical themes. Manuel Castells's *The Information Age* (3 vols., 1996–98) is an ambitious attempt to theorize change in the late twentieth century. B. L. Turner II et al., eds., *The Earth as Transformed by Human Action* (1990), attempts to quantify the extent of human impacts on the environment, while the yearly volumes of Lester Brown et al., *State of the World* (1984–), offer ecological statistics. David Held et al., eds., *Global Transformations* (1999), offers a thorough discussion of aspects of globalization; while the books by Paul Harrison, *Inside the Third World* (1981) and *The Third Revolution* (1992), provide many insights into the realities of life in the third world. Paul Kennedy's *Preparing for the Twenty-First Century* (1994) looks at many long trends.

PART VI

PERSPECTIVES ON THE FUTURE

15

FUTURES

This book started out examining very large structures and huge timescales. But its focus has narrowed—first to a single planet, then to the history of a single species, and finally to a single century in the history of that species. Now we must move back up the temporal and spatial scales once more as we look toward the future.

THINKING ABOUT THE FUTURE

> We are all in a situation that resembles driving a fast vehicle at night over unknown terrain that is rough, full of gullies, with precipices not far off. Some kind of headlight, even a feeble and flickering one, may help to avoid some of the worst disasters.

It may seem foolish to discuss the future. After all, the future really is unpredictable.

It is not just that we don't know enough. Some nineteenth-century scientists believed that reality was both deterministic and predictable. They thought that if we knew enough about the position and motion of everything around us, we could predict the future with great precision. It is now clear that this is not so. Quantum physics shows that *it is in the nature of reality to be unpredictable*. At the smallest levels, reality has something fuzzy about it. There appears to be a limit to the precision with which we can measure the movements of subatomic particles. It is as if they were in some sense smeared out over space and time, so that the best we can do is to estimate the probability of their being at a particular place at a particular time. This type of unpredictability is often described as *chaos*, because chaos theory has shown that billions of tiny uncertainties can accumulate through long chains of causation until, in the large-scale world that humans

Figure 15.1.   Earthrise as seen from the Moon. This famous photograph was taken from *Apollo 8* in December 1968. It has become a powerful symbol of our growing awareness of human unity and fragility. William Anders, one of the three astronauts on the mission, probably actually took the photograph. In a 1998 interview, he said, "All of the views of the Earth from the moon have led the human race, and its political leaders, and its environmental leaders, and its citizenry, [to] realize that we're all jammed together on one really kind of dinky little planet, and we'd better treat it, and ourselves, better, or we're not gonna be here very long." As Fred Spier has pointed out, the Earthrise photo also provides an ironic symbol of the fragility of human maps of reality, for the accounts by the three astronauts of who took the photograph, and when, are utterly contradictory (Fred Spier, "The Apollo 8 Earthrise Photo," 2000 <http://www.i2o.uva.nl/inhoud/gig/Apollo%208%20US.pdf> [accessed April 2003]). Photo courtesy of NASA.

occupy, they create considerable large-scale unpredictability. In the 1990s, rigorous mathematical proofs found that chaotic behavior is more than a matter of ignorance or imprecision: it is just the way things are. Even if changes take place according to precise, deterministic rules, we can never know the starting point of change with enough accuracy to forecast its future course exactly. Thus, even if reality is deterministic, it need not be predictable.

But there is a second kind of uncertainty. Understanding how a particular object works may not help us predict its behavior when it is combined with other objects into a larger system. Interacting systems with different elements appear to function according to emergent rules that we cannot always deduce simply by knowing how their components work. Understanding hydrogen and oxygen does not tell us much about water, which is formed by their chemical combination.[1] Ricard Solé and Brian Goodwin observe, "With chaos, it is sensitivity to initial conditions that makes the dynamics unpredictable. With emergent properties, it is the general inability of observers to predict the behavior of nonlinear systems from an understanding of their parts and interactions."[2]

We have seen both types of unpredictability at work in evolution and in human history. Many possible futures are compatible with the same rules of natural selection or cultural change. So change is always, to some degree, open-ended. There really is a difference between past and future, which makes prediction a dangerous game. Peter Stearns reminds us how dangerous it is by listing some of the more spectacular failed predictions made in the United States in the twentieth century: "[E]lectronic impulses from a supersonic alarm clock enter your brain directly to wake you up (1955); electronic brains will decide who marries whom, making more happy marriages (1952); only 10 percent of the population will work, while the rest are paid to be idle (1966 and recurrently); within a few decades, communicable diseases and also heart disease will be wiped out (again, 1966, clearly a banner year for optimistic technologists)."[3] For all these reasons, historians normally avoid considering the future entirely. R. G. Collingwood wrote, sternly: "The historian's business is to know the past, not to know the future, and whenever historians claim to be able to determine the future in advance of its happening we may know with certainty that something has gone wrong with their fundamental conception of history."[4]

Despite these cautions, we cannot entirely avoid the challenge of trying to predict. There are at least two types of situation in which we can and *must* attempt forecasts. The first is when we are dealing with entities that change slowly or simply. There are degrees of open-endedness, for even chaotic processes will generally confine their unpredictability within limits. Thus for some processes and at some scales, change is reasonably simple and fairly easy to anticipate. These are the types of change that determinists once thought were typical of *all* change. For example, chemists can normally foresee the exact result of mixing defined quantities of simple chemicals at particular temperatures. This does not mean that prediction is easy, but it is sometimes possible if we take enough care about it. When firing a canon, it matters greatly where the projectile lands; for a gunner, the mathematics of ballistics is worth mastering, because it may make the difference between winning and losing a battle. Deterministic thinking also works pretty well when change is slow. For such processes, the present moment appears to stretch out, reaching well into what we think of as the future. The rise and fall of a single breath may last only a second or two, but the rise and fall of a mountain may take millions of years. So we can say with some confidence that Mount Everest will be around in 1,000 years' time.

It is also worth thinking hard about the future when we are dealing with complex processes whose outcomes matter to us and over which we have some influence. Choosing which stocks to buy and which horse to back at the races are good examples. These are not deterministic processes, so we cannot predict them with the confidence of a gunner. But they are not totally open-ended. If change is utterly random, it is a waste of energy to attempt prediction; tossing a coin is as rational as any other way of making decisions. But where there is even a slight element of predictability in systems that matter to us, it is worth thinking hard about what is going on—and such situations are all around us. In handling them, prediction becomes a game of percentages. Those who carefully consider the variables involved in these types of change may find, over time, that they predict with slightly more success than those who make no effort. Some gamblers do make money. In such situations, the effort put into prediction matters, and it matters profoundly. Animals constantly have to make predictions about the likelihood of, say, finding dangerous predators in a particular place. Those that predict best will survive, and those that don't won't; in this way, skill in such predicting eventually gets built into the genetic makeup of most species. Choices whose outcomes matter even though they are neither deterministic nor completely random surround us all the time. So it is not surprising that in all human societies, entire professions have been based on the making of such predictions—think of astrologers, stockbrokers, professional gamblers, weather forecasters, or . . . politicians.

Making predictions of these two kinds, and making them as well as possible, is something living creatures do all the time, whether they are eagles swooping for the kill or investors buying shares. Indeed, action is impossible without prediction. Properly understood, prediction is as inevitable as breathing.

In thinking about the future at the scales of big history, we face both types of prediction. This chapter will begin by discussing our near future, on a scale of about one hundred years. At this scale, change is complex and unstable, but we have no reason to think it is totally random. Besides, we have to predict at this scale because our predictions will affect our actions, and our actions will help shape the lives of our children and grandchildren. So, trying to predict the shape of the next century is a serious task. In the "middle future," a scale of several hundred to several thousand years, serious prediction about the future of our species is almost impossible. We have little influence over these scales, and there are too many possible futures. Our ability to predict is so limited that it is not worth putting much effort into the task. Yet when we shift to the remote future, to larger timescales and larger objects, such as whole planets, or stars, or galaxies, or even the universe itself, prediction becomes easier again. This is because at these scales, we are dealing with slower and more predictable types of change, so that deterministic thinking comes into its own once more. Even here, there is no certainty, but the range of possibilities narrows.

THE NEAR FUTURE: THE NEXT HUNDRED YEARS

"Things happened very slowly and we didn't notice them at first," Jean-Marie explained. "At the beginning of an illness, you don't realize it can do you harm. It's only when you can no longer walk that you realize you are really sick. When we saw that the land was dying, we knew we had to do something. But we didn't know what to do." [Jean-Marie Sawadogo, 55, head of a family living near Ouagadougo, capital of Burkina Faso]

What we now call the plains of Pheleus [Plato's homeland in Attica], were once covered in rich soil, and there was abundant timber on the mountains, of which traces may still be seen. Some of our mountains at present will only support bees. But not so very long ago trees fit for the roofs of vast buildings were felled there, and the rafters are still in existence. There were also many other lofty cultivated trees which provided unlimited fodder for beasts. The soil got the benefit of the yearly "water from Zeus." This was not lost, as it is today, by running off a barren ground to the sea. A plentiful supply was received into the soil and stored up in the layers of clay. The moisture absorbed in the higher regions percolated to the hollows, and so all quarters were lavishly provided with springs and rivers. To this day the sanctuaries at their former sources survive. By comparison with the original territory, what is left now is like the skeleton of a body wasted by disease. The rich, soft soil has been carried off. Only the bare framework of the district is left.

The scale of a single century is strategic because it will be shaped by people living today, and it will affect the lives of our children and grandchildren. It is the scale we must consider if we want to pass the world on in good shape to our heirs. Furthermore, the accelerating transformations of the twentieth century make it socially and politically irresponsible *not* to consider futures on this scale, for things may change very fast. Besides, at this scale, political will and creativity may count for as much as prediction. Thus our predictions may themselves shape the future. We must learn to step outside the modern creation story, and accept that we are the collective authors of its next chapter.

But prediction at this scale is extremely difficult, more like forecasting the weather than plotting the trajectory of a missile. To play this game of percentages well, we must look first at the large trends we have considered in earlier chapters, because these, like geological processes, are likely to continue at least some distance into the future. But we must also consider the possibility that these trends may be changing direction, or could take sudden, random turns. And we need to be disciplined in our thinking, in order to make our maps of the future as plausible as possible. The emerging discipline known as *futurology*, whose roots lie in attempts to forecast technological developments during the Second World War, has been dominated by attempts to model futures with a focus on technologies, military outcomes (as in Herman Kahn's 1960 book, *On Thermonuclear War*), and ecological impacts (as in the models of Donella Meadows and her colleagues at the Massachusetts Institute of Technology, beginning with the 1972 volume *The Limits to Growth*).[5] But despite the sophistication of some of these models, those who construct them, from stockbrokers to meteorologists, know that the best they can hope for is a slightly better percentage of right guesses than their rivals. So, the basic rules of serious futurology are (a) look for the large trends and analyze how they work, (b) construct models to suggest how different trends may interact, and (c) be alert for countertrends or other factors that might falsify or cut across the predictions suggested by long trends and simple modeling. Beyond that, all we can do is prepare for the likelihood that many of our predictions will fail. This may not seem much of a claim for futurology, but it is better than doing nothing at all, just as studying the form at a racetrack is better than tossing a coin. In the long run, you will end up with more money if you study the form.

Some of the trends described in the previous chapter, including the accelerating pace of change itself, are worrying. These anxieties have been captured well by Clive Ponting in his remarkable *Green History of the World* (1992).[6] In the first chapter of that book, Ponting offers a striking parable for human history as a whole, drawn from the history of one of the remotest places on earth, Rapa Nui. That island lies in the Pacific Ocean, 3,500 kilometers west of Chile; the nearest inhabited place is Pitcairn Island, 2,000 kilometers to the west. Rapa Nui is known to Westerners as Easter Island because it was first encountered by Europeans on a Dutch ship, *Arena*, on Easter Day, 1722. The crew of the *Arena* found about 3,000 people living on the island in poor reed huts or caves. They seemed to be engaged in almost constant warfare over scarce food resources. All in all, it seemed a desperately impoverished place. Yet the visitors also found more than 600 huge stone statues, most more than 6 meters high. These were astonishingly elegant and beautifully carved, and many had heavy stone topknots (some weighing 10 tons) resting on their heads. Carving, transporting, and setting up these statues must have required considerable technical and managerial sophistication, but there was no sign of such skills among the Easter Islanders of the eighteenth century. Moreover, it was hard to understand how such an impoverished environment could have supported a society capable of such monumental construction. In the eighteenth century, the islands had only one species of wild tree and one wild shrub. (The wild tree went extinct in the twentieth century, but was later reintroduced from specimens kept in a botanical garden in Sweden.) The only source of animal foods appeared to be chickens, as the inhabitants' lack of boats prevented them from fishing.

The puzzle of Easter Island has been partly unraveled using modern techniques, such as the study of pollen remains, that can help archaeologists reconstruct ancient environments and landscapes. What has emerged is a sad story. The occupation of Easter Island was one of the final phases in the settlement of the Pacific, the fourth world zone of the Holocene era. (It is not impossible that there was an earlier population, of South American origin, but this remains unproven.) Easter Island was probably settled about 1,500 years ago by a boatload of twenty to thirty migrants from the Marquesas Islands in what is today French Polynesia. The small size of Easter Island, and its limited resources, ensured that colonizing it would not be easy. The island is only 22.5 kilometers long and about 11 kilometers wide. There were no indigenous mammals, and fish stocks in the waters around it were limited. The settlers brought chickens and rats with them; they soon found that of the crops they were used to, such as yam, taro, banana, and coconut, only one, the sweet potato, would really thrive there. So chicken and sweet potato became the basis of their diet. The good news was that it didn't take much effort to make a living from these basic foodstuffs. The island was well forested, and there were fertile volcanic soils.

Over time, populations increased, and a number of separate villages emerged, scattered across the island. Competition between the villages and their chiefs may have taken the form of warfare, but it also took a recognizably modern form: competitive monument building. From as early as 700 CE, villages began to erect large stone courtyards, or *ahus*, with statues on them. These may have been monuments to living or dead local leaders, as some certainly contain tombs. Similar monuments can be found in many parts of Polynesia, but none as grand as those built on Easter Island. As these societies flourished, material and political hierarchies developed, and the managerial and technological skills of the islanders increased. Many of the *ahus* appear to be aligned with the stars in ways that suggest detailed astronomical knowledge, something to be expected of a people descended from seafarers. The islanders may even have created a simple form of writing.

The main puzzle for archaeologists was to figure out how the carvings were transported and placed in position. The answer seems to be that they were carried on rollers made from tree trunks. By about 500 years ago, the population of the island had grown to perhaps 7,000 people, and competition between villages was fierce. The building and transportation of more and more statues meant that more and more trees were cut down—until, eventually, the last tree fell. Quite suddenly, the society collapsed. The abruptness of the cataclysm is apparent from the presence in the island's main quarry of unfinished statues, half carved from the volcanic rock. The effects of deforestation were devastating, for wood was needed not just to transport the statutes but also to build fishing boats and houses, to make nets and cloth (from fibers of the paper mulberry tree), and to provide fuel for cooking and heating. People could no longer fish, make cloth, or build houses, so their diets became impoverished and they began to live in caves or reed huts. Deforestation also led to erosion, reducing soil fertility and crop yields. Chickens became the most important item in their diets, and the population was reduced to the miserable strategy of building stone chicken fortresses, which they defended in grotesque and bloody chicken wars. Cannibalism sometimes made up for lack of animal protein. Political structures broke down as the ceremonies surrounding statue construction could no longer be carried out. Indeed, the old traditions died out so thoroughly that two centuries later, the inhabitants had little idea of the past of their island or the significance of the statues. In short, population growth and increasing consumption of resources, driven by political and economic competition, led to sudden environmental and social collapse.

The most horrifying aspect of this story is that the islanders and their leaders must have seen it coming. They must have known as they felled the last trees that they were destroying their own future and that of their children. And yet they cut the trees down. Does Rapa Nui provide an appropriate parable for thinking about the larger trajectory of human history? After all, the creation of degraded environments after periods of rapid change, whether caused by megafaunal extinctions in the Stone Age or overirrigation in Mesopotamia in the third millennium BCE or in the Mayan lands just over a thousand years ago, has been a recurring theme in human history.

There are disturbing parallels between the trends described in the previous chapter and the history of Rapa Nui. As global inequalities increase, resources are being consumed in ever-increasing amounts to sustain the vast hierarchical structures of modern capitalist societies. Modern societies have their own forms of competitive monument building. Resources, from fresh water to timber, are being used faster than they can be replenished; and wastes, from plastics to carbon emissions, are being disposed of faster than they can be absorbed by natural ecological cycles. Yet populations continue to increase, and politicians the world over argue that economic growth must continue and even accelerate in order to alleviate the poverty of poorer countries and sustain the living standards of richer ones. But is growth really sustainable? If existing consumption levels are already dangerous, then the idea of a world in which the entire population consumes resources and produces wastes at the rate of the richer industrial nations is terrifying. Gandhi understood the problem as early as 1928, when he wrote: "God forbid that India should ever take to industrialism after the manner of the West. . . . If an entire nation of 300 million took to similar economic exploitation, it would strip the world bare like locusts."[7] Nevertheless, capitalism, now the dominant force in economic development, thrives on growth; and the political and business leaders who hold the greatest power today respond to the demands of local constituencies with short-term plans and projects, as did the statue-building chiefs of Rapa Nui. As on Rapa Nui, we appear incapable of stopping processes that threaten the future of our children and grandchildren.

But perhaps we can do better than the Easter Islanders.[8] The most important reason for hope may be that collective learning now operates on a larger scale and more efficiently than ever before. If there are solutions to be found, both for humans and for the biosphere as a whole, the global information networks of modern humans can surely find them. These networks gave us the technologies that helped us mold the biosphere as we wished, and modern, electronically driven networks of collective learning have helped us understand the dangers of our increasing ecological power. In broad terms, the challenge is clear. To avoid a global replay of the catastrophes that overtook Easter Island, we must find more sustainable ways of living. We must use water, timber, energy, and raw materials at rates that can be supported for centuries, not decades; and we must produce waste products only in amounts that can be absorbed safely so we do not damage the environment and our fellow creatures. Can we do these things?

If populations keep growing at the rates typical of the late twentieth century, there is no hope. Here, though, we have reason for optimism, for global rates of population growth appear to be slowing, not only in the more affluent countries but also in some of the world's poorer countries. The evidence for this demographic transition is now very strong. For most of the agrarian era, rates of population growth were governed by high birth rates and death rates; these encouraged parents to have many children, because they knew that some would die before adulthood. In the wealthier countries today, population growth is governed by a very different regime, dominated by lower death rates and birth rates and by improved welfare services. More children survive, and people expect to live longer; but because children are no longer the only source of support in old age, there is less need to have babies as a form of long-term insurance. The result is that birth rates have fallen, and population growth has declined—in some countries, to zero. The rapid population growth of recent decades and centuries was caused by a regime halfway between these two extremes, in which death rates fell (because of better medical care and increased food production), while birth rates remained high. The key to stabilizing global populations in the next century will be to reduce birth rates in the poorer countries, where they remain highest. The factors most likely to achieve this result are increased affluence, urbanization, improved infant health, and increased education, particularly of third world women (and particularly about contraception and health). Investment in improving health care and women's education in poorer countries could have a dramatic impact on growth rates in the next few decades. Birth rates are already falling sharply in many poorer countries, so it is likely that sometime in the next century, global rates of population growth will stabilize. By 1998, thirty-three countries had zero population growth.[9] The most optimistic estimates suggest that global populations will stabilize at about 9 to 10 billion. Feeding, clothing, and housing 3 or 4 billion more people will be a huge challenge, particularly as most will be born in the countries least able to provide for them; but given the rapid increases in food production in the twentieth century and the immense resources available in the richer countries, it should not be impossible. The graph in figure 15.2 suggests one likely scenario for population growth over the next century, in both richer and poorer countries.

Can consumption be similarly stabilized? To do so, we must take two crucial steps, both of which have begun to be implemented in small ways. The first is to shift from use of virgin resources to recycling. The second is to rely more heavily on sustainable and nonpolluting energy supplies. The necessary technologies already exist to exploit solar power, wind power, and fuel

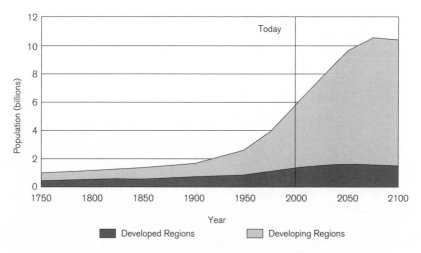

Figure 15.2.   A modern "Malthusian cycle," 1750–2100? This graph includes estimates of future population growth in both the developed and developing regions of the world. Today, most demographers are agreed that the rapid growth of the last two centuries will slow, and world populations should stabilize by the year 2100. But, as this graph shows, for a time, growth will continue in those regions least able to support larger populations. Adapted from Paul Kennedy, *Preparing for the Twenty-First Century* (London: Fontana, 1994), p. 23.

cells powered by hydrogen, though in present global markets (which do not factor in the environmental costs of different sources of energy) they cannot compete commercially with the fossil fuels that still power the Modern Revolution. But technologies for the cheap exchange of information are already with us as a result of the electronics revolution of the late twentieth century. In principle, we have the technologies needed to build a sustainable global economy without drastically reducing average living standards in the wealthier countries. But we may find, as in Rapa Nui, that the most difficult problems turn out to be political and educational rather than technological.

The political problems are indeed formidable. The political and business leaders who have the greatest power to make decisions on such issues are all accountable to particular regional or economic interest groups, and the workings of the political process encourage them to think on timescales too short to deal effectively with global ecological and social issues. They will be supported in their resistance to change by the affluent populations of the richer countries, for whom ecological crisis remains a distant and uncertain threat rather than the catastrophe it already is in many poorer countries.

Besides, capitalism itself seems to depend on continued growth for its existence. Does this mean that capitalism must be overthrown? Sadly, the Communist revolutions of the twentieth century suggest that overthrowing capitalism may be an extremely destructive project, and one that is not in any case likely to create societies that are notably egalitarian or ecologically sensitive.

But politically, too, some indications are promising. One positive sign is the rapid emergence of a new global awareness of ecological issues and their interconnection with social and economic issues. Twenty years ago, few governments had ministries that specialized in environmental issues—now most governments take such issues seriously, and so do the electorates that choose them. The "Earth Summit"—the United Nations Conference on Environment and Development, held in Rio de Janeiro in 1992—was an important symbolic gesture toward sustainability, and it included a general agreement that the richer countries would have to help the poorer countries to develop in ways that were "environmentally sound." For the first time, an international agreement argued that growth must be balanced against sustainability. Here, at least, was a rhetorical victory; a second conference followed, ten years later, in Johannesburg.

There have also been some examples of international cooperation, particularly on issues on which it has been easier to achieve a broad consensus. In the 1970s, evidence began to accumulate that the ozone layer was thinning because of the use of chlorofluorocarbons (CFCs).[10] These were used in refrigeration, in air-conditioning, and as cleaners and solvents. In 1977, several developed nations urged UNEP (the United Nations Environment Program) to consider the issue, and a conference held that year adopted a plan for global action. At that point, no one took the issue seriously enough to act, in part because the scientific evidence remained ambiguous. In the early 1980s, the United States, which accounted for 30 percent of total emissions, took a leading position on reducing use of CFCs, partly because substitutes were available, and partly because of internal pressures from the emerging environmental lobby. But several other countries— including a number of countries in the European Community (EC), which produced 45 percent of global output—argued against regulation. Several developing countries, including China and India, also resisted regulation, as they were planning on increasing their production of CFCs. Clearly, an international agreement would be meaningless without the cooperation of these major current or potential producers. Some poorer countries maintained that they would need international funding to help them move away from dependence on CFCs. In the mid-1980s, the scientific evidence became clearer, and several "lead states" pushed for an international convention coupled with specific, binding protocols on the issue. In 1985, the Vienna Convention for the Protection of the Ozone Layer was signed, but it did little more than require international monitoring of CFC emissions. Then, at UNEP's Montreal conference in 1987, under pressure from the lead states (including the United States), struggling with internal divisions, and facing superb negotiators, the EC agreed to a 50 percent cut by 1999. The Montreal Protocol on Substances That Deplete the Ozone Layer allowed developing countries to increase production for a time, but set caps on eventual output. Unfortunately, a veto by the United States and Japan kept money from being made available to help developing countries adjust. Within months, however, new scientific discoveries, including the discovery of a large hole in the ozone layer over the Antarctic, made the issue appear much more urgent. By May 1989, eighty nations had come out in support of the total elimination of CFCs by 2000. In 1990, a fund was created to help developing countries adjust, and thirty-two industrialized countries put about $1 billion into it. There are still loopholes in these agreements, but on the whole they have been extraordinarily successful. Production of CFCs fell from ca. 1.1 million tons in 1986 to 160,000 tons in 1996, and there is evidence that the hole in the ozone layer has started to shrink.

The international response to the ozone crisis shows that cooperation is possible. Nations, like individuals, can sometimes work together to solve common problems. And where evidence about the seriousness of a problem is clear, cooperation can be organized quickly and efficiently, even if it threatens some regional interests. The international mechanisms of cooperation that exist are clumsy and cumbersome, but they may be able to do the job in a crisis. The response to the thinning of the ozone layer is not the only example of their effectiveness, as Lester Brown points out: "Air pollution in Europe, for instance, has been reduced dramatically as a result of the 1979 treaty on transboundary air pollution. Global chlorofluorocarbon (CFC) emissions have dropped 60 per cent from their peak in 1988 following the 1987 treaty on ozone depletion and its subsequent amendments. The killing of elephants has plummeted in Africa because of the 1990 ban on commercial trade in ivory under the Convention on International Trade in Endangered Species of Wild Flora and Fauna."[11]

But there is an even deeper problem. We have seen that capitalism is the driving force of innovation in the modern world, and capitalist economies depend on increasing production and sales. Is that growth incompatible with sustainability? The answer is unclear, but there are reasons for thinking that capitalism may well manage to coexist with at least some of the early stages of a transition to sustainability. One is that capitalist economies need increasing profits more than increasing production—and profits can be made in many ways, some of which are compatible with a sustainable economy. In principle, the recycling of resources or the sale of information and services rather than goods can generate profits as effectively as the exploitation of virgin resources. If governments were to start taxing unsustainable production methods more harshly, investment would soon move into more sustainable activities, where large profits could then be made. There is no absolute contradiction between capitalism and sustainability. Markets can be steered, as governments have known ever since John Maynard Keynes made this point in the 1930s. And some of the most effective methods of steering them include the use of taxes and subsidies to alter costs and direct economic activity in new directions. As Brown has argued forcefully, contemporary capitalism is ecologically destructive in part because it has no way of accounting for ecological values. For example, modern accounting methods cannot properly assess the services provided by forests in preventing floods, absorbing excess carbon dioxide, preventing soil erosion, and maintaining biodiversity. It is perfectly feasible, in principle, to use taxes and subsidies to build these costs into economic transactions. Indeed, governments routinely use these mechanisms today. An obvious example of how they could steer markets in more sustainable directions would be to introduce taxes on the use of fossil fuels—paid for, perhaps, by reductions in income taxes. Such taxes could transform the current balance of profitability between fossil fuels and less damaging energy sources such as wind power and fuel cells, because in a market economy, price signals can rapidly transform the behaviors of millions of consumers and producers.

But does the political will for taking such actions exist? For the answer to be "yes," two things must happen: ecological dangers must become apparent to those who wield power in the modern world (governments can respond rapidly to crises once there is no doubt about their seriousness and magnitude), and popular attitudes, particularly in the richer countries, must change. Attitudes are critical. The widespread belief that continued growth in production is a good in itself poses one of the main barriers to reform. Such a belief will persist as long as good living is conceived in the ways we have been taught by consumer capitalism—as the never-ending consumption of more and better material goods. Changing definitions of what makes a good life may turn out to be one of the crucial steps toward a more sustainable relationship with the environment.

The other major challenges are ethical and political. Are the huge inequalities of the modern world tolerable? Will they not generate conflicts that guarantee the eventual use of the destructive military technologies now available to us? After all, the information networks of the modern world can disseminate knowledge about the manufacture of nuclear and biological weapons as well as about solar power cells. Thus it is a good bet that in the next few decades, more nations will have access to destructive weapons, and so will increasing numbers of guerrilla organizations such as al Qaeda that see themselves as defenders of the dispossessed and disempowered. Here, prediction is harder, because political changes are so dependent on the decisions and actions of individuals. Will the governments of wealthier countries decide that reducing global poverty can increase their own security? Perhaps forces less obvious but no less fundamental will encourage politicians to tackle the poverty of the world's poorest countries. Capitalist economies need markets, and we have seen that consumer capitalism is distinguished from the system's earlier forms by levels of productivity so high that it has to sell goods to its own workforce, to the subordinate classes that Marx called the proletariat. The same pressures will surely lead, eventually, to a rise in the living standards of subordinate classes in even the world's poorest countries. And in this way, as global capitalism assumes less predatory forms, it may begin to raise living standards beyond the industrialized heartland. Thus, if a maturing world capitalist system can avoid the dangers of global overconsumption that Gandhi warned against, there is hope that even if relative inequalities continue to grow, the material living standards of subordinate classes in many other countries may be raised in the next century, generating new markets and reducing global political and military conflicts. Such a course of action might reduce the more abject forms of poverty, though inequality in general is bound to survive as long as capitalism remains the dominant shaper of economic change.

If some of the ecological and political problems of the twentieth century are genuinely addressed in this century, there is a chance that the gains of the Modern Revolution will be passed on to future generations. If not, there is a real danger that the Modern Revolution will spin out of control, causing military and ecological catastrophes that will leave our children and grandchildren a world as degraded as Easter Island, but with its devastation on a far larger scale.

THE MIDDLE FUTURE: THE NEXT CENTURIES AND MILLENNIA

When we think about more distant futures, say the next millennium or two, the open-endedness of historical change defeats us. Peter Stearns rightly describes "millennial forecasting" as a "nonstarter."[12] At this scale, alternative futures proliferate so fast that anything we say can be little more than guesswork. Besides, at the millennial scale, unlike the hundred-year scale, our capacity to shape the future also dwindles to insignificance, so we are under less pressure to predict.

It is easy to imagine catastrophic scenarios brought about by nuclear or biological warfare, or ecological disaster, or perhaps even a collision with a large asteroid. If caused by human action, such endings to human history might suggest that our species overreached itself, that what we think of today as progress was, in fact, the beginning of the end. Icarus will then seem the most appropriate metaphor for human ambition and creativity. It is almost as easy to imagine utopian scenarios, in which most of the modern world's problems are overcome—in which humans learn to construct ecologically sustainable economies, inequalities between different groups and regions have diminished significantly, and human technological prowess is used to provide a majority of the world's population with a better life rather than with more and more material goods. Such an outcome would vindicate those who see human history as a story of progress.

But it is the in-between scenarios that are both most likely and most difficult to imagine. The best we can do here is to consider some of the larger trends shaping the modern world and assume that they will continue some way into the future.

If current demographic trends are sustained for a century or more, population growth will grind to a halt; human population numbers will stabilize or even decline, while average ages will rise. But another trend, technological innovation, shows no sign of slowing. There may well be eras of technological stagnation in the future, as there have been in the past, but the present burst of technological creativity seems set to continue, perhaps for a few centuries more. Stable populations and accelerating innovation in information technology, genetic engineering, and control of new energy sources (possibly including hydrogen fusion) ought to mean that increased productivity can be used not just to maintain minimum standards for ever-greater numbers of people but to raise the real living standards of everyone. Social and economic trends over the past 5,000 years offer little hope for a significant reduction in economic and political inequality. On the contrary, they suggest that gradients of wealth will get steeper, and the difference between the weakest and the most powerful will grow. But, as we have seen, the evolution of consumer capitalism during the past century indicates that the living standards of those at the bottom of these gradients may rise, if only because the poor are numerous enough to provide valuable markets for capitalist economies whose search for new consumers will become more frantic as populations stabilize while productivity keeps rising.

If environmental constraints do not bring the capitalist world system crashing down—if, instead, it manages to find new markets by selling to the poor as well as the rich, by seeking profits in ecologically sustainable production, and by trading more in services and information than in materials—then we can envisage further transformations generated by technologies we can only glimpse at present. Biotechnology may create new ways of feeding, clothing, and equipping a world of 10 to 12 billion people. It may also enable more and more of them to live longer and healthier lives. Nanotechnology and new, faster microchips may surround us with intelligent robots of all sizes, some of which may behave in ways that are hard to distinguish from human intelligence. Meanwhile, new energy sources should increase the energy available to us. Finally, the space technologies envisioned by the Russian schoolteacher Konstantin Tsiolkovsky, which enabled the first human to leave Earth on 12 April 1961 and the first human to land on another heavenly body on 21 July 1969, will surely lead, eventually, to a new migratory phase in human history. In this phase, the networked world of today will be torn apart once more into separate regional webs. What makes such ideas more than science fiction is the knowledge that 500 years ago, no one had any idea of the speed and significance of the changes that would transform North America, a region of foragers and small-scale farming societies, into today's superpower.

Colonization of other worlds may begin with the industrial exploitation of the Moon, nearby planets, and asteroids. It will continue with the planting of settlements on planets within our solar system. Both the industrial exploitation of asteroids and the initial colonization of Mars may be feasible within a century. More speculative (and more complex ethically) are plans for the "terraforming" of Mars—that is, the modification of Mars's atmosphere and temperature to make it habitable for humans and other living organisms from Earth.[13] Several plans of this kind already exist, though the changes they envisage could take a thousand years to complete. If they succeed, humans will have learned how to "domesticate" entire planets as they once domesticated large herbivores. Should humans start migrating from their home planet in larger numbers, the human history described so far in this book will eventually appear as merely the first chapter in a history, most of which will take place beyond the earth. In some ways, migrations to other planets will be reminiscent of the great migrations of the Stone Age that took members of our species into new environments within Africa, and then into the undiscovered lands of Australia, Siberia, and the Americas. Or perhaps a better analogy is with the great sea voyages that colonized the Pacific. But surviving beyond our earth will require all the technological ingenuity humans can muster. Migrants of the future will have to create entirely new ways of living, probably in totally artificial environments. And, like the Easter Islanders, they will not always succeed. Even on our nearest celestial neighbor, the Moon, they will live in a barren desert, enduring horrifying temperature extremes under a totally black sky.

Travel beyond our solar system is a different proposition, because of the huge distances involved and Einstein's rule that nothing can travel faster than light.[14] Light takes more than 4 years to travel to the nearest star, Proxima Centauri, while it takes almost 30,000 years to travel to the center of our own galaxy. And at present, we have no idea how to build a spaceship that could travel merely one-tenth of the speed of light, the lowest speed compatible with return journeys shorter than a human lifetime. As yet, even the most optimistic proposals do not envisage the making of such journeys for a few more centuries. Journeys of colonization, taken by travelers who, like Polynesian colonists, do not expect to return home, may be more realistic. These could rely on larger, slower spacecraft that would take hundreds of years to reach their destinations. Unlike Polynesian ships, "space arks" could become permanent homes, more comfortable and more attractive than any planets they might stumble across (see figure 15.3). Instead of traveling the universe as we do today, aboard naturally created planets whose movements we cannot control, humans of the future may travel in artificial planets that they can steer. In that case, the human future will lie not in the colonization of thousands of other planets but in the creation of thousands, even millions, of space arks, which periodically dip down to nearby planets to replenish stocks of fuel and raw materials. It has been estimated that successive waves of interstellar colonizers, traveling at relatively low speeds, could take a hundred million years to reach the most distant parts of our own galaxy; our existing knowledge barely enables us to begin to visualize journeys to other galaxies. But even space arks will not be built in the near future.

If humans do begin to travel beyond our solar system, there may again emerge a human society broken into separate worlds, like the many societies of the Pacific, each with its own distinctive history, for contact will be intermittent and slow. According to Arthur C. Clarke, "The finite velocity of light will, inevitably, divide the human race once more into scattered communities, sundered by barriers of space and time. We will be one with our remote ancestors, who lived in a world of immense and often insuperable distances, for we are moving out into a universe vaster than all their dreams."[15] If the separation lasts for long enough, the networks that have linked humans for most of their history will fray. The cultural networks will snap first, but then the genetic links that define modern humans as a single

Figure 15.3.   A possible design for a space colony: a future for humans in the cosmos? Is this how a majority of humans will live in two or three centuries from now? Will humans repeat the epic migrations of the Paleolithic era, but now on the scale of the solar system, or even in interstellar space? This illustration is based on ideas for a space colony to explore the solar system developed by the Princeton physicist Gerard K. O'Neill in the 1970s and 1980s. Each cylinder would be up to 30 kilometers long, and could contain populations of thousands, or hundreds of thousands, of people. Each of the three strips ("countries"?) would enjoy daylight for one-third of the colony's "day." Adapted from Nikos Prantzos, *Our Cosmic Future: Humanity's Fate in the University* (Cambridge: Cambridge University Press, 2000), p. 42.

species will thin and at some points break. Humans, like the finches of the Galápagos Islands, will start to evolve into innumerable separate and diverging species, each adjusted to a particular local environment.

Evolutionary change is unavoidable, whether or not humans colonize other worlds. Few mammal species last more than a few million years without evolving into new species. Humans count as a young species, with a potential future of hundreds of thousands or perhaps even a few million years. But modern genetic technologies may soon enable humans to start consciously manipulating their own genetic makeup. With the decoding of the human genome at the end of the twentieth century, we already know the blueprint on which humans are constructed, even if we do not yet understand all the intricate ways in which different parts of that blueprint interact with each other. It is thus likely that in the next few centuries, humans will begin to engineer their own bodies, without waiting for the slower processes of natural selection to do their work.[16] Will it make sense any longer to think of these people as *us* or as *our* descendants?

And will these descendants ever meet other intelligent, networked beings? There are good reasons to think they will not, at least not within our own galaxy. Observations of planets orbiting nearby stars and discoveries of living organisms in what were once thought to be impossibly harsh environments—in volcanic vents in the sea, and frozen deep within rocks—both suggest that life may be common, at least where stars and planets exist. Moreover, the speed with which the first life-forms appeared on Earth indicates that life can form rapidly where the conditions are right. But intelligent life-forms that can share information as humans do may be extraordinarily rare. It took 4 billion years to evolve networked, large-brained creatures on this earth, and that was a chancy business and could easily have taken longer; the evolutionary pathways leading to large brains seem to be extraordinarily narrow. There thus is no certainty that anything like our species would evolve even after huge periods of time. Besides, if intelligent, information-sharing creatures were common, the absence of any clear evidence for their existence would be puzzling. On a visit to Los Alamos in 1950, the physicist Enrico Fermi put this argument in the form of a simple question: "But where are they?" If such species were common, then there should be many intelligent, networked communities with technologies much more advanced than ours, and we should have come across signals of some kind from *some* of them.[17] If humans ever reach planets near other stars, they may find, like Polynesian travelers who made their way across the Pacific, that there are no other creatures as complex or technologically sophisticated as themselves.

But at this point we are moving into pure speculation, as is inevitable when we make any guesses about the nature of human societies in a thousand years' time. We can remind ourselves how speculative such ideas may be by remembering that the dinosaurs, as a group, appeared to be flourishing before they were destroyed in a geological instant by an asteroid impact 65 million years ago.

THE REMOTE FUTURE: THE FUTURE OF THE SOLAR SYSTEM, THE GALAXY, AND THE UNIVERSE

Oddly, the obscurity lifts at the largest scales, for astronomers deal with larger but simpler objects than historians, objects that change very slowly over huge time periods. Astronomers are confident that they have a good idea of what is in store for planets and stars, and even for the universe itself.

The ultimate fate of the biosphere will be determined by the evolution of the earth and its sun. Though these are large systems, they are simpler than the biosphere or human society, so their future evolution is more predictable. Our sun is about halfway through its life cycle, giving it another 4 billion or so years to live. But life on Earth will die out well before the Sun dies. As it ages, the Sun will heat up until eventually the surface of Earth begins to heat up, too. The biosphere may evolve in ways that slow the impact of these changes, but eventually those organisms still living on Earth will run out of options. In 3 billion years' time, Earth will receive as much heat from the Sun as Venus does today; the oceans will boil, and their steam will contribute to massive global warming. Earth will become uninhabitable.[18] Eventually, it will be as barren as the Moon is today.

When the Sun burns up all its hydrogen, it will become unstable. It will eject material from its outer layers, and its inner core, freed from the pressure of these outer layers, will expand until it reaches where the earth is now. However, the Sun's reduced mass and gravitational pull will allow the earth to drift out to a more distant orbit, 60 million kilometers away. Nikos Prantzos describes the resulting view from Earth: "If an observer could survive the fiery furnace on its surface, at temperatures approaching 2000° C, he or she would see a sight worthy of Dante's inferno. The Sun's disk would occupy more than three quarters of the sky."[19] If anyone is watching as the Sun engulfs our earth, they may be visitors from farther out in the solar system; for a time, the moons of Jupiter and Saturn, such as Titan and Europa, may become habitable. Then the Sun will shrink once more, as it starts to burn helium in its core, but only for about 100 million years. When it runs out of helium, it will flare up again and start manufacturing oxygen and carbon. At this stage, even the outer planets will become uninhabitable. Then, the furnaces at the center of the Sun will finally die down and it will shrink into a white dwarf: an extremely dense, brilliantly hot mass of material that, because it has no internal heat engine, will gradually cool and darken during an afterlife lasting many times longer than its fusion phase.

The hundreds of billions of stars in the Milky Way will not notice its passing—though perhaps they should, for it will offer a small portent of the galaxy's distant future. About 90 percent of the material from which stars can be manufactured has already been used, so the era of star formation is drawing to a close. In just a few tens of billions of years from now, star formation will cease; then, when the surviving stars start dying, the lights will dim and begin to go out. In a cold, dark universe, energy gradients will no longer be steep enough to create complex entities; the universe will become simpler and simpler, and the second law of thermodynamics will assert its bleak authority more and more effectively. But this will not happen quickly, and not without reverses: the smaller stars, like remnants of a once-powerful guerrilla army, will live for many times the age of the existing universe. Then, a few thousand billion years from now, even they will shut down, and the universe will be dark again, as it was in its early days. But now it will be like a huge cosmic junkyard, full of cold, dark objects such as brown dwarfs, dead planets, asteroids, neutron stars, and black holes.[20]

And what will happen next? We do not know for sure, but we know some of the likely scripts. The future depends largely on the balance between expansion, which drives the universe apart, and the force of gravity, which draws it together. If there is enough mass/energy to slow the expansion of the universe to a halt, then, after perhaps a few thousand billion years, the universe must start contracting. The contraction phase will not be a mere reversal of the expansion phase, as some once believed. It was even supposed at one time that "big crunches" might be followed by new big bangs, in a scenario of bouncing universes that some saw as a modern version of cyclical cosmologies, such as those of the Maya.[21] Such ideas encouraged astronomers to attempt a detailed census of the amount of matter/energy in the universe. At first it seemed that there was far too little matter to halt the expansion of the universe, but it gradually became clear that there is a huge amount of matter or energy that we cannot see. And, as various indirect methods were used to estimate the amount of dark matter, it began to appear that gravity and expansion were extraordinarily finely balanced, making the universe's ultimate fate uncertain. In the late 1990s, however, the discovery of so-called vacuum energy offered a resolution to these debates— in part because the vacuum energy could itself account for much of the missing matter/energy, and in part because it seemed to guarantee that the expansion of the universe would not slow but would instead increase, for vacuum energy appears to be gently accelerating the rate at which the universe expands.

Currently, most astrophysicists believe that the universe will keep expanding forever. It is, in their jargon, "open" rather than "closed." As it gets bigger, the spaces between galaxies will increase, and the universe will get simpler, colder, and lonelier in an infinitely slow diminuendo. The good times will be over for good. With the reduction in temperature difference between hot and cold objects, entropy will increase, making the formation of complex entities increasingly difficult, though the continued expansion of the universe will ensure that it never reaches a state of perfect thermodynamic equilibrium. As the universe ages, light will come only from rare flare-ups, as cold lumps of matter collide randomly to form a few new stars. These lonely beacons of light will find themselves in a colossal galactic graveyard, surrounded by billions of stellar corpses. Gravitational forces will push some of the corpses out into empty space, where each will endure a lonely purgatory as it travels farther and farther away from anything else, until finally it perishes in its own private universe. Those star corpses that stay within the former galaxies will be pulled together by gravity until they merge into huge galactic black holes. Any matter left outside them will also begin to decay if (as some modern theories suggest) even protons are not forever. From perhaps 10^{30} years after the big bang onward, the universe will be a dark, cold place, filled only with black holes and stray subatomic particles that wander light-years apart from each other.

But as Stephen Hawking showed in the early 1970s, even black holes lose energy, and over unimaginable periods of time they, too, will disappear. Their deaths by quantum evaporation will last billions of times longer than all the eras that passed before, so long that each billion years will count as no more than a single grain of sand on an earthly beach (see table 15.1). On these scales, according to Prantzos, the 10^{30} years before black holes began to dominate the universe "will look even shorter than the Planck time does for us today!"[22] What will the dying black holes leave behind? Very little: Paul Davies imagines "an inconceivably dilute soup of photons, neutrinos, and a dwindling number of electrons and positrons, all slowly moving farther and farther apart. As far as we know, no further basic physical processes would ever happen. No significant event would occur to interrupt the bleak sterility of a universe that has run its course yet still faces eternal life—perhaps eternal death would be a better description."[23]

To an imaginary observer watching the death agony of the last black holes, the few billion years considered in this book will seem like a dazzling flash of creativity at the beginning of time, a split second in which huge and chaotic energies challenged the second law of thermodynamics and conjured up the menagerie of exotic and complex entities that make up our world. In that fleeting springtime, before it cooled and darkened, the universe was bursting with creativity. And in at least one obscure galaxy, there appeared a networked, intelligent species capable of contemplating the universe as a whole and of reconstructing much of its past.[24]

It is tempting to think that this flash of creativity was laid on for humans—the ultimate justification, perhaps, for the universe's creation from nothing. Modern science offers no good reason for believing in such an-

TABLE 15.1. A CHRONOLOGY OF THE COSMIC FUTURE IN AN OPEN UNIVERSE

Time Since Big Bang (years)	Significant Events
10^{14}	Most stars are dead; the universe is dominated by cold objects, black dwarfs, neutron stars, dead planets/asteroids, and stellar black holes; surviving matter is isolated as universe keeps expanding.
10^{20}	Many objects have drifted away from galaxies; those remaining have collapsed into galactic black holes.
10^{32}	Protons have largely decayed, leaving a universe of energy, leptons, and black holes.
10^{66}–10^{106}	Stellar and galactic black holes evaporate.
10^{1500}	Through quantum "tunneling," remaining matter is transformed into iron.
$10^{10^{76}}$	Remaining matter is transformed into neutronic matter, then into black holes, which evaporate.

SOURCE: Adapted from Nikos Prantzos, *Our Cosmic Future: Humanity's Fate in the Universe* (Cambridge: Cambridge University Press, 2000), p. 263.

thropocentrism. Instead, it seems, we are one of the more exotic creations of a universe in the most youthful, exuberant, and productive phase of a very long life. Though we no longer see ourselves as the center of the universe or the ultimate reason for its existence, this may still be grandeur enough for many of us.

SUMMARY

Predicting the future is a chancy business, because the universe is inherently unpredictable. But in some situations we have to try. It is worth thinking hard about the next century, because what we do today may have a significant impact on the lives of those who live a century from now. If our predictions are not too far from the mark, and we act intelligently in the light of those predictions, we may be able to avoid disaster. Such disasters could take several forms, including severe ecological degradation and military conflicts generated by growing inequalities in access to resources. The two issues are linked; and, with intelligent management, it may be possible to steer the world toward a more sustainable relationship with the environment and create a global economy that raises the living conditions of the poor, even if it remains biased toward the wealthy. On scales of several centuries, the possibilities multiply so rapidly that it is hardly worth the effort of trying to make predictions. But there are large trends, particularly in technology, that may hint at some plausible futures. Humans may migrate to planets or moons within the solar system, and perhaps even farther afield; and they may learn to control genetic processes with great precision. But any particular predictions could, of course, be derailed by unexpected crises, whether caused by humans or by geological or astronomical phenomena such as asteroid impacts. At cosmological scales, our predictions become more confident once again. The Sun and our solar system will die within 4 billion years, but the universe will survive much longer. Recent evidence suggests that the expansion of the universe will continue forever. If this is so, then we can use contemporary understanding of fundamental physical and astronomical processes to describe how, as the universe keeps expanding, it will also decay. From the standpoint of an inconceivably distant future, when the universe contains no more than a depressingly thin sprinkling of photons and subatomic particles, the 13 billion years covered in this book will seem like a brief, exuberant springtime.

FURTHER READING

Peter Stearns, *Millennium III, Century XXI* (1996), discusses the history of futurology; and Yorick Blumenfeld, ed., *Scanning the Future* (1999), brings together some essays on futurology. In *Signs of Life* (2000), Ricard Solé and Brian Goodwin offer a good discussion of the problems of prediction and the nature of unpredictability. On the ecological future, some of the most accessible works are Lester Brown, *Eco-Economy* (2001) (though it has been subjected to a tough statistical critique in Bjørn Lomborg, *The Skeptical Environmentalist* [2001]), and Paul Kennedy, *Preparing for the Twenty-First Century* (1994). The middle future is best represented in works of fiction. Brian Stableford and David Langford's *The Third Millennium* (1985) is a fascinating and moderately optimistic "history" of the next thousand years, while Walter Miller's *A Canticle for Leibowitz* (1959), written at the height of the cold war, portrays a future in which human creativity and rationality lead only to periodic nuclear holocausts. On even larger scales, science comes into its own again. Nikos Prantzos, *Our Cosmic Future* (2000), discusses possibilities for space travel, and also explores the most remote cosmological futures, as does Paul Davies, *The Last Three Minutes* (1995).

:ation myth, as in any story, there is a time-
1eline constructed? And how can we begin to
t scales?

CONSTRUCTING A MODEL... ...INE

One of the most astonishing features of the modern creation story is that it confidently describes events that happened billions of years before humans existed. Many of these chronological details have come into focus only in the past few decades, so the timeline behind the story told in this book is, in many parts, very recent. How was it constructed?

Where written records exist, dating is not a great problem, and modern historians have relied mainly on written records to construct their accounts of the past. But things are different when we deal with larger time spans not covered by written records. Even fifty years ago, attempts to construct such timelines would have been vastly more difficult than they are today. Until the middle of the twentieth century, it seemed that precise knowledge of the distant past was impossible. We might be able to determine the relative order of events (such as the sequence in which particular rocks had been laid down), but there seemed no way of determining absolute dates.

In the Christian world, the Bible was regarded as a primary source for ancient dates until as late as the nineteenth century. Estimates for the moment of creation were established by adding up the dates of all the generations listed in the Bible. Such calculations suggested that God had made the Earth about 6,000 years ago. In the seventeenth century, as noted in chapter 1, one British scholar concluded that humans had been created at 9:00 AM on 23 October 4004 BCE. But even in the seventeenth century, some scholars with an interest in geology realized that the earth had to be older than that. For example, they became aware that fossil objects found high in mountain regions seemed to be the remains of ancient fish, which suggested that the mountains they were in must have risen up from sea level. Such changes, the scholars thought, must have taken longer than 6,000 years. By the nineteenth century, geologists were getting used to the idea of much larger timescales, and they were growing very skilled at identifying relative dates. They could tell which layer of rock had been laid down first, knowledge that in turn enabled them to put fossils in sequence to describe the rough stages of evolutionary history. But there seemed to be no precise way of determining absolute dates. One influential attempt to fix the age of the earth was that of William Thompson (Lord Kelvin). He argued in the 1860s that the earth had existed for less than 100 million years, and perhaps as little as 20 million years, based on his assumption that the earth and Sun had once been molten balls of matter that had cooled to their present temperatures. To estimate their age, Lord Kelvin calculated how long this cooling would have taken. He was wrong, because he did not understand the role of radioactivity, which maintained the internal heat of both bodies (though in different ways). Indeed, it was an understanding of radioactivity that eventually made it possible to determine precise absolute dates for the modern creation story.[1]

Radiometric dating techniques exploit a feature of all radioactive materials, including many isotopes of normally stable chemicals, such as carbon.[2] Many radioactive elements contain large numbers of protons and neutrons in the nuclei of their atoms. Because protons have a positive charge, they repel each other electrically; the more of them there are crammed into a single nuclei, the greater the repulsive force. Eventually, these repulsive forces can weaken the strong nuclear forces that hold nuclei together; for this reason, large nuclei tend to be more fragile than small nuclei. But even smaller nuclei can be unstable in certain configurations. Periodically, the nuclei of radioactive elements just start falling apart. They eject small numbers of protons and neutrons, and sometimes a single electron or positron; in so doing, they turn into different elements. This process, known as *radioactive decay*, continues until the original material has been transmuted, step by radioactive step, into a stable element, such as lead. Such decay occurs with great statistical regularity; though we can never predict when a particular nucleus will break down (just as we can never predict the outcome of a particular coin toss), we can be very precise about the behavior of large numbers of radioactive events. Thus we can estimate how rapidly large amounts of material will decay. This rate is normally calculated in *half-lives*. For example, the half-life of uranium 238 (the most common isotope of uranium) is ca. 4.5 billion years, or slightly less than the age of the earth. This means that if we start with a newly formed lump of uranium 238 (created, perhaps, in a supernova), after 4.5 billion years about half of it will have broken down into other elements. (The fact that so much of the uranium on Earth appears to be about 4.56 billion years old is one reason for thinking that a supernova exploded in our part of the galaxy just before our solar system formed.) The half-lives of radioactive elements vary greatly. For example, the half-life of carbon 14 (a rare isotope of carbon) is 5,715 years, which is why archaeologists use it to date events that occurred up to ca. 40,000 years ago.[3] For earlier dates, too little of the original carbon 14 is left for accurate analysis, leading to anomalous results; so other methods have to be used.

The statistical regularity of radioactive decay enables us to calculate when a particular lump of matter containing radioactive material was formed. By using such techniques, we can say, for example, that the earth was formed 4.56 billion years ago, or that the Cambrian era lasted from about 570 to 510 million years ago. The technical details are complex, but the general principle is simple enough. If you take a lump of radioactive material, you can measure what proportion of it has broken down into other elements, and from that figure you can calculate how long the lump has existed. There is always a certain amount of unreliability in these calculations, but even the degree of reliability can be estimated with some precision. The principles of radiometric dating were first developed in the United States by Willard Libby, in the 1950s. Since that time, the techniques have been improved greatly. As a result, since the mid–twentieth century, archaeologists, geologists, paleontologists, and astronomers have been able to calculate precise absolute dates for many major events in the remote past of our planet and solar system. Radiometric dating techniques provide many of the more important dates for the modern timeline.

Molecular dating is a more recent technique, developed first in the 1980s; it is used mainly to determine the evolutionary distance between two related species (see chapter 6). It works by comparing similar genetic material (such as DNA) from two organisms and then estimating the difference between the two samples. The calculation of many estimates of this kind has shown that much genetic change is statistically random; thus, like the breakdown of radioactive materials, it can function as a sort of clock. Scientists first used molecular clocks to determine when the human and ape lines diverged, and they came up with the scandalously short result of ca. 5 to 7 million years ago. The rapid acceptance of this date by paleontologists greatly enhanced the credibility of the technique, which is now used to date many other important processes, such as the date of human migrations to different parts of the world.

The big bang poses its own chronological problems. Edwin Hubble showed that the universe was expanding, and he also showed that it was possible, in principle, to calculate the rate of expansion. To undertake this calculation, he first had to determine the distance between galaxies and the rate at which they were moving apart. Neither task is easy, and the problem is further complicated because the rate of expansion has probably changed over time, influenced by gravity or perhaps (as recent studies suggest) by some form of "vacuum energy." Hubble's first attempt to calculate the rate of expansion (the Hubble constant) suggested that the universe was only 2 billion years old—clearly an impossible figure, as Earth itself was thought to be at least 4.5 billion years old. Modern estimates put the origin of the universe at about 13 billion years ago. This date is (just) consistent with dates for the oldest known stars (about 12 billion years old), and with the absence of any older radiometric dates. The most recent studies, based on evidence released in 2003 using the Wilkinson Microwave Anisotropy Probe (WMAP), give the extraordinarily precise date of 13.7 billion years ago for the big bang. The same studies suggest that the first stars lit up a mere 200 million years later, so it is no surprise that the estimated ages of the oldest stars have sometimes seemed dangerously close to estimates for the age of the universe itself.

UNDERSTANDING LARGE TIMESCALES

Grasping the scales of the modern creation myth is extremely difficult for those not used to dealing with large timescales. But this is not a problem unique to the modern creation story. Some Hindu and Buddhist chronologies of the universe's history are even more extravagant than those of modern science.

> Suppose, O Monks—the Buddha once told his followers—there
> was a huge rock of one solid mass, one mile long, one mile wide, one
> mile high, without split or flaw. And at the end of every hundred
> years a man should come and rub against it once with a silken cloth.
> Then that huge rock would wear off and disappear quicker than a
> world-period [kalpa]. But of such world-periods, O Monks, many
> have passed away, many hundreds, many thousands, many hundred
> thousands.[4]

To really grasp the scale of the modern creation myth, we need to make a similar imaginative effort. This appendix includes several timelines that may help the reader become more familiar with the different timescales of the modern creation story.

The earliest parts of the book figure dates in relation to the present. Thus, the universe was probably created about 13 billion years ago, and the earth about 4.6 billion years ago, while the earliest evidence for multicellular organisms appears about 600 million years ago, and the earliest skeletal evidence for the existence of hominines (the bipedal primates from which modern humans are descended) appears about 4 million years ago (though recent finds are pushing these dates back to about 6 million years). As we approach the era of human history, we use this system more formally, adopting the archaeologists' terminology of dates BP (before the present). Strictly speaking, such dates, if based on radiometric dating techniques, are calculated "before 1950." For all dates after ca. 5,000 years ago (from chapter 9 on, generally speaking), I use the more familiar system of BCE (before the common era) and CE (common era), which are equivalent to the traditional Christian dating system of BC and AD. To translate from dates BP into dates BCE, simply subtract 2,000 years. Thus, 5000 BP is equivalent to 3000 BCE.

What follows is a brief summary of the modern story of creation, and three different chronologies that may help the reader keep track of the book's vast chronological scale. Scattered through the book are eight timelines at different scales, which may also help familiarize the reader with the multiple timescales of this story.

The Core Story

What follows is just one possible attempt to summarize the story told in the rest of this book.

Thirteen billion (13,000,000,000) years ago there was nothing. There wasn't even emptiness. Time did not exist, nor did space. In this nothing, there occurred an explosion, and within a split second, something did exist. The early universe was fantastically hot—a searing cloud of energy and matter, much hotter than the interior of a sun. For a trillionth of a second it expanded faster than the speed of light, growing from the size of an atom to the size of a galaxy. Then the rate of expansion slowed, but the universe has continued expanding to the present day. As the early universe expanded, its temperature dropped. After about 300,000 years, it was cool enough for atoms of hydrogen and helium to form. Within about a billion years, huge clouds of hydrogen and helium began to gather and then collapse in on themselves under the pressure of gravity. As the center of these clouds heated up, atoms fused together violently like vast hydrogen bombs, and the first stars lit up. Hundreds of billions of stars appeared, gathered in the huge communities we call *galaxies*. The early universe consisted of little more than hydrogen and helium, but inside stars, and in the violent death agonies of large stars, new elements were created. And over time, more complex elements began to appear in interstellar space. Our own sun was formed about 4.5 billion years ago from a cloud of gas and matter that contained many of these new elements, in addition to hydrogen and helium. The planets of our solar system were formed at the same time as the Sun, from the debris left over from the Sun's creation.

The early earth was a dangerous place, bombarded by meteorites and so hot that much of it was molten. Over a billion years, however, it began to cool, and as it did so, water rained down on its surface to create the first seas. By 3.5 billion years ago, complex chemical reactions, probably taking place around deep-sea volcanoes, had created simple forms of life. Over the next 3.5 billion years, these simple, single-cell organisms became more and more diverse, evolving through natural selection. Quite early, some learned to extract energy from sunlight through the process of photosynthesis. As other organisms began to feed on the photosynthesizers, sunlight became the main "battery" of life on earth. Powered by the Sun, living organisms spread through the seas and eventually over the land, creating an interconnected web of life that had a profound impact on the atmosphere, the land, and the sea. From about 600 million years ago, there began to appear larger organisms, each made up of billions of individual cells. A mere 250,000 years ago, our own species appeared, having evolved from apelike ancestors through the same unpredictable processes of natural selection.

Though they evolved in the same way as other animals, humans turned out to be unusually good at extracting resources from the environment. Their advantage lay in their ability to share information and ideas with a precision that no other animal could match. And over time, their shared knowledge accumulated, enabling each generation to build on the knowledge of earlier generations. The number of humans grew as they learned how to live in more and more diverse environments, first in Africa; then in Eurasia, Australia, and the Americas; and eventually in the myriad islands of the Pacific. These global migrations took many tens of thousands of years. Eventually, beginning a mere ten thousand years ago, humans in some parts of the world began to manipulate their environments so successfully that they could produce ever greater amounts of food from a given area of land. Using the technologies we refer to as *agriculture*, they began to settle down in small village communities. As populations grew, the number and size of villages also grew until the first large cities appeared, about 5,000 years ago. These large, dense settlements required new and complex forms of regulation to prevent disputes and coordinate the activities of many people living at close quarters. In this way there appeared the first states, groups of powerful individuals capable of regulating the activities of the community as a whole. Conflict appeared both within and between communities, as different groups competed for resources and power. But as communities also exchanged information, the technological resources available to humanity as a whole continued to accumulate. Over several thousand years, the size, the reach, and the populations of societies with states expanded, until eventually most humans were living within state-based societies with cities and some form of agriculture. As their numbers and technological skills grew, so did their impact on the biosphere—the community of other organisms on earth. In some regions, the impact of human activities such as irrigation or deforestation proved so damaging that the local environment could no longer support large human populations, and entire civilizations collapsed.

As technologies of communication and transportation improved, more and more communities came into contact with each other. About 500 years ago, for the first time, these changes connected human communities in all parts of the world. For many communities this coming together was disastrous; it brought conquest, disease, and exploitation, sometimes of the most brutal kind. But the merging of regional communities also helped trigger new technological breakthroughs that could now be shared throughout the world. In the past two centuries, new technologies, beginning with the harnessing of steam power, have given human societies access to the vast sources of energy locked up in fossil fuels such as coal and oil. Human populations have grown more rapidly than ever before, and the problems of administering these huge communities, and coping with conflicts between them, have demanded the creation of even more powerful and complex state systems. Today, human numbers are so great, and the impact of humans on the biosphere is so significant, that we are in real danger of doing serious damage to the environment that is our home. Such damage could lead to a global collapse of human civilizations and have devastating effects on other organisms as well. At the same time, the ability of humans to share knowledge is now greater than ever before, and it may be that new technologies and new ways of organizing human societies will enable us to avoid the dangers created by our ecological virtuosity.

A Chronology for the Whole of Time

The first chronology gives a list of (approximate) dates. These cover some of the fundamental changes and transitions dealt with in the text.

History of the Universe before Our Sun (from 13 Billion Years to 4.5 Billion Years Ago)

- ca. 13,000,000,000 (13 billion) years ago: The big bang, the origin of the universe; the universe expands to the size of a galaxy; many vital events occur during the next few seconds; protons and electrons appear in the first second.
- ca. 300,000 years later: The universe has cooled to about a few thousand degrees Celsius, and electrons are captured by protons to form the first (electrically neutral) atoms, of hydrogen and helium; cosmic background radiation (CBR) is released as the universe becomes electrically neutral (the detection of CBR in 1964 led to general acceptance of the big bang theory of the origins of the universe).
- ca. 1,000,000,000 (1 billion) years after the big bang: The first stars light up as hydrogen atoms begin to fuse into helium atoms at the center of huge clouds of gas, under the pressure of gravity; billions of stars cluster into galaxies; new elements form in the interior of stars (all elements up to iron, with 26 protons) or in the vast explosions of dying stars known as *supernovae* (elements up to uranium, with 92 protons).
- ca. 4,600,000,000 (4.6 billion) years ago: The Sun, Earth, and solar system form from clouds of stardust containing the debris of older stars.

History of Earth and Life on Earth (from 4.5 Billion Years Ago)

- ca. 3,500,000,000 (3.5 billion) years ago: The first living organisms on earth appear; DNA is the basis of reproduction, and still present in every cell of every living thing (it reproduces by making almost perfect copies of itself; change and evolution are possible because the copies are not absolutely perfect, and when imperfect copies manage to survive, their descendants eventually form new species); early life consists of prokaryotes, little more than strands of DNA floating inside a protective container, or *cell;* photosynthesizing cells use the energy of sunlight and produce oxygen.
- ca. 2,500,000,000 (2.5 billion) years ago: Free oxygen, produced by photosynthesizing organisms, begins to change the earth's atmosphere.
- ca. 1,500,000,000 (1.5 billion) years ago: The first complex cells or eukaryotes appear, which have nuclei containing DNA and complex internal organelles (all complex life-forms evolve from eukaryotes); groups of cells start clustering together into large colonies to form the first multicelled creatures; with sexual reproduction, in which two not quite identical organisms swap their DNA to form a new creature, different from either parent, the pace of change accelerates.
- ca. 600,000,000 (600 million) years ago: The first fossils of larger, multicellular creatures appear during the Cambrian era; formation of an

ozone layer, from oxygen high up in the atmosphere, makes it easier for life to evolve on land, as it shields the surface from the harmful ultraviolet rays of the Sun, but not from its warmth and light; life spreads to the land and air, as well as multiplying and diversifying within the seas.

- ca. 65,000,000 (65 million) years ago: The extinction of the dinosaurs occurs, probably as the result of an asteroid impact, whose effects were similar to those of a nuclear war; mammals begin to replace dinosaurs as dominant large land animals; the first primates appear, tree-dwelling mammals with larger brains, dexterous hands, and stereoscopic vision.

The Paleolithic Era of Human History (from ca. 7 million to ca. 10,000 Years Ago)

- ca. 7,000,000 (7 million) years ago: The first hominines evolve from apes, distinguished by bipedalism.
- ca. 4,000,000 (4 million) years ago: Australopithecines emerge.
- ca. 2–1.5 million years ago: *Homo habilis,* the first member of our genus, appears.
- ca. 1.8 million years ago: *Homo ergaster/erectus* develops.
- ca. 1 million years ago: Members of the species *Homo erectus* migrate to southern parts of Eurasia.
- ca. 250,000 years ago: The appearance of the first modern humans, probably with fully developed language: *Homo sapiens.*
- ca. 100,000 years ago: Modern humans move into the Near East, where they probably encounter Neanderthals.
- ca. 60,000 years ago: The first colonization of Sahul/Australia by modern humans.
- ca. 25,000 years ago: Modern humans move into Siberia; Neanderthals, the only remaining nonhuman hominines, become extinct.
- ca. 13,000 years ago: The first clear evidence of colonization of the Americas, across the Bering Straits.

The Holocene Era of Human History (the Past 10,000 Years)

- ca. 10,000–5,000 years ago: The last ice age ends; intensive foraging technologies, some sedentary societies appear, along with early forms of agriculture; populations begin to grow rapidly; there are early signs of complexity and hierarchy, as large communities require new, more complex forms of organization.
- ca. 5,000 years ago: The first cities, states, and agrarian civilizations emerge; powerful elites control resources through tribute-taking; these elites organize warfare, large-scale worship, and monument building; writing is invented; agrarian civilizations spread and become the most populous and powerful of human communities.

The Modern Era (the Past 500 Years–the Future)

- ca. 500 years ago: Afro-Eurasia and the Americas come together, forming the largest "world zones" on earth; the first global system of exchange is created.
- ca. 200 years ago: The first capitalist societies emerge in western Europe; the Industrial Revolution exploits fossil fuels; there is huge increase in the power, wealth, and influence of European states; European imperialism dominates the globe.
- ca. 100 years ago: The Industrial Revolution begins to spread more widely; conflict breaks out between leading capitalist states; a communist backlash occurs.
- ca. 50 years ago: The first use of a nuclear weapon takes place (humans learn to use the explosive power present at the origin of the universe, and are in danger of destroying themselves and the rest of the biosphere).
- ca. 4–5,000 million years into the future: The Sun begins to die.
- Many billions of years further into the future: The universe will decay into a state of featureless equilibrium.

Thirteen Billion Years in Thirteen Years

The second chronology also covers 13 billion years. However, it collapses the timescales of modern cosmology by a factor of one billion, reducing 13 billion years to 13 years. Doing this may make it easier to grasp the crucial differences between different types of timescale.

History of the Universe before Our Sun: From 13 to ca. 4.5 Years Ago

- The big bang occurs ca. 13 years ago.
- The first stars and galaxies appear by about 12 years ago.
- The Sun and solar system form about 4.5 years ago.

History of Earth and Life on Earth: From 4 Years to ca. 3 Weeks Ago

- The first living organisms appear about 4 years ago.
- The first multicelled organisms appear about 7 months ago.
- Pangaea forms about 3 months ago.
- Dinosaurs are driven to extinction after a meteor impact about 3 weeks ago; mammals flourish.

The Paleolithic Era of Human History: From 3 Days Ago to 6 Minutes Ago

- First hominines evolve in Africa about 3 days ago.
- First *Homo sapiens* evolve about 50 minutes ago in Africa.
- First humans reach Papua New Guinea/Australia about 26 minutes ago.
- First humans reach Americas about 6 minutes ago.

The Holocene Era of Human History: From 6 Minutes Ago to 15 Seconds Ago

- First agricultural communities flourish about 5 minutes ago.
- First literate urban civilizations appear about 3 minutes ago.
- Classical civilizations of China, Persia, India, and the Mediterranean and the first agrarian civilizations in Americas emerge about 1 minute ago.
- Mongol Empire briefly unites much of Eurasia about 24 seconds ago; Black Death.

The Modern Era: The Past 15 Seconds

- Human communities are linked into a single "world system" about 15 seconds ago.
- The Industrial Revolution occurs about 6 seconds ago.
- Industrial Revolution spreads to Europe about 6 seconds ago.
- The First World War is fought about 2 seconds ago.
- Human populations reach 5, then 6 billion; the first atomic weapons are used; humans walk on the Moon; and the electronic revolution occurs, all within the last second.

So, at the end of 13 years, the universe would have existed for 13 years, and the earth for fewer than 5. Complex, multicelled organisms would have existed for about 7 months, hominines for a mere 3 days, and our own species, *Homo sapiens,* for a mere 50 minutes. Agricultural societies would have existed for only 5 minutes, and the entire recorded history of civilization for 3 minutes. The modern industrial civilizations that dominate the world today would have existed for 6 seconds.

The Geological Timescale

The third chronology will be familiar to students of geology. It is the geological timescale. You will undoubtedly come across references to it, so it is worth getting used to its main features. Table A1 presents one, highly simplified version. Don't worry too much if the dates seem to vary slightly from version to version; it's the large picture that matters.

TABLE A1. THE GEOLOGICAL TIMESCALE

Geological Era	Period	Starting Date (years BP)	Major Events
Hadean		4.6 billion	Formation of solar system; Moon; meltdown and "differentiation," oldest rocks, early atmosphere
Archean		4.0 billion	Earliest life; Prokaryota
Proterozoic		2.5 billion	Increasing oxygen; Eukaryota
	Ediacaran	590 million	Earliest multicelled organisms
Paleozoic	Cambrian	570 million	Earliest organisms with shells
	Ordovician	510 million	First corals, vertebrates
	Silurian	439 million	First bony fishes, first trees
	Devonian	409 million	First sharks, amphibians
	Carboniferous	363 million	First reptiles, winged insects; coal formation
	Permian	290 million	Mass extinctions
Mesozoic	Triassic	250 million	First dinosaurs, lizards, mammals
	Jurassic	208 million	First birds
	Cretaceous	146 million	First flowering plants, marsupials
Cenozoic (Tertiary)	Paleocene	65 million	Asteroid impact; dinosaur extinction, radiation of mammals, flowering plants; first primates
	Eocene	57 million	First apes
	Oligocene	36 million	Early hominoids
	Miocene	23 million	Separation of hominid and ape lines
	Pliocene	5.2 million	Australopithecines, *Homo habilis*
(Quaternary)	Pleistocene	1.6 million	*Homo erectus*, modern humans
	Holocene	10,000	Post–Ice Age human history

APPENDIX 2
CHAOS AND ORDER

In this appendix, I will argue that there are objects that recur at all the different scales discussed in this book. Though not essential to understanding the book's argument, the appendix may clarify some details and may help the reader see more clearly some of the links between different parts of the modern creation story.

Of all the patterns that occur at many different scales, the most fundamental is the existence of pattern itself.[1] Wherever we look, we see organized structures, or regimes. We don't see unrelated bits and pieces, like a sort of cosmic static; patterns that are too simple and repetitive also tend to fade into the background. What we notice are complex patterns that combine structure and diversity. These are the patterns that stand out against a background of disorder or extreme simplicity, and that have histories. If there are general rules of historical change, they concern the ways in which these patterns are created and evolve.

We see complex structures in part because we are built to see complex structures. All living organisms have to map their environment in order to survive. They have to be able to detect seasonal changes, the movements of the Sun and Moon, the movements of prey and predators. Thus they have to be pattern detectors, tracking how the bits and pieces of their environment fall into larger and more predictable shapes. Humans, too, are constantly distinguishing between those parts of the environment that have structure and those that do not. We are necessarily more interested in the stars than in the vast reaches of near-empty space that lie between them. We have also learned how to track many patterns that are not immediately accessible to our senses, such as the patterns of deep time. Order and chaos shape all our attempts to understand our world.

But the patterns we detect are really there, and their existence is one of the great puzzles of the universe. Why is there order of any kind? And what rules allow the creation and evolution of ordered structures? Creating disorder seems to be much easier than creating order. Think of a pack of cards. Shuffled randomly, it will rarely produce an ordered sequence—say, an ordered run of thirteen hearts. And if it does, that sequence will vanish after a few more shuffles. But when we study the universe as a whole, we find complex and durable patterns at many different scales, from clusters of galaxies stretching over millions of light-years, to the complex social structures of human history, to the even more durable patterns that lock quarks into the subatomic particles we call protons and neutrons.

Many religions solve the problem of explaining these complex and durable patterns by claiming that complex entities such as ourselves were created by an intelligent creator or deity. For modern science this can be no solution, for it merely raises the further problem of how such a deity might have been created. Can we explain complexity without introducing a hypothesis that begs so many more questions? At present, no entirely satisfactory answers to these questions exist; the following paragraphs can merely hint at some modern approaches to a solution.

One thing is immediately clear: creating and maintaining patterns requires work. A pack of cards has many more disordered than ordered states, so most shuffles will yield a disordered state. The universe seems to work similarly, with a natural tendency toward disorder and chaos. Creating and sustaining patterns means working against this apparently universal tendency toward disorder; it means helping unlikely things to happen and keep happening.

Understanding patterns therefore means understanding how energy does work. In the nineteenth century, studies of the efficiency with which energy was used in steam engines led the French engineer Sadi Carnot to the conclusion that energy never vanishes; it simply changes the forms in which it exists. Thus as heat is used to generate steam, whose pressure can generate the mechanical energy of a steam engine, energy itself seems to be conserved. The law of the conservation of energy is often called the *first law of thermodynamics*. The *second law of thermodynamics* seems, at first sight, to contradict the first: it says that in a closed system (which the universe seems to be), the amount of *free* energy, or energy capable of doing work, tends to dissipate over time. A waterfall can drive a turbine because the water at its top has been lifted to higher levels, and the energy used to lift it (supplied by the Sun, which evaporated water vapor and lifted it into the clouds) is returned as water falls toward the sea. By the time the water has reached the sea, it can no longer do work, because all the water at sea level has about the same amount of available energy; it is in a state of thermodynamic equilibrium. Usable or free energy, energy that can do work, requires a gradient, a slope, some form of difference. The second law predicts that over colossal periods of time, and in a closed system, all differentials will diminish; as they do so, there will be ever-diminishing amounts of free energy available to do the hard work of creating and sustaining complex entities. This seems to mean that the whole universe will eventually become less and less ordered as it tends toward a state of thermodynamic equilibrium. In the nineteenth century, this depressing idea was described as the "heat death" of the universe. A German scientist, Rudolf Clausius, labeled the steadily increasing pile of unusable energy *entropy*. In the very long run, it seems, entropy must increase, and complexity must diminish.[2] In the end, *everything* must become background noise. The second law apparently implies that everything in the universe is riding down the same escalator toward chaos.

These ideas are fundamental to modern physics, but they raise two deep problems. First, how is order possible at all? Why do we not find ourselves in a universe of total disorder, in which the second law has fulfilled its deadly mission? Did the universe start out with a stock of free energy on which all ordered entities have drawn ever since? If so, where did that energy capital come from and how long will it be before it runs out? Something (or someone?) must have done some heavy lifting in the early days of the universe to create the gradients and differences that create and sustain the patterns we see around us.[3] If it was not a creator god who did this work, then how was it done? The ultimate source of free energy (and therefore of order) remains one of the great puzzles of modern cosmology, because, as far as we can tell, the early universe was remarkably homogenous.

That early universe was apparently dense and incredibly hot, in a state of thermodynamic equilibrium. But as it expanded it cooled, and as it cooled, its symmetry was broken. The first differences appeared, and the first gradients of temperature and pressure. At first, there seemed little distinction between forces such as electricity and gravity. They all appeared to blend together in the violent energies of a universe that was nearly infinitely hot and infinitely dense. As it expanded and cooled, however, the different fundamental forces each assumed its own particular form. For example, before about 300,000 years after the big bang, the electromagnetic force was too weak to bind electrons and protons together into atoms. But after that time, the universe was cool enough for electricity to begin sculpting the atomic structures studied in modern physics and chemistry. At this point, matter and energy also became quite distinct.

As the universe expanded, tiny initial differences multiplied, and each of these forces began to operate in distinctive ways. Gravity operated at large scales, and shaped the large structures of the universe. Because matter was slow moving and heavy, it could be herded together by gravity more easily than energy, which was light and fast moving. So, as energy and matter separated, gravity got to work shaping matter into large, complex structures, while energy, for the most part, escaped its influence, except in extreme regions such as areas near black holes. First, gravity gathered hydrogen and helium into huge clouds. Then it began to squeeze each cloud into smaller and smaller spaces, until pressure and temperature built up, particularly at the center. When the cores reached about 10 million degrees, fusion reactions began, and stars lit up. The fusion reactions at the center of all stars counterbalanced the crushing force of gravity, negotiating a sort of cosmic truce that is the foundation of every star. Once created, stars provided stable, long-lived energy differentials, which offered durable stores of free energy, or *negentropy*. Stars created stable hot spots, dotted throughout the cooling early universe like raisins in a dough. Today, the cosmic background radiation is just a few degrees above absolute zero—this is the base temperature of the universe. But at their centers, stars must have enormously high temperatures for fusion to start—and in large stars, they can rise much higher than 10 million degrees. In the regions near these hot points, complex entities could begin to form, exploiting the huge temperature differentials between stars and surrounding space, just as early life on Earth formed around volcanic vents deep in the sea. As Paul Davies puts it: "Matter and energy in far-from-equilibrium open systems have a propensity to seek out higher and higher levels of organization and complexity."[4]

On Earth, the temperature differential between our sun and surrounding space provides the free energy needed to create most forms of complexity, including ourselves; energies created early in the history of our solar system drive the internal heat battery of Earth, which drives plate tectonics. These differentials enable energy to flow, and energy flows make patterns possible. And given enough time, the mere possibility of pattern makes it likely that patterns of many different kinds will eventually appear.

According to this line of argument, the expansion of the universe, which allowed the early universe to cool and diversify, is the ultimate source of all temperature and pressure differentials, and therefore of the free energy needed to create order. We can put this argument slightly differently. At its moment of origin, the universe was so small and so homogenous that there were few possible disordered states; it was like a pack of cards with just one card. Expansion created a larger space and new possibilities for disorder, and these possibilities multiplied as the universe kept expanding. As a general rule, the larger a system, the greater the possible entropy; just as, to continue the analogy, a larger pack of cards increases the number of possible disordered states.[5] Thus, while the second law of thermodynamics suggests that entropy will always increase, the expansion of the universe seems to ensure that there will always be more steps inserted on the path down the thermodynamic escalator toward the state of complete disorder that lies at the bottom. Whatever caused the universe to expand is also in some sense the source of order and pattern.

After the first problem—explaining how order of any kind is possible—is addressed, the second problem remains. How did complex entities emerge, and, once they had emerged, how did they sustain themselves long enough to be noticed by us (or to *be* us)? Paradoxically, the tendency toward increasing entropy—the drive toward disorder—may itself be the engine that creates order. It creates order *on the way* to creating disorder. Poetically, we can think of the steady increase in entropy as an attempt by the universe to return to its initial state of thermodynamic equilibrium; many creation myths similarly describe the breaking of an original unity, whose divided parts try to return to their initial state. In Plato's *Symposium*, one of the proposed explanations of love between men and women is that it was created when the gods split a hermaphroditic being into two different creatures, whose attempts to reunite created all future generations of humans. The drive toward disorder seems to create new forms of order, just as the energy of falling water can cause droplets of water to splash upward, or a river's current can create eddies in which small amounts of water flow against the main current.

At a local scale, and in the short run, complex entities seem to reverse the workings of the second law of thermodynamics by increasing order. But viewed within the larger environment from which they draw free energy, they clearly actually increase entropy by speeding up the transformation of free energy into unusable forms of heat. Thus complexity is, in a sense, a cunning way for the second law of thermodynamics to work more efficiently toward its bleak goal of a universe without order.[6] Ilya Prigogine and Isabelle Stengers have applied the curious term *dissipative* to the complex structures described here.[7] What complex structures do is to handle huge flows of energy and, in the process, dissipate large amounts of free energy, thereby increasing entropy overall. Though they appear to reduce entropy momentarily and locally, they in fact generate entropy more effectively than do simpler structures, facilitating the deadly workings of the second law.

Nevertheless, creating order is not easy. Somehow or other, significant flows of energy need to be concentrated and focused in ways that generate pockets of increased order. Complex phenomena require a constant throughput of energy to help them climb entropy's remorseless down escalator. So the existence of stable differentials to guarantee a steady supply of energy, such as the temperature and pressure differentials available near stars, is an essential precondition for complexity. What is not clear is whether there exist mechanisms that actively seek out complexity. Does the presence of differentials and disequilibria actively drive matter and energy toward complexity? Or do they merely make it possible? Does complexity work like natural selection, through the random generation of structures that, once they have appeared, become locked into place simply because they fit their environment well? Or does the second law create complexity through a devious cosmological cunning of its own?

Whatever the source of order, its creation, whether in the Sun or the stock exchange, requires creating structures that can channel and control large flows of energy without falling apart. This is an extremely difficult trick. And that difficulty explains why ordered entities are fragile and rare, and why they stand out against a background that is much simpler. Roughly speaking, the more complex a phenomenon is, the denser the energy flows it must juggle and the more likely it is to break down. So we should expect that as entities become more complex, they become less stable, shorter-lived, and rarer. Perhaps even a slight increase in complexity can sharply increase their fragility and, therefore, their scarcity. Of all complex chemicals in existence, only a minute fraction have ever formed living organisms; of all living organisms, an even smaller proportion has formed into intelligent, networked species such as ourselves. (Table 4.1 offers some evidence for these generalizations.) But it is also clear that the likelihood of complex entities' appearing would be greatly increased if instead of relying on random changes to generate such structures by accident, we could identify laws that tended to actively create such structures. At present, we simply do not know whether there are such laws, though the emerging science of complexity is attempting to identify them.

What we can do is to describe some of the ways in which complex structures emerge. The fundamental rule seems to be that complexity normally emerges step by step, linking already existing patterns into larger and more complex patterns at different scales. Once achieved, some patterns seem to lock their constituents into new arrangements that are more stable and more durable than the simpler arrangements from which they are created. Such processes create the hierarchy of different levels of complexity that we observe in the universe, because at each scale, new rules of construction and change seem to come into play. These are known as *emergent* properties, because they do not seem to be derived from the properties of the original components; instead, they apparently emerge as these components are assembled into a larger structure. The word *universe* is a verbal structure consisting of eight letters. But its meaning cannot be deduced just by knowing the letters used to construct it. Its meaning is an emergent property. Similarly in chemistry, the properties of water cannot be explained by describing how hydrogen and oxygen behave, yet water is formed by combining molecules of hydrogen and oxygen. Its properties emerge only when atoms of hydrogen and oxygen are combined into water molecules.[8] The myriad different ways in which these rules play out at different scales and at different degrees of complexity provide the subject matter for the various disciplines of modern knowledge. Each deals with the rules that emerge at new levels of complexity, from particle physics to chemistry, biology, ecology, and history.

Being complex creatures ourselves, we know from personal experience how hard it is to climb the down escalator, to work against the universal slide into disorder, so we are inevitably fascinated by other entities that appear to do the same thing. Thus this theme—the achievement of order despite, and perhaps with the aid of, the second law of thermodynamics—is woven through all parts of the story told here. The endless waltz of chaos and complexity provides one of this book's unifying ideas.

NOTES

Chapter epigraphs: Fernand Braudel, *On History*, trans. Sarah Matthews (Chicago: University of Chicago Press, 1980), p. viii, from the 1969 preface; Leopold von Ranke, quoted in Arthur Marwick, *The Nature of History* (London: Macmillan, 1970), p. 38; Edward Fitzgerald, *The Rubáiyát of Omar Khayyám of Naishápúr*, stanza 47, in *The Norton Anthology of Poetry*, 3rd ed., ed. Alexander W. Allison et al. (New York: W. W. Norton, 1970), p. 688.

Section epigraphs, p. 1: James Joyce, *Finnegans Wake*; quoted in Joseph Campbell, *The Masks of God*, vol. 1, *Primitive Mythology* (1959; reprint, Harmondsworth: Penguin, 1976), p. xx; p. 5: Mark Twain, "The Damned Human Race," in *Letters from the Earth*, ed. Bernard De Voto (New York: Harper and Row, 1962), pp. 215–16; quoted in Lynn Margulis and Dorion Sagan, *Microcosmos: Four Billion Years of Microbial Evolution* (London: Allen and Unwin, 1987), p. 194.

1. David Christian, "The Case for 'Big History'," *Journal of World History* 2, no. 2 (fall 1991): 223–38; the essay is reprinted in *The New World History: A Teacher's Companion*, ed. Ross E. Dunn (Boston: Bedford/St. Martin's, 2000), pp. 575–87. I originally used the label "big history" somewhat flippantly, and only later realized how overblown it might sound; however, I have stuck with it, as it now provides a convenient shorthand for the project of attempting to view history on the largest possible scale.

2. Two important modern attempts at unified accounts of the past—written, respectively, from a cosmological and a geological perspective—are Eric J. Chaisson, *Cosmic Evolution: The Rise of Complexity in Nature* (Cambridge, Mass.: Harvard University Press, 2001), and Preston Cloud, *Cosmos, Earth, and Man: A Short History of the Universe* (New Haven: Yale University Press, 1978).

3. Murray Gell-Mann, "Transitions to a More Sustainable World," in *Scanning the Future: Twenty Eminent Thinkers on the World of Tomorrow*, ed. Yorick Blumenfeld (London: Thames and Hudson, 1999), pp. 61–62.

4. Fred Spier, *The Structure of Big History: From the Big Bang until Today* (Amsterdam: Amsterdam University Press, 1996).

5. Edward O. Wilson, *Consilience: The Unity of Knowledge* (London: Abacus, 1998).

6. William H. McNeill, "History and the Scientific Worldview," *History and Theory*, 37, no. 1 (1998): 12–13.

7. Erwin Schrödinger, *What Is Life?* in *What Is Life? The Physical Aspect of the Living Cell;* with, *Mind and Matter;* and *Autobiographical Sketches* (Cambridge: Cambridge University Press, 1992), p. 1 (first published in 1944).

8. Spier, *The Structure of Big History*.

9. In Lyotard's famous formulation, postmodernism is, above all, "incredulity towards metanarratives" (Jean-François Lyotard, *The Postmodern Condition: A Report on Knowledge*, trans. Geoff Bennington and Brian Massumi [Minneapolis: University of Minnesota Press, 1984], p. xxiv); Keith Jenkins defines grand narratives as "overarching philosophies of history like the Enlightenment story of the steady progress of reason and freedom, or Marx's drama of the forward march of human productive capacities via class conflict culminating in proletarian revolution" (Jenkins, ed., *The Postmodern History Reader* [London: Routledge, 1997], p. 7).

10. George E. Marcus and Michael M. J. Fischer, *Anthropology as Cultural Critique: An Experimental Moment in the Human Sciences* (Chicago: University of Chicago Press, 1986), p. 15 (and elsewhere).

11. Natalie Zemon Davis, commenting on a symposium titled "Cultural Encounters between the Continents over the Centuries," in *Nineteenth International Congress of Historical Sciences* (Oslo: Nasjonalbiblioteket, 2000), p. 47.

12. William Cronon, "A Place for Stories: Nature, History, and Narrative," *Journal of American History* 78, no. 4 (March 1992): 1349.

13. Ernest Gellner, *Plough, Sword, and Book: The Structure of Human History* (London: Paladin, 1991), pp. 12–13.

14. Patrick O'Brien, "Is Universal History Possible?" in *Nineteenth International Congress of Historical Sciences*, p. 13.

15. The metaphor of knowledge as a collection of maps of reality is deliberately balanced between instrumentalist and realist theories of knowledge. There is an elegant discussion of history writing as mapping in John Lewis Gaddis, *The Landscape of History: How Historians Map the Past* (Oxford: Oxford University Press, 2002), particularly pp. 31–34. For a difficult but up-to-date account of the instrumentalist/realist debate within the philosophy of science, see Stathis Psillos, *Scientific Realism: How Science Tracks Truth* (London: Routledge, 1999), which tracks the borderline between the two approaches, while ultimately opting for a realist position.

CHAPTER 1. THE FIRST 300,000 YEARS

Chapter epigraph: William Shakespeare, *Twelfth Night* 1.2; quoted in John Lewis Gaddis, *The Landscape of History: How Historians Map the Past* (Oxford: Oxford University Press, 2002), p. 16. Gaddis adds, "The first lines Shakespeare has Viola speak, filled as they are with intelligence, curiosity, and some dread, could well be the starting point for any historian contemplating the landscape of history."

1. Deborah Bird Rose, *Nourishing Terrains: Australian Aboriginal Views of Landscape and Wilderness* (Canberra: Australian Heritage Commission, 1996), p. 23.

2. *The Rig Veda*, ed. and trans. Wendy Doniger O'Flaherty (Harmondsworth: Penguin, 1981), no. 10.129, pp. 25–26.

3. *Popol Vuh: The Mayan Book of the Dawn of Life*, trans. Dennis Tedlock, rev. ed. (New York: Simon and Schuster, 1996), p. 64.

4. Barbara Sproul, *Primal Myths: Creation Myths around the World* (1979; reprint, San Francisco: HarperSanFrancisco, 1991), p. 15.

5. In one of the best modern accounts of time, *Time: An Essay*, trans. Edmund Jephcott (Oxford: Blackwell, 1992), Norbert Elias argues that our modern sense of time is shaped largely by the need of people in complex societies to coordinate their activities precisely.

6. Tony Swain describes such a place-based ontology among some Aboriginal Australian communities; see *A Place for Strangers: Towards a History of Australian Aboriginal Being* (Cambridge: Cambridge University Press, 1993), chap. 1.

7. Quoted from Sproul, *Primal Myths*, pp. 137–38.

8. Peter White, "The Settlement of Ancient Australia," in *The Illustrated History of Humankind*, ed. Göran Burenhult, vol. 1, *The First Humans: Human Origins and History to 10,000 BC* (St. Lucia: University of Queensland Press, 1993), p. 148.

9. Stephen Hawking, *The Universe in a Nutshell* (New York: Bantam, 2001), p. 85.

10. The concept of the "Dreaming," or "Dreamtime," first entered the English language in reports from an 1894 expedition to Central Australia, as a translation of the Arunta word *altyerre*, "to dream"; see Rhys Jones, "Folsom and Talgai: Cowboy Archaeology in Two Continents," in *Approaching Australia: Papers from the Harvard Australian Studies Symposium*, ed. Harold Bolitho and Chris Wallace-Crabbe (Cambridge, Mass.: Harvard University Press, 1997), p. 20.

11. Mircea Eliade, *The Myth of the Eternal Return, or, Cosmos and History*, trans. Willard R. Trask (New York: Pantheon, 1954).

12. Lee Smolin, *The Life of the Cosmos* (London: Phoenix, 1998), particularly chap. 7.

13. It is easy to be amused by these calculations, but, as Timothy Ferris points out, the idea of measuring the moment of origin is quite modern in spirit; and anyway, Archbishop Ussher, whose calculation Lightfoot later refined, was off

merely by a factor of one million—which in modern cosmology may not be that bad (*The Whole Shebang: A State-of-the-Universe(s) Report* [New York: Simon and Schuster, 1997], p. 172).

14. There is a short and up-to-date account of these processes in Charles Lineweaver, "Our Place in the Universe," in *To Mars and Beyond: Search for the Origins of Life*, ed. Malcolm Walter (Canberra: National Museum of Australia, 2002), pp. 88–99.

15. Evidence from the Wilkinson Microwave Anisotropy Probe (WMAP) released by NASA in February 2003 suggests the most precise date calculated so far for the big bang: about 13.7 billion years ago. See "Imagine the Universe News," 12 February 2003 <http://imagine.gsfc.nasa.gov/docs/features/news/12feb03.html> (accessed April 2003).

16. Martin Rees, *Just Six Numbers: The Deep Forces That Shape the Universe* (New York: Basic Books, 2000), p. 133, makes this point: "The leap back from 10^{-14} seconds to 10^{-35} seconds is . . . bigger (in that it spans more factors of ten) than the timespan between the three minute threshold when helium was formed . . . and the present time (10^{37} seconds, or ten billion years)."

17. Richard P. Feynman, *Six Easy Pieces: The Fundamentals of Physics Explained* (London: Penguin, 1998), p. 5.

18. That the expansion was crucial for the emergence of complex entities is the argument of Eric Chaisson, *Cosmic Evolution: The Rise of Complexity in Nature* (Cambridge, Mass.: Harvard University Press, 2001); see p. 126.

19. Ferris, *The Whole Shebang*, p. 78. One of the most accessible accounts of the idea of inflation can be found in Paul Davies, *The Last Three Minutes* (London: Phoenix, 1995), pp. 28–35; for help with exponential notation, see "Note on Exponential Notation" at the end of this chapter.

20. Rees, *Just Six Numbers*, pp. 93–97.

21. Chaisson, *Cosmic Evolution*, p. 112.

22. The results from NASA's WMAP suggest that this transition, which released the cosmic background radiation, occurred about 380,000 years after the big bang. See "Imagine the Universe News," 12 February 2003.

23. Chaisson, *Cosmic Evolution*, p. 113.

24. Feynman, *Six Easy Pieces*, p. 34; Ferris writes that if the nucleus were a golf ball, the farthest electrons would orbit two miles away from it (*The Whole Shebang*, p. 108).

25. Chaisson, *Cosmic Evolution*, p. 2.

26. We are in complex symbolic territory here. As the Vietnamese Zen master Thich Nhat Hanh explains: "Form is the wave and emptiness is the water" (*The Heart of Understanding: Commentaries on the Prajñaparamita Heart Sutra*, ed. Peter Levitt [Berkeley: Parallax, 1988], p. 15). The Heart Sutra is quoted in *Heart of Understanding*, p. 1.

27. Wendy L. Freedman, "The Expansion Rate and Size of the Universe," *Scientific American*, spring 1998, pp. 92–97; Ken Croswell, "Uneasy Truce," *New Scientist*, 30 May 1998, pp. 42–46. But see note 15 for the most recent estimates.

28. Max Tegmark, quoted in James Glanz, "In the Big Bang's Echoes: Clues

to the Cosmos," a "Science Times" column, *New York Times*, 6 February 2001, p. D1.

29. For NASA's website on WMAP, see "Wilkinson Microwave Anisotropy Probe," 14 March 2003 <http://map.gsfc.nasa.gov/> (accessed April 2003); for the probe's most recent results, announced in February 2003, see note 15.

30. Lynn Margulis and Dorion Sagan, *Microcosmos: Four Billion Years of Microbial Evolution* (London: Allen and Unwin, 1987), p. 41.

31. Peter Coles, *Cosmology: A Very Short Introduction* (Oxford: Oxford University Press, 2001), pp. 91–92; Hawking, *Universe in a Nutshell*, pp. 96–99, describes the theory of "vacuum energy."

32. The explanation of exponential notation in the text is based on the extremely lucid description in Cesare Emiliani, *The Scientific Companion: Exploring the Physical World with Facts, Figures, and Formulas*, 2nd ed. (New York: John Wiley, 1995), pp. 5–10.

CHAPTER 2. ORIGINS OF THE GALAXIES AND STARS: THE BEGINNINGS OF COMPLEXITY

Chapter epigraph: Martin Rees, *Just Six Numbers: The Deep Forces That Shape the Universe* (London: Weidenfeld and Nicolson, 2000), p. 126.

1. Timothy Ferris, *The Whole Shebang: A State-of-the-Universe(s) Report* (New York: Simon and Schuster, 1997), pp. 151–52.

2. There is a good, up-to-date discussion of the problem of dark matter in Ferris, "The Black Taj, " chap. 5 of *The Whole Shebang;* see also Rees, *Just Six Numbers,* chap. 6. The most recent estimates suggest that radiation may make up as little as 0.005 percent of the mass of the universe; particles such as neutrinos may make up 0.3 percent; ordinary matter—things made from particles such as protons and electrons—may account for as little as 5 percent; "cold dark matter," consisting of particles whose existence has been predicted by theory but not yet detected in practice, may account for as much as 25 percent; and the remaining 70 percent may be contributed by "dark energy." See David B. Cline, "The Search for Dark Matter," *Scientific American*, March 2003, pp. 50–59, particularly the table on p. 53.

3. Evidence from WMAP released by NASA in February 2003 suggests that the first stars may have appeared as early as 200 million years after the big bang. See "Imagine the Universe News," 12 February 2003 <http://imagine.gsfc.nasa .gov/docs/features/news/12feb03.html> (accessed April 2003).

4. Rees, *Just Six Numbers*, p. 53; the only way to convert 100 percent of mass to energy is to bring matter and antimatter together.

5. Ferris, *The Whole Shebang*, pp. 79–80.

6. Lee Smolin, *The Life of the Cosmos* (London: Phoenix, 1998), particularly chap. 7, "Did the Universe Evolve?" Theories of "universal Darwinism" claim that any system containing replicators (in this case, universes and black holes) may be able to create complex entities through blind, algorithmic processes analogous to natural selection; see, for example, Henry Plotkin, *Evo-*

lution in Mind: An Introduction to Evolutionary Psychology (London: Penguin, 1997), pp. 251–52.

7. Charles Lineweaver, "Our Place in the Universe," in *To Mars and Beyond: Search for the Origins of Life*, ed. Malcolm Walter (Canberra: National Museum of Australia, 2002), p. 95. See also note 2.

8. John Wilford Noble, "Cosmic Players That Could've Been Stars," *New York Times*, 8 June 2001, p. A21.

9. See Armand Delsemme, *Our Cosmic Origins: From the Big Bang to the Emergence of Life and Intelligence* (Cambridge: Cambridge University Press, 1998), p. 61, for a chart summarizing the different lifestyles of stars of different masses; details of a supernova explosion are described well in Paul Davies, *The Last Three Minutes* (London: Phoenix, 1995), pp. 41–45.

10. See Delsemme, *Our Cosmic Origins*, pp. 74–75.

11. Ken Croswell, *The Alchemy of the Heavens* (Oxford: Oxford University Press, 1996), pp. 47–48.

12. E. O. Wilson, *Consilience* (London: Abacus, 1999), p. 49.

13. I first heard these thought experiments in lectures given in the early 1990s by the late David Allen, an English astronomer who lived and worked in Sydney.

14. Cesare Emiliani, *The Scientific Companion: Exploring the Physical World with Facts, Figures, and Formulas*, 2nd ed. (New York: John Wiley, 1995), p. 9.

15. Croswell, *Alchemy of the Heavens*, p. 182.

CHAPTER 3. ORIGINS AND HISTORY OF THE EARTH

1. Ross Taylor, "The Solar System: An Environment for Life?" in *To Mars and Beyond: Search for the Origins of Life*, ed. Malcolm Walter (Canberra: National Museum of Australia, 2002), pp. 59–60.

2. Nigel Hawkes, "First Sight of a Planet outside Our Solar System," *Times* (London), 29 May 1998, p. 5.

3. Current estimates of the likelihood of encountering other life-forms in the near future are assessed in Ian Crawford, "Where Are They?" *Scientific American*, July 2000, pp. 38–43.

4. Armand Delsemme, *Our Cosmic Origins: From the Big Bang to the Emergence of Life and Intelligence* (Cambridge: Cambridge University Press, 1998), pp. 116–21, argues that comets played this important role.

5. For a review of dating techniques, see appendix 1; see also Delsemme, *Our Cosmic Origins*, p. 285; Neil Roberts, *The Holocene: An Environmental History*, 2nd ed. (Oxford: Blackwell, 1998), chap. 2; and Nigel Calder, *Timescale: An Atlas of the Fourth Dimension* (London: Chatto and Windus, 1983).

6. There is a good chronology of the earth's history in Lynn Margulis and Dorion Sagan, *What Is Life?* (Berkeley: University of California Press, 1995), pp. 64–80.

7. Ian W. D. Dalziel, "Earth before Pangea," *Scientific American*, January 1995, pp. 38–43.

CHAPTER 4. THE ORIGINS OF LIFE AND THE THEORY OF EVOLUTION

1. Erwin Schrödinger, *What Is Life?* in *What Is Life? The Physical Aspect of the Living Cell;* with, *Mind and Matter;* and *Autobiographical Sketches* (Cambridge: Cambridge University Press, 1992), p. 77 (first published in 1944).

2. Eric Chaisson calls this quantity Φ_m, the "free energy rate density," and he measures it in "units of energy per time per mass"; he adds that the notion is "familiar to astronomers as the luminosity-to-mass ratio, to physicists as the power density, to geologists as the specific radiant flux, to biologists as the specific metabolic rate, and to engineers as the power-to-mass ratio" (*Cosmic Evolution: The Rise of Complexity in Nature* [Cambridge, Mass.: Harvard University Press, 2001], p. 134).

3. Schrödinger, *What Is Life?*, p. 73.

4. Martin Rees, "Exploring Our Universe and Others," *Scientific American*, December 1999, p. 46.

5. Another naturalist, Alfred Russel Wallace, stumbled on the same idea at about the same time, and he and Darwin first presented the theory in two papers published in 1858 in the *Journal of the Linnaean Society*.

6. Charles Darwin, *The Origin of Species by Means of Natural Selection: The Preservation of Favored Races in the Struggle for Life* (1859), ed. and intro. J. W. Burrow (Harmondsworth: Penguin, 1968), p. 82.

7. Darwin, *Origin of Species*, p. 90.

8. That blind algorithmic processes could generate extraordinary complexity is the fundamental insight that Daniel Dennett has described as "Darwin's dangerous idea" in his modern classic, *Darwin's Dangerous Idea: Evolution and the Meaning of Life* (London: Allen Lane, 1995).

9. Darwin, *The Origin of Species*, pp. 441–42, 115; quoted in Tim Megarry, *Society in Prehistory: The Origins of Human Culture* (Basingstoke: Macmillan, 1995), pp. 33–34.

10. Hubert Reeves, Joël de Rosnay, Yves Coppens, and Dominique Simonnet, *Origins: Cosmos, Earth and Mankind* (New York: Arcade Publishing, 1998), p. 138 (section by Coppens).

11. Armand Delsemme, *Our Cosmic Origins: From the Big Bang to the Emergence of Life and Intelligence* (Cambridge: Cambridge University Press, 1998), p. 135. Stuart Kauffman discusses the intriguing possibility that this delicate balance between order and chaos may itself be a product of evolutionary processes; see *At Home in the Universe: The Search for Laws of Complexity* (London: Viking, 1995), p. 90.

12. The material in this section draws on the lectures of David Briscoe, given at Macquarie University since 1989; see also Lynn Margulis and Dorion Sagan, *What Is Life?* (Berkeley: University of California Press, 1995), pp. 64–69.

13. Lynn Margulis and Dorion Sagan, *Microcosmos: Four Billion Years of Microbial Evolution* (London: Allen and Unwin, 1987), p. 48.

14. Amino acids are simple organic molecules containing an amino group ($-NH_2$) and a carboxyl group ($-COOH$), as well as a variable group of other atoms, all of which are attached to a single carbon atom; nucleotides are equally

simple, consisting of a sugar molecule, a phosphate group, and one of four *bases* (compounds containing nitrogen). Linked together in huge chains, nucleotides form the two main types of *nucleic acids,* DNA and RNA, so named because they were first isolated in cell nuclei; DNA and RNA are the keys to heredity in all living organisms.

15. These issues are discussed well by a pioneering researcher in the field of early life, Malcolm Walter, in *The Search for Life on Mars* (Sydney: Allen and Unwin, 1999).

16. Darwin, quoted in Paul Davies, *The Fifth Miracle: The Search for the Origin of Life* (Harmondsworth: Penguin, 1999), p. 54.

17. A. G. Cairns-Smith, *Seven Clues to the Origins of Life* (Cambridge: Cambridge University Press, 1985).

18. Reeves, de Rosnay, Coppens, and Simonnet, *Origins,* p. 92 (section by de Rosnay).

19. For a detailed discussion of these discoveries about archaebacteria and their implications, see Davies, *The Fifth Miracle,* particularly chap. 7.

20. Karen L. Von Damm, "Lost City Found," *Nature,* 12 July 2001, pp. 127–28.

21. Birger Rasmussen, "Filamentous Microfossils in a 3,235-Million-Year-Old Volcanogenic Massive Sulphide Deposit," *Nature,* 8 June 2000, pp. 676–79, announces the oldest fossil finding of an "extremophile," age 3.2 billion years, from the Pilbara region in Australia. These bacteria probably "ate" chemicals and lived near sulfurous vents deep in the ocean; previously, the oldest remains of archaebacteria were only 500 million years old.

22. There is a clear discussion of this question of priority in Freeman Dyson, *Origins of Life,* 2nd ed. (Cambridge: Cambridge University Press, 1999).

23. Cesare Emiliani, *The Scientific Companion: Exploring the Physical World with Facts, Figures, and Formulas,* 2nd ed. (New York: John Wiley, 1995), p. 151. Fred Hoyle famously compared the chances of constructing even the simplest bacteria by purely random processes to the chances of a whirlwind constructing "a Boeing 747 from scraps in a junkyard" (quoted in Delsemme, *Our Cosmic Origins,* p. 151).

24. For a description of these proposed deep mechanisms, see Kauffman, *At Home in the Universe,* and Paul Davies, *The Cosmic Blueprint* (London: Unwin, 1989).

25. There is a description of the "RNA world" in John Maynard Smith and Eörs Szathmáry, *The Origins of Life: From the Birth of Life to the Origins of Language* (Oxford: Oxford University Press, 1999), chaps. 3 and 4.

26. Dyson, *Origins of Life,* p. 40.

27. Dyson, *Origins of Life,* chap. 3.

CHAPTER 5. THE EVOLUTION OF LIFE AND THE BIOSPHERE

1. Lynn Margulis and Dorion Sagan, *Microcosmos: Four Billion Years of Microbial Evolution* (London: Allen and Unwin, 1987).

2. Stephen Jay Gould, *Life's Grandeur: The Spread of Excellence from Plato to Darwin* (London: Jonathan Cape, 1996); see chap. 14 for arguments supporting the dominance of bacteria, including the possibility that bacteria make up more than half of the total mass of living organisms on earth; see also Margulis and Sagan, *Microcosmos*.

3. John Maynard Smith and Eörs Szathmáry, *The Origins of Life: From the Birth of Life to the Origins of Language* (Oxford: Oxford University Press, 1999), p. 15. For another account that sees evolution as part of a process of increasing complexity, see Hubert Reeves, Joël de Rosnay, Yves Coppens, and Dominique Simonnet, *Origins: Cosmos, Earth and Mankind* (New York: Arcade Publishing, 1998).

4. This story of life's increasing complexity is told superbly in Maynard Smith and Szathmáry, *The Origins of Life;* for a sense of how modern biologists argue about complexity, see Roger Lewin, *Complexity: Life on the Edge of Chaos* (London: Phoenix, 1993), chap. 7.

5. Maynard Smith and Szathmáry, *The Origins of Life*, p. 62.

6. For a more detailed discussion of these finds of bacteria by one of the researchers who worked most closely on them, see Malcolm Walter, *The Search for Life on Mars* (Sydney: Allen and Unwin, 1999), chap. 3. There remains some possibility that such evidence is flawed; if so, the earliest certain evidence for the existence of life dates to less than 2 billion years ago.

7. Paul Davies, *The Fifth Miracle: The Search for the Origin of Life* (Harmondsworth: Penguin, 1999), particularly chap. 10.

8. The following account is based on John Snyder and C. Leland Rodgers, *Biology*, 3rd ed. (New York: Barron's, 1995), chaps. 5 and 6.

9. Margulis and Sagan, *Microcosmos*, p. 97.

10. Margulis and Sagan, *Microcosmos*, p. 88.

11. Margulis and Sagan, *Microcosmos*, p. 93.

12. The figures on oxygen are given in Armand Delsemme, *Our Cosmic Origins: From the Big Bang to the Emergence of Life and Intelligence* (Cambridge: Cambridge University Press, 1998), p. 168; and see graph, p. 170.

13. Margulis and Sagan, *Microcosmos*, chap. 6.

14. J. E. Lovelock, *Gaia: A New Look at Life on Earth* (1979; reprint, Oxford: Oxford University Press, 1987), p. 69.

15. Margulis and Sagan, *Microcosmos*, p. 93.

16. Margulis and Sagan, *Microcosmos*, p. 114.

17. Lynn Margulis and Dorion Sagan, *What Is Life?* (Berkeley: University of California Press, 1995), p. 73.

18. See Maynard Smith and Szathmáry, *The Origins of Life*, chap. 6.

19. Margulis and Sagan, *Microcosmos*, p. 115.

20. Margulis and Sagan, *Microcosmos*, p. 119; and on the delicate balance between cooperation and competition in evolution, see pp. 123–25.

21. Margulis and Sagan, *Microcosmos*, p. 142.

22. Maynard Smith and Szathmáry, *The Origins of Life*, chap. 7.

23. Richard A. Fortey, *Life: An Unauthorised Biography: A Natural His-*

tory of the First Four Thousand Million Years of Life on Earth (London: Flamingo, 1998), p. 89.

24. Reeves, de Rosnay, Coppens, and Simonnet, *Origins,* p. 110 (section by de Rosnay).

25. Maynard Smith and Szathmáry, *The Origins of Life,* pp. 125–29.

26. Maynard Smith and Szathmáry, *The Origins of Life,* p. 28.

27. These examples are taken from Stuart Kauffman, *At Home in the Universe: The Search for Laws of Complexity* (London: Viking, 1995), p. 109, but with a reduced estimate of the number of genes in humans.

28. Ernst Mayr, *The Growth of Biological Thought* (Cambridge, Mass.: Harvard University Press, 1982), p. 273; quoted in Tim Megarry, *Society in Prehistory: The Origins of Human Culture* (Basingstoke: Macmillan, 1995), p. 19.

29. This categorizing of human beings derives from the modern classification described in Roger Lewin, *Human Evolution: An Illustrated Introduction,* 4th ed. (Oxford: Blackwell, 1999), p. 43.

30. Stephen Jay Gould, *Wonderful Life: The Burgess Shale and the Nature of History* (London: Hutchinson, 1989).

31. Margulis and Sagan, *Microcosmos,* p. 187.

32. Margulis and Sagan, *Microcosmos,* pp. 174–75.

33. Nicholas Humphrey, *A History of the Mind* (London: Chatto and Windus, 1992), p. 97; Humphrey adds (p. 195) that consciousness in this simple sense may have appeared first in the more complex brains of mammals and birds, but there is a case for detecting it among much simpler organisms.

34. Terrence W. Deacon, *The Symbolic Species: The Co-evolution of Language and the Brain* (Harmondsworth: Penguin, 1997), p. 455.

35. Deacon, *The Symbolic Species,* p. 450.

36. See Luann Becker, "Repeated Blows," *Scientific American,* March 2002, pp. 76–83; a possible site for this asteroid's impact has now turned up at Bedout, in the sea just off northwestern Australia.

37. Dinosaurs have gripped the human imagination with great power. To give just one striking illustration, it seems likely that ancient Inner Asian legends about huge birdlike dragons or "griffins," which guarded hoards of gold, were based on fossils of dinosaurs such as protoceratops, which have often been found in the Tien Shan mountains of Xinjiang, in China, near deposits of gold. See Jeannine Davis-Kimball, with Mona Behan, *Warrior Women: An Archaeologist's Search for History's Hidden Heroines* (New York: Warner, 2002), chap. 6.

38. There is a vivid account of this asteroid collision in Tim Flannery, *The Eternal Frontier: An Ecological History of North America and Its Peoples* (New York: Atlantic Monthly Press, 2001), chap. 1.

39. On the vexed question of whether evolution is biased toward increasing complexity, see the good, brief discussion in Davies, *The Fifth Miracle,* pp. 219–25.

40. Alvares, quoted in Jeanne M. Sept and George E. Brooks, "Reports of Chimpanzee Natural History, Including Tool Use, in Sixteenth- and Seventeenth-Century Sierra Leone," *International Journal of Primatology* 15, no. 6

(December 1994): 872 (bracketed addition theirs); my thanks to George Brooks for giving me an offprint of this article.

41. James Lovelock, *The Ages of Gaia: A Biography of Our Living Earth* (Oxford: Oxford University Press, 1988), p. 19.

42. The term *biosphere* was created, by analogy with terms such as *lithosphere* or *atmosphere* (the regions of rocks or air), by an Austrian geologist, Edward Suess (1831–1914), and popularized by the Russian Vladimir Vernadsky (1863–1945), who regarded it as the sphere of all "living matter" (Margulis and Sagan, *What Is Life?*, pp. 48 ff.).

43. Lynn Hunt, "Send in the Clouds," *New Scientist*, 30 May 1998, pp. 28–33; for more on the Gaia hypothesis, see the writings of James Lovelock.

44. Margulis and Sagan, *Microcosmos*, p. 91.

45. Lewin, *Human Evolution*, p. 27. It is also possible that these major extinction events, like that of the late Cretaceous, were caused by asteroid impacts.

46. Lewin, *Human Evolution*, pp. 22–24.

47. Lewin, *Human Evolution*, p. 22.

48. How this pattern worked in the lands of Australasia is described vividly in Tim Flannery, *The Future Eaters: An Ecological History of the Australasian Lands and People* (Chatswood, N.S.W.: Reed, 1995), passim; for a summary, see p. 344.

49. "Leibig's Law of the Minimum . . . states that populations will be limited by critical resources (e.g., water) that are in shortest supply" (Allen W. Johnson and Timothy Earle, *The Evolution of Human Societies*, 2nd ed. [Stanford: Stanford University Press, 2000], pp. 14–15).

CHAPTER 6. THE EVOLUTION OF HUMANS

1. See, for example, John Maynard Smith and Eörs Szathmáry, *The Origins of Life: From the Birth of Life to the Origins of Language* (Oxford: Oxford University Press, 1999).

2. The best short history of human energy use is Vaclav Smil, *Energy in World History* (Boulder, Colo.: Westview Press, 1994).

3. Paul Ehrlich, *The Machinery of Nature* (New York: Simon and Schuster, 1986), p. 287; the figures for NPP appropriated by humans on land are taken from I. G. Simmons, *Changing the Face of the Earth: Culture, Environment, History*, 2nd ed. (Oxford: Blackwell, 1996), p. 361, which are in turn adapted from J. M. Diamond, "Human Use of World Resources," *Nature*, 6 August 1987, pp. 479–80.

4. *World Resources, 2000–2001: People and Ecosystems: The Fraying Web of Life* (Washington, D.C.: World Resources Institute, 2000), pp. 246, 248.

5. See, for example, Richard Leakey and Roger Lewin, *The Sixth Extinction: Patterns of Life and the Future of Humankind* (New York: Doubleday, 1995).

6. A. J. McMichael, *Planetary Overload: Global Environmental Change and*

the Health of the Human Species (Cambridge: Cambridge University Press, 1993), p. 33.

7. Maynard Smith and Szathmáry offer the following description of new forms of complexity in the biological realm: "Entities that were capable of independent replication before the transition could afterwards replicate only as part of a larger whole" (*The Origins of Life*, p. 19).

8. Craig Stanford, *The Hunting Apes: Meat Eating and the Origins of Human Behavior* (Princeton: Princeton University Press, 1999), pp. 28–29.

9. John A. Mears, "Agricultural Origins in Global Perspective," in *Agricultural and Pastoral Societies in Ancient and Classical History*, ed. Michael Adas (Philadelphia: Temple University Press, 2001), p. 65. For a general discussion of the importance in human history of non-zero-sum games (i.e., games in which sharing is more fruitful than competing), see Robert Wright, *Nonzero: The Logic of Human Destiny* (New York: Random House, 2000).

10. McMichael, *Planetary Overload*, p. 34.

11. Derek Bickerton, *Language and Species* (Chicago: University of Chicago Press, 1990), p. 157; cited from William H. Calvin, *How Brains Think: Evolving Intelligence, Then and Now* (London: Phoenix, 1998), p. 82.

12. Terrence W. Deacon, *The Symbolic Species: The Co-evolution of Language and the Brain* (Harmondsworth: Penguin, 1997), p. 397.

13. Chris Stringer and Robin McKie, *African Exodus* (London: Cape, 1996), p. 150.

14. The seminal essay by Vincent Sarich and Alan Wilson was titled, with the lack of poetry common to many great scientific articles, "Immunological Time Scale for Hominid Evolution"; it was published in *Science*, 1 December 1967, pp. 1200–1203.

15. On the limitations of "molecular systematics," see Roger Lewin, *Human Evolution: An Illustrated Introduction*, 4th ed. (Oxford: Blackwell, 1999), pp. 41–45.

16. Recent finds of earlier hominine remains mean that paleontologists may in the future be able to speak of that common ancestor with more authority. Meanwhile there is, of course, plenty of speculation; see Lewin, *Human Evolution*, pp. 84–85.

17. On environment and human evolution, see the brief summary in Calvin, *How Brains Think*, pp. 69–81, and William H. Calvin, *The Ascent of Mind: Ice Age Climates and the Evolution of Intelligence* (New York: Bantam, 1991).

18. Two recent books by Craig Stanford, *The Hunting Apes* and *Significant Others: The Ape-Human Continuum and the Quest for Human Nature* (New York: Basic Books, 2001), give a fascinating, if controversial, overview of recent research in this field.

19. I use the term *foragers* rather than the more common expression, *hunters and gatherers,* because it has become increasingly apparent that in such societies the gathering of plant foods is normally more important, at least from a dietary point of view, than the hunting of meat. The terminology was proposed by Richard Lee in his influential study of the !Kung San, *The !Kung San:*

Men, Women, and Work in a Foraging Society (Cambridge: Cambridge University Press, 1979); for an introduction to foraging technologies, see the account in Allen W. Johnson and Timothy Earle, *The Evolution of Human Societies*, 2nd ed. (Stanford: Stanford University Press, 2000), chap. 3.

20. See Robert Foley, "In the Shadow of the Modern Synthesis? Alternative Perspectives on the Last Fifty Years of Paleoanthropology," *Evolutionary Anthropology* 10, no. 1 (2001): 5–15, for a survey of hominine history that focuses on such adaptive radiations.

21. See Lewin, *Human Evolution*, chap. 10.

22. Lewin, *Human Evolution*, p. 55.

23. Ann Gibbons, "In Search of the First Hominids," *Science*, 15 February 2002, pp. 1214–19; the team's less striking name for the species is *Orrorin tugenensis*.

24. Yohannes Halle-Selassie, "Late Miocene Hominids from the Middle Awash, Ethiopia," *Nature*, 12 July 2001, pp. 178–81.

25. It is possible that bipedalism is not confined to hominines; recently, a 9-million-year-old fossil of an apelike creature that seems to have been bipedal has been discovered on a Mediterranean island (Stanford, *The Hunting Apes*, p. 220).

26. See Lewin, "Origin of Bipedalism," chap. 17 in *Human Evolution*.

27. Hubert Reeves, Joël de Rosnay, Yves Coppens, and Dominique Simonnet, *Origins: Cosmos, Earth and Mankind* (New York: Arcade Publishing, 1998), pp. 152–56 (section by Coppens); Lewin, *Human Evolution*, pp. 108–9.

28. Reeves, de Rosnay, Coppens, and Simonnet, *Origins*, p. 156 (section by Coppens).

29. In this and the next two chapters, dates will be given, as archaeologists prefer, "before the present." Strictly speaking, the "present" used as a reference point in radiometric dating techniques is the year 1950, but for most purposes this discrepancy can be ignored.

30. For a fascinating, if controversial, account of the evolution of early hominines by the discoverer of Lucy, see Donald Johanson and James Shreeve, *Lucy's Child: The Discovery of a Human Ancestor* (Harmondsworth: Penguin, 1989).

31. See Stanford, *The Hunting Apes*.

32. On the significance of and evidence for sexual dimorphism among hominines, see Walter Leutenegger, "Sexual Dimorphism: Comparative and Evolutionary Perspectives," in *The Illustrated History of Humankind*, ed. Göran Burenhult, vol. 1, *The First Humans: Human Origins and History to 10,000 BC* (San Francisco: HarperSanFrancisco, 1993), p. 41.

33. Ian Tattersall, *Becoming Human: Evolution and Human Uniqueness* (New York: Harcourt Brace, 1998), pp. 133–34.

34. Recently, it has been argued that there were two species of *habilis: Homo rudolfensis*, which was bigger-brained and generally more heavily built, and *H. habilis*, which was smaller but had a more modern jaw and teeth (Lewin, *Human Evolution*, p. 124).

35. In *The Hunting Apes,* Craig Stanford has proposed a modified version of the "man the hunter" hypothesis by exploring the nutritional and social importance of meat in primate societies.

36. On hunting by chimps, see Stanford, *The Hunting Apes.*

37. It is becoming common to refer to African examples of these species as *ergaster* and non-African species as *erectus;* I will refer to *erectus* only when specifically discussing species that lived outside of Africa.

38. There is a good, short account of *Homo ergaster* in Paul Ehrlich, *Human Natures: Genes, Cultures, and the Human Prospect* (Washington, D.C.: Island Press, 2000), pp. 92–96.

39. Johan Goudsblom has argued that the use of fire marks a fundamental transition in human history; see Goudsblom, *Fire and Civilization* (Harmondsworth: Allen Lane, 1992).

40. Deacon, *The Symbolic Species,* p. 358.

41. Steven Mithen, *The Prehistory of the Mind: A Search for the Origins of Art, Religion, and Science* (London: Thames and Hudson, 1996), pp. 179 ff.

42. The significance of the absence of *erectus* from northern Eurasia is explored in David Christian, *A History of Russia, Central Asia, and Mongolia,* vol. 1, *Inner Eurasia from Prehistory to the Mongol Empire* (Oxford: Blackwell, 1998), chap. 2.

43. Edward O. Wilson, *Consilience: The Unity of Knowledge* (London: Abacus, 1998), p. 107; chap. 6 of this book is a superb brief introduction to modern understandings of the human brain. The other turning points listed by Wilson are the origins of life, the appearance of eukaryotic cells, and the appearance of multicellular organisms.

44. Robert Lewin, *Complexity: Life on the Edge of Chaos* (Phoenix: London, 1993), p. 163.

45. Such a feedback loop is argued for in work by Nicholas Humphrey, summarized in Calvin, *How Brains Think,* pp. 66–68.

46. See Jared Diamond, *Why Is Sex Fun? The Evolution of Human Sexuality* (London: Weidenfeld and Nicolson, 1997), and Donna J. Haraway, *Simians, Cyborgs, and Women: The Reinvention of Nature* (New York: Routledge, 1991), p. 107, on attempts to explain the significance of the lack of a clear estrous cycle in modern humans.

47. Robert Foley, *Humans before Humanity* (Oxford: Blackwell, 1995), pp. 165–71.

48. Ehrlich, *Human Natures,* p. 96.

CHAPTER 7. THE BEGINNINGS OF HUMAN HISTORY

1. Henry Plotkin, *Evolution in Mind: An Introduction to Evolutionary Psychology* (London: Penguin, 1997), p. 248.

2. There is a good, general account of such theories of language acquisition in Steven Mithen, *The Prehistory of the Mind* (London: Thames and Hudson, 1996), whose argument is summarized briefly in John Maynard Smith and Eörs

Szathmáry, *The Origins of Life: From the Birth of Life to the Origins of Language* (Oxford: Oxford University Press, 1999), pp. 143–45. On modularity in language, see Steven Pinker, *The Language Instinct: The New Science of Language and Mind* (New York: Penguin, 1994).

3. Mithen, *The Prehistory of the Mind.*

4. See Terrence W. Deacon, *The Symbolic Species: The Co-evolution of Language and the Brain* (Harmondsworth: Penguin, 1997), esp. chap. 10; Steven Pinker has argued that if there are indeed distinct mental modules or "organs," they probably look more like roadkill than like the organs we are more familiar with, such as hearts or lungs (*How the Mind Works* [New York: W. W. Norton, 1997], p. 30).

5. "[I]cons are mediated by a similarity between sign and object, indices are mediated by some physical or temporal connection between sign and object, and symbols are mediated by some formal or merely agreed upon link irrespective of any physical characteristics of either sign or object" (Deacon, *The Symbolic Species*, p. 70). Quotations from this book are hereafter cited parenthetically in the text.

6. Deacon, *The Symbolic Species*, pp. 322–24, 345.

7. Deacon, *The Symbolic Species*, pp. 84–92.

8. On the impact of language on the sense of time, see John McCrone, *The Ape That Spoke* (Basingstoke: Macmillan, 1990), and *How the Brain Works: A Beginner's Guide to the Mind and Consciousness* (London: Dorling Kindersley, 2002), particularly pp. 56–58.

9. Deacon, *The Symbolic Species*, pp. 310–18.

10. Deacon, *The Symbolic Species*, pp. 340, 353.

11. A useful, if hostile, summary, is given in Chris Stringer and Robin McKie, *African Exodus* (London: Cape, 1996), pp. 48 ff.; see also Alan G. Thorne and Milford H. Wolpoff, "The Multiregional Evolution of Humans," *Scientific American*, April 1992, pp. 28–33.

12. For a discussion of some of the genetic evidence for this model, see Luigi Luca Cavalli-Sforza and Francesco Cavalli-Sforza, *The Great Human Diasporas: The History of Diversity and Evolution*, trans. Sarah Thorne (Reading, Mass.: Addison-Wesley, 1995).

13. Sally McBrearty and Alison S. Brooks, "The Revolution That Wasn't: A New Interpretation of the Origin of Modern Human Behavior," *Journal of Human Evolution* 39 (2000): 453–563. Quotations from this article are hereafter cited parenthetically in the text.

14. McBrearty and Brooks, "The Revolution That Wasn't," p. 497.

15. McBrearty and Brooks, "The Revolution That Wasn't," pp. 493–94.

16. Robert Wright, *Nonzero: The Logic of Human Destiny* (New York: Random House, 2000), p. 51.

17. The metaphor of "networks" has not been used as much as it might have been in world history. William McNeill's *The Rise of the West: A History of the Human Community* (Chicago: University of Chicago Press, 1963) argued that interactions between different human communities were the major driving force

of change in world history. In his most recent work, written with John McNeill, the metaphor of "webs of interaction" is developed with great virtuosity. See J. R. McNeill and William H. McNeill, *The Human Web: A Bird's-Eye View of World History* (New York: W. W. Norton, 2003).

18. "The American architect R. Buckminster Fuller (1895–1983) applied 'synergy' (from Greek *synergos*, working together) to describe entities that behave as more than the sum of their parts," an elegant definition of the term from Lynn Margulis and Dorion Sagan, *What Is Life?* (Berkeley: University of California Press, 1995), p. 8.

19. Wright, *Nonzero*, p. 52; chap. 4 of his book, "The Invisible Brain," discusses the general rule that increased population density tends to stimulate innovation.

20. For a brief survey of the decline in the number of languages in the course of world history, see Frances Karttunen and Alfred W. Crosby, "Language Death, Language Genesis, and World History," *Journal of World History* 6, no. 2 (fall 1995): 157–74; for the figure for California in 1750, see p. 159; for Papua New Guinea today, see p. 173.

21. For an example of some of the ways in which studies of modern foragers can help us think about Paleolithic lifeways, and of some of the limitations of such analogies, see Allen W. Johnson and Timothy Earle, *The Evolution of Human Societies*, 2nd ed. (Stanford: Stanford University Press, 2000), particularly chaps. 2 and 3.

22. Colin Renfrew, *Archaeology and Language: The Puzzle of Indo-European Origins* (Harmondsworth: Penguin, 1989), p. 125, gives the lower estimates of population density; Massimo Livi-Bacci, *A Concise History of World Population*, trans. Carl Ipsen (Oxford: Blackwell, 1992), pp. 26–27, gives the higher.

23. Marshall Sahlins, "The Original Affluent Society," in *Stone Age Economics* (London: Tavistock, 1972), pp. 1–39; quotation, p. 1.

24. Sahlins, "The Original Affluent Society," p. 16.

25. Johnson and Earle, *The Evolution of Human Societies*, p. 14.

26. Eric R. Wolf's models of "kin-ordered," "tribute-taking," and "capitalist" societies are described in chap. 2 of *Europe and the People without History* (Berkeley: University of California Press, Berkeley, 1982); for another account of foraging lifeways, which also focuses on kinship structures, see Johnson and Earle, "The Family-Level Group," part 1 of *The Evolution of Human Societies*.

27. The economic rules of reciprocity have been explored in the work of Karl Polanyi and his followers; see, for example, Karl Polanyi, Conrad M. Arensberg, and Harry W. Pearson, eds., *Trade and Market in the Early Empires: Economies in History and Theory* (Glencoe, Ill.: Free Press, 1957), or, for an introduction to Polanyi's ideas, see S. C. Humphrey, "History, Economics, and Anthropology: The Work of Karl Polanyi," *History and Theory* 8 (1969): 165–212.

28. Quoted in Wright, *Nonzero*, p. 20.

29. Richard Lee, *The Dobe !Kung* (New York: Holt, Rinehart, and Winston, 1984), p. 96; quoted in Johnson and Earle, *The Evolution of Human Societies*, p. 75.

30. Irenäus Eibl-Eibesfeldt, "Aggression and War: Are They Part of Being Human?" in *The Illustrated History of Humankind*, ed. Göran Burenhult, vol. 1, *The First Humans: Human Origins and History to 10,000 BC* (St. Lucia: University of Queensland Press, 1993), pp. 26–27.

31. Christopher Chase-Dunn and Thomas D. Hall, *Rise and Demise: Comparing World-Systems* (Boulder, Colo.: Westview Press, 1997), p. 138.

32. There is a superb discussion of the dominant role of a sense of place in Aboriginal religious and cosmological thought in Tony Swain, *A Place for Strangers: Towards a History of Australian Aboriginal Being* (Cambridge: Cambridge University Press, 1993). See also Deborah Bird Rose, *Nourishing Terrains: Australian Aboriginal Views of Landscape and Wilderness* (Canberra: Australian Heritage Commission, 1996). My thanks to Frank Clarke for both these references.

33. Hobbles Danaiyarri, quoted in Rose, *Nourishing Terrains*, p. 9.

34. Some idea of how such societies perceived change may be suggested in Mircea Eliade's difficult but important work, *The Myth of the Eternal Return, or, Cosmos and History*, trans. Willard R. Trask (New York: Harper, 1959).

35. Much of this section draws on Stringer and McKie, *African Exodus*.

36. Clive Gamble, *Timewalkers: The Prehistory of Global Colonization* (Harmondsworth: Penguin, 1995), p. 215. On the redating of the Lake Mungo skeleton by Alan Thorne, see Alan Thorne et al., "Australia's Oldest Human Remains: Age of the Lake Mungo 3 Skeleton," *Journal of Human Evolution* 36 (June 1999): 591–612; Richard G. Roberts, "Thermoluminescence Dating," in Burenhult, ed., *The First Humans*, gives a picture of this rock shelter (p. 153) and a grindstone from it (p. 156). But see also John Mulvaney and Johan Kamminga, *Prehistory of Australia* (Sydney: Allen and Unwin, 1999), pp. 130–46; they are skeptical of dates earlier than about 50,000 BP (on Malakunanja, see pp. 140–42).

37. Cavalli-Sforza and Cavalli-Sforza, *The Great Human Diasporas*, p. 123.

38. Paul Ehrlich, *Human Natures: Genes, Cultures, and the Human Prospect* (Washington, D.C.: Island Press, 2000), p. 166.

39. Rhys Jones, "Fire Stick Farming," *Australian Natural History*, September 1969, pp. 224–28.

40. Leigh Dayton, "Mass Extinctions Pinned on Ice Age Hunters," *Science*, 8 June 2001, p. 1819.

41. Stephen Pyne, *Fire in America: A Cultural History of Wildland and Rural Fire* (Princeton: Princeton University Press, 1982), p. 3.

42. Neil Roberts, *The Holocene: An Environmental History*, 2nd ed. (Oxford: Blackwell, 1998), p. 112; he cites P. Mellars, "Fire Ecology, Animal Populations, and Man: A Study of Some Ecological Relationships in Prehistory," *Proceedings of the Prehistoric Society* 42 (1975): 15–45. On fire stick farming in Australia, see Tim Flannery, *The Future Eaters: An Ecological History of the Australasian Lands and People* (Chatswood, N.S.W.: Reed, 1995), pp. 217–36, which suggests, controversially, that the increased role of fire was an indirect consequence of the extinction of large herbivores that had previously consumed large amounts of dead plant matter.

43. Andrew Goudie, *The Human Impact on the Natural Environment*, 5th ed. (Oxford: Blackwell, 2000), pp. 38–41.

44. I. G. Simmons, *Environmental History: A Concise Introduction* (Oxford: Blackwell, 1993), p. 74 .

45. Johan Goudsblom, *Fire and Civilization* (Harmondsworth: Allen Lane, 1992).

46. Peter Bogucki, *The Origins of Human Society* (Oxford: Blackwell, 1999), p. 92; and see Elizabeth Wayland Barber, *Women's Work: The First 20,000 Years: Women, Cloth, and Society in Early Times* (New York: W. W. Norton, 1994), chap. 2.

47. Richard G. Klein, *Ice Age Hunters of the Ukraine* (Chicago: University of Chicago Press, 1973), p. 110; and see Olga Soffer, "Sungir: A Stone Age Burial Site," in Burenhult, ed., *The First Humans*, pp. 138–39.

48. Olga Soffer, "The Middle to Upper Palaeolithic Transition on the Russian Plain," in *The Human Revolution*, ed. Paul Mellars and Chris Stringer (Edinburgh: Edinburgh University Press, 1989), 1:736.

49. Z. A. Abramovo, "Two Models of Cultural Adaptation," *Antiquity* 63 (1989): 789; and see Roland Fletcher, "Mammoth Bone Huts," in Burenhult, ed., *The First Humans*, pp. 134–35.

50. Brian M. Fagan, *The Journey from Eden: The Peopling of Our World* (London: Thames and Hudson, 1990), p. 186; N. D. Praslov, "Late Palaeolithic Adaptations to the Natural Environment on the Russian Plain," *Antiquity* 63 (1989): 786.

51. Olga Soffer, "Patterns of Intensification as Seen from the Upper Paleolithic of the Central Russian Plain," in *Prehistoric Hunter-Gatherers: The Emergence of Cultural Complexity*, ed. T. Douglas Price and James A. Brown (Orlando, Fla.: Academic Press, 1985), p. 243, and Soffer, "Storage, Sedentism, and the Eurasian Palaeolithic Record," *Antiquity* 63 (1989): 726.

52. Timothy Champion et al., *Prehistoric Europe* (London: Academic Press, 1984), p. 81. Similar figures have been found in Siberia, at the sites of Mal'ta and Buret'; see A. P. Okladnikov, "Inner Asia at the Dawn of History," in *Cambridge History of Early Inner Asia*, ed. Denis Sinor (Cambridge: Cambridge University Press, 1990), p. 56. Clive Gamble, *The Palaeolithic Settlement of Europe* (Cambridge: Cambridge University Press, 1986), p. 326, shows the distribution of finds of Venus figurines; and see Chris Stringer and Clive Gamble, *In Search of the Neanderthals: Solving the Puzzle of Human Origins* (London: Thames and Hudson, 1993), p. 210.

53. Mulvaney and Kamminga, *Prehistory of Australia*, pp. 28–31.

54. Stringer and McKie, *African Exodus*, p. 150.

55. Göran Burenhult, "The Rise of Art," in Burenhult, ed., *The First Humans*, p. 100.

56. See the figures cited in Stringer and McKie, *African Exodus*, p. 150 (for 100,000 BP), and Livi-Bacci, *A Concise History of World Population*, p. 31; and see the good survey in Thomas M. Whitmore et al., "Long-Term Population Change," in *The Earth as Transformed by Human Action: Global and Regional*

Changes in the Biosphere over the Past 300 Years, ed. B. L. Turner II et al. (Cambridge: Cambridge University Press, 1990), pp. 25–39.

57. A fascinating, if controversial, book on the impact of fire stick farming in Australia and New Zealand is Flannery, *The Future Eaters.*

58. Recent fossil finds show that a dwarf species of mammoth survived on isolated Wrangel Island, in the Arctic Ocean, until perhaps 4,500 years ago (Roberts, *The Holocene,* p. 86).

59. Dayton, "Mass Extinctions Pinned on Ice Age Hunters," provides evidence that humans were using fire stick farming techniques as early as 45,000 years ago; and see Tim Flannery, *The Eternal Frontier: An Ecological History of North America and Its Peoples* (New York: Atlantic Monthly Press, 2001), pp. 189–91.

60. Alfred Wallace, quoted in Flannery, *The Future Eaters,* p. 181.

61. See Richard G. Roberts, Timothy F. Flannery, Linda K. Ayliffe, Hiroyuki Yoshida, et al., "New Ages for the Last Australian Megafauna: Continent-wide Extinction about 46,000 Years Ago," *Science,* 8 June 2001, pp. 1888–92, and John Alroy, "A Multispecies Overkill Simulation of the End-Pleistocene Megafaunal Mass Extinction," *Science,* 8 June 2001, pp. 1893–96. These two articles, by dating the extinction events more precisely and using more sophisticated computer models of likely human impacts on Paleolithic megafauna, offer powerful support for the notion that early humans were largely responsible; in Australia, it now seems, all land species larger than 100 kg disappeared about 46,000 years ago, soon after the best dates for the arrival of humans.

62. Roberts, *The Holocene,* p. 83, citing Paul S. Martin and Richard G. Klein, eds., *Quaternary Extinctions* (Tucson: University of Arizona Press, 1984).

63. See Flannery, *The Future Eaters,* pp. 164–207, for a good recent discussion of this debate, which argues forcefully for the role of human action, and Jared Diamond, *Guns, Germs, and Steel: The Fates of Human Societies* (London: Vintage, 1998), pp. 46–47, who argues strongly for the critical importance of these extinctions. See also Mulvaney and Kamminga, *Prehistory of Australia,* pp. 124–29, for an account of the Australian evidence that is skeptical of the human overkill thesis.

64. Stringer and McKie, *African Exodus,* pp. 101–4.

65. Burenhult, "The Rise of Art," p. 104.

66. On the possible survival of hominines distinct from modern humans in Java as late as 53,000–27,000 BP, see Richard G. Klein, *The Human Career: Human Biological and Cultural Origins,* 2nd ed. (Chicago: University of Chicago Press, 1999), p. 395; on the survival of Neanderthals in western Europe to perhaps 30,000 BP, see pp. 477 ff.

CHAPTER 8. INTENSIFICATION AND THE ORIGINS OF AGRICULTURE

Chapter epigraph: Lester R. Brown, *Eco-Economy: Building an Economy for the Earth* (New York: W. W. Norton, 2001), p. 93.

1. See Colin Renfrew, *Archaeology and Language: The Puzzle of Indo-*

European Origins (Harmondsworth: Penguin, 1989), p. 125, for the lower estimate, and Massimo Livi-Bacci, *A Concise History of World Population*, trans. Carl Ipsen (Oxford: Blackwell, 1992), pp. 26–27, for the higher.

2. J. R. Biraben, "Essai sur l'évolution du nombre des hommes," *Population* 34 (1979): 23.

3. The description in this paragraph is based on Neil Roberts, *The Holocene: An Environmental History*, 2nd ed. (Oxford: Blackwell, 1998), chap. 4.

4. Robert Wright, *Nonzero: The Logic of Human Destiny* (New York: Random House, 2000), p. 29; Wright also suggests, quite rightly (p. 52), that Tasmania could also be treated as an entirely distinct world.

5. On Southeast Asian influences on Australia, see Josephine Flood, *Archaeology of the Dreamtime* (Sydney: Collins, 1983), pp. 222–93.

6. Of the Americas, John Kicza comments: "No compelling evidence indicates any contact other than incidental with societies from outside the Americas until Columbus's 1492 voyage" ("The Peoples and Civilizations of the Americas before Contact," in *Agricultural and Pastoral Societies in Ancient and Classical History*, ed. Michael Adas [Philadelphia: Temple University Press, 2001], p. 183).

7. Flood, *Archaeology of the Dreamtime*, pp. 236–37.

8. See Ben Finney, "The Other One-Third of the Globe," *Journal of World History* 5, no. 2 (fall 1994): 273–98; J. R. McNeill, "Of Rats and Men: A Synoptic Environmental History of the Island Pacific," *Journal of World History* 5, no. 2 (fall 1994): 299–349; and Tim Flannery, *The Future Eaters: An Ecological History of the Australasian Lands and People* (Chatswood, N.S.W.: Reed, 1995).

9. Robert J. Wenke, *Patterns in Prehistory: Humankind's First Three Million Years*, 3rd ed. (New York: Oxford University Press, 1990), p. 208; and see Joseph Greenberg and Merritt Ruhlen, "Linguistic Origins of Native Americans," *Scientific American*, November 1992, p. 94.

10. The comparison between these worlds has been explored with great virtuosity by Jared Diamond in *Guns, Germs, and Steel: The Fates of Human Societies* (London: Vintage, 1998); the argument in this section relies at many points on both Diamond's questions and his answers.

11. Jared Diamond, *The Rise and Fall of the Third Chimpanzee* (London: Vintage, 1991), p. 165.

12. "Several species of domestic animals have smaller brains and less developed sense organs than their wild ancestors, because they no longer need the bigger brains and more developed sense organs on which their ancestors depended to escape from wild predators" (Diamond, *Guns, Germs, and Steel*, p. 159).

13. Bruce D. Smith, *The Emergence of Agriculture* (New York: Scientific American Library, 1995), p. 18.

14. There is a good discussion of the impact of domestication on plants in Diamond, *Guns, Germs, and Steel*, chap. 7.

15. Flood, *Archaeology of the Dreamtime*, p. 219.

16. The earliest remains of dogs—that is, of domesticated wolves—have been found in Iraq, and dated to ca. 10,000 to 12,000 BCE; see Charles B. Heiser, *Seed to Civilization: The Story of Food*, new ed. (Cambridge, Mass.: Harvard University Press, 1990), p. 37.

17. Smith, *Emergence of Agriculture*, pp. 67, 72.

18. Smith, *Emergence of Agriculture*, pp. 85–86, 61, 57, 64–65.

19. On pigs, see Clive Ponting, *A Green History of the World* (Harmondsworth: Penguin, 1992), p. 44; on cattle, see Heiser, *Seed to Civilization*, p. 43; Wenke, *Patterns in Prehistory*, p. 248.

20. Brian M. Fagan, *People of the Earth: An Introduction to World Prehistory*, 10th ed. (Upper Saddle River, N.J.: Prentice Hall, 2001), p. 244. On the last aurochs, see Heiser, *Seed to Civilization*, pp. 43–44.

21. Jared Diamond has also argued persuasively that the gap reflects a smaller number of really valuable potential plant domesticates as well; see *Guns, Germs, and Steel*, chaps. 8 and 9.

22. Smith, *Emergence of Agriculture*, pp. 159, 181, 197–98; Diamond, *Guns, Germs, and Steel*, pp. 150–51.

23. There is a good, if slighted dated, discussion of the problem of explaining the transition to agriculture in Mark Cohen, *The Food Crisis in Prehistory* (New Haven: Yale University Press, 1977), chap. 1.

24. Cohen, *The Food Crisis in Prehistory*, p. 5.

25. Marek Zvelebil, "Mesolithic Prelude and Neolithic Revolution," in *Hunters in Transition: Mesolithic Societies of Temperate Eurasia and Their Transition to Farming*, ed. Marek Zvelebil (Cambridge: Cambridge University Press, 1986), pp. 11–13.

26. Mark Cohen, *Health and the Rise of Civilization* (New Haven: Yale University Press, 1989), pp. 112–13.

27. Diamond, *Germs, Guns, and Steel*, pp. 206–10.

28. Cohen, *Health and the Rise of Civilization*, pp. 132, 139.

29. Cohen, *Health and the Rise of Civilization*, p. 139.

30. John H. Coatsworth, "Welfare," *American Historical Review* 101, no. 1 (February 1996): 2; my thanks to Tom Passananti for this reference.

31. There is a brief survey of different explanations for the origins of agriculture in Fagan, *People of the Earth*, pp. 232–35.

32. The explanation given in the text owes much to the models offered by Bruce Smith in *Emergence of Agriculture* and by Donald O. Henry in *From Foraging to Agriculture: The Levant at the End of the Ice Age* (Philadelphia: University of Pennsylvania Press, 1989).

33. Cohen, *The Food Crisis in Prehistory*, pp. 19 ff.

34. Henry, *From Foraging to Farming*, p. 231.

35. Flood, *Archaeology of the Dreamtime*, pp. 187–90.

36. Flood, *Archaeology of the Dreamtime*, p. 195. There is an interesting comparison of intensification in Australia and Papua New Guinea in chap. 15 of Diamond, *Guns, Germs, and Steel*.

37. Flood, *Archaeology of the Dreamtime*, p. 205.

38. Flood, *Archaeology of the Dreamtime*, pp. 204–7.

39. Flood, *Archaeology of the Dreamtime*, pp. 226–28.

40. Wenke, *Patterns in Prehistory*, pp. 254–56.

41. Fagan, *People of the Earth*, pp. 216, 218.

42. P. M. Dolukhanov, "The Late Mesolithic and the Transition to Food Production in Eastern Europe," in Zvelebil, ed., *Hunters in Transition*, p. 116.

43. Roland Oliver, *The African Experience*, 2nd ed. (Boulder, Colo.: Westview Press, 2000), p. 35.

44. Roberts, *The Holocene*, pp. 147–48.

45. On the way in which such practices preadapted human communities to agriculture, see David Rindos's marvelous, if difficult, study, *Origins of Agriculture: An Evolutionary Perspective* (New York: Academic Press, 1984).

46. Diamond argues for the crucial role played by the availability of such species in *Guns, Germs, and Steel*, chap. 8.

47. Diamond, *Guns, Germs and Steel*, pp. 134–38.

48. Cohen, *The Food Crisis in Prehistory*, p. 65. See also Ester Boserup, *The Conditions of Agricultural Growth: The Economics of Agrarian Change under Population Pressure* (Chicago: Aldine, 1965).

49. Cohen, *The Food Crisis in Prehistory*, p. 85.

50. Paul Bairoch, *Cities and Economic Development: From the Dawn of History to the Present*, trans. Christopher Braider (Chicago: University of Chicago Press, 1988), p. 7; citing Fekri A. Hassan, *Demographic Archaeology* (New York: Academic Press, 1981).

51. James Dawson, an 1881 account quoted in John Mulvaney and Johan Kamminga, *Prehistory of Australia* (Sydney: Allen and Unwin, 1999), p. 94.

52. Andrew Sherratt, "Reviving the Grand Narrative: Archaeology and Long-Term Change," *Journal of European Archaeology* 3, no. 1 (1995): 20–21.

53. Henry, *From Foraging to Farming*, pp. 49–51.

54. The pioneering role of women is an argument developed in Margaret Ehrenberg, *Women in Prehistory* (Norman: University of Oklahoma Press, 1989), pp. 80–85; see also Elizabeth Wayland Barber, *Women's Work: The First 20,000 Years: Women, Cloth, and Society in Early Times* (New York: W. W. Norton, 1994), chap. 3.

55. Fagan, *People of the Earth*, pp. 257–59.

56. Fagan, *People of the Earth*, p. 257.

57. See the discussion in Smith, *Emergence of Agriculture*, pp. 210–14.

58. Smith, *Emergence of Agriculture*, p. 213.

59. On semisedentary communities of the Americas, see Kicza, "Peoples and Civilizations of the Americas before Contact," particularly pp. 212–17.

60. Elman R. Service, *Primitive Social Organization: An Evolutionary Perspective*, 2nd ed. (New York: Random House, 1971), passim; see also the discussion of typologies of human societies in Allen W. Johnson and Timothy Earle, *The Evolution of Human Societies: From Foraging Group to Agrarian State*, 2nd ed. (Stanford: Stanford University Press, 2000), pp. 32–35.

61. Marija Gimbutas's ideas are summarized in *The Civilization of the God-*

dess: The World of Old Europe, ed. Joan Marler (San Francisco: Harper and Row, 1991).

62. I. G. Simmons, *Changing the Face of the Earth: Culture, Environment, History,* 2nd ed. (Oxford: Blackwell, 1996), p. 94.

63. Andrew Goudie, *The Human Impact on the Natural Environment,* 5th ed. (Oxford: Blackwell, 2000), p. 188. On average, under natural conditions, soils form at less than 0.1 mm/year, and ca. 0.05–2 mm is removed a year. Under cultivation, ca. 5–10 mm/year is removed; under pasture, ca. 1 mm/year. But where the soil is left bare, rates of loss can be as high as 25–100 mm/year. Thus human activity can rapidly destroy soils that have built up over thousands of years.

64. On the limited ecological impact of early farmers in Europe, see Roberts, *The Holocene,* pp. 154–58.

CHAPTER 9. FROM POWER OVER NATURE TO POWER OVER PEOPLE: CITIES, STATES, AND "CIVILIZATIONS"

1. Anthony Giddens refers to power over things and over people, respectively, as "allocative" and "authoritative" power (*The Nation-State and Violence,* vol. 2 of *A Contemporary Critique of Historical Materialism* [Cambridge: Polity Press, 1985], p. 7).

2. Marvin Harris, "The Origin of Pristine States," in *Cannibals and Kings,* ed. Marvin Harris (New York: Vintage, 1978), p. 102.

3. A good recent survey of models of the emergence of cities and states can be found in Allen W. Johnson and Timothy Earle, *The Evolution of Human Societies,* 2nd ed. (Stanford: Stanford University Press, 2000); the typology they propose for all preindustrial societies is summarized on pp. 32, 36. See also Brian M. Fagan, *People of the Earth: An Introduction to World Prehistory,* 10th ed. (Upper Saddle River, N.J.: Prentice Hall, 2001), pp. 368–85.

4. That population growth is the central factor in the evolution of more complex societies is the central argument of Johnson and Earle in *The Evolution of Human Societies;* e.g., "Although we will see that its precise role is hotly contested, population growth is undeniably central to the process of sociocultural evolution, because of its clear consequences for how people meet their basic needs. In any environment, population growth creates problems in technology, the social organization of production, and political regulation that must be solved. We will show how the solutions to these problems bring about the changes we know as sociocultural evolution" (p. 2).

5. Lynn Margulis and Dorion Sagan, *Microcosmos: Four Billion Years of Microbial Evolution* (London: Allen and Unwin, 1987), p. 130.

6. William H. McNeill, *The Pursuit of Power: Technology, Armed Force, and Society since A.D. 1000* (Oxford: Blackwell, 1982), p. vii.

7. Lewis Thomas, *Societies as Organisms* (London: Viking, 1974); quoted in C. Tickell, "The Human Species: A Suicidal Success?" in *The Human Impact Reader: Readings and Case Studies,* ed. Andrew Goudie (Oxford: Blackwell, 1997), p. 450.

8. Johnson and Earle define a slightly simpler feedback loop: "we identify the process of feedback between population and technology as the engine of the evolutionary process" (*Evolution of Human Societies*, p. 14). Robert Wright explores a similar feedback loop in *Nonzero: The Logic of Human Destiny* (New York: Random House, 2000), chap. 4, esp. p. 50.

9. Neil Roberts, *The Holocene: An Environmental History*, 2nd ed. (Oxford: Blackwell, 1998), p. 112.

10. Andrew Goudie, *The Human Impact on the Natural Environment*, 5th ed. (Oxford: Blackwell, 2000), p. 82; Goudie points out that modern experiments in Denmark used an unsharpened 4,000-year-old chert ax successfully to fell more than 100 trees.

11. Goudie, *The Human Impact*, p. 52.

12. Andrew Sherratt, "Plough and Pastoralism: Aspects of the Secondary Products Revolution," in *Patterns of the Past: Studies in Honour of David Clarke*, ed. Ian Hodder, Glynn Isaac, and Norman Hammond (Cambridge: Cambridge University Press, 1981), pp. 261–305; and see the updated version of his argument in "The Secondary Exploitation of Animals in the Old World (1983, revised)," in *Economy and Society in Prehistoric Europe: Changing Perspectives* (Princeton: Princeton University Press, 1997), pp. 199–228.

13. I. G. Simmons, *Changing the Face of the Earth: Culture, Environment, History*, 2nd ed. (Oxford: Blackwell, 1996), p. 94.

14. See David Christian, *A History of Russia, Central Asia, and Mongolia*, vol. 1, *Inner Eurasia from Prehistory to the Mongol Empire* (Oxford: Blackwell, 1998), esp. chap. 4, and Christian, "Silk Roads or Steppe Roads? The Silk Roads in World History," *Journal of World History* 11, no. 1 (spring 2000): 1–26.

15. Margaret Ehrenberg, *Women in Prehistory* (Norman: University of Oklahoma Press, 1989), p. 99; see pp. 99–107 for a general discussion of the links between the secondary products revolution and patriarchy. See also Elizabeth Wayland Barber, *Women's Work: The First 20,000 Years: Women, Cloth, and Society in Early Times* (New York: W. W. Norton, 1994), pp. 97–98.

16. The significance of this contrast between livestock-rich Afro-Eurasia and the livestock-poor Americas is explored well in Jared Diamond, *Guns, Germs, and Steel: The Fates of Human Societies* (London: Vintage, 1998); see in particular chap. 18.

17. John McNeill has pointed out that even as late as 1800 CE, ca. 70 percent of all energy was generated by human muscle power (J. R. McNeill, *Something New under the Sun: An Environmental History of the Twentieth-Century World* [New York: W. W. Norton, 2000], p. 11).

18. Robert J. Wenke, *Patterns in Prehistory: Humankind's First Three Million Years*, 3rd ed. (New York: Oxford University Press, 1990), p. 336.

19. John H. Coatsworth, "Welfare," *American Historical Review* 101, no. 1 (February 1996): 6–7.

20. Christian, *A History of Russia, Central Asia, and Mongolia*, 1:129–31.

21. Humans convert ca. 18 percent of their food energy into energy, horses

only 10 percent; so, to their owners, slaves were a peculiarly efficient way of storing energy (J. R. McNeill, *Something New under the Sun*, pp. 11–12).

22. Barber, *Women's Work*, pp. 29–33, develops the argument that in agrarian societies, women's roles were restricted primarily by the demands of child rearing.

23. This section borrows from ideas that I first heard in the elegant lectures on the sociology of power given at Macquarie University by Bob Norton in the early 1990s.

24. Michael Mann draws a similar distinction between "distributive" power (power of domination of A over B) and "collective" power (power based on cooperation); distributive power tends toward the coercive and the illegitimate, while collective tends toward the voluntary and the legitimate, but in reality the two overlap and their relationship is dialectical. See Mann, *The Sources of Social Power*, vol. 1, *A History of Power from the Beginning to A.D. 1760* (Cambridge: Cambridge University Press, 1986).

25. Mann makes the same point in slightly different terminology, arguing that "collective power antedated distributive power" (*The Sources of Social Power*, 1:53).

26. Harris, "The Origin of Pristine States," p. 106.

27. Malinowski, quoted in Harris, "The Origin of Pristine States," p. 109.

28. My account of the development of cities in Mesopotamia is based largely on Hans Jörg Nissen, *The Early History of the Ancient Near East, 9000–2000 B.C.*, trans. Elizabeth Lutzeier, with Kenneth J. Northcott (Chicago: University of Chicago Press, 1988); see also Susan Pollock, *Ancient Mesopotamia: The Eden That Never Was* (Cambridge: Cambridge University Press, 1999), and D. T. Potts, *Mesopotamian Civilization: The Material Foundations* (Ithaca, N.Y.: Cornell University Press, 1997).

29. Andrew Sherratt, "Reviving the Grand Narrative: Archaeology and Long-Term Change," *Journal of European Archaeology* 3, no. 1 (1995): 17.

30. Sherratt, "Reviving the Grand Narrative," p. 19: "the element of 'sparking across the gap' between resource-rich areas is essential for initiating a labor-intensive mode of production. . . . Textiles are particularly useful in this respect, and have almost always been associated with the development of urbanism." This observation about textiles holds later, with the Industrial Revolution, too.

31. Christopher Ehret, "Sudanic Civilization," in *Agricultural and Pastoral Societies in Ancient and Classical History*, ed. Michael Adas (Philadelphia: Temple University Press, 2001), pp. 244–45.

32. Nissen, *The Early History of the Ancient Near East*, pp. 55–61, 67–73.

33. Anthony Giddens, *A Contemporary Critique of Historical Materialism*, 2nd ed. (Basingstoke: Macmillan, 1995), p. 96, refers in this context to Lewis Mumford's notion of towns as containers and concentrators of power.

34. The distinction between city-states and territorial states can be observed in all regions of early state formation; it is developed in Bruce G. Trigger, *Early*

Civilizations: Ancient Egypt in Context (Cairo: American University in Cairo Press, 1993), pp. 8–14.

35. The state may not achieve the monopoly over legitimate control of the means of violence that Max Weber regarded as the essence of a state, but it certainly aims at it; Giddens defines a state as "a political organization [i.e., an organization that can exercise power] whose rule is territorially ordered and which is able to mobilize the means of violence to sustain that rule" (*The Nation-State and Violence*, p. 20).

36. Harris, "The Origin of Pristine States," pp. 113–15.

37. Valerie Hansen, *The Open Empire: A History of China to 1600* (New York: W. W. Norton, 2000), p. 35.

38. *The Russian Primary Chronicle: Laurentian Text*, trans. and ed. Samuel Hazzard Cross and Olgerd P. Sherbovitz-Wetzor (Cambridge, Mass.: Mediaeval Academy of America, 1953), p. 122 (year 994–96 CE).

39. Constantine Porphyrogenitus, *De Administrando Imperio*, ed. G. Moravcsik, trans. R. J. H. Jenkins, rev. ed. (Washington, D.C.: Dumbarton Oaks Center for Byzantine Studies, 1967), 1.63.

40. Charles Tilly, *Coercion, Capital, and European States, A.D. 990–1992*, rev. ed. (Cambridge, Mass.: Blackwell, 1992), pp. 1–2. I do not, however, accept Tilly's assertion that such entities can be found as early as Jericho or Çatal Hüyük.

41. J. R. McNeill, *Something New under the Sun*, p. 12.

42. "Military force, or the threat of its use, was normally a highly important basis of the traditional state because the state lacked the means to 'directly administrate' the regions subject to its domination" (Giddens, *Nation-State and Violence*, p. 58).

43. Giddens, *A Contemporary Critique of Historical Materialism*, p. 104.

44. Giddens has made this point about the weakness of traditional states forcefully; see, for example, *Nation-State and Violence*, p. 57, for a short summary of his claims about the limits of state power in what he calls "class-divided" societies; and see also his summary and generalization from the work of historians of modern France on levels of rural violence (p. 60).

45. Nissen, *Early History of the Ancient Near East*, p. 80.

46. Nissen, *Early History of the Ancient Near East*, pp. 83, 84 (picture of symbol).

47. Giddens stresses the link between writing and power, as well as the notion of writing as stored information; see, e.g., *A Contemporary Critique of Historical Materialism*, pp. 94–95. In *The Nation-State and Violence*, he writes: "Writing did not originate as an isomorphic representation of speech, but as a mode of administrative notation, used to keep records or tallies" (p. 41).

48. Hansen, *The Open Empire*, pp. 17–28.

49. Trigger, *Early Civilizations*, p. 44.

50. See Eric R. Wolf, *Europe and the People without History* (Berkeley: University of California Press, 1982), chap. 3.

51. Giddens, *A Contemporary Critique of Historical Materialism*, p. 112.

52. Wenke, *Patterns in Prehistory*, pp. 480–81, 534.

53. Michael D. Coe, *Mexico: From the Olmecs to the Aztecs*, 4th ed. (New York: Thames and Hudson, 1994), p. 71.

54. Coe, *Mexico*, pp. 75–76.

55. Fagan, *People of the Earth*, pp. 540–41.

CHAPTER 10. LONG TRENDS IN THE ERA OF AGRARIAN "CIVILIZATIONS"

1. Robert Wright, *Nonzero: The Logic of Human Destiny* (New York: Random House, 2000), p. 108.

2. For some recent discussions of such language dispersals in Eurasia, see Colin Renfrew, *Archaeology and Language: The Puzzle of Indo-European Origins* (Harmondsworth: Penguin, 1989), and J. P. Mallory, *In Search of the Indo-Europeans: Language, Archaeology, and Myth* (London: Thames and Hudson, 1989). For a critique of theories of "Bantu" migrations in Africa, see Jan Vansina, "New Linguistic Evidence and 'the Bantu Expansion,'" *Journal of African History* 36, no. 2 (1995): 173–95; my thanks to Heike Schmidt for this reference.

3. Jared Diamond, *Guns, Germs, and Steel: The Fates of Human Societies* (London: Vintage, 1998), chaps. 2 and 8; Ben Finney, "The Other One-Third of the Globe," *Journal of World History* 5, no. 2 (fall 1994): 273–98; and J. R. McNeill, "Of Rats and Men: A Synoptic Environmental History of the Island Pacific," *Journal of World History* 5, no. 2 (fall 1994): 299–349. Tim Flannery, *The Future Eaters: An Ecological History of the Australasian Lands and People* (Chatswood, N.S.W.: Reed, 1995), pp. 164–65, gives later dates for the settlement of New Zealand.

4. Diamond, *Guns, Germs, and Steel*, chap. 2.

5. David Christian, "Silk Roads or Steppe Roads? The Silk Roads in World History," *Journal of World History* 11, no. 1 (spring 2000): 1–26.

6. It is because of the importance and pervasiveness of this gap between the masses and the elites that Anthony Giddens refers to such societies as "class-divided"; see, for example, *A Contemporary Critique of Historical Materialism*, 2nd ed. (Basingstoke: Macmillan, 1995), p. 159.

7. Eric R. Wolf, *Europe and the People without History* (Berkeley: University of California Press, 1982), particularly chap. 3.

8. William McNeill, "The Disruption of Traditional Forms of Nurture," in *The Disruption of Traditional Forms of Nurture* (Amsterdam: Het Spinhuis, 1998), pp. 7–8, 29–53; my thanks to Professor McNeill for this reference.

9. The best analysis of these relationships is in Thomas J. Barfield, *The Perilous Frontier: Nomadic Empires and China* (Oxford: Blackwell, 1989); and see Nicola di Cosmo, "State Formation and Periodization in Inner Asian History," *Journal of World History* 10, no. 1 (spring 1999): 1–40.

10. Janet Abu-Lughod, *Before European Hegemony: The World System, A.D. 1250–1350* (New York: Oxford University Press, 1989); and Andre Gunder Frank and Barry K. Gills, eds., *The World System: Five Hundred Years or Five Thousand?* (London: Routledge, 1992).

11. Christopher Chase-Dunn and Thomas D. Hall, *Rise and Demise: Comparing World Systems* (Boulder, Colo.: Westview Press, 1997).

12. Michael Mann, *The Sources of Social Power*, vol. 1, *A History of Power from the Beginning to A.D. 1760* (Cambridge: Cambridge University Press, 1986).

13. "The use of multiple bounding criteria often will result in nested levels of system boundedness. Generally, bulk goods will compose the smallest regional interaction net. Political/military interaction will compose a larger net that may include more than one bulk-goods net, and prestige-goods exchanges will link even larger regions that may contain one or more political/military nets. We expect the information net to be of the same order of size as the prestige goods net: sometimes larger, sometimes smaller" (Chase-Dunn and Hall, *Rise and Demise*, p. 53).

14. Hans Jörg Nissen, *The Early History of the Ancient Near East* (Chicago: University of Chicago Press, 1988), pp. 167–68.

15. Charles L. Redman, "Mesopotamia in the First Cities," in *The Illustrated History of Humankind*, ed. Göran Burenhult, vol. 3, *Old World Civilizations: The Rise of Cities and States* (St. Lucia: University of Queensland Press, 1994), p. 32.

16. A Sumerian poet, trans. S. N. Kramer, in *Ancient Near Eastern Texts Relating to the Old Testament*, ed. James B. Pritchard, 3rd ed. (Princeton: N.J.: Princeton University Press, 1969), pp. 647–48; quoted in A. Bernard Knapp, *The History and Culture of Ancient Western Asia and Egypt* (Chicago: Dorsey Press, 1988), p. 87. The exact site of Agade remains unknown.

17. This is the claim of Barry K. Gills and Andre Gunder Frank, "World System Cycles, Crises, and Hegemonic Shifts, 1700 BC to 1700 AD," in Frank and Gills, eds., *The World System*, pp. 143–99, particularly pp. 153–55.

18. Paul Bairoch, *Cities and Economic Development: From the Dawn of History to the Present*, trans. Christopher Braider (Chicago: University of Chicago Press, 1988), p. 27.

19. See Christian, "Silk Roads or Steppe Roads?"

20. See Barry K. Gills and Andre Gunder Frank, "The Cumulation of Accumulation," in Frank and Gills, eds., *The World System*, pp. 81–114, particularly 86, and Christian, "Silk Roads or Steppe Roads?" for two discussions that emphasize the significance of early trans-Eurasian exchanges.

21. Lynda Shaffer, "Southernization," in *Agricultural and Pastoral Societies in Ancient and Classical History*, ed. Michael Adas (Philadelphia: Temple University Press, 2001), pp. 308–24; originally published in *Journal of World History* 5, no. 1 (spring 1994): 1–21.

22. Shaffer, "Southernization."

23. On the histories of Kush and Axum, see Stanley M. Burstein, ed., *Ancient African Civilizations: Kush and Axum* (Princeton: Markus Wiener, 1998); and see Christopher Ehret, "Sudanic Civilization," in Adas, ed., *Agricultural and Pastoral Societies in Ancient and Classical History*, pp. 224–74.

24. I have discussed the Eurasian steppes at greater length in David Christian, *A History of Russia, Central Asia and Mongolia*, vol. 1, *Inner Eurasia from Prehistory to the Mongol Empire* (Oxford: Blackwell, 1998).

25. Some have claimed that Chinese naval expeditions may have crossed the Pacific in the first millennium BCE and influenced the Mayan cultures of Mesoamerica and the Chavin culture in Peru, but the evidence is indirect and circumstantial; see Louise Levathes, *When China Ruled the Seas: The Treasure Fleet of the Dragon Throne, 1405–1433* (New York: Simon and Schuster, 1994), chap. 1, and Joseph Needham and Lu Gwei-djen, *Trans-Pacific Echoes and Resonances: Listening Once Again* (Singapore: World Scientific, 1984).

26. Robert J. Wenke, *Patterns in Prehistory: Humankind's First Three Million Years*, 3rd ed. (New York: Oxford University Press, 1990), p. 498.

27. The account is quoted in John E. Kicza, "The Peoples and Civilizations of the Americas before Contact," in Adas, ed., *Agricultural and Pastoral Societies in Ancient and Classical History*, p. 190.

28. Bernal Díaz, *The Conquest of New Spain*, trans. J. M. Cohen (Harmondsworth: Penguin, 1963), p. 214.

29. Wenke, *Patterns in Prehistory*, p. 515.

30. William R. Thompson, "The Military Superiority Thesis and the Ascendancy of Western Eurasia in the World System," *Journal of World History* 10, no. 1 (1999): 172, citing Rein Taagepera, "Expansion and Contraction Patterns of Large Polities: Context for Russia," *International Studies Quarterly* 41(1997): 475–504.

31. Robert S. Lopez, *The Commercial Revolution of the Middle Ages, 950–1350* (Englewood Cliffs, N.J.: Prentice-Hall, 1971), p. 1.

32. Emmanuel Le Roy Ladurie, *The Peasants of Languedoc*, trans. John Day (Urbana: University of Illinois Press, 1974), p. 4.

33. E. A. Wrigley, *Population and History* (London: Weidenfeld and Nicolson, 1969), pp. 55–57.

34. See also the conclusion of Thomas M. Whitmore et al., "Long-Term Population Change," in *The Earth as Transformed by Human Action: Global and Regional Changes in the Biosphere over the Past 300 Years*, ed. B. L. Turner et al. (Cambridge: Cambridge University Press, 1990), p. 37: "Examination of long-term population records at the regional scale reveals a common pattern of significant episodic growth and decline."

35. I. G. Simmons, *Environmental History: A Concise Introduction* (Oxford: Blackwell, 1993), p. 13. See also Whitmore et al., "Long-Term Population Change."

36. Michael D. Coe, *The Maya* (New York: Praeger, 1966), p. 128.

37. E. L. Jones, *The European Miracle: Environments, Economies, and Geopolitics in the History of Europe and Asia*, 2nd ed. (Cambridge: Cambridge University Press, 1987), chap. 3.

38. Diamond, *Guns, Germs, and Steel*, chap. 11, particularly pp. 212–14.

39. William H. McNeill, *Plagues and People* (Oxford: Blackwell, 1977), p. 102.

40. William H. McNeill, *Rise of the West* (Chicago: University of Chicago Press, 1963), chap. 7. See also McNeill, "Confluence of the Civilized Disease Pools of Eurasia, 500 B.C. to A.D. 1200," chap. 3 in *Plagues and Peoples,* and Kenneth F. Kiple, ed., *The Cambridge World History of Human Disease* (Cambridge: Cambridge University Press, 1993).

41. William McNeill, *Plagues and Peoples,* p. 111.

42. Kenneth F. Kiple, introduction to Kiple, ed., *Cambridge World History of Human Disease,* p. 3.

43. William McNeill, *Plagues and Peoples,* p. 119.

44. William McNeill, *Plagues and Peoples,* pp. 116–29.

45. See Henri J. M. Claessen and Peter Skalnik, eds., *The Early State* (The Hague: Mouton, 1978).

46. Taagepera's figures are cited in William Eckhardt, "A Dialectical Evolutionary Theory of Civilizations, Empires, and Wars," in *Civilizations and World Systems: Studying World-Historical Change,* ed. Stephen K. Sanderson (Walnut Creek, Calif.: Altamira Press, 1995), pp. 80–81. Eckhardt's essay relies heavily on Rein Taagepera, "Size and Duration of Empires: Systematics of Size," *Social Science Research* 7 (1978): 108–27.

47. Thompson, "The Military Superiority Thesis," p. 172, citing Taagepera, "Expansion and Contraction Patterns of Large Polities."

48. There is a helpful discussion of religion in early civilizations in Bruce G. Trigger, *Early Civilizations: Ancient Egypt in Context* (Cairo: American University in Cairo Press, 1993), chap. 4.

49. See Karl Jaspers, *The Origin and Goal of History,* trans. Michael Bullock (New Haven: Yale University Press, 1953), chap. 1.

50. The prince is quoted in Christian, *A History of Russia, Central Asia, and Mongolia,* 1:306.

51. Joel Mokyr, *The Lever of Riches: Technological Creativity and Economic Progress* (New York: Oxford University Press, 1990), p. 20.

52. Marx's classic formulation is as follows: "In all forms where the actual worker himself remains the 'possessor' of the means of production and the conditions of labor needed for the production of his own means of subsistence, the property relationship must appear at the same time as a direct relationship of domination and servitude, and the direct producer therefore as an unfree person—an unfreedom which may undergo a progressive attenuation from serfdom with statute-labor down to a mere tribute obligation. . . . Under these conditions, the surplus labor for the nominal landowner can only be extorted from them by extra-economic compulsion, whatever the form this might assume" (Karl Marx, *Capital: A Critique of Political Economy,* vol. 3, trans. David Fernbach [Harmondsworth: Penguin, 1981], p. 926).

53. M. I. Finley, "Empire in the Greco-Roman World," *Greece and Rome,* 2nd ser., 25, no. 1 (April 1978): 1. There is an interesting analysis of the "benefits" of conquest in G. D. Snooks, *The Dynamic Society: Exploring the Sources of Global Change* (London: Routledge, 1996), chap. 10.

54. That profit-maximizing was not one of the central preoccupations of

rulers in the classical era is a central theme of M. I. Finley's *The Ancient Economy* (London: Chatto and Windus, 1973).

55. Niccolò Machiavelli, *The Prince*, trans. George Bull (Harmondsworth: Penguin, 1961), pp. 87, 88.

56. Nizam al-Mulk, *The Book of Government, or Rules for Kings*, trans. Hubert Darke, 2nd ed. (London: Routledge, 1978), pp. 131–32; the French writer is quoted in C. Warren Hollister, *Medieval Europe: A Short History*, 5th ed. (New York, 1982), p. 163.

57. R. Bin Wong, *China Transformed: Historical Change and the Limits of European Experience* (Ithaca, N.Y.: Cornell University Press, 1997), pp. 129, 134.

58. Mokyr, *The Lever of Riches*, p. 175.

59. This synergy is analogous but not identical to the type of growth that Adam Smith associated with increasing exchanges; Smith focused on the productivity gains generated by increased specialization, which, in turn, was made possible by larger markets.

60. The data in this paragraph are drawn from Stephen K. Sanderson, "Expanding World Commercialization: The Link between World Systems and Civilizations," in Sanderson, ed., *Civilizations and World Systems*, p. 267.

61. Referring to the ideas of the German sociologist Max Weber, Anthony Giddens defines economic exchange as "any non-coerced agreement offering an existing or future utility against another or others given in return" (*The Nation-State and Violence*, vol. 2 of *A Contemporary Critique of Historical Materialism* [Cambridge: Polity Press, 1985], pp. 123–24).

62. William H. McNeill, *The Pursuit of Power: Technology, Armed Force, and Society since A.D. 1000* (Oxford: Blackwell, 1982), p. 5.

63. For a brief survey of the Hellenistic era, see Stanley M. Burstein, "The Hellenistic Period in World History," in Adas, ed., *Agricultural and Pastoral Societies in Ancient and Classical History*, pp. 275–307.

64. María Eugenia Aubet, *Phoenicians and the West: Politics, Colonies, and Trade*, trans. Mary Turton (Cambridge: Cambridge University Press, 1993), pp. 85–87; cited in Snooks, *The Dynamic Society*, p. 345.

65. William McNeill, *Pursuit of Power*, pp. 22–23, mentions cuneiform archives from Anatolia in 1800 BCE containing correspondence of traders in tin who used donkey caravans.

66. Knapp, *The History and Culture of Ancient Western Asia and Egypt*, pp. 141–43.

67. Knapp, *The History and Culture of Ancient Western Asia and Egypt*, pp. 95, 142; Valerie Hansen, *The Open Empire: A History of China to 1600* (New York: W. W. Norton, 2000), p. 92.

68. Christian, "Silk Roads or Steppe Roads?"

69. Mokyr, *The Lever of Riches*, p. 26.

70. Wong, *China Transformed*, pp. 45–46, 90.

71. Giddens, *The Nation-State and Violence*, pp. 40–41, 79–81.

72. On the classical Greek contributions to warfare, William McNeill writes: "It was no accident that the major period of weapons development in the an-

cient Mediterranean world occurred in the centuries when competing rulers applied commercial principles to the tasks of military mobilization" (*Pursuit of Power,* p. 70).

CHAPTER 11. APPROACHING MODERNITY

1. Anthony Giddens, *The Nation-State and Violence,* vol. 2 of *A Contemporary Critique of Historical Materialism* (Cambridge: Polity Press, 1985), p. 33.

2. An argument for the planetary importance of this transformation is made persuasively in J. R. McNeill, *Something New under the Sun: An Environmental History of the Twentieth-Century World* (New York: W. W. Norton, 2000).

3. Eric R. Wolf, *Europe and the People without History* (Berkeley: University of California Press, 1982), pp. 24–72. For a similar tour, in the year 1000 CE, see John Man's marvelous *Atlas of the Year 1000* (Cambridge, Mass.: Harvard University Press, 1999).

4. Wolf, *Europe and the People without History,* p. 71.

5. Brian M. Fagan, *People of the Earth: An Introduction to World Prehistory,* 10th ed. (Upper Saddle River, N.J.: Prentice Hall, 2001), p. 362.

6. Wolf, *Europe and the People without History,* p. 42.

7. Michael D. Coe, *Mexico: From the Olmecs to the Aztecs,* 4th ed. (London: Thames and Hudson, 1994), p. 158.

8. On pastoralism, the best introductions are Thomas J. Barfield, *The Nomadic Alternative* (Englewood Cliffs, N.J.: Prentice-Hall, 1993), and Anatoly M. Khazanov, *Nomads and the Outside World,* trans. Julia Crookenden, 2nd ed. (Madison: University of Wisconsin Press, 1994).

9. Pseudo-Hippocrates, "Airs, Waters, Places," quoted from *Hippocratic Writings,* ed. and intro. G. E. R. Lloyd, trans. J. Chadwick and W. N. Mann (Harmondsworth: Penguin, 1978), p. 163.

10. See David Christian, *A History of Russia, Central Asia, and Mongolia,* vol. 1, *Inner Eurasia from Prehistory to the Mongol Empire* (Oxford: Blackwell, 1998).

11. Christian, *A History of Russia, Central Asia and Mongolia,* 1:85–94, 149–57, and chap. 8; Nicola di Cosmo, "State Formation and Periodization in Inner Asian History," *Journal of World History* 10, no. 1 (spring 1999): 1–40.

12. The ambassador is quoted in Terence Armstrong, *Russian Settlement in the North* (Cambridge: Cambridge University Press, 1965), p. 36; and see James Forsyth's superb history of Siberia, *A History of the Peoples of Siberia: Russia's North Asian Colony, 1581–1990* (Cambridge: Cambridge University Press, 1992), particularly pp. 10–16.

13. Wolf, *Europe and the People without History,* p. 65.

14. Allen W. Johnson and Timothy Earle, *The Evolution of Human Societies,* 2nd ed. (Stanford: Stanford University Press, 2000), p. 9.

15. Lynn Margulis and Dorion Sagan, *Microcosmos: Four Billion Years of Microbial Evolution* (London: Allen and Unwin, 1987), p. 228.

16. Carlo M. Cipolla, *The Economic History of World Population*, 6th ed. (Harmondsworth: Penguin, 1974), pp. 114–15.

17. David S. Landes, *The Unbound Prometheus: Technological Change and Industrial Development in Western Europe from 1750 to the Present* (London: Cambridge University Press, 1969), p. 6.

18. Joel Mokyr, *The Lever of Riches: Technological Creativity and Economic Progress* (New York: Oxford University Press, 1990), p. 99.

19. E. A. Wrigley, *Continuity, Chance, and Change: The Character of the Industrial Revolution in England* (Cambridge: Cambridge University Press, 1988), pp. 54–55.

20. McNeill, *Something New under the Sun*, pp. 14–15; and see chap. 6.

21. McNeill, *Something New under the Sun*, p. 15.

22. Paul Bairoch, *Cities and Economic Development: From the Dawn of History to the Present*, trans. Christopher Braider (Chicago: University of Chicago Press, 1988), p. 513.

23. There is a fine analysis of the rise of clock time in Norbert Elias, *Time: An Essay*, trans. Edmund Jephcott (Oxford: Blackwell, 1992); Elias argues that the need to coordinate increasingly complex networks of dependence is the fundamental cause of the increasing precision and generality of modern time scheduling.

24. Charles Tilly, *Coercion, Capital, and European States, AD 990–1992*, rev. ed. (Cambridge, Mass.: Blackwell, 1992), p. 69; as Tilly points out, the United States today constitutes a partial exception to the rule that modern states monopolize the means of violence (p. 68).

25. Norbert Elias, *The Civilizing Process*, vol. 1, *The History of Manners*, trans. Edmund Jephcott (New York: Pantheon, 1978), p. 186.

26. David T. Courtwright, *Forces of Habit: Drugs and the Making of the Modern World* (Cambridge, Mass.: Harvard University Press, 2002).

27. On the sense of place in premodern cosmologies, see the rich and suggestive account in Tony Swain, *A Place for Strangers: Towards a History of Australian Aboriginal Being* (Cambridge: Cambridge University Press, 1993).

28. Recently, a number of historians have attempted the necessary comparative studies: see the superb studies by R. Bin Wong, *China Transformed: Historical Change and the Limits of European Experience* (Ithaca, N.Y.: Cornell University Press, 1997); Kenneth Pomeranz, *The Great Divergence: China, Europe, and the Making of the Modern World Economy* (Princeton: Princeton University Press, 2000); and Andre Gunder Frank, *ReOrient: Global Economy in the Asian Age* (Berkeley: University of California Press, 1998). See also the fine brief survey of these debates in Robert B. Marks, *The Origins of the Modern World: A Global and Ecological Narrative* (Lanham, Md.: Rowman and Littlefield, 2002).

29. There is a good short survey of these debates over modernity in Craig Lockard, "Global Historians and the Great Divergence," *World History Bulletin* 17, no. 1 (fall 2000): 17, 32–34.

30. Mokyr, *The Lever of Riches*, p. 148.

31. Daniel Headrick, "Technological Change," in *The Earth as Transformed by Human Action: Global and Regional Changes in the Biosphere over the Past 300 Years*, ed. B. L. Turner II et al. (Cambridge: Cambridge University Press, 1990), p. 59.

32. There is a good short survey of theories of growth in J. L. Anderson, *Explaining Long-Term Economic Change* (Basingstoke: Macmillan, 1991); and see the survey in Mokyr, *The Lever of Riches*, chap. 7 ("Understanding Technological Progress").

33. See, for example, Ester Boserup, *Population and Technology* (Oxford: Blackwell, 1981).

34. Though the shortage of wood in China was not *that* much worse than in Britain; see the discussion in Pomeranz, *The Great Divergence*, pp. 220–36.

35. The importance of coal is particularly stressed in Wrigley, *Continuity, Chance, and Change*, and in Pomeranz, *The Great Divergence*. Geographical factors also loom large in the highly influential studies by E. L. Jones, *The European Miracle: Environments, Economies, and Geopolitics in the History of Europe and Asia*, 2nd ed. (Cambridge: Cambridge University Press, 1987), and *Growth Recurring: Economic Change in World History* (Oxford: Clarendon, 1988).

36. Mokyr, *The Lever of Riches*, p. 162.

37. T. S. Ashton, quoted in Gary Hawke, "Reinterpretations of the Industrial Revolution," in *The Industrial Revolution and British Society*, ed. Patrick O'Brien and Roland Quinault (Cambridge: Cambridge University Press, 1993), p. 55.

38. See T. S. Ashton, *The Industrial Revolution, 1760–1830* (London: Oxford University Press, 1948).

39. Max Weber, *The Protestant Ethic and the Spirit of Capitalism*, trans. Talcott Parsons (1930; reprint, New York: Scribners, 1958).

40. A subtle recent reworking of the argument that ideas were a significant prime mover in the Modern Revolution is offered in Margaret Jacob, *Scientific Culture and the Making of the Industrial West* (New York: Oxford University Press, 1997). I will make use of Jacob's argument in chapter 12.

41. Adam Smith, *An Inquiry into the Nature and Causes of the Wealth of Nations*, 5th ed., ed. Edwin Cannan (New York: Modern Library, 1937), pp. 1, 13.

42. See Mokyr, *The Lever of Riches*, p. 5: "Economic growth caused by an increase in trade may be termed *Smithian growth*."

43. Frank, *ReOrient*, pp. 173, 166. On productivity levels in China, see also Pomeranz, *The Great Divergence*, and Wong, *China Transformed*. For a different view, arguing that European economic superiority dates from as early as the fifteenth century, see Angus Maddison, *The World Economy: A Millennial Perspective* (Paris: OECD, 2001).

44. See Wolf, *Europe and the People without History*, chap. 3.

45. Anthony Giddens, *A Contemporary Critique of Historical Materialism*, 2nd ed. (Basingstoke: Macmillan, 1995), p. 1. My quotation of this passage im-

plies not agreement with everything in Giddens's critique of Marx but rather approval of his willingness to salvage what is still of value in Marx. Fernand Braudel has also argued that Marx's models of society, if unfrozen and handled with more flexibility and nuance, should still be of great value to historians; see Braudel, "History and the Social Sciences," in *On History*, trans. Sarah Matthews (Chicago: University of Chicago Press, 1980), p. 51.

46. Daniel C. Dennett, *Consciousness Explained* (London: Penguin, 1993), p. 204, refers to *memes* as entities that infest human brains rather like parasites. Richard Dawkins coined the term in the first edition of *The Selfish Gene* (Oxford : Oxford University Press, 1976), applying it to any intellectual or cultural information that can be transferred from one person to another by imitation. The idea of "memes" is explored thoroughly enough to see its limitations in Susan Blackmore, *The Meme Machine* (Oxford: Oxford University Press, 1999).

47. Pomeranz, *The Great Divergence*; Wong, *China Transformed*.

48. For this argument about the processes of accumulation, see Pomeranz, *The Great Divergence*.

49. One of the best accounts of the economic history of the new hub region is still Ralph Davis, *The Rise of the Atlantic Economies* (Ithaca, N.Y.: Cornell University Press, 1973).

50. Andrew Sherratt, like Marx, has stressed the economic significance of the changing topology of global exchange networks: "The capital concentration possible from such a huge catchment area, whose nodal points linked routes from every continent, permitted an investment in machinery, labour discipline, and settlement agglomeration which permitted a new scale of added-value manufacturing" (Sherratt, "Reviving the Grand Narrative: Archaeology and Long-Term Change," *Journal of European Archaeology* 3, no. 1 [1995]: 21). These arguments are important and familiar, but new information networks may have been equally significant in helping to explain changing rates of innovation.

51. Wong, *China Transformed*, p. 151.

CHAPTER 12. GLOBALIZATION, COMMERCIALIZATION, AND INNOVATION

Chapter epigraph: Tim Flannery, *The Future Eaters: An Ecological History of the Australasian Lands and People* (Chatswood, N.S.W.: Reed, 1995), p. 334.

1. The outstanding study of the ecological advantages enjoyed by Afro-Eurasians and the team of animals, plants, and bugs they captained is Alfred W. Crosby, *Ecological Imperialism: The Biological Expansion of Europe, 900–1900* (Cambridge: Cambridge University Press, 1986).

2. Crosby, *Ecological Imperialism*, p. 200.

3. See, for example, Kenneth Pomeranz, *The Great Divergence: China, Europe, and the Making of the Modern World Economy* (Princeton: Princeton University Press, 2000); Andre Gunder Frank, *ReOrient: Global Economy in the Asian Age* (Berkeley: University of California Press, 1998); and R. Bin Wong, *China Transformed: Historical Change and the Limits of European Experience* (Ithaca, N.Y.: Cornell University Press, 1997).

4. Adam Smith, quoted in Frank, *ReOrient*, p. 13.

5. Marshall G. S. Hodgson, "The Great Western Transmutation," in his *Rethinking World History: Essays on Europe, Islam, and World History*, ed. Edmund Burke III (Cambridge: Cambridge University Press, 1993), p. 47.

6. Valerie Hansen, *The Open Empire: A History of China to 1600* (New York: W. W. Norton, 2000), p. 263.

7. Asa Briggs, *A Social History of England*, 2nd ed. (Harmondsworth: Penguin, 1987), pp. 74–75.

8. Paul Bairoch, *Cities and Economic Development: From the Dawn of History to the Present*, trans. Christopher Braider (Chicago: University of Chicago Press, 1988), p. 159.

9. al-Muqaddasi, quoted in W. Barthold, *Turkestan down to the Mongol Invasion*, 4th ed., trans. T. Minorsky, ed. C. E. Bosworth (London: E. J. W. Gibb Memorial Trust, 1977), pp. 103–4.

10. Mark Elvin, *The Pattern of the Chinese Past* (Stanford: Stanford University Press, 1973), p. 177.

11. Janet Abu-Lughod, *Before European Hegemony: The World System, A.D. 1250–1350* (New York: Oxford University Press, 1989), pp. 337–39.

12. Abu-Lughod, *Before European Hegemony*, p. 331, referring to Jacques Gernet, *Daily Life in China on the Eve of the Mongol Invasion, 1250–1276*, trans. H. M. Wright (London: Allen and Unwin, 1962), p. 87.

13. Elvin, *The Pattern of the Chinese Past*, p. 177.

14. The poem is quoted in Elvin, *Pattern of the Chinese Past*, p. 169.

15. Carlo M. Cipolla, *Before the Industrial Revolution: European Society and Economy, 1000–1700*, 2nd ed. (London: Methuen, 1981), p. 143. See also Robert S. Lopez, *The Commercial Revolution of the Middle Ages, 950–1350* (Englewood Cliffs, N.J.: Prentice-Hall, 1971).

16. Abu-Lughod, *Before European Hegemony*.

17. Thomas T. Allsen's latest work, *Culture and Conquest in Mongol Eurasia* (Cambridge: Cambridge University Press, 2001), focuses on exchanges between China and Ilkhanid Persia in the late thirteenth and early fourteenth centuries.

18. Ibn Battuta's travels are described in Ross E. Dunn's classic study, *The Adventures of Ibn Battuta: A Muslim Traveler of the Fourteenth Century* (Berkeley: University of California Press, 1989).

19. The classic study of these trading networks, and of the networks of merchants they depended on, is Philip Curtin, *Cross-Cultural Trade in World History* (Cambridge: Cambridge University Press, 1985).

20. Andrew M. Watson, *Agricultural Innovation in the Early Islamic World: The Diffusion of Crops and Farming Techniques, 700–1100* (Cambridge: Cambridge University Press, 1983).

21. See Wong, *China Transformed*.

22. Lynda Shaffer, "Southernization," in *Agricultural and Pastoral Societies in Ancient and Classical History*, ed. Michael Adas (Philadelphia: Temple Uni-

versity Press, 2001), pp. 308–24; originally published in *Journal of World History* 5, no. 1 (spring 1994): 1–21.

23. al Jahiz, quoted in Shaffer, "Southernization," p. 312. See also James E. McClellan III and Harold Dorn, *Science and Technology in World History: An Introduction* (Baltimore: Johns Hopkins University Press, 1999), pp. 145–54.

24. Shaffer, "Southernization," p. 316.

25. Thomas A. Brady, "Rise of Merchant Empires, 1400–1700: A European Counterpoint," in *The Political Economy of Merchant Empires: State Power and World Trade, 1350–1750*, ed. James D. Tracy (Cambridge: Cambridge University Press, 1991), p. 150.

26. Barbarossa's uncle is quoted in Cipolla, *Before the Industrial Revolution*, p. 148.

27. Hansen, *The Open Empire*, p. 135.

28. Sima Qian, quoted in Elvin, *Pattern of the Chinese Past*, p. 164.

29. Chao Cuo, early second century BCE, quoted in Elvin, *Pattern of the Chinese Past*, p. 164.

30. S. A. M. Adshead, *China in World History*, 2nd ed. (Basingstoke: Macmillan, 1995), p. 117.

31. Archibald R. Lewis, *Nomads and Crusaders, A.D. 1000–1360* (Bloomington: Indiana University Press, 1991), pp. 109, 130, 161. According to Wong, at times under the Song commercial revenues may have exceeded half of total government revenues (*China Transformed*, p. 95).

32. On Chinese shipbuilding techniques, see Arnold Pacey, *Technology in World Civilization* (Cambridge, Mass.: MIT Press, 1990), pp. 65–66.

33. Hansen, *The Open Empire*, p. 266.

34. Elvin, *Pattern of the Chinese Past*, p. 118. There is a good, short survey of economic growth under the Song in Abu-Lughod, *Before European Hegemony*, chap. 10.

35. Hansen, *The Open Empire*, p. 264.

36. J. R. McNeill, *Something New under the Sun: An Environmental History of the Twentieth-Century World* (New York: W. W. Norton, 2000), p. 56; McNeill adds that the other preindustrial spike of this kind occurred after the introduction of copper coinage in the Mediterranean in the first millennium BCE.

37. Elvin, *Pattern of the Chinese Past*, p. 88; Pacey, *Technology in World Civilization*, p. 47.

38. Pacey, *Technology in World Civilization*, pp. 24–26.

39. Hansen, *The Open Empire*, pp. 266–67, 270–71.

40. See Joseph Needham, *Science and Civilisation in China*, 7 vols. (Cambridge: Cambridge University Press, 1954–2003).

41. Wong, *China Transformed*, p. 131.

42. There is a good popular account of these voyages in Louise Levathes, *When China Ruled the Seas: The Treasure Fleet of the Dragon Throne, 1405–1433* (New York: Simon and Schuster, 1994).

43. Karl Marx, *Capital: A Critique of Political Economy*, vol. 1, trans. Ben

Fowkes (Harmondsworth: Penguin, 1976), part 2, p. 247 (the opening paragraph). The idea that the creation of a world market was critical to the emergence of modernity has been central in Marxist historiography. See Immanuel Waller- stein, "World-System," in *A Dictionary of Marxist Thought*, ed. Tom Botto- more, 2nd ed. (Oxford: Blackwell, 1991), pp. 590–91.

44. As Dennis O. Flynn and Arturo Giráldez have pointed out, strictly speak- ing, a global system of exchanges did not exist before 1571, the year when the Pacific was also bridged by the beginning of regular trade between the Ameri- cas and Manila. See Flynn and Giráldez, "Cycles of Silver: Global Economic Unity through the Mid–Eighteenth Century," *Journal of World History* 13, no. 2 (fall 2002): 393.

45. For an introduction to these critical flows of silver, see two articles by Dennis O. Flynn and Arturo Giráldez: "Born with a 'Silver Spoon': The Origin of World Trade in 1571," *Journal of World History* 6, no. 2 (fall 1995): 201–21, and "Cycles of Silver."

46. See, for example, Massimo Livi-Bacci's summary account of the trans- mission of disease across world zones in *A Concise History of World Popula- tion*, trans. Carl Ipsen (Oxford: Blackwell, 1992), pp. 50–56.

47. Based on figures in Angus Maddison, *The World Economy: A Millen- nial Perspective* (Paris: UNESCO, 2001), p. 235. There is much debate about pre-Columbian population figures and therefore about the relative decline in populations in the sixteenth century. Estimates of the population decline in Mexico alone range from 15 to 90 percent. The figures in table 11.1 (from J. R. Biraben, "Essai sur l'évolution du nombre des hommes," *Population* 34 [1979]: 16) suggest that Latin American populations (excluding North America) de- clined by about 75 percent between 1500 and 1600, dropping from about 39 million to about 10 million. On the more conservative estimates of Angus Mad- dison, Latin American populations declined by about 50 percent, from ca. 17.5 million to ca. 8.6 million; the older estimates of Woodrow Borah and Sher- burne F. Cook, in *The Aboriginal Population of Central Mexico on the Eve of the Spanish Conquest* (Berkeley: University of California Press, 1963), suggest that populations in 1500 may have been as high as 100 million, and the decline may have been as much as 90 to 95 percent. See the discussions in Maddison, *The World Economy*, pp. 233–36, and Massimo Livi-Bacci, *A Concise History of World Population*, pp. 50–56. My thanks to Bruce Castleman for some of these references.

48. The native American is quoted in Alfred W. Crosby, *The Columbian Ex- change: Biological and Cultural Consequences of 1492* (Westport, Conn.: Greenwood Press, 1972), p. 36.

49. Thomas Hariot, quoted in Crosby, *The Columbian Exchange*, pp. 40–41 (from David B. Quinn, ed., *The Roanoke Voyages, 1584–1590*, 2 vols. [London: Hakluyt Society, 1955], 1:387).

50. In *Ecological Imperialism*, Crosby describes this process of settlement as the creation of "neo-Europes."

51. Crosby, *Columbian Exchange*, p. 170.

52. Crosby, *Columbian Exchange,* p. 199.

53. Frank, *ReOrient,* p. 60.

54. Joel Mokyr, *The Lever of Riches: Technological Creativity and Economic Progress* (New York: Oxford University Press, 1990), p. 70.

55. Crosby, *Columbian Exchange,* pp. 185, 199–201.

56. See Mokyr, *The Lever of Riches,* chap. 4, for a survey of rates of innovation.

57. Peter N. Stearns, *The Industrial Revolution in World History* (Boulder, Colo.: Westview Press, 1993), p. 18.

58. Frank, *ReOrient,* pp. 168, 172.

59. Frank, *ReOrient,* p. 166.

60. Curtin, *Cross-Cultural Trade,* p. 149.

61. Pomeranz, *The Great Divergence,* particularly chaps. 1, 2, and 3.

62. Wong, *China Transformed,* p. 17.

63. Pomeranz, *The Great Divergence,* makes a similar argument.

64. There is a fine description of how such processes of commercial exchange worked through the North American fur trade in Eric R. Wolf, *Europe and the People without History* (Berkeley: University of California Press, 1982).

65. David Christian, *Living Water: Vodka and Russian Society on the Eve of Emancipation* (Oxford: Clarendon, 1990), pp. 33, 384–88.

66. Elvin, *Pattern of the Chinese Past,* p. 167.

67. Wong, *China Transformed,* p. 45.

68. The petition is quoted in John Bushnell, *Mutiny amid Repression: Russian Soldiers in the Revolution of 1905–1906* (Bloomington: Indiana University Press, 1985), p. 180.

69. For a survey of Europe's role as middleman, see Frank, *ReOrient.*

70. Andrew Sherratt, "Reviving the Grand Narrative: Archaeology and Long-Term Change," *Journal of European Archaeology* 3, no. 1 (1995): 13.

71. Sherratt, "Reviving the Grand Narrative," p. 21.

72. Christopher Chase-Dunn and Thomas D. Hall have also explored the distinctive role played by "semi-peripheral" regions in the long-term history of agrarian civilizations; see Chase-Dunn and Hall, *Rise and Demise: Comparing World Systems* (Boulder, Colo.: Westview Press, 1997), chap. 5.

73. Wong, *China Transformed,* p. 129.

74. On European innovations in the medieval period, see Mokyr, *The Lever of Riches,* pp. 31–56.

75. Margaret Jacob, *The Cultural Meaning of the Scientific Revolution* (Philadelphia: Temple University Press, 1988), p. 109.

76. Sherratt, "Reviving the Grand Narrative," p. 25.

77. Steven Shapin, *The Scientific Revolution* (Chicago: University of Chicago Press, 1996), pp. 79–80.

78. Frank, *ReOrient,* p. 256: "The structural similarity of the Mongols and the Europeans is that both were peoples in (semi)marginal or peripheral areas who were attracted to and made incursions into the 'core' areas and economies, which were principally in East Asia and secondarily in West Asia."

79. The story is told in Dava Sobel, *Longitude: The True Story of a Lone Genius Who Solved the Greatest Scientific Problem of His Time* (New York: Walker, 1995). On the role of European states in technological innovation, see also Mokyr, *The Lever of Riches*, pp. 78–79.

80. Charles Loyseau, *Traité des Ordres* (1613); quoted in Henry Kamen, *European Society, 1500–1700* (London: Hutchinson, 1984), p. 99.

81. Charles Tilly, *Coercion, Capital, and European States, AD 990–1992*, rev. ed. (Cambridge, Mass.: Blackwell, 1992), p. 30; the argument of the entire book turns on the distinction between these three routes to the modern nation-state.

82. Tilly, *Coercion, Capital, and European States*, p. 14 and chap. 3.

83. Aldeni's conversation is cited in Tilly, *Coercion, Capital, and European States*, p. 128.

84. Nicola di Cosmo, "European Technology and Manchu Power: Reflections on the 'Military Revolution' in Seventeenth Century China," paper presented at the International Congress of Historical Sciences, Oslo, August 2000.

85. Tilly, *Coercion, Capital, and European States*, p. 29.

86. Robert de Balsac, quoted in Tilly, *Coercion, Capital, and European States*, p. 84.

87. See Geoffrey Parker, *The Military Revolution: Military Innovation and the Rise of the West, 1500–1800*, 2nd ed. (Cambridge: Cambridge University Press, 1996), and William H. McNeill, *The Pursuit of Power: Technology, Armed Force, and Society since A.D. 1000* (Oxford: Blackwell, 1982).

88. Elvin, *Pattern of the Chinese Past*, p. 88; Pacey, *Technology in World Civilization*, p. 47.

89. Christian, *Living Water*, pp. 5, 383, 385; in the late eighteenth century, revenues from vodka taxes normally covered 50–60 percent of the costs of defense; in the nineteenth century, they covered, on average, ca. 70 percent of the defense budget.

90. Catharina Lis and Hugo Soly, *Poverty and Capitalism in Pre-Industrial Europe*, [trans. James Coonan] (Atlantic Highlands, N.J.: Humanities Press, 1979), p. 15.

91. Maxine Berg, *The Age of Manufactures, 1700–1820: Industry, Innovation, and Work in Britain*, 2nd ed. (London: Routledge, 1994), pp. 98–99.

92. Berg, *The Age of Manufactures*, p. 99; quoting Keith Wrightson, *English Society, 1580–1680* (London: Hutchinson, 1982), p. 139.

93. George Huppert, *After the Black Death: A Social History of Early Modern Europe* (Bloomington: Indiana University Press, 1986), p. 72.

94. David Christian, "Accumulation and Accumulators: The Metaphor Marx Muffed," *Science and Society* 54, no. 2 (summer 1990): 219–24.

95. In *Capital*, Marx wrote: "Accumulation of wealth at one pole is, therefore, at the same time accumulation of misery, the torment of labour, slavery, ignorance, brutalization and moral degradation at the opposite pole, i.e. on the side of the class that produces its own product as capital" (*Capital*, vol. 1, p. 799; from chap. 25, "The General Law of Capitalist Accumulation").

96. Marx describes the impact of "Absolute Surplus-Value" vividly in *Cap-*

ital, vol. 1, part 3; see also Jan de Vries, "The Industrial Revolution and the Industrious Revolution," *Journal of Economic History* 54, no. 2 (June 1994): 249–70.

97. Karl Marx, *Grundrisse: Foundations of the Critique of Political Economy*, trans. Martin Nicolaus (Harmondsworth: Penguin, 1973), p. 463.

98. N. F. R. Crafts, *British Economic Growth during the Industrial Revolution* (Oxford: Clarendon, 1985), pp. 13–14, for the percentage employed in agriculture in 1688; Lis and Soly, *Poverty and Capitalism*, p. 100, for the estimate of landownership.

99. Peter Mathias, *The First Industrial Nation: An Economic History of Britain, 1700–1914*, 2nd ed. (London: Methuen, 1983), p. 26.

100. Crafts, *British Economic Growth*, pp. 13–14.

101. Mathias, *The First Industrial Nation*, p. 29.

102. Huppert, *After the Black Death*, p. 59.

103. Tilly, *Coercion, Capital, and European States*, p. 17.

CHAPTER 13. BIRTH OF THE MODERN WORLD

1. Daniel R. Headrick, *The Tools of Empire: Technology and European Imperialism in the Nineteenth Century* (New York: Oxford University Press, 1981), p. 3.

2. Patrick O'Brien, "Introduction: Modern Conceptions of the Industrial Revolution," in *The Industrial Revolution and British Society*, ed. Patrick O'Brien and Roland Quinault (Cambridge: Cambridge University Press, 1993), p. 2; for historical "pointillism," see p. 5. See also R. Bin Wong, *China Transformed: Historical Change and the Limits of European Experience* (Ithaca, N.Y.: Cornell University Press, 1997), p. 279: "Much effort has gone into downplaying the rupture caused by the Industrial Revolution. But the world of material possibilities was dramatically altered between 1780 and 1880. No previous century witnessed such changes."

3. That priority made Britain atypical was, for example, the message of Alexander Gerschenkron's pioneering studies of comparative industrialization in *Economic Backwardness in Historical Perspective, a Book of Essays* (Cambridge, Mass.: Harvard University Press, Belknap Press, 1962).

4. Patrick O'Brien and Caglar Keyder, *Economic Growth in Britain and France, 1780–1914: Two Paths to the Twentieth Century* (London: Allen and Unwin, 1978), p. 196.

5. Gary Hawke, "Reinterpretations of the Industrial Revolution," in O'Brien and Quinault, eds., *The Industrial Revolution and British Society*, p. 54.

6. N. F. R. Crafts, *British Economic Growth during the Industrial Revolution* (Oxford: Clarendon, 1985), p. 115.

7. For some examples of such exploitation of others' technological discoveries, see Joel Mokyr, *The Lever of Riches: Technological Creativity and Economic Progress* (New York: Oxford University Press, 1990), pp. 100–109.

8. See Patrick O'Brien, "Political Preconditions for the Industrial Revolu-

tion," in O'Brien and Quinault, eds., *The Industrial Revolution and British Society*, pp. 124–55.

9. Catharina Lis and Hugo Soly, *Poverty and Capitalism in Pre-Industrial Europe* [trans. James Coonan] (Atlantic Highlands, N.J.: Humanities Press, 1979), p. 108.

10. English writers of 1640s, quoted in Lis and Soly, *Poverty and Capitalism*, p. 108; research based on the estimates of Gregory King, cited from Crafts, *British Economic Growth*, p. 13.

11. Figures from Gregory King, summarized in Lis and Soly, *Poverty and Capitalism*, p. 111.

12. Crafts, *British Economic Growth*, pp. 13, 16.

13. W. G. Hoskins, *The Midland Peasant: The Economic and Social History of a Leicestershire Village* (London: Macmillan, 1965), p. 269; quoted in Maxine Berg, *The Age of Manufactures, 1700–1820: Industry, Innovation, and Work in Britain*, 2nd ed. (London: Routledge, 1994), p. 85.

14. Asa Briggs, *A Social History of England*, 2nd ed. (Harmondsworth: Penguin, 1987), p. 206.

15. Berg, *The Age of Manufactures*, p. 80.

16. E. J. Hobsbawm, *Industry and Empire* (Harmondsworth: Penguin, 1969), pp. 28–29.

17. Crafts, *British Economic Growth*, pp. 62–63.

18. Crafts, *British Economic Growth*, pp. 62, 121.

19. Crafts gives growth rates for total factor productivity of 0.2–0.3 percent per year for the eighteenth century, rising to 0.7 percent for 1801–30 and 1.0 percent for 1831–60 (*British Economic Growth*, pp. 2, 76–77, 81).

20. James E. McClellan III and Harold Dorn, *Science and Technology in World History: An Introduction* (Baltimore: Johns Hopkins University Press, 1999), p. 279.

21. Arnold Pacey, *Technology in World Civilization* (Cambridge, Mass.: MIT Press, 1990), p. 113.

22. McClellan and Dorn, *Science and Technology*, pp. 280–81.

23. Mokyr, *The Lever of Riches*, pp. 84–85.

24. For more detailed descriptions of these spinning devices, see Mokyr, *The Lever of Riches*, pp. 96–98.

25. Mokyr, *The Lever of Riches*, p. 111.

26. See the discussion of the factory system in Anthony Giddens, *A Contemporary Critique of Historical Materialism*, 2nd ed. (Basingstoke: Macmillan, 1995), pp. 124–25.

27. Both Marx and Weber remarked on the importance of experience with modern armies as a preparation for the factory system, as Giddens notes (*A Contemporary Critique of Historical Materialism*, p. 125).

28. Pacey, *Technology in World Civilization*, pp. 106, 117–19.

29. Joseph Needham, *Clerks and Craftsmen in China and the West* (Cambridge: Cambridge University Press, 1970), p. 202; quoted in George Basalla, *The Evolution of Technology* (Cambridge: Cambridge University Press, 1988), p. 40.

30. Peter Mathias, *The First Industrial Nation: An Economic History of Britain, 1700–1914*, 2nd ed. (London: Methuen, 1983), pp. 124–25; and see McClellan and Dorn, *Science and Technology*, pp. 287–89.

31. Margaret Jacob has rightly stressed the indirect significance of widespread scientific knowledge in *Scientific Culture and the Making of the Industrial West* (New York: Oxford University Press, 1997); and Mokyr has stressed the creativity of particular engineers (*The Lever of Riches*, pp. 111–12).

32. Hobsbawm, *Industry and Empire*, pp. 50–51.

33. James Watt, quoted in Mokyr, *The Lever of Riches*, p. 87.

34. Charles Tilly, *Coercion, Capital, and European States, AD 990–1992*, rev. ed. (Cambridge, Mass.: Blackwell, 1992), p. 96.

35. See Charles Tilly, "How War Made States, and Vice Versa," chap. 3 of *Coercion, Capital and European States*.

36. Tilly, *Coercion, Capital, and European States*, pp. 103–4.

37. Tilly, *Coercion, Capital, and European States*, pp. 106–7.

38. Anthony Giddens, *A Contemporary Critique of Historical Materialism*, and *The Nation-State and Violence*, vol. 2 of *A Contemporary Critique of Historical Materialism* (Cambridge: Polity Press, 1985), passim. Giddens borrows the term *surveillance* from the work of Michel Foucault.

39. Tilly, *Coercion, Capital, and European States*, p. 110.

40. Giddens, *The Nation-State and Violence*, p. 152.

41. The protection of royal forests from poachers is a central theme of E. P. Thompson, *Whigs and Hunters: The Origin of the Black Act* (London: Allen Lane, 1975), and the defense of entrepreneurial property against labor radicalism is a major theme of *The Making of the English Working Class* (London: Victor Gollancz, 1968).

42. Karl Polanyi, *The Great Transformation: The Political and Economic Origins of Our Time* (Boston: Beacon, 1957).

43. See the lively account of the exchanges in madrassas in John Merson, *Roads to Xanadu: East and West in the Making of the Modern World* (French's Forest, N.S.W.: Child and Associates, 1989), pp. 83 ff.

44. Merson quotes in full a letter from Leonardo to the duke of Milan listing the various types of military invention he has to sell (*Roads to Xanadu*, p. 70).

45. Jacob, *The Cultural Meaning of the Scientific Revolution*, p. 221.

46. Crafts, *British Economic Growth*, p. 98.

47. Mokyr, *The Lever of Riches*, pp. 134–35.

48. Gerschenkron, *Economic Backwardness in Historical Perspective*.

49. For a good, short summary of industrialization, see Daniel R. Headrick, "Technological Change," in *The Earth as Transformed by Human Action: Global and Regional Changes in the Biosphere over the Past 300 Years*, ed. B. L. Turner II et al. (Cambridge: Cambridge University Press, 1990), pp. 55–67.

50. For a typology of modern production units of various kinds, see Richard Barff, "Multinational Corporations and the New International Division of Labour," in *Geographies of Global Change: Remapping the World in the Late*

Twentieth Century, ed. R. J. Johnston, Peter J. Taylor, and Michael J. Watts (Oxford: Blackwell, 1995), p. 51.

51. See, for example, Mike Davis, *Late Victorian Holocausts: El Niño Famines and the Making of the Third World* (London: Verso, 2001), p. 51.

52. Davis, *Late Victorian Holocausts*, p. 16; and see chap. 9.

53. Davis, *Late Victorian Holocausts*, p. 115, and passim.

54. I have argued this point at greater length in David Christian, *Imperial and Soviet Russia: Power, Privilege, and the Challenge of Modernity* (Basingstoke: Macmillan, 1997).

CHAPTER 14. THE GREAT ACCELERATION OF THE TWENTIETH CENTURY

Chapter epigraph: Yehudi Menuhin, quoted in E. J. Hobsbawm, *The Age of Extremes* (London: Weidenfeld and Nicolson, 1994), p. 2, from Paola Agosti and Giovanna Borgese, *Mi pare un secolo: Ritratti e parole di centosei protagonisti del Novecento* (Turin, 1992).

1. J. R. McNeill, *Something New under the Sun: An Environmental History of the Twentieth-Century World* (New York: W. W. Norton, 2000), p. 4.

2. Robert Wright, *Nonzero: The Logic of Human Destiny* (New York: Random House, 2000), p. 51.

3. D. J. Bradley, cited in Andrew Cliff and Peter Haggett, "Disease Implications of Global Change," in *Geographies of Global Change: Remapping the World in the Late Twentieth Century*, ed. R. J. Johnston, Peter J. Taylor, and Michael J. Watts (Oxford: Blackwell, 1995), pp. 206–23, data from p. 207; chart from p. 208.

4. Walter Benjamin, "Theses on the Philosophy of History," in *Illuminations*, ed. Hannah Arendt, trans. Harry Zohn (London: Jonathan Cape, 1970), no. IX, pp. 259–60.

5. Hobsbawm refers to the "snapping of the links between generations, that is to say, between past and present" (*Age of Extremes*, p. 15).

6. Robert W. Kates, B. L. Turner II, and William C. Clark, "The Great Transformation," in *The Earth as Transformed by Human Action: Global and Regional Changes in the Biosphere over the Past 300 Years*, ed. R. L. Turner II et al. (Cambridge: Cambridge University Press, 1990), p. 11.

7. The information in this paragraph based on Lester R. Brown et al., *State of the World, 1999: A Worldwatch Institute Report on Progress toward a Sustainable Society* (London: Earthscan Publications, 1999), pp. 115–16.

8. Here, I continue to follow the periodization of Daniel R. Headrick, "Technological Change," in Turner et al., eds., *The Earth as Transformed by Human Action*, pp. 55–67.

9. Richard Barff, "Multinational Corporations and the New International Division of Labor," in Johnston, Taylor, and Watts, eds., *Geographies of Global Change*, p. 51.

10. Manuel Castells makes his argument in three volumes of *The Information Age: Economy, Society and Culture* (Oxford: Blackwell): vol. 1, *The Rise*

of the Network Society (1996); vol. 2, *The Power of Identity* (1997); and vol. 3, *End of Millennium* (1998).

11. Headrick, "Technological Change," p. 59.

12. Brown et al., *State of the World, 1999*, graph, p. 10.

13. These observations on changes in human bodies are drawn from the work of Richard and Lee Meadows Jantz, of the University of Tennessee, Knoxville, cited in J. J. Stambaugh, "Human Bodies Have Changed since 1800s, Study Shows," *San Diego Union-Tribune*, 22 December 2001, p. A21.

14. Susan Christopherson, "Changing Women's Status in a Global Economy," in Johnston, Taylor, and Watts, eds., *Geographies of Global Change*, p. 202. For a brief survey of these changes in women's status, see Hobsbawm, *Age of Extremes*, pp. 310–19.

15. Brown et al., *State of the World, 1999*, p. 10.

16. Lester R. Brown et al., *State of the World, 1995: A Worldwatch Institute Report on Progress Toward a Sustainable Society* (London: Earthscan Publications, 1995), p. 176.

17. Paul Kennedy, *Preparing for the Twenty-First Century* (London: Fontana, 1994), p. 215.

18. Brown et al., *State of the World, 1999*, p. 10.

19. Paul Harrison, *Inside the Third World: The Anatomy of Poverty*, 2nd ed. (Harmondsworth: Penguin, 1981), p. 261.

20. Brown et al., *State of the World, 1999*, p. 11.

21. *Encyclopaedia Britannica CD 98: Multimedia Edition* (Chicago: Encyclopedia Britannica, Britannica Centre, 1994–97), s.v. "Urbanization."

22. Hobsbawm, *Age of Extremes*, p. 289; more generally, see pp. 289–91.

23. Harrison, *Inside the Third World*, p. 67.

24. Brown et al., *State of the World, 1995*, p. 12.

25. The emperor is quoted in Immanuel C. Y. Hsü, *The Rise of Modern China*, 2nd ed. (New York: Oxford University Press, 1975), p. 213.

26. The official is quoted in Arnold Pacey, *Technology in World Civilization* (Cambridge, Mass.: MIT Press, 1990), p. 143. See also David T. Courtwright, *Forces of Habit: Drugs and the Making of the Modern World* (Cambridge, Mass.: Harvard University Press, 2002), pp. 31–36.

27. I have developed this argument in *Imperial and Soviet Russia: Power, Privilege, and the Challenge of Modernity* (Basingstoke: Macmillan, 1997).

28. Robert Lewis, "Technology and the Transformation of the Soviet Economy," in *The Economic Transformation of the Soviet Union, 1913–1945*, ed. R. W. Davies, Mark Harrison and S. G. Wheatcroft (Cambridge: Cambridge University Press, 1994), pp. 182–97; information from p. 194 (and see table 41, p. 310).

29. Mikhail Gorbachev, *Perestroika: New Thinking for Our Country and the World* (New York: Harper and Row, 1987), pp. 18–19.

30. See also the graphs in Charles Tilly, *Coercion, Capital, and European States, AD 990–1992*, rev. ed. (Cambridge, Mass.: Blackwell, 1992), p. 73, which present casualties of European states.

31. Brown et al., *State of the World, 1999*, pp. 154–55.

32. Tilly, *Coercion, Capital, and European States*, p. 67.

33. Brown et al., *State of the World, 1999*, pp. 155–56.

34. Brown et al., *State of the World, 1999*, pp. 159, 163.

35. John F. Richards, "Editorial Introduction," in Turner et al., eds., *The Earth as Transformed by Human Action*, p. 21.

36. Lester R. Brown et al., *Vital Signs, 1998–99: The Trends That Are Shaping Our Future* (London: Earthscan, 1998), p. 128.

37. Marine organisms provide the fullest and therefore the most precise evidence of such changes; see Richard Leakey and Roger Lewin, *The Sixth Extinction: Patterns of Life and the Future of Humankind* (New York: Doubleday, 1995), p. 45.

38. Brown et al., *State of the World, 1999*, pp. 116–17, 123.

39. Lester R. Brown, *Eco-Economy: Building an Economy for the Earth* (New York: W. W. Norton, 2001), p. 93.

40. Kennedy, *Preparing for the Twenty-First Century*, p. 112. In 2001, the people of the Pacific island nation of Tuvalu decided to leave their homeland because of rising water levels.

41. Brown et al., *State of the World, 1999*, p. 11.

42. Brown et al., *State of the World, 1999*, p. 116.

43. Kates, Turner, and Clark, "The Great Transformation," p. 12.

CHAPTER 15. FUTURES

Section epigraphs, p. 467: Murray Gell-Mann, "Transitions to a More Sustainable World," in *Scanning the Future: Twenty Eminent Thinkers on the World of Tomorrow*, ed. Yorick Blumenfeld (London: Thames and Hudson, 1999), p. 79. Page 471: Paul Harrison, *The Third Revolution: Population, Environment and a Sustainable World* (London: Penguin, 1993), p. 149; Plato, *Critias* 111A–D, quoted in Harrison, *Third Revolution*, p. 115.

1. This example and the distinction between these two kinds of unpredictability both come from Ricard Solé and Brian Goodwin, *Signs of Life: How Complexity Pervades Biology* (New York: Basic Books, 2000), chap. 1.

2. Solé and Goodwin, *Signs of Life*, p. 20.

3. Peter N. Stearns, *Millennium III, Century XXI: A Retrospective on the Future* (Boulder, Colo.: Westview Press, 1996), p. 158.

4. R. G. Collingwood, *The Idea of History* (New York: Oxford University Press, 1956), p. 54; quoted from John Lewis Gaddis, *The Landscape of History: How Historians Map the Past* (Oxford: Oxford University Press, 2002), p. 58.

5. See Yorick Blumenfeld, introduction to Blumenfeld, ed., *Scanning the Future*, pp. 7–23. See also Herman Kahn, *On Thermonuclear War* (Princeton: Princeton University Press, 1960), and Donella H. Meadows et al., *The Limits to Growth: A Report for the Club of Rome's Project on the Predicament of Mankind* (New York: Universe Books, 1972). For a recent discussion of some of the more terrifying possibilities that face us, see Martin Rees, *Our Final*

Hour: A Scientist's Warning: How Terror, Error, and Environmental Disaster Threaten Humankind's Future in This Century—on Earth and Beyond (New York: Basic Books, 2003).

6. Clive Ponting, *A Green History of the World* (Harmondsworth: Penguin, 1992).

7. Ghandi, quoted in J. R. McNeill, *Something New under the Sun: An Environmental History of the Twentieth-Century World* (New York: W. W. Norton, 2000), p. 330.

8. One of the best recent attempts to think through the problem of building a sustainable economy is Lester R. Brown, *Eco-Economy: Building an Economy for the Earth* (New York: W. W. Norton, 2001); see also the short discussion by Gell-Mann in "Transitions to a More Sustainable World," pp. 61–79.

9. Lester R. Brown and Jennifer Mitchell, "Building a New Economy," in Lester R. Brown et al., *State of the World, 1998: A Worldwatch Institute Report on Progress toward a Sustainable Society,* (London: Earthscan Publications, 1998), p. 174.

10. Gareth Porter, Janet Welsh Brown, and Pamela S. Chasek, *Global Environmental Politics,* 3rd ed. (Boulder, Colo.: Westview Press, 2000), pp. 87–93.

11. Lester R. Brown et al., *State of the World, 1995: A Worldwatch Institute Report on Progress toward a Sustainable Society* (London: Earthscan Publications, 1995), p. 172.

12. Stearns, *Millennium III, Century XXI,* p. 74.

13. Nikos Prantzos, *Our Cosmic Future: Humanity's Fate in the Universe* (Cambridge: Cambridge University Press, 2000), pp. 56, 73; on plans for the terraforming of Mars, see pp. 75–80.

14. On interstellar travel, see Prantzos, *Our Cosmic Future,* chap. 2.

15. Arthur C. Clarke, quoted in Blumenfeld, ed., introduction to *Scanning the Future,* p. 19.

16. Some of the possibilities of genetic engineering are explored interestingly in a fascinating history of the future written in the mid-1980s by Brian Stableford and David Langford, *The Third Millennium: A History of the World, AD 2000–3000* (London: Sidgwick and Jackson, 1985).

17. Prantzos, *Our Cosmic Future,* pp. 162–69; as Prantzos points out (p. 164), Fermi's question had already been raised by the French scientist Fontenelle in the eighteenth century. For a more optimistic assessment of the question of intelligent life elsewhere in the universe, see Armand Delsemme, *Our Cosmic Origins: From the Big Bang to the Emergence of Life and Intelligence* (Cambridge: Cambridge University Press, 1998), pp. 236–44.

18. Prantzos, *Our Cosmic Future,* pp. 209 ff.

19. Prantzos, *Our Cosmic Future,* p. 214.

20. Prantzos, *Our Cosmic Future,* pp. 225–29.

21. The idea of cyclical universe creation was first put forward by the physicist John Wheeler; see Ken Croswell, *The Alchemy of the Heavens* (Oxford: Oxford University Press, 1996), p. 216; Stephen Hawking once flirted with the idea that the arrow of time might reverse itself in the contracting phase as the sec-

ond law of thermodynamics went into reverse, but later rejected the notion as erroneous; see *A Brief History of Time: From the Big Bang to Black Holes* (New York: Bantam, 1988), pp. 150–51.

22. Prantzos, *Our Cosmic Future*, p. 263.

23. Paul Davies, *The Last Three Minutes* (London: Phoenix, 1995), pp. 98–99.

24. The image of a cosmic springtime is borrowed from Arthur C. Clarke, *Profiles of the Future* (1962), cited in Prantzos, *Our Cosmic Future*, p. 225.

APPENDIX 1. DATING TECHNIQUES, CHRONOLOGIES, AND TIMELINES

1. There are good surveys of modern dating techniques in Armand Delsemme, *Our Cosmic Origins: From the Big Bang to the Emergence of Life and Intelligence* (Cambridge: Cambridge University Press, 1998), p. 285; Neil Roberts, *The Holocene: An Environmental History*, 2nd ed. (Oxford: Blackwell, 1998), chap. 2; and Nigel Calder, *Timescale: An Atlas of the Fourth Dimension* (London: Chatto and Windus, 1983).

2. Isotopes have the same number of protons but varying numbers of neutrons. Thus, C^{14} is unstable, while C^{13} and C^{12} are not, but all have 6 protons (a fact that defines them all as atoms of carbon); C^{12} and C^{13} account respectively for 98.9 percent and 1.1 percent of all carbon atoms, while C^{14} exists in tiny amounts.

3. See Roberts, *The Holocene*, pp. 11–25, for a good survey of the evolution of radiocarbon dating methods.

4. Nyanatiloka, *Buddhist Dictionary: Manual of Buddhist Terms and Doctrines*, 3rd ed. (Colombo [Sri Lanka]: Frewin, 1972), s.v. "Kappa." The metaphor of a rock wearing away seems to be widespread, for the same source cites, under the same heading, a German tale told by the brothers Grimm that contains the following: "In Farther Pommerania there is the diamond-mountain, one hour high, one hour wide, one hour deep. There every hundred years a little bird comes and whets its little beak on it. And when the whole mountain is ground off, then the first second of eternity has passed."

APPENDIX 2. CHAOS AND ORDER

1. This argument about pattern has been greatly influenced by two recent attempts to identify patterns at the many different scales of big history: Fred Spier, *The Structure of Big History: From the Big Bang until Today* (Amsterdam: Amsterdam University Press, 1996), and Eric J. Chaisson, *Cosmic Evolution: The Rise of Complexity in Nature* (Cambridge, Mass.: Harvard University Press, 2001). These, in turn, owe much to the pioneering discussions in Erwin Schrödinger, *What Is Life?* in *What Is Life? The Physical Aspect of the Living Cell;* with, *Mind and Matter;* and *Autobiographical Sketches* (Cambridge: Cambridge University Press, 1992) (first published in 1944). On the emergence of complexity, see also Ilya Prigogine and Isabelle Stengers, *Order out of Chaos: Man's New Dialogue with Nature* (London: Heinemann, 1984); Paul Davies, *The Cosmic Blueprint* (London: Unwin, 1989); Ricard Solé and Brian Goodwin,

Signs of Life: How Complexity Pervades Biology (New York: Basic Books, 2000); Stuart Kauffman, *At Home in the Universe: The Search for Laws of Complexity* (London: Viking, 1995); and Roger Lewin, *Complexity: Life on the Edge of Chaos* (London: Phoenix, 1993).

2. "Entropy: defines the amount of unusable energy; entropy can never diminish in a closed system" (Armand Delsemme, *Our Cosmic Origins: From the Big Bang to the Emergence of Life and Intelligence* [Cambridge: Cambridge University Press, 1998], pp. 299–300). However, recent hints that the rate of expansion of the universe is accelerating may undermine this idea if, as I will argue later in this appendix, expansion is itself a source of negative entropy, or *negentropy*; see Nikos Prantzos, *Our Cosmic Future: Humanity's Fate in the Universe* (Cambridge: Cambridge University Press, 2000), pp. xi, 241–42.

3. For a sample of recent discussions of how order is possible, see Roger Penrose, *The Emperor's New Mind: Concerning Computers, Minds, and the Laws of Physics* (London: Vintage, 1990); Chaisson, *Cosmic Evolution;* and Martin Rees, *Just Six Numbers: The Deep Forces That Shape the Universe* (New York: Basic Books, 2000); Prantzos, *Our Cosmic Future*, pp. 239–42.

4. Davies, *The Cosmic Blueprint*, p. 119.

5. Prantzos, *Our Cosmic Future*, p. 241.

6. This view of complexity is Rod Swenson's idea, as summarized in Lynn Margulis and Dorion Sagan, *What Is Life?* (Berkeley: University of California Press, 1995), p. 16. See also Delsemme, *Our Cosmic Origins*, p. 300: "Living organisms can diminish their entropy because they can reject the unusable energy to the outside world."

7. See Prigogine and Stengers, *Order out of Chaos*.

8. The example of the word *universe* is suggested by Hubert Reeves, Joël de Rosnay, Yves Coppens, and Dominique Simonnet, *Origins: Cosmos, Earth, and Mankind* (New York: Arcade Publishing, 1998), p. 35, where Reeves compares the "primitive puree" of the early universe to alphabet soup; the example of water comes from Solé and Goodwin, *Signs of Life*, p. 13.

BIBLIOGRAPHY

This bibliography includes three main categories of references, under two headings—General Reference and Other Works. First, it lists a number of valuable texts that are relevant to several parts of this book, which may or may not have been cited. Second, it gives full references for all works appearing in the notes; some of these sources are general synoptic studies, and some are more specialized. Third, it gives full references for all the works cited under "Further Reading" at the end of each chapter. These consist almost entirely of works that should be accessible to the general reader (though not necessarily with ease); most are synoptic or introductory in nature.

GENERAL REFERENCE

Asimov, Isaac. *Asimov's New Guide to Science.* Rev. ed. Harmondsworth: Penguin, 1987.

———. *Beginnings: The Story of Origins—of Mankind, Life, the Earth, the Universe.* New York: Walker, 1987.

Barraclough, Geoffrey, ed. *Times Concise Atlas of World History.* 5th ed. London: Times Books, 1994.

Bentley, Jerry H., and Herbert F. Ziegler. *Traditions and Encounters: A Global Perspective on the Past.* 2 vols. 2nd ed. Boston: McGraw-Hill, 2003.

Brown, Lester R., et al. *State of the World, 1995: A Worldwatch Institute Report on Progress toward a Sustainable Society.* London: Earthscan Publications, 1995.

———. *State of the World, 1999: A Worldwatch Institute Report on Progress toward a Sustainable Society.* London: Earthscan Publications, 1999. [Series began in 1984.]

Burenhult, Göran, ed. *The Illustrated History of Humankind.* 5 vols. San Francisco: HarperSanFrancisco, 1993–94. [A good, up-to-date, and well-illustrated world history from an archaeological perspective.]

Calder, Nigel. *Timescale: An Atlas of the Fourth Dimension.* London: Chatto and Windus, 1983. [A remarkable chronology for the whole of time, now slightly dated.]

Cambridge Encyclopaedia of Archaeology. Edited by Andrew Sherratt. Cambridge: Cambridge University Press, 1980.

Cambridge Encyclopedia of Earth Sciences. Edited by David G. Smith. Cambridge: Cambridge University Press, 1982.

Cambridge Encyclopedia of Human Evolution. Edited by Steven Jones, Robert Martin, and David Pilbeam. Cambridge: Cambridge University Press, 1992.

Clark, Robert P. *The Global Imperative: An Interpretive History of the Spread of Humankind.* Boulder, Colo.: Westview Press, 1997. [An attempt to theorize human history, building on the notion of entropy.]

Cowan, C. Wesley, and Patty Jo Watson, eds. *The Origins of Agriculture: An International Perspective.* Washington, D.C.: Smithsonian Institution Press, 1992.

Dunn, Ross E., ed. *The New World History: A Teacher's Companion.* Boston: Bedford/St. Martin's, 2000. [A collection of essays on world history.]

Dunn, Ross E., and David Vigilante, eds. *Bring History Alive! A Sourcebook for Teaching World History.* Los Angeles: National Center for History in the Schools, UCLA, 1996. [A collection of recent essays on world history.]

Emiliani, Cesare. *The Scientific Companion: Exploring the Physical World with Facts, Figures, and Formulas.* 2nd ed. New York: John Wiley, 1995.

Livi-Bacci, Massimo. *A Concise History of World Population.* Translated by Carl Ipsen. Oxford: Blackwell, 1992.

Manning, Patrick. *Navigating World History: Past, Present, and Future of a Global Field.* Basingstoke: Palgrave Macmillan, 2003.

Mazlish, Bruce, and Ralph Buultjens, eds. *Conceptualizing Global History.* Boulder, Colo.: Westview Press, 1993.

McEvedy, Colin, and Richard Jones. *Atlas of World Population History.* Harmondsworth: Penguin, 1978.

Moore, R. L. "World History." In *Companion to Historiography,* edited by Michael Bentley, pp. 941–59. New York: Routledge, 1997.

Morrison, Philip, and Phylis Morrison. *Powers of Ten: A Book about the Relative Size of Things in the Universe and the Effect of Adding Another Zero.* Redding, Conn.: Scientific American Library; San Francisco: dist. by W. H. Freeman, 1982. [On scales from the very small to the very large.]

Myers, Norman, ed. *Gaia Atlas of First Peoples.* Harmondsworth: Penguin, 1990.

———. *Gaia Atlas of Future Worlds.* Harmondsworth: Penguin, 1990.

———. *The Gaia Atlas of Planet Management.* 2nd ed. London: Pan, 1995. [A superb overview of the state of the planet today.]

Past Worlds: The Times Atlas of Archaeology. London: Times Books, 1988. [Magnificent!]

Penguin Atlas of World History. Edited by Hermann Kinder and Werner Hilge-
mann. 2 vols. Harmondsworth: Penguin, 1978. [Cheap and accessible, with
superb maps and a detailed chronology for most of recorded history.]

Reilly, Kevin, and Lynda Norene Shaffer. "World History." In *The American
Historical Association's Guide to Historical Literature,* edited by Mary Beth
Norton, 1:42–45. 3rd ed. New York: Oxford University Press, 1995.

Renfrew, Colin. *Archaeology and Language: The Puzzle of Indo-European Origins.* Harmondsworth: Penguin, 1989.

Renfrew, Colin, and Paul Bahn. *Archaeology.* London: Thames and Hudson,
1992. [A superb introduction to archaeology.]

UNESCO. *History of Humanity: Scientific and Cultural Development.* Vol. 1,
Prehistory and the Beginnings of Civilization. Edited by S. J. De Laet. London: Routledge, 1994.

————. *History of Humanity: Scientific and Cultural Development.* Vol. 2, *From
the Third Millennium to the Seventh Century BC.* Edited by A. H. Dani and
J-.P. Mohen London: Routledge, 1996.

————. *History of Humanity: Scientific and Cultural Development.* Vol. 3, *From
the Seventh Century BC to the Seventh Century AD.* Edited by Joachim Her-
rmann and Erik Zürcher. London: Routledge, 1996.

OTHER WORKS

Abramovo, Z. A. "Two Models of Cultural Adaptation." *Antiquity* 63 (1989):
789–91.

Abu-Lughod, Janet. *Before European Hegemony: The World System, A.D.
1250–1350.* New York: Oxford University Press, 1989.

Adams, Robert M. *The Evolution of Urban Society: Early Mesopotamia and
Prehispanic Mexico.* Chicago: Aldine, 1966.

————. *Paths of Fire: An Anthropologist's Inquiry into Western Technology.*
Princeton: Princeton University Press, 1996.

Adas, Michael, ed. *Agricultural and Pastoral Societies in Ancient and Classical
History.* Philadelphia: Temple University Press, 2001.

————. *Islamic and European Expansion: The Forging of a Global Order.*
Philadelphia: Temple University Press, 1993.

Adshead, S. A. M. *China in World History.* 2nd ed. Basingstoke: Macmillan, 1995.

Allsen, Thomas T. *Culture and Conquest in Mongol Eurasia.* Cambridge: Cambridge University Press, 2001.

Alroy, John. "A Multispecies Overkill Simulation of the End-Pleistocene Megafaunal Mass Extinction." *Science,* 8 June 2001, pp. 1893–96.

Amin, Samir. "The Ancient World-Systems versus the Modern Capitalist
World-System." In *The World System: From Hundred Years or Five Thou-
sand?,* edited by Andre Gunder Frank and Barry K. Gills, pp. 247–77. London:
Routledge, 1992.

Anderson, J. L. *Explaining Long-Term Economic Change.* Basingstoke: Macmillan, 1991.

Armstrong, Terence. *Russian Settlement in the North.* Cambridge: Cambridge University Press, 1965.

Ashton, T. S. *The Industrial Revolution, 1760–1830.* London: Oxford University Press, 1948.

Aubet, María Eugenia. *The Phoenicians and the West: Politics, Colonies, and Trade.* Translated by Mary Turton. Cambridge: Cambridge University Press, 1993.

Bahn, Paul, and John Flenley. *Easter Island, Earth Island.* London: Thames and Hudson, 1992.

Bairoch, Paul. *Cities and Economic Development: From the Dawn of History to the Present.* Translated by Christopher Brauder. Chicago: University of Chicago Press, 1988.

———. "International Industrialization Levels from 1705 to 1980." *Journal of European Economic History* 11 (1982): 269–333.

Barber, Elizabeth Wayland. *Women's Work: The First 20,000 Years: Women, Cloth, and Society in Early Times.* New York: W. W. Norton, 1994.

Barff, Richard. "Multinational Corporations and the New International Division of Labour." In *Geographies of Global Change: Remapping the World in the Late Twentieth Century,* edited by R. J. Johnston, Peter J. Taylor, and Michael J. Watts, pp. 50–62. Oxford: Blackwell, 1995.

Barfield, Thomas J. *The Nomadic Alternative.* Englewood Cliffs, N.J.: Prentice-Hall, 1993.

———. *The Perilous Frontier: Nomadic Empires and China.* Oxford: Blackwell, 1989.

Barnett, S. Anthony. *The Science of Life: From Cells to Survival.* Sydney: Allen and Unwin, 1998.

Barraclough, Geoffrey. *An Introduction to Contemporary History.* 1965. Reprint, Harmondsworth: Penguin, 1967.

Barrow, John D. *The Origin of the Universe.* London: Weidenfeld and Nicolson, 1994.

———. *Theories of Everything: The Quest for Ultimate Explanation.* Oxford: Clarendon, 1991.

Barthold, W. *Turkestan down to the Mongol Invasion.* Translated by T. Minorsky. Edited by C. E. Bosworth. 4th ed. London: E. J. W. Gibb Memorial Trust, 1977.

Basalla, George. *The Evolution of Technology.* Cambridge: Cambridge University Press, 1988.

Bawden, Stephen, Stephen Dovers, and Megan Shirlow. *Our Biosphere under Threat: Ecological Realities and Australia's Opportunities.* Melbourne: Oxford University Press, 1990.

Bayley, Chris. *The Birth of the Modern World: Global Connections and Comparisons, 1780–1914.* Oxford: Blackwell, 2003.

Becker, Luann. "Repeated Blows." *Scientific American,* March 2002, pp. 76–83.

Bellwood, Peter. *Man's Conquest of the Pacific: The Prehistory of Southeast Asia and Oceania.* New York: Oxford University Press, 1979.

———. *The Polynesians: Prehistory of an Island People.* Rev. ed. London: Thames and Hudson, 1987.

Bentley, Jerry. "Cultural Encounters between the Continents over the Centuries." In *Nineteenth International Congress of Historical Sciences,* pp. 29–45. Oslo: Nasjonalbiblioteket, 2000.

———. *Old World Encounters: Cross-Cultural Contacts and Exchanges in Pre-Modern Times.* New York: Oxford University Press, 1993.

———. *Shapes of World History in Twentieth-Century Scholarship.* Washington, D.C.: American Historical Association, 1996. (Reprinted in *Agricultural and Pastoral Societies in Ancient and Classical History,* edited by Michael Adas [Philadelphia: Temple University Press, 2001], pp. 3–35.)

Berg, Maxine. *The Age of Manufactures, 1700–1820: Industry, Innovation, and Work in Britain.* 2nd ed. London: Routledge, 1994.

Berry, Thomas. *The Dream of the Earth.* San Francisco: Sierra Club Books, 1988.

Biraben, J. R. "Essai sur l'évolution du nombre des hommes." *Population* 34 (1979): 13–25.

Black, Jeremy. *War and the World: Military Power and the Fate of Continents, 1450–2000.* New Haven: Yale University Press, 1998.

Blackmore, Susan. *The Meme Machine.* Oxford: Oxford University Press, 1999.

Blank, Paul W., and Fred Spier, eds. *Defining the Pacific: Constraints and Opportunities.* Aldershot, Hants.: Ashgate, 2002. [A survey of Pacific history on scales up to those of big history.]

Blaut, J. M. *The Colonizer's Model of the World: Geographical Diffusionism and Eurocentric History.* London: Guildford Press, 1993.

Blumenfeld, Yorick, ed. *Scanning the Future: Twenty Eminent Thinkers on the World of Tomorrow.* London: Thames and Hudson, 1999.

Bogucki, Peter. *The Origins of Human Society.* Oxford: Blackwell, 1999.

Borah, Woodrow, and Sherburne F. Cook. *The Aboriginal Population of Central Mexico on the Eve of the Spanish Conquest.* Berkeley: University of California Press, 1963.

Boserup, Ester. *The Conditions of Agricultural Growth: The Economics of Agrarian Change under Population Pressure.* Chicago: Aldine, 1965.

———. *Population and Technology.* Oxford: Blackwell, 1981.

Bottomore, Tom, ed. *A Dictionary of Marxist Thought.* 2nd ed. Oxford: Blackwell, 1991.

Boyden, S. *Biohistory: The Interplay between Human Society and the Biosphere.* Man and the Biosphere Series, ed. J. N. R. Jeffers, vol. 8. Paris: UNESCO; Park Ridge, N.J.: Parthenon, 1992.

Brady, Thomas A. "Rise of Merchant Empires, 1400–1700: A European Counterpoint." In *The Political Economy of Merchant Empires: State Power and World Trade, 1350–1750,* edited by James D. Tracy, pp. 117–60. Cambridge: Cambridge University Press, 1991.

Braudel, Fernand. *Civilization and Capitalism, Fifteenth–Eighteenth Century.* 3 vols. London: Collins, 1981–84.

———. *On History.* Translated by Sarah Matthews. Chicago: University of Chicago Press, 1980.

Briggs, Asa. *A Social History of England.* 2nd ed. Harmondsworth: Penguin, 1987.

Brown, Lester R. *Eco-Economy: Building an Economy for the Earth.* New York: W. W. Norton, 2001.

Brown, Lester R., and Jennifer Mitchell. "Building a New Economy." In *State of the World, 1998: A Worldwatch Institute Report on Progress toward a Sustainable Society,* by Lester R. Brown et al., pp. 168–87. London: Earthscan Publications, 1998.

Brown, Lester R., et al. *Vital Signs, 1998–99: The Trends That Are Shaping Our Future.* London: Earthscan, 1998.

Budiansky, Stephen. *The Covenant of the Wild.* New York: Morrow, 1992. [Popular account of animal domestication.]

Bulliet, Richard, et al. *The Earth and Its Peoples: A Global History.* Boston: Houghton Mifflin, 1997.

Burenhult, Göran. "The Rise of Art." In *The Illustrated History of Humankind,* edited by Göran Burenhult. Vol. 1, *The First Humans: Human Origins and History to 10,000 BC,* pp. 97–121. St. Lucia: University of Queensland Press, 1993.

Burstein, Stanley M. "The Hellenistic Period in World History." In *Agricultural and Pastoral Societies in Ancient and Classical History,* edited by Michael Adas, pp. 275–307. Philadelphia: Temple University Press, 2001.

———, ed. *Ancient African Civilizations: Kush and Axum.* Princeton: Markus Wiener, 1998.

Bushnell, John. *Mutiny amid Repression: Russian Soldiers in the Revolution of 1905–1906.* Bloomington: Indiana University Press, 1985.

Bynum, W. F., and Roy Porter, eds. *Companion Encyclopedia of the History of Medicine.* London: Routledge, 1993.

Cairns-Smith, A. G. *Evolving the Mind: On the Nature of Matter and the Origin of Conscious.* Cambridge: Cambridge University Press, 1996.

———. *Seven Clues to the Origin of Life.* Cambridge: Cambridge University Press, 1985.

Calvin, William H. *The Ascent of Mind: Ice Age Climates and the Evolution of Intelligence.* New York: Bantam, 1991.

———. *How Brains Think: Evolving Intelligence, Then and Now.* London: Phoenix, 1998.

Campbell, Joseph. *The Hero with a Thousand Faces.* Bollingen no. 17. Princeton: Princeton University Press, 1959.

———. *The Masks of God.* Vol. 1, *Primitive Mythology.* 1959. Reprint, Harmondsworth: Penguin, 1976.

Campbell, Joseph, with Bill Moyers. *The Power of Myth.* New York: Doubleday: 1988.

Cardwell, Donald. *The Fontana History of Technology.* London: Fontana, 1994.

Carneiro, Robert. "Political Expansion as an Expression of the Principle of Competitive Exclusion." In *Origins of the State: The Anthropology of Political Evolution,* edited by Ronald Cohen and Elman R. Service, pp. 205–20. Philadelphia: Institute for the Study of Human Issues, 1978.

Castells, Manuel. *End of Millennium.* Vol. 3 of *The Information Age: Economy, Society and Culture.* Oxford: Blackwell, 1998.

———. *The Power of Identity.* Vol. 2 of *The Information Age: Economy, Society and Culture.* Oxford: Blackwell, 1997.

———. *The Rise of the Network Society.* Vol. 1 of *The Information Age: Economy, Society and Culture.* Oxford: Blackwell, 1996.

Cattermole, Peter, and Patrick Moore. *The Story of the Earth.* Cambridge: Cambridge University Press, 1986.

Cavalli-Sforza, Luigi Luca, and Francesco Cavalli-Sforza. *The Great Human Diasporas.* Translated by Sarah Thorne. Reading, Mass.: Addison-Wesley, 1995.

Chaisson, Eric J. *Cosmic Evolution: The Rise of Complexity in Nature.* Cambridge, Mass.: Harvard University Press, 2001.

———. *The Life Era: Cosmic Selection and Conscious Evolution.* New York: W. W. Norton, 1987.

———. *Universe: An Evolutionary Approach to Astronomy.* Englewood Cliffs, N.J.: Prentice-Hall, 1988.

Champion, Timothy, et al. *Prehistoric Europe.* London: Academic Press, 1984.

Chandler, Tertius. *Four Thousand Years of Urban Growth: An Historical Census.* Lewiston, N.Y.: St. David's University Press, 1987.

Chase-Dunn, Christopher. *Global Formation: Structures of the World-Economy.* Oxford: Blackwell, 1989.

Chase-Dunn, Christopher, and Thomas D. Hall. "Cross-World System Comparisons: Similarities and Differences." In *Civilizations and World Systems: Studying World-Historical Change,* edited by Stephen K. Sanderson, pp. 109–35. Walnut Creek, Calif.: Altamira, 1995.

———. *Rise and Demise: Comparing World Systems.* Boulder, Colo.: Westview Press, 1997.

———, eds. *Core/Periphery Relations in Precapitalist Worlds.* Boulder, Colo.: Westview Press, 1991.

Chaudhuri, K. N. *Asia before Europe: Economy and Civilization of the Indian Ocean from the Rise of Islam to 1750.* Cambridge: Cambridge University Press, 1990.

Chew, Sing C. *World Ecological Degradation: Accumulation, Urbanization, and Deforestation, 3000 B.C.–A.D. 2000.* Lanham, Md.: Rowman and Littlefield, 2001.

Childe, V. Gordon. *Man Makes Himself.* London: Watts, 1936.

———. *What Happened in History?* Harmondsworth: Penguin, 1942.

Christian, David. "Accumulation and Accumulators: The Metaphor Marx Muffed." *Science and Society* 54, no. 2 (summer 1990): 219–24.

———. "Adopting a Global Perspective." In *The Humanities and a Creative Na-*

tion: Jubilee Essays, edited by D. M. Schreuder, pp. 249–62. Canberra: Australian Academy of the Humanities, 1995.

———. "The Case for 'Big History.'" *Journal of World History* 2, no. 2 (fall 1991): 223–38. [Reprinted in *The New World History: A Teacher's Companion*, edited by Ross E. Dunn (Boston: Bedford/St. Martin's, 2000), pp. 575–87.]

———. *A History of Russia, Central Asia, and Mongolia*. Vol. 1, *Inner Eurasia from Prehistory to the Mongol Empire*. Oxford: Blackwell, 1998.

———. *Imperial and Soviet Russia: Power, Privilege, and the Challenge of Modernity*. Basingstoke: Macmillan, 1997.

———. *Living Water: Vodka and Russian Society on the Eve of Emancipation*. Oxford: Clarendon, 1990.

———. "The Longest Durée: A History of the Last 15 Billion Years." *Australian Historical Association Bulletin*, nos. 59–60 (August–November 1989): 27–36.

———. "Maps of Time: Human History and Terrestrial History." In *Symposium ter Gelegenheid van het 250-jarig Jubileum*, pp. 33–63. Haarlem: Koninklijke Hollandsche Maatschappij der Wetenschappen, 2002.

———. "Science in the Mirror of 'Big History.'" In *The Changing Image of the Sciences*, edited by I. H. Stamhuis, T. Koetsier, C. de Pater, and A. van Helden, pp. 143–71. Dordrecht: Kluwer Academic Publishers, 2002.

———. "Silk Roads or Steppe Roads? The Silk Roads in World History." *Journal of World History* 11, no. 1 (spring 2000): 1–26.

Christopherson, Susan. "Changing Women's Status in a Global Economy." In *Geographies of Global Change: Remapping the World in the Late Twentieth Century*, edited by R. J. Johnston, Peter J. Taylor, and Michael J. Watts, pp. 191–205. Oxford: Blackwell, 1995.

Cipolla, Carlo M. *Before the Industrial Revolution: European Society and Economy, 1000–1700*. 2nd ed. London: Methuen, 1981.

———. *The Economic History of World Population*. 6th ed. Harmondsworth: Penguin, 1974. [Dated, particularly on kin-ordered societies, but remains an interesting overview of human history.]

Claessen, Henri J. M., and Peter Skalnik, eds. *The Early State*. The Hague: Mouton, 1978.

Cliff, Andrew, and Peter Haggett. "Disease Implications of Global Change." In *Geographies of Global Change: Remapping the World in the Late Twentieth Century*, edited by R. J. Johnston, Peter J. Taylor, and Michael J. Watts, pp. 206–23. Oxford: Blackwell, 1995.

Cline, David B. "The Search for Dark Matter." *Scientific American*, March 2003, pp. 50–59.

Cloud, Preston. *Cosmos, Earth, and Man: A Short History of the Universe*. New Haven: Yale University Press, 1978.

———. *Oasis in Space: Earth History from the Beginning*. New York: W. W. Norton, 1988.

Clutton-Brock, Juliet. *Domesticated Animals from Early Times*. London: British Museum, 1981.

Coatsworth, John H. "Welfare." *American Historical Review* 101, no. 1 (February 1996): 1–17.

Coe, Michael D. *The Maya*. New York: Praeger, 1966.

———. *Mexico: From the Olmecs to the Aztecs*. 4th ed. New York: Thames and Hudson, 1994.

Cohen, H. Floris. *The Scientific Revolution: A Historiographical Inquiry*. Chicago: University of Chicago Press, 1994.

Cohen, Mark. *The Food Crisis in Prehistory*. New Haven: Yale University Press, 1977.

———. *Health and the Rise of Civilization*. New Haven: Yale University Press, 1989.

Cohen, Ronald, and Elman R. Service, eds. *Origins of the State: The Anthropology of Political Evolution*. Philadelphia: Institute for the Study of Human Issues, 1978.

Coles, Peter. *Cosmology: A Very Short Introduction*. Oxford: Oxford University Press, 2001.

Collins, Randall. *Macrohistory: Essays in the Sociology of the Long Run*. Stanford: Stanford University Press, 1999.

Constantine Porphyrogenitus. *De Administrando Imperio*. Edited by G. Moravcsik. Translated by R. J. H. Jenkins. Rev. ed. Washington, D.C.: Dumbarton Oaks Center for Byzantine Studies, 1967.

Costello, Paul. *World Historians and Their Goals: Twentieth-Century Answers to Modernism*. De Kalb: Northern Illinois University Press, 1994.

Courtwright, David T. *Forces of Habit: Drugs and the Making of the Modern World*. Cambridge, Mass.: Harvard University Press, 2002.

Crafts, N. F. R. *British Economic Growth during the Industrial Revolution*. Oxford: Clarendon, 1985.

Crawford, Ian. "Where Are They?" *Scientific American*, July 2000, pp. 38–43.

Cronon, William. "A Place for Stories: Nature, History, and Narrative." *Journal of American History* 78, no. 4 (March 1992): 1347–76.

Crosby, Alfred W. *The Columbian Exchange: Biological and Cultural Consequences of 1492*. Westport, Conn.: Greenwood Press, 1972.

———. *Ecological Imperialism: The Biological Expansion of Europe, 900–1900*. Cambridge: Cambridge University Press, 1986.

———. *The Measure of Reality: Quantification in Western Europe, 1250–1600*. Cambridge University Press, 1997.

Croswell, Ken. *The Alchemy of the Heavens*. Oxford: Oxford University Press, 1996.

———. "Uneasy Truce." *New Scientist*, 30 May 1998, pp. 42–46.

Curtin, Philip D. *Cross-Cultural Trade in World History*. Cambridge: Cambridge University Press, 1985.

Dalziel, Ian W. D. "Earth before Pangea." *Scientific American*, January 1995, pp. 38–43.

Darwin, Charles. *The Origin of Species by Means of Natural Selection: The*

Preservation of Favored Races in the Struggle for Life. Edited and with an introduction by J. W. Burrow. Harmondsworth: Penguin, 1968. [First published in 1859.]

Davies, Norman. *Europe: A History.* 1996. Reprint, London: Pimlico, 1997.

Davies, Paul. *About Time.* London: Viking, 1995.

———. *The Cosmic Blueprint.* London: Unwin, 1989.

———. *The Fifth Miracle: The Search for the Origin of Life.* Harmondsworth: Penguin, 1999.

———. *The Last Three Minutes.* London: Phoenix, 1995.

Davis, Mike. *Late Victorian Holocausts: El Niño Famines and the Making of the Third World.* London: Verso, 2001.

Davis, Natalie Zemon. Discussant's comment on "Cultural Encounters between the Continents over the Centuries." In *Nineteenth International Congress of Historical Sciences,* pp. 46–47. Oslo: Nasjonalbiblioteket, 2000.

Davis, Ralph. *The Rise of the Atlantic Economies.* Ithaca, N.Y.: Cornell University Press, 1973.

Davis-Kimball, Jeannine, with Mona Behan. *Warrior Women: An Archaeologist's Search for History's Hidden Heroines.* New York: Warner, 2002.

Dawkins, Richard. *River out of Eden: A Darwinian View of Life.* New York: Bantam, 1995.

———. *The Selfish Gene.* 2nd ed. Oxford: Oxford University Press, 1989.

Dayton, Leigh. "Mass Extinctions Pinned on Ice Age Hunters." *Science,* 8 June 2001, p. 1819.

Deacon, Terrence W. *The Symbolic Species: The Co-evolution of Language and the Brain.* Harmondsworth: Penguin, 1997.

Delsemme, Armand. *Our Cosmic Origins: From the Big Bang to the Emergence of Life and Intelligence.* Cambridge: Cambridge University Press, 1998.

Denemark, Robert A., et al., eds. *World System History: The Social Science of Long-Term Change.* London: Routledge, 2000.

Dennell, Robin C. *European Economic Prehistory: A New Approach.* New York: Academic Press, 1983.

Dennett, Daniel C. *Consciousness Explained.* London: Penguin, 1993.

———. *Darwin's Dangerous Idea: Evolution and the Meaning of Life.* London: Allen Lane, 1995.

———. *Kinds of Minds: Toward an Understanding of Consciousness.* London: Weidenfeld, 1997.

DeVries, B., and J. Goudsblom, eds. *Mappae Mundi: Humans and Their Habitats in a Long-Term Socio-Ecological Perspective.* Amsterdam: Amsterdam University Press, 2002.

de Vries, Jan. "The Industrial Revolution and the Industrious Revolution." *Journal of Economic History* 54, no. 2 (June 1994): 249–70.

Diamond, Jared. *Guns, Germs, and Steel: The Fates of Human Societies.* London: Vintage, 1998.

———. "Human Use of World Resources." *Nature,* 6 August 1987, pp. 479–80.

———. *The Rise and Fall of the Third Chimpanzee.* London: Vintage, 1991.

————. *Why Is Sex Fun? The Evolution of Human Sexuality.* London: Weidenfeld and Nicolson, 1997.

Díaz, Bernal. *The Conquest of New Spain.* Translated by J. M. Cohen. Harmondsworth: Penguin, 1963.

di Cosmo, Nicola. "European Technology and Manchu Power: Reflections on the 'Military Revolution' in Seventeenth Century China." Paper presented at the International Congress of Historical Sciences, Oslo, August 2000.

————. "State Formation and Periodization in Inner Asian History." *Journal of World History* 10, no. 1 (spring 1999): 1–40.

Diesendorf, Mark, and Clive Hamilton, eds. *Human Ecology, Human Economy.* Sydney: Allen and Unwin, 1997.

Dingle, Tony. *Aboriginal Economy.* Fitzroy, Vic.: McPhee Gribble/Penguin, 1988.

Dolukhanov, P. M. "The Late Mesolithic and the Transition to Food Production in Eastern Europe." In *Hunters in Transition: Mesolithic Societies of Temperate Eurasia and Their Transition to Farming,* edited by Marek Zvelebil, pp. 109–20. Cambridge: Cambridge University Press, 1986.

Dunn, Ross E. *The Adventures of Ibn Battuta: A Muslim Traveler of the Fourteenth Century.* Berkeley: University of California Press, 1986.

Dyson, Freeman. *Origins of Life.* 2nd ed. Cambridge: Cambridge University Press, 1999.

Earle, Timothy. *How Chiefs Come to Power: The Political Economy in Prehistory.* Stanford: Stanford University Press, 1997.

Eckhardt, William. "A Dialectical Evolutionary Theory of Civilizations, Empires, and Wars." In *Civilizations and World Systems: Studying World-Historical Change,* edited by Stephen K. Sanderson, pp. 79–82. Walnut Creek, Calif.: Altamira Press, 1995.

Ehrenberg, Margaret. *Women in Prehistory.* Norman: University of Oklahoma Press, 1989.

Ehret, Christopher. *An African Classical Age: Eastern and Southern Africa in World History, 1000 B.C. to A.D. 400.* Charlottesville: University Press of Virginia, 1998.

————. "Sudanic Civilization." In *Agricultural and Pastoral Societies in Ancient and Classical History,* edited by Michael Adas, pp. 224–74. Philadelphia: Temple University Press, 2001.

Ehrlich, Paul. *Human Natures: Genes, Cultures, and the Human Prospect.* Washington, D.C.: Island Press, 2000.

————. *The Machinery of Nature.* New York: Simon and Schuster, 1986.

Ehrlich, Paul R., and Anne H. Ehrlich. *The Population Explosion.* New York: Simon and Schuster, 1990.

Eibl-Eibesfeldt, Irenäus. "Aggression and War: Are They Part of Being Human?" In *The Illustrated History of Humankind,* edited by Gören Burenhult. Vol. 1, *The First Humans: Human Origins and History to 10,000 BC,* pp. 26–29. St. Lucia: University of Queensland Press, 1993.

Eliade, Mircea. *The Myth of the Eternal Return, or, Cosmos and History.* Translated by Willard R. Trask. New York: Harper, 1954.

Elias, Norbert. *The Civilizing Process.* Vol. 1, *The History of Manners.* Translated by Edmund Jephcott. Oxford: Blackwell, 1978.

———. *The Civilizing Process.* Vol. 2, *State Formation and Civilization.* Translated by Edmund Jephcott. Oxford: Blackwell, 1982.

———. *The Civilizing Process: Sociogenetic and Psychogenetic Investigations.*
Translated by Edmund Jephcott. Edited by Eric Dunning, Johan Goudsblom,
and Stephen Mennell. 2nd ed. Oxford: Blackwell, 2000.

———. *Norbert Elias on Civilization, Power, and Knowledge: Selected Writings.* Edited by Stephen Mennell and Johan Goudsblom. Chicago: University of Chicago Press, 1998.

———. *The Norbert Elias Reader: A Biographical Selection.* Edited by Johan
Goudsblom and Stephen Mennell. Oxford: Blackwell, 1998.

———. *Time: An Essay.* Translated by Edmund Jephcott. Oxford: Blackwell,
1992.

Elvin, Mark. *The Pattern of the Chinese Past.* Stanford: Stanford University
Press, 1973.

Emiliani, Cesare. *Planet Earth: Cosmology, Geology, and the Evolution of Life
and Environment.* Cambridge: Cambridge University Press, 1992.

Evans, L. T. *Feeding the Ten Billion: Plants and Population Growth.* Cambridge:
Cambridge University Press, 1998.

Fagan, Brian M. *Floods, Famines, and Emperors: El Niño and the Fate of Civilizations.* New York: Basic Books, 1999.

———. *The Journey from Eden: The Peopling of Our World.* London: Thames
and Hudson, 1990.

———. *People of the Earth: An Introduction to World Prehistory.* 10th ed. Upper Saddle River, N.J.: Prentice Hall, 2001. [A good, comprehensive, and up-
to-date textbook on prehistory.]

Ferris, Timothy. *Coming of Age in the Milky Way.* New York: William Morrow, 1988.

———. *The Whole Shebang: A State-of-the-Universe(s) Report.* New York: Simon and Schuster, 1997.

Feynman, Richard P. *Six Easy Pieces: The Fundamentals of Physics Explained.*
London: Penguin, 1998. [A very good introduction to basic concepts of modern physics by one of its pioneers.]

Finley, M. I. *The Ancient Economy.* London: Chatto and Windus, 1973.

———. "Empire in the Greco-Roman World." *Greece and Rome,* 2nd ser., 25,
no. 1 (April 1978): 1–15.

Finney, Ben. "The Other One-Third of the Globe." *Journal of World History*
5, no. 2 (fall 1994): 273–98.

Flannery, Tim. *The Eternal Frontier: An Ecological History of North America
and Its Peoples.* New York: Atlantic Monthly Press, 2001.

———. *The Future Eaters: An Ecological History of the Australasian Lands
and People.* Chatswood, N.S.W.: Reed, 1995.

Fletcher, Roland. *The Limits of Settlement Growth: A Theoretical Outline.* Cambridge: Cambridge University Press, 1995.

———. "Mammoth Bone Huts." In *The Illustrated History of Humankind*, edited by Göran Burenhult. Vol. 1, *The First Humans: Human Origins and History to 10,000 BC*, pp. 134–35. St. Lucia: University of Queensland Press, 1993.

Flood, Josephine. *Archaeology of the Dreamtime*. Sydney: Collins, 1983.

Floud, Roderick, and Donald McCloskey, eds. *The Economic History of Britain since 1700*. 2nd ed. Cambridge: Cambridge University Press, 1994.

Flynn, Dennis O., and Arturo Giráldez. "Born with a 'Silver Spoon': The Origin of World Trade in 1571." *Journal of World History* 6, no. 2 (fall 1995): 201–21.

———. "Cycles of Silver: Global Economic Unity through the Mid–Eighteenth Century." *Journal of World History* 13, no. 2 (fall 2002): 391–427.

———. *Metals and Monies in an Emerging Global Economy*. Brookfield, Vt.: Variorum, 1997.

Fodor, Jerry A. *The Modularity of Mind: An Essay on Faculty Psychology*. Cambridge, Mass.: MIT Press, 1983.

Foley, Robert. *Humans before Humanity*. Oxford: Blackwell, 1995.

———. "In the Shadow of the Modern Synthesis? Alternative Perspectives on the Last Fifty Years of Paleoanthropology." *Evolutionary Anthropology* 10, no. 1 (2001): 5–15.

Foltz, Richard. *Religions of the Silk Road: Overland Trade and Cultural Exchange from Antiquity to the Fifteenth Century*. New York: St. Martin's Press, 1999.

Forsyth, James. *A History of the Peoples of Siberia: Russia's North Asian Colony, 1581–1990*. Cambridge: Cambridge University Press, 1992.

Fortey, Richard A. *Life: An Unauthorised Biography: A Natural History of the First Four Thousand Million Years of Life on Earth*. London: Flamingo, 1998.

Frank, Andre Gunder. *ReOrient: Global Economy in the Asian Age*. Berkeley: University of California Press, 1998.

Frank, Andre Gunder, and Barry K. Gills, eds. *The World System: Five Hundred Years or Five Thousand?* London: Routledge, 1992.

Freedman, Wendy L. "The Expansion Rate and Size of the Universe." *Scientific American*, November 1992, p. 54.

Gaddis, John Lewis. *The Landscape of History: How Historians Map the Past*. Oxford: Oxford University Press, 2002.

Gamble, Clive. *The Paleolithic Settlement of Europe*. Cambridge: Cambridge University Press, 1986.

———. *Timewalkers: The Prehistory of Global Colonization*. Harmondsworth: Penguin, 1995.

Gell-Mann, Murray. "Transitions to a More Sustainable World." In *Scanning the Future: Twenty Eminent Thinkers on the World of Tomorrow*, edited by Yorick Blumenfeld, pp. 61–79. London: Thames and Hudson, 1999. [Extracts from *The Quark and the Jaguar: Adventures in the Simple and the Complex* (1994).]

Gellner, Ernest. *Plough, Sword, and Book: The Structure of Human History*. London: Paladin, 1991.

Gerschenkron, Alexander. *Economic Backwardness in Historical Perspective, a Book of Essays.* Cambridge, Mass.: Harvard University Press, Belknap Press, 1962.

Gibbons, Ann. "In Search of the First Hominids." *Science,* 15 February 2002, pp. 1214–19.

Giddens, Anthony. *Beyond Left and Right: The Future of Radical Politics.* Cambridge: Polity, 1994.

———. *A Contemporary Critique of Historical Materialism.* 2nd ed. Basingstoke: Macmillan, 1995.

———. *The Nation-State and Violence.* Vol. 2 of *A Contemporary Critique of Historical Materialism.* Cambridge: Polity Press, 1985. [Taken together, these three volumes by Giddens offer a theory of the nature of modernity and modern society.]

Gills, Barry K., and Andre Gunder Frank. "The Cumulation of Accumulation." In *The World System: Five Hundred Years or Five Thousand?,* edited by Andre Gunder Frank and Barry K. Gills, pp. 81–114. London: Routledge, 1992.

———. "World System Cycles, Crises, and Hegemonic Shifts, 1700 BC to 1700 AD." In *The World System: Five Hundred Years or Five Thousand?,* edited by Andre Gunder Frank and Barry K. Gills, pp. 143–99. London: Routledge, 1992.

Gimbutas, Marija. *The Civilization of the Goddess: The World of Old Europe.* Edited by Joan Marler. San Francisco: Harper and Row, 1991.

Gleick, James. *Chaos: Making a New Science.* New York: Penguin, 1988.

Golden, Peter B. "Nomads and Sedentary Societies in Eurasia." In *Agricultural and Pastoral Societies in Ancient and Classical History,* edited by Michael Adas, pp. 71–114. Philadelphia: Temple University Press, 2001.

Goldstone, Jack A. *Revolution and Rebellion in the Early Modern World.* Berkeley: University of California Press, 1991.

Goody, Jack. *The East in the West.* Cambridge: Cambridge University Press, 1996.

Gorbachev, Mikhail. *Perestroika: New Thinking for Our Country and the World.* New York: Harper and Row, 1987.

Goudie, Andrew. *The Human Impact on the Natural Environment.* 5th ed. Oxford: Blackwell, 2000.

———, ed. *The Human Impact Reader: Readings and Case Studies.* Oxford: Blackwell, 1997.

Goudie, Andrew, and Heather Viles, eds. *The Earth Transformed: An Introduction to Human Impacts on the Environment.* Oxford: Blackwell, 1997.

Goudsblom, Johan. *Fire and Civilization.* Harmondsworth: Allen Lane, 1992.

Goudsblom, Johan, Eric Jones, and Stephen Mennell. *The Course of Human History: Economic Growth, Social Process, and Civilization.* Armonk, N.Y.: M. E. Sharpe, 1996.

Gould, Stephen Jay. *Ever Since Darwin: Reflections in Natural History.* New York: W. W. Norton, 1977.

———. *Life's Grandeur: The Spread of Excellence from Plato to Darwin.* London: Jonathan Cape, 1996. [The U.S. edition, which has the same subtitle, is titled *Full House.*]

————. *The Mismeasure of Man*. New York: W. W. Norton, 1981.

————. *The Panda's Thumb: More Reflections in Natural History*. Harmondsworth: Penguin, 1980.

————. *Time's Arrow, Time's Cycle: Myth and Metaphor in the Discovery of Geological Time*. Cambridge, Mass.: Harvard University Press, 1987.

————. *Wonderful Life: The Burgess Shale and the Nature of History*. London: Hutchinson, 1989.

Greenberg, Joseph, and Merritt Ruhlen. "Linguistic Origins of Native Americans." *Scientific American*, November 1992, pp. 94–99.

Griaule, Marcel. *Conversations with Ogotemmêli*. 1965. Reprint, London: Oxford University Press for the International African Institute, 1975.

Gribbin, John. *Genesis: The Origins of Man and the Universe*. New York: Delta, 1981. [A scientist's introduction to the history of the universe, the stars, and the Earth.]

————. *In Search of the Big Bang: Quantum Physics and Cosmology*. London: Corgi, 1987.

Halle-Selassie, Yohannes. "Late Miocene Hominids from the Middle Awash, Ethiopia." *Nature*, 12 July 2001, pp. 178–81.

Hansen, Valerie. *The Open Empire: A History of China to 1600*. New York: W. W. Norton, 2000.

Haraway, Donna J. *Simians, Cyborgs, and Women: The Reinvention of Nature*. New York: Routledge, 1991.

Harris, David, and Gordon Hillman, eds. *Foraging and Farming: The Evolution of Plant Exploitation*. London: Unwin Hyman, 1989.

Harris, Marvin. *Culture, People, Nature*. 5th ed. New York: Harper and Row, 1988. [A clear, simple, but opinionated introduction to anthropology.]

————. "The Origin of Pristine States." In *Cannibals and Kings*, edited by Marvin Harris, pp. 101–23. New York: Vintage, 1978.

Harrison, Paul. *Inside the Third World: The Anatomy of Poverty*. 2nd ed. Harmondsworth: Penguin, 1981.

————. *The Third Revolution: Population, Environment, and a Sustainable World*. London: I. B. Tauris, 1992.

al-Hassan, Ahmand Y., and Donald R. Hill. *Islamic Technology: An Illustrated History*. Cambridge: Cambridge University Press; Paris: UNESCO, 1986.

Haub, Carl. "How Many People Have Ever Lived on Earth?" *Population Today*, February 1995, p. 4.

Hawke, Gary. "Reinterpretations of the Industrial Revolution." In *The Industrial Revolution and British Society*, edited by Patrick O'Brien and Roland Quinault, pp. 54–78. Cambridge: Cambridge University Press, 1993.

Hawking, Stephen. *A Brief History of Time: From the Big Bang to Black Holes*. New York: Bantam, 1988.

————. "The Direction of Time." *New Scientist*, 9 July 1987, pp. 46–49.

————. "The Edge of Spacetime." In *The New Physics*, edited by Paul Davies, pp. 61–69. Cambridge: Cambridge University Press, 1989.

————. *The Universe in a Nutshell*. New York: Bantam, 2001.

Headrick, Daniel R. "Technological Change." In *The Earth as Transformed by Human Action: Global and Regional Changes in the Biosphere over the Past 300 Years*, edited by B. L. Turner II et al., pp. 55–67. Cambridge: Cambridge University Press, 1990.

———. *The Tentacles of Progress: Technology Transfer in the Age of Imperialism, 1850–1940*. New York: Oxford University Press, 1988.

———. *The Tools of Empire: Technology and European Imperialism in the Nineteenth Century*. New York: Oxford University Press, 1981.

Heiser, Charles B. *Seed to Civilization: The Story of Food*. New ed. Cambridge, Mass.: Harvard University Press, 1990.

Held, David, and Anthony McGrew, eds. *The Global Transformations Reader: An Introduction to the Globalization Debate*. Cambridge: Polity Press, 2000.

Held, David, Anthony McGrew, David Goldblatt, and Jonathan Perraton. *Global Transformations: Politics, Economics and Culture*. Cambridge: Polity Press, 1999.

Henry, Donald O. *From Foraging to Agriculture: The Levant at the End of the Ice Age*. Philadelphia: University of Pennsylvania Press, 1989.

Hippocratic Writings. Edited and with an introduction by G. E. R. Lloyd. Translated by J. Chadwick and W. N. Mann. Harmondsworth: Penguin, 1978.

Hobsbawm, E. J. *The Age of Capital*. London: Abacus, 1977.

———. *The Age of Empire*. London: Weidenfeld and Nicolson, 1987.

———. *The Age of Extremes*. London: Weidenfeld and Nicolson, 1994.

———. *The Age of Revolution, 1789–1848*. 1962. Reprint, New York: New American Library, [1964].

———. *Industry and Empire*. Harmondsworth: Penguin, 1969.

Hodgson, Marshall G. S. *Rethinking World History: Essays on Europe, Islam, and World History*. Edited by Edmund Burke III. Cambridge: Cambridge University Press, 1993.

———. *The Venture of Islam: Conscience and History in a World Civilization*. 3 vols. Chicago: University of Chicago Press, 1974.

Hollister, C. Warren. *Medieval Europe: A Short History*. 5th ed. New York: John Wiley, 1982.

Hsü, Immanuel C. Y. *The Rise of Modern China*. 2nd ed. New York: Oxford University Press, 1975.

Hudson, Pat. *The Industrial Revolution*. London: Routledge, 1992.

Hughes, J. Donald. *An Environmental History of the World: Humankind's Changing Role in the Community of Life*. London: Routledge, 2001.

———, ed. *The Face of the Earth: Environment and World History*. Armonk, N.Y.: M. E. Sharpe, 1999. [Essays on an environmental approach to world history.]

Hughes, Sarah Shaver, and Brady Hughes. *Women in World History*. 2 vols. Armonk, N.Y.: M. E. Sharpe, 2000.

Hughes-Warrington, Marnie. "Big History." *Historically Speaking*, November 2002, pp. 16–20.

———. *Fifty Key Thinkers on History*. London: Routledge, 2000.

Humphrey, Nicholas. *A History of the Mind.* London: Chatto and Windus, 1992.

Humphrey, S. C. "History, Economics, and Anthropology: The Work of Karl Polanyi." *History and Theory* 8 (1969): 165–212.

Hunt, Lynn. "Send in the Clouds." *New Scientist,* 30 May 1998, pp. 28–33.

Huppert, George. *After the Black Death: A Social History of Early Modern Europe.* Bloomington: Indiana University Press, 1986.

Independent Commission on International Development. *Common Crisis North-South: Cooperation for World Recovery.* London: Pan, 1983.

———. *Issues North-South: A Programme for Survival: The Report of the Independent Commission on International Development Issues.* London: Pan, 1980.

Irwin, Geoffrey. *The Prehistoric Exploration and Colonisation of the Pacific.* Cambridge: Cambridge University Press, 1992.

Jacob, Margaret C. *The Cultural Meaning of the Scientific Revolution.* Philadelphia: Temple University Press, 1988.

———. *Scientific Culture and the Making of the Industrial West.* New York: Oxford University Press, 1997.

Jantsch, Erich. *The Self-Organizing Universe: Scientific and Human Implications of the Emerging Paradigm of Evolution.* Oxford: Pergamon Press, 1980.

Jaspers, Karl. *The Origin and Goal of History.* Translated by Michael Bullock. New Haven: Yale University Press, 1953.

Jenkins, Keith, ed. *The Postmodern History Reader.* London: Routledge, 1997.

Johanson, Donald C., and Maitland A. Edey. *Lucy: The Beginnings of Humankind.* New York: Simon and Schuster, 1981.

Johanson, Donald, and James Shreeve. *Lucy's Child: The Discovery of a Human Ancestor.* Harmondsworth: Penguin, 1989.

Johnson, Allen W., and Timothy Earle. *The Evolution of Human Societies: From Foraging Group to Agrarian State.* 2nd ed. Stanford: Stanford University Press, 2000.

Johnston, R. J., Peter J. Taylor, and Michael J. Watts, eds. *Geographies of Global Change: Remapping the World in the Late Twentieth Century.* Oxford: Blackwell, 1995.

Jones, E. L. *The European Miracle: Environments, Economies, and Geopolitics in the History of Europe and Asia.* 2nd ed. Cambridge: Cambridge University Press, 1987.

———. *Growth Recurring: Economic Change in World History.* Oxford: Clarendon, 1988.

Jones, Eric, Lionel Frost, and Colin White. *Coming Full Circle: An Economic History of the Pacific Rim.* Boulder, Colo.: Westview Press, 1993.

Jones, Rhys. "Fire Stick Farming." *Australian Natural History,* September 1969, pp. 224–28.

———. "Folsom and Talgai: Cowboy Archaeology in Two Continents." In *Approaching Australia: Papers from the Harvard Australian Studies Symposium,* edited by Harold Bolitho and Chris Wallace-Crabbe, pp. 3–50. Cambridge, Mass.: Harvard University Press, 1997.

Jones, Steve. *Almost Like a Whale: The Origin of Species Updated.* London: Anchor, 2000.

Kahn, Herman. *On Thermonuclear War.* Princeton: Princeton University Press, 1960.

Kamen, Henry. *European Society, 1500–1700.* London: Hutchinson, 1984.

Karttunen, Frances, and Alfred W. Crosby. "Language Death, Language Genesis, and World History." *Journal of World History* 6, no. 2 (fall 1995): 157–74.

Kates, Robert W., B. L. Turner II, and William C. Clark. "The Great Transformation." In *The Earth as Transformed by Human Action: Global and Regional Changes in the Biosphere over the Past 300 Years,* edited by R. L. Turner II et al., pp. 1–17. Cambridge: Cambridge University Press, 1990.

Kauffman, Stuart. *At Home in the Universe: The Search for Laws of Complexity.* London: Viking, 1995.

Kennedy, Paul. *Preparing for the Twenty-First Century.* London: Fontana, 1994.

———. *The Rise and Fall of the Great Powers: Economic Change and Military Conflict from 1500 to 2000.* London: Unwin Hyman, 1988.

Khazanov, Anatoly M. *Nomads and the Outside World.* Translated by Julia Crookenden. 2nd ed. Madison: University of Wisconsin Press, 1994.

Kicza, John E. "The Peoples and Civilizations of the Americas before Contact." In *Agricultural and Pastoral Societies in Ancient and Classical History,* edited by Michael Adas, pp. 183–222. Philadelphia: Temple University Press, 2001.

Kiple, Kenneth F. Introduction to *The Cambridge World History of Human Disease,* edited by Kenneth F. Kiple, pp. 1–7. Cambridge: Cambridge University Press, 1993.

———, ed. *The Cambridge World History of Human Disease.* Cambridge: Cambridge University Press, 1993.

Klein, Richard G. *The Human Career: Human Biological and Cultural Origins.* 2nd ed. Chicago: University of Chicago Press, 1999.

———. *Ice Age Hunters of the Ukraine.* Chicago: University of Chicago Press, 1973.

Knapp, A. Bernard. *The History and Culture of Ancient Western Asia and Egypt.* Chicago: Dorsey Press, 1988.

Knudtson, Peter, and David Suzuki. *Wisdom of the Elders.* New York: Bantam, 1992.

Kohl, Philip L., ed. *The Bronze Age Civilization of Central Asia: Recent Soviet Discoveries.* Armonk, N.Y.: M. E. Sharpe, 1981.

Kuhn, Thomas. *The Structure of Scientific Revolutions.* 2nd ed. Chicago: University of Chicago Press, 1970.

Kuppuram, G., and K. Kumudamani. *History of Science and Technology in India.* 12 vols. Delhi: Sundeep Prakashan, 1990.

Kutter, G. Siegfried. *The Universe and Life: Origins and Evolution.* Boston: Jones and Bartlett, 1987.

Lambert, David. *The Cambridge Guide to Prehistoric Man.* Cambridge: Cambridge University Press, 1987.

———. *The Cambridge Guide to the Earth.* Cambridge: Cambridge University Press, 1988.

Landes, David S. *Revolution in Time: Clocks and the Making of the Modern World.* Cambridge, Mass.: Harvard University Press, Belknap Press, 1983.

———. *The Unbound Prometheus: Technological Change and Industrial Development in Western Europe from 1750 to the Present.* London: Cambridge University Press, 1969.

———. *The Wealth and Poverty of Nations: Why Some Are So Rich and Some Are So Poor.* New York: Little, Brown, 1998.

Leakey, R. E. *The Making of Mankind.* London: M. Joseph, 1981. [Revised, with Roger Lewin, as *Origins Reconsidered.*]

———. *The Origin of Humankind.* New York: Basic Books, 1994. [Superb introductions to human origins by a pioneer.]

Leakey, Richard, and Roger Lewin. *Origins Reconsidered.* London: Abacus, 1992.

———. *The Sixth Extinction: Patterns of Life and the Future of Humankind.* New York: Doubleday, 1995.

Lee, Richard. *The !Kung San: Men, Women, and Work in a Foraging Society.* Cambridge: Cambridge University Press, 1979.

Le Roy Ladurie, Emmanuel. *The Peasants of Languedoc.* Translated by John Day. Urbana: University of Illinois Press, 1974.

Leutenegger, Walter. "Sexual Dimorphism: Comparative and Evolutionary Perspectives." In *The Illustrated History of Humankind,* edited by Göran Burenhult. Vol. 1, *The First Humans: Human Origins and History to 10,000 BC,* p. 41. St. Lucia: University of Queensland Press, 1993.

Levathes, Louise. *When China Ruled the Seas: The Treasure Fleet of the Dragon Throne, 1405–1433.* New York: Simon and Schuster, 1994.

Lewin, Roger. *Complexity: Life on the Edge of Chaos.* London: Phoenix, 1993.

———. *Human Evolution: An Illustrated Introduction.* 4th ed. Oxford: Blackwell, 1999.

Lewis, Archibald R. *Nomads and Crusaders, A.D. 1000–1368.* Bloomington: Indiana University Press, 1991.

Lewis, Martin W., and Kären E. Wigen. *The Myth of Continents: A Critique of Metageography.* Berkeley: University of California Press, 1997.

Lewis, Robert. "Technology and the Transformation of the Soviet Economy." In *The Economic Transformation of the Soviet Union, 1913–1945,* edited by R. W. Davies, Mark Harrison, and S. G. Wheatcroft, pp. 182–97. Cambridge: Cambridge University Press, 1994.

Liebes, Sidney, Elisabet Sahtouris, and Brian Swimme. *A Walk through Time: From Stardust to Us: The Evolution of Life on Earth.* New York: John Wiley, 1998.

Lineweaver, Charles. "Our Place in the Universe." In *To Mars and Beyond: Search for the Origins of Life,* edited by Malcolm Walter, pp. 88–99. Canberra: National Museum of Australia, 2002.

Lis, Catharina, and Hugo Soly. *Poverty and Capitalism in Pre-Industrial Europe.* [Translated by James Coonan.] Atlantic Highlands, N.J.: Humanities Press, 1979.

Liu, Xinru. "The Silk Road: Overland Trade and Cultural Interactions in Eurasia." In *Agricultural and Pastoral Societies in Ancient and Classical History*, edited by Michael Adas, pp. 151–79. Philadelphia: Temple University Press, 2001.

Livingston, John A. *Rogue Primate: An Exploration of Human Domestication.* Boulder, Colo.: Roberts Rinehart, 1994.

Lloyd, Christopher. "Can There Be a Unified Theory of Cosmic-Ecological World History? A Critique of Fred Spier's Construction of 'Big History.'" *Focaal*, no. 29 (1997): 171–80.

———. *The Structures of History.* Oxford: Blackwell, 1993.

Lockard, Craig. "Global Historians and the Great Divergence." *World History Bulletin* 17, no. 1 (fall 2000): 17, 32–34.

Lomborg, Bjørn. *The Skeptical Environmentalist: Measuring the Real State of the World.* Cambridge: Cambridge University Press, 2001.

Long, Charles H. *Alpha: The Myths of Creation.* 1963. Reprint, Chico, Calif.: Scholars Press and the American Academy of Religion, 1983. [One of the best and most readily available anthologies of creation myths in English.]

Lopez, Robert S. *The Commercial Revolution of the Middle Ages, 950–1350.* Englewood Cliffs, N.J.: Prentice-Hall, 1971.

Lourandos, Harry. *Continent of Hunter-Gatherers.* Cambridge: Cambridge University Press, 1997.

Lovelock, J. E. *The Ages of Gaia: A Biography of Our Living Earth.* Oxford: Oxford University Press, 1988.

———. *Gaia: A New Look at Life on Earth.* 1979. Reprint, Oxford: Oxford University Press, 1987.

———. *Gaia: The Practical Science of Planetary Medicine.* London: Unwin, 1991. [Lovelock's books provide a rich, if controversial, theory about the role of life in the history of the planet.]

Lunine, Jonathan I. *Earth: Evolution of a Habitable World: New Perspectives in Australian Prehistory.* Cambridge: Cambridge University Press, 1999.

Lyotard, Jean-François. *The Postmodern Condition: A Report on Knowledge.* Translated by Geoff Bennington and Brian Massumi. Minneapolis: University of Minnesota Press, 1984.

Macdougall, J. D. *A Short History of Planet Earth: Mountains, Mammals, Fire, and Ice.* New York: John Wiley, 1996.

MacNeish, Richard S. *The Origins of Agriculture and Settled Life.* Norman: University of Oklahoma Press, 1992.

Maddison, Angus. *The World Economy: A Millennial Perspective.* Paris: OECD, 2001.

Maisels, Charles Keith. *The Emergence of Civilization: From Hunting and Gathering to Agriculture, Cities, and the State in the Near East.* London: Routledge, 1990.

Mallory, J. P. *In Search of the Indo-Europeans: Language, Archaeology, and Myth.* London: Thames and Hudson, 1989.

Man, John. *Atlas of the Year 1000*. Cambridge, Mass.: Harvard University Press, 1999.

Mandel, Ernst. *Late Capitalism*. Translated by Joris De Bres. [Rev. ed.] London: Verso, 1978.

Mann, Michael. *The Sources of Social Power*. Vol. 1, *A History of Power from the Beginning to A.D. 1760*. Cambridge: Cambridge University Press, 1986.

Marcus, George E., and Michael M. J. Fischer. *Anthropology as Cultural Critique: An Experimental Moment in the Human Sciences*. Chicago: University of Chicago Press, 1986.

Margulis, Lynn, and Dorion Sagan. *Microcosmos: Four Billion Years of Microbial Evolution*. London: Allen and Unwin, 1987.

———. *What Is Life?* Berkeley: University of California Press, 1995.

Marks, Robert B. *The Origins of the Modern World: A Global and Ecological Narrative*. Lanham, Md.: Rowman and Littlefield, 2002.

Marwick, Arthur. *The Nature of History*. London: Macmillan, 1970.

Marx, Karl. *Capital: A Critique of Political Economy*. Vol. 1. Translated by Ben Fowkes. Harmondsworth: Penguin, 1976.

———. *Capital: A Critique of Political Economy*. Vol. 3. Translated by David Fernbach. Harmondsworth: Penguin, 1981.

———. *Grundrisse: Foundations of the Critique of Political Economy*. Translated by Martin Nicolaus. Harmondsworth: Penguin, 1973.

Mathias, Peter. *The First Industrial Nation: An Economic History of Britain, 1700–1914*. 2nd ed. London: Methuen, 1983.

Mathias, Peter, and John A. Davis, eds. *The First Industrial Revolutions*. Oxford: Blackwell, 1989.

Maynard Smith, John. *The Theory of Evolution*. 3rd ed. New York: Penguin, 1975.

Maynard Smith, John, and Eörs Szathmáry. *The Origins of Life: From the Birth of Life to the Origins of Language*. Oxford: Oxford University Press, 1999.

Mayr, Ernst. *One Long Argument: Charles Darwin and the Genesis of Modern Evolutionary Thought*. London: Penguin, 1991.

Mazlish, Bruce, and Ralph Buultjens, eds. *Conceptualizing Global History*. Boulder, Colo.: Westview Press, 1993.

McBrearty, Sally, and Alison S. Brooks. "The Revolution That Wasn't: A New Interpretation of the Origin of Modern Human Behavior." *Journal of Human Evolution* 39 (2000): 453–563.

McClellan, James E., III, and Harold Dorn. *Science and Technology in World History: An Introduction*. Baltimore: Johns Hopkins University Press, 1999.

McCrone, John. *The Ape That Spoke*. Basingstoke: Macmillan, 1990.

———. *How the Brain Works: A Beginner's Guide to the Mind and Consciousness*. London: Dorling Kindersley, 2002.

McKeown, Thomas. *The Origins of Human Disease*. Oxford: Oxford University Press, 1998.

McMichael, A. J. *Planetary Overload: Global Environmental Change and the*

Health of the Human Species. Cambridge: Cambridge University Press, 1993.

McNeill, J. R. "Of Rats and Men: A Synoptic Environmental History of the Island Pacific." *Journal of World History* 5, no. 2 (fall 1994): 299–349.

———. *Something New under the Sun: An Environmental History of the Twentieth-Century World.* New York: W W. Norton, 2000.

McNeill, J. R., and William H. McNeill. *The Human Web: A Bird's-Eye View of World History.* New York: W. W. Norton, 2003.

McNeill, William H. *The Disruption of Traditional Forms of Nurture.* Amsterdam: Het Spinhuis, 1998.

———. *A History of the Human Community.* 3rd ed. Englewood Cliffs, N.J.: Prentice-Hall, 1990.

———. "History and the Scientific Worldview." *History and Theory* 37, no. 1 (1998): 1–13.

———. *Keeping Together in Time: Dance and Drill in Human History.* Cambridge, Mass.: Harvard University Press, 1995.

———. *Mythistory and Other Essays.* Chicago: University of Chicago Press, 1985.

———. *Plagues and People.* Oxford: Blackwell, 1977.

———. *The Pursuit of Power: Technology, Armed Force, and Society since A.D. 1000.* Oxford: Blackwell, 1982.

———. *The Rise of the West: A History of the Human Community.* Chicago: University of Chicago Press, 1963. [Still perhaps the best one-volume world history, less Eurocentric than its title suggests; more up-to-date, but less interesting, is McNeill's textbook, *A History of the Human Community*.]

McSween, Harry Y., Jr. *Fanfare for Earth: The Origin of Our Planet and Life.* New York: St. Martin's Press, 1997.

Meadows, Donella H., Dennis L. Meadows, and Jørgen Randers. *Beyond the Limits: Confronting Global Collapse, Envisioning a Sustainable Future.* Post Mills, Vt.: Chelsea Green, 1992.

Meadows, Donella H., et al. *The Limits to Growth: A Report for the Club of Rome's Project on the Predicament of Mankind.* New York: Universe Books, 1972. [Both this book and the preceding are on modeling futures.]

Mears, John. "Agricultural Origins in Global Perspective." In *Agricultural and Pastoral Societies in Ancient and Classical History,* edited by Michael Adas, pp. 36–70. Philadelphia: Temple University Press, 2001.

Megarry, Tim. *Society in Prehistory: The Origins of Human Culture.* Basingstoke: Macmillan, 1995.

Merson, John. *Roads to Xanadu: East and West in the Making of the Modern World.* French's Forest, N.S.W.: Child and Associates, 1989.

Miller, Walter M. *A Canticle for Leibowitz.* 1959. Reprint, New York: Bantam, 1997.

Mithen, Steven. *The Prehistory of the Mind: A Search for the Origins of Art, Religion, and Science.* London: Thames and Hudson, 1996.

Modelski, George, and William R. Thompson. *Leading Sectors and World Pow-*

ers: The Coevolution of Global Politics and Economics. Columbia: University of South Carolina Press, 1996. [An attempt to define Kondratieff cycles for the past millennium.]

Mokyr, Joel. *The Lever of Riches: Technological Creativity and Economic Progress.* New York: Oxford University Press, 1990.

Morrison, Philip, and Phylis Morrison. *Powers of Ten: A Book about the Relative Size of Things in the Universe and the Effect of Adding Another Zero.* Redding, Conn.: Scientific American Library; San Francisco: dist. by W. H. Freeman, 1982.

al-Mulk, Nizam. *The Book of Government, or Rules for Kings.* Translated by Hubert Darke. 2nd ed. London: Routledge, 1978.

Mulvaney, John, and Johan Kamminga. *Prehistory of Australia.* Sydney: Allen and Unwin, 1999.

Myers, Norman. *The Sinking Ark: A New Look at the Problem of Disappearing Species.* Oxford: Pergamon Press, 1979. [A classic statement about extinction and biodiversity loss.]

Needham, Joseph. *Science and Civilisation in China.* 7 vols. Cambridge: Cambridge University Press, 1954–2003.

Needham, Joseph, and Lu Gwei-djen. *Trans-Pacific Echoes and Resonances: Listening Once Again.* Singapore: World Scientific, 1984. [Summarizes the slender evidence on trans-Pacific contacts before Columbus.]

Nhat Hanh, Thich. *The Diamond That Cuts through Illusion: Commentaries on the Prajñaparamita Diamond Sutra.* Translated by Anh Huong Nguyen. Berkeley: Parallax, 1992.

———. *The Heart of Understanding: Commentaries on the Prajñaparamita Heart Sutra.* Edited by Peter Levitt. Berkeley: Parallax, 1988.

Nisbet, E. G. *Living Earth—A Short History of Life and Its Home.* London: HarperCollins Academic Press, 1991.

Nissen, Hans Jörg. *The Early History of the Ancient Near East, 9000–2000 B.C.* Translated by Elizabeth Lutzeier, with Kenneth J. Northcott. Chicago: University of Chicago Press, 1988.

Nitecki, Matthew H., and Doris V. Nitecki, eds. *History and Evolution.* Albany: State University of New York Press, 1992.

North, Douglass C. *Structure and Change in Economic History.* New York: W. W. Norton, 1981.

North, Douglass C., and Robert Paul Thomas. *The Rise of the Western World.* Cambridge: Cambridge University Press, 1973.

Nyanatiloka. *Buddhist Dictionary: Manual of Buddhist Terms and Doctrines.* 3rd ed. Colombo [Sri Lanka]: Frewin, 1972.

Oates, David, and Joan Oates. *The Rise of Civilization.* Oxford: Elsevier Phaidon, 1976.

O'Brien, Patrick. "Introduction: Modern Conceptions of the Industrial Revolution." In *The Industrial Revolution and British Society,* edited by Patrick O'Brien and Roland Quinault, pp. 1–30. Cambridge: Cambridge University Press, 1993.

———. "Is Universal History Possible?" *Nineteenth International Congress of Historical Sciences*, pp. 3–18. Oslo: Nasjonalbiblioteket, 2000.

———. "Political Preconditions for the Industrial Revolution." In *The Industrial Revolution and British Society*, edited by Patrick O'Brien and Roland Quinault, pp. 124–55. Cambridge: Cambridge University Press, 1993.

O'Brien, Patrick, and Caglar Keyder. *Economic Growth in Britain and France, 1780–1914: Two Paths to the Twentieth Century*. London: Allen and Unwin, 1978.

O'Brien, Patrick, and Roland Quinault, eds. *The Industrial Revolution and British Society*. Cambridge: Cambridge University Press, 1993.

Ogilvie, Sheilagh, and Markus Cerman, eds. *European Proto-Industrialization: An Introductory Handbook*. Cambridge: Cambridge University Press, 1996.

Okladnikov, A. P. "Inner Asia at the Dawn of History." In *Cambridge History of Early Inner Asia*, edited by Denis Sinor, pp. 41–96. Cambridge: Cambridge University Press, 1990.

Oliver, Roland. *The African Experience: From Olduvai Gorge to the Twenty-First Century*. 2nd ed. Boulder, Colo.: Westview Press, 2000.

Overton, Mark. *Agricultural Revolution in England: The Transformation of the Agrarian Economy, 1500–1850*. Cambridge: Cambridge University Press, 1996.

Pacey, Arnold. *Technology in World Civilization*. Cambridge, Mass.: MIT Press, 1990.

Packard, Edward. *Imagining the Universe: A Visual Journey*. New York: Perigee Books, 1994.

Parker, Geoffrey. *The Military Revolution: Military Innovation and the Rise of the West, 1500–1800*. 2nd ed. Cambridge: Cambridge University Press, 1996.

———, ed. *The World: An Illustrated History*. New York: Harper and Row, 1986. [Beautifully illustrated.]

Pearson, M. N. "Merchants and States." In *The Political Economy of Merchant Empires: State Power and World Trade, 1350–1750*, edited by James D. Tracy, pp. 41–116. Cambridge: Cambridge University Press, 1991.

Penrose, Roger. *The Emperor's New Mind: Concerning Computers, Minds, and the Laws of Physics*. London: Vintage, 1990.

Pinker, Steven. *How the Mind Works*. New York: W. W. Norton, 1997.

———. *The Language Instinct: The New Science of Language and Mind*. New York: Penguin, 1994.

Plotkin, Henry. *Evolution in Mind: An Introduction to Evolutionary Psychology*. London: Penguin, 1997.

Polanyi, Karl. *The Great Transformation: The Political and Economic Origins of Our Time*. Boston: Beacon, 1957.

Polanyi, Karl, Conrad M. Arensberg, and Harry W. Pearson, eds. *Trade and Market in the Early Empires: Economies in History and Theory*. Glencoe, Ill.: Free Press, 1957.

Pollock, Susan. *Ancient Mesopotamia: The Eden That Never Was*. Cambridge: Cambridge University Press, 1999.

Pomeranz, Kenneth. *The Great Divergence: China, Europe, and the Making of the Modern World Economy.* Princeton: Princeton University Press, 2000.

Pomeranz, Kenneth, and Steven Topik. *The World That Trade Created: Society, Culture, and the World Economy, 1400 to the Present.* Armonk, N.Y.: M. E. Sharpe, 1999.

Pomper, Philip, Richard H. Elphick, and Richard T. Vann, eds. *World History: Ideologies, Structures, and Identities.* Oxford: Blackwell, 1998.

Ponting, Clive. *A Green History of the World.* Harmondsworth: Penguin, 1992. [The best short introduction to the history of human impact on the environment.]

————. *World History: A New Perspective.* London: Chatto and Windus, 2000.

Poole, Ross. *Nation and Identity.* London: Routledge, 1999.

Popol Vuh: The Mayan Book of the Dawn of Life. Translated by Dennis Tedlock. Rev. ed. New York: Simon and Schuster, 1996.

Porter, Gareth, Janet Welsh Brown, and Pamela S. Chasek. *Global Environmental Politics.* 3rd ed. Boulder, Colo.: Westview Press, 2000.

Potts, D. T. *Mesopotamian Civilization: The Material Foundations.* Ithaca, N.Y.: Cornell University Press, 1997.

Potts, Malcolm, and Roger Short. *Ever Since Adam and Eve: The Evolution of Human Sexuality.* Cambridge: Cambridge University Press, 1999.

Prantzos, Nikos. *Our Cosmic Future: Humanity's Fate in the Universe.* Cambridge: Cambridge University Press, 2000.

Praslov, N. D. "Late Palaeolithic Adaptations to the Natural Environment on the Russian Plain." *Antiquity* 63 (1989): 784–87.

Priem, H. N. A. *Aarde en Leven: Het leven in relatie tot zijn planetaire omgeving/ Earth and Life: Life in Relation to Its Planetary Environment.* Dordrecht: Kluwer, 1993.

Prigogine, Ilya, and Isabelle Stengers. *Order out of Chaos: Man's New Dialogue with Nature.* London: Heinemann, 1984.

Psillos, Stathis. *Scientific Realism: How Science Tracks Truth.* London: Routledge, 1999.

Pyne, Stephen. *Fire in America: A Cultural History of Wildland and Rural Fire.* Princeton: Princeton University Press, 1982.

————. *Vestal Fire: An Environmental History.* Seattle: University of Washington Press, 1997.

Rahman, Abdur, ed. *Science and Technology in Indian Culture: A Historical Perspective.* New Delhi: National Institute of Science, Technology, and Development Studies, 1984.

Rasmussen, Birger. "Filamentous Microfossils in a 3,235-Million-Year-Old Volcanogenic Massive Sulphide Deposit." *Nature,* 8 June 2000, pp. 676–79.

Redman, Charles L. "Mesopotamia and the First Cities." In *The Illustrated History of Humankind,* edited by Göran Burenhult. Vol. 3, *Old World Civilizations: The Rise of Cities and States,* pp. 17–36. St. Lucia: University of Queensland Press, 1994.

Rees, Martin. *Just Six Numbers: The Deep Forces That Shape the Universe.* New York: Basic Books, 2000.

Reeves, Hubert, Joël de Rosnay, Yves Coppens, and Dominique Simonnet. *Origins: Cosmos, Earth, and Mankind.* New York: Arcade Publishing, 1998.

Renfrew, Colin. *Archaeology and Language: The Puzzle of Indo-European Origins.* Harmondsworth: Penguin, 1987.

Renfrew, Colin, and Stephen Shennan, eds. *Ranking, Resource, and Exchange: Aspects of the Archaeology of Early European Society.* Cambridge: Cambridge University Press, 1982.

Ridley, Matt. *Evolution.* Oxford: Blackwell, 1993. [An introduction to modern neo-Darwinianism.]

———. *Genome: The Autobiography of a Species in Twenty-three Chapters.* London: Fourth Estate, 1999. [A superb series of essays on aspects of modern genetics.]

The Rig Veda: An Anthology: One Hundred and Eight Hymns. Selected, edited, and translated by Wendy Doniger O'Flaherty. Harmondsworth: Penguin, 1981.

Rindos, David. *Origins of Agriculture: An Evolutionary Perspective.* New York: Academic Press, 1984.

Ringrose, David R. *Expansion and Global Interaction, 1200–1700.* New York: Longman, 2001.

Roberts, J. M. *The Pelican History of the World.* Rev. ed. Harmondsworth: Penguin, 1988.

Roberts, Neil. *The Holocene: An Environmental History.* 2nd ed. Oxford: Blackwell, 1998.

Roberts, Richard G. "Thermoluminescence Dating." In *The Illustrated History of Humankind,* edited by Göran Burenhult. Vol. 1, *The First Humans: Human Origins and History to 10,000 BC,* pp. 152–53. St. Lucia: University of Queensland Press, 1993.

Roberts, Richard G., Timothy F. Flannery, Linda K. Ayliffe, Hiroyuki Yoshida, et al. "New Ages for the Last Australian Megafauna: Continent-wide Extinction about 46,000 Years Ago." *Science,* 8 June 2001, pp. 1888–92.

Rose, Deborah Bird. *Nourishing Terrains: Australian Aboriginal Views of Landscape and Wilderness.* Canberra: Australian Heritage Commission, 1996.

Rose, Steven, ed. *From Brains to Consciousness? Essays on the New Sciences of the Mind.* London: Penguin, 1999.

Rowlands, Michael. "Centre and Periphery: A Review of a Concept." In *Centre and Periphery in the Ancient World,* edited by Michael Rowlands, Mogens Larsen, and Kristian Kristiansen, pp. 1–11. Cambridge: Cambridge University Press, 1987.

The Russian Primary Chronicle: Laurentian Text. Translated and edited by Samuel Hazzard Cross and Olgerd P. Sherbovitz-Wetzor. Cambridge, Mass.: Mediaeval Academy of America, 1953.

Sabloff, Jeremy A., and C. C. Lamberg-Karlovsky, eds. *Ancient Civilization and Trade.* Albuquerque: University of New Mexico Press, 1975.

Sahlins, Marshall. "The Original Affluent Society." In *Stone Age Economics,* pp. 1–39. London: Tavistock, 1972. [This essay is superb; the others are also well worth reading.]

———. *Tribesmen.* Englewood Cliffs, N.J.: Prentice-Hall, 1968.

Salmon, Wesley C. *Scientific Explanation and the Causal Structure of the World.* Princeton: Princeton University Press, 1984.

Sanderson, Stephen K. "Expanding World Commercialization: The Link between World Systems and Civilizations." In *Civilizations and World Systems: Studying World-Historical Change,* edited by Stephen K. Sanderson, pp. 261–72. Walnut Creek, Calif.: Altamira Press, 1995.

———. *Social Transformations: A General Theory of Historical Development.* London: Blackwell, 1995.

———, ed. *Civilizations and World Systems: Studying World-Historical Change.* Walnut Creek, Calif.: Altamira Press, 1995.

Sarich, Vincent, and Alan Wilson. "Immunological Time Scale for Hominid Evolution." *Science,* 1 December 1967, pp. 1200–1203.

Schneider, Stephen H. *Laboratory Earth: The Planetary Gamble We Can't Afford to Lose.* London: Phoenix, 1997.

Schrire, Carmel, ed. *Past and Present in Hunter Gatherer Studies.* Orlando, Fla.: Academic Press, 1985.

Schrödinger, Erwin. *What Is Life? The Physical Aspect of the Living Cell;* with, *Mind and Matter;* and *Autobiographical Sketches.* Cambridge: Cambridge University Press, 1992. [*What Is Life?* was first published in 1944.]

Schumpeter, Joseph A. *Business Cycles: A Theoretical, Historical, and Statistical Analysis of the Capitalist Process.* New York: McGraw-Hill, 1939.

Scott, Joan W. "Gender: A Useful Category of Historical Analysis." *American Historical Review* 75, no. 5 (December 1986): 1053–75.

Sept, Jeanne M., and George E. Brooks. "Reports of Chimpanzee Natural History, Including Tool Use, in Sixteenth- and Seventeenth-Century Sierra Leone." *International Journal of Primatology* 15, no. 6 (December 1994): 867–77.

Service, Elman R. *Primitive Social Organization: An Evolutionary Perspective.* 2nd ed. New York: Random House, 1971. [1st ed. 1962.]

Shaffer, Lynda. "Southernization." In *Agricultural and Pastoral Societies in Ancient and Classical History,* edited by Michael Adas, pp. 308–24. Philadelphia: Temple University Press, 2001. [Originally published in *Journal of World History* 5, no. 1 (spring 1994): 1–21.]

Shannon, Thomas R. *An Introduction to the World-System Perspective.* 2nd ed. Boulder, Colo.: Westview Press, 1996.

Shapin, Steven. *The Scientific Revolution.* Chicago: University of Chicago Press, 1996.

Shapiro, Robert. *Origins: A Skeptic's Guide to the Creation of Life on Earth.* London: Penguin, 1986.

Sherratt, Andrew. *Economy and Society in Prehistoric Europe: Changing Perspectives.* Princeton: Princeton University Press, 1997.

————. "Plough and Pastoralism: Aspects of the Secondary Products Revolution." In *Patterns of the Past: Studies in Honour of David Clarke*, edited by Ian Hodder, Glynn Isaac, and Norman Hammond, pp. 261–305. Cambridge: Cambridge University Press, 1981.

————. "Reviving the Grand Narrative: Archaeology and Long-Term Change." *Journal of European Archaeology* 3, no. 1 (1995): 1–32.

————. "The Secondary Exploitation of Animals in the Old World (1983, revised)." In *Economy and Society in Prehistoric Europe: Changing Perspectives*, pp. 199–228. Princeton: Princeton University Press, 1997.

Silk, Joseph. *The Big Bang: The Creation and Evolution of the Universe.* San Francisco: W. H. Freeman, 1980.

Simmons, I. G. *Changing the Face of the Earth: Culture, Environment, History.* 2nd ed. Oxford: Blackwell, 1996.

————. *Environmental History: A Concise Introduction.* Oxford: Blackwell, 1993.

Sinor, Denis, ed. *The Cambridge History of Early Inner Asia.* Cambridge: Cambridge University Press, 1990.

Smil, Vaclav. *Energy in World History.* Boulder, Colo.: Westview Press, 1994.

Smith, Adam. *An Inquiry into the Nature and Causes of the Wealth of Nations*, edited by Edwin Cannan. 5th ed. New York: Modern Library, 1937.

Smith, Bonnie. *The Gender of History: Men, Women, and Historical Practice.* Cambridge, Mass.: Harvard University Press, 1998.

Smith, Bruce D. *The Emergence of Agriculture.* New York: Scientific American Library, 1995.

Smolin, Lee. *The Life of the Cosmos.* London: Phoenix, 1998.

Snooks, G. D. *The Dynamic Society: Exploring the Sources of Global Change.* London: Routledge, 1996.

————. *The Ephemeral Civilization: Exploding the Myth of Social Evolution.* London: Routledge, 1997.

————, ed. *Was the Industrial Revolution Necessary?* London: Routledge, 1994.

Snyder, John, and C. Leland Rodgers. *Biology.* 3rd ed. New York: Barron's, 1995.

Snyder, Lee Daniel. *Macro-History: A Theoretical Approach to Comparative World History.* Lewiston, N.Y.: Edwin Mellen Press, 1999.

Sobel, Dava. *Longitude: The True Story of a Lone Genius Who Solved the Greatest Scientific Problem of His Time.* New York: Walker, 1995.

Soffer, Olga. "The Middle to Upper Paleolithic Transition on the Russian Plain." In *The Human Revolution*, edited by Paul Mellars and Chris Stringer, 1:714–42. Edinburgh: Edinburgh University Press, 1989.

————. "Patterns of Intensification as Seen from the Upper Paleolithic of the Central Russian Plain." In *Prehistoric Hunter-Gatherers: The Emergence of Cultural Complexity*, edited by T. Douglas Price and James A. Brown, pp. 235–70. Orlando: Academic Press, 1985.

————. "Storage, Sedentism, and the Eurasian Palaeolithic Record." *Antiquity* 63 (1989): 719–32.

————. "Sungir: A Stone Age Burial Site." In *The Illustrated History of Hu-*

mankind, edited by Göran Burenhult. Vol. 1, *The First Humans: Human Origins and History to 10,000 BC,* pp. 138–39. St. Lucia: University of Queensland Press, 1993.

Solé, Ricard, and Brian Goodwin. *Signs of Life: How Complexity Pervades Biology.* New York: Basic Books, 2000.

Spier, Fred. *The Structure of Big History: From the Big Bang until Today.* Amsterdam: Amsterdam University Press, 1996.

Spodek, Howard. *The World's History.* 2nd ed. Upper Saddle River, N.J.: Prentice Hall, 2001.

Sproul, Barbara. *Primal Myths: Creation Myths around the World.* 1979. Reprint, San Francisco: HarperSanFrancisco, 1991.

Stableford, Brian, and David Langford. *The Third Millenium: A History of the World, AD 2000–3000.* London: Sidgwick and Jackson, 1985.

Stanford, Craig B. *The Hunting Apes: Meat Eating and the Origins of Human Behavior.* Princeton: Princeton University Press, 1999.

———. *Significant Others: The Ape-Human Continuum and the Quest for Human Nature.* New York: Basic Books, 2001.

Stanley, Steven M. *Children of the Ice Age: How a Global Catastrophe Allowed Humans to Evolve.* 1996. Reprint, New York: W. H. Freeman, 1998.

———. *Earth and Life through Time.* New York: W. H. Freeman, 1986.

Stavrianos, L. S. *Lifelines from Our Past: A New World History.* New York: W. H. Freeman, 1989. [An interpretive essay by one of the pioneers of world history; he uses Eric Wolf's typology of human societies in simplified form.]

Stearns, Peter N. *The Industrial Revolution in World History.* Boulder, Colo.: Westview Press, 1993.

———. *Millennium III, Century XXI: A Retrospective on the Future.* Boulder, Colo.: Westview Press, 1996.

Stearns, Peter N., and John H. Hinshaw. *The ABC-CLIO World History Companion to the Industrial Revolution.* Santa Barbara, Calif.: ABC-CLIO, 1996.

Stokes, Gale. "The Fates of Human Societies: A Review of Recent Macrohistories." *American Historical Review* 106, no. 2 (April 2001): 508–25.

Stringer, Chris, and Clive Gamble. *In Search of the Neanderthals: Solving the Puzzle of Human Origins.* London: Thames and Hudson, 1993.

Stringer, Chris, and Robin McKie. *African Exodus.* London: Cape, 1996.

Suzuki, David, with Amanda McConnell. *The Sacred Balance: Rediscovering Our Place in Nature.* St. Leonards, N.S.W.: Allen and Unwin, 1997.

Swain, Tony. *A Place for Strangers: Towards a History of Australian Aboriginal Being.* Cambridge: Cambridge University Press, 1993.

Sweezey, Paul, et al. *The Transition from Feudalism to Capitalism.* Rev. ed. London: New Left Books; Atlantic Highlands, N.J.: Humanities Press, 1976.

Swimme, Brian, and Thomas Berry. *The Universe Story: From the Primordial Flaring Forth to the Ecozoic Era: A Celebration of the Unfolding of the Cosmos.* San Francisco: HarperSanFrancisco, 1992.

Taagepera, Rein. "Expansion and Contraction Patterns of Large Polities: Context for Russia." *International Studies Quarterly* 41 (1997): 475–504.

————. "Size and Duration of Empires: Growth-Decline Curves, 3000 to 600 BC." *Social Science Research* 7 (1978): 180–96.

————. "Size and Duration of Empires: Growth-Decline Curves, 600 BC to 600 AD." *Social Science Research* 3 (1979): 115–38.

————. "Size and Duration of Empires: Systematics of Size." *Social Science Research* 7 (1978): 108–27.

Tattersall, Ian. *Becoming Human: Evolution and Human Uniqueness.* New York: Harcourt Brace, 1998.

Taylor, Stuart Ross. "The Solar System: An Environment for Life?" In *To Mars and Beyond: Search for the Origins of Life,* edited by Malcolm Walter, pp. 56–67. Canberra: National Museum of Australia, 2002.

Thompson, E. P. *The Making of the English Working Class.* London: Victor Gollancz, 1963.

————. *Whigs and Hunters: The Origin of the Black Act.* London: Allen Lane, 1975.

Thompson, William R. "The Military Superiority Thesis and the Ascendancy of Western Eurasia in the World System." *Journal of World History* 10, no. 1 (1999): 143–78.

Thorne, Alan G., and Milford H. Wolpoff. "The Multiregional Evolution of Humans." *Scientific American,* April 1992, pp. 28–33.

Thorne, Alan, et al. "Australia's Oldest Human Remains: Age of the Lake Mungo 3 Skeleton." *Journal of Human Evolution* 36 (June 1999): 591–612.

Tickell, C. "The Human Species: A Suicidal Success?" In *The Human Impact Reader: Readings and Case Studies,* edited by Andrew Goudie, pp. 450–59. Oxford: Blackwell, 1997.

Tilly, Charles. *As Sociology Meets History.* New York: Academic Press, 1981.

————. *Big Structures, Large Processes, Huge Comparisons.* New York: Russell Sage Foundation, 1984.

————. *Coercion, Capital, and European States, AD 990–1992.* Rev. ed. Cambridge, Mass.: Blackwell, 1992.

Toynbee, Arnold. *A Study of History.* Oxford: Oxford University Press, 1946.

Tracy, James D., ed. *The Political Economy of Merchant Empires: State Power and World Trade, 1350–1750.* Cambridge: Cambridge University Press, 1991.

————. *The Rise of Merchant Empires: Long-Distance Trade in the Early Modern World, 1350–1750.* Cambridge: Cambridge University Press, 1990.

Trigger, Bruce G. *Early Civilizations: Ancient Egypt in Context.* Cairo: American University in Cairo Press, 1993.

Tudge, Colin. *The Time before History: Five Million Years of Human Impact.* New York: Scribner, 1996.

Turner, II, B. L., et al., eds. *The Earth as Transformed by Human Action: Global and Regional Changes in the Biosphere over the Past 300 Years.* Cambridge: Cambridge University Press, 1990.

Van Creveld, Martin L. *Technology and War: From 2000 B.C. to the Present.* New York: Free Press; London: Collier Macmillan, 1989.

Vansina, Jan. "New Linguistic Evidence and 'the Bantu Expansion.'" *Journal of African History* 36, no. 2 (1995): 173–95.

Voll, John O. "Islam as a Special World-System." In *The New World History: A Teacher's Companion,* edited by Ross E. Dunn, pp. 276–86. Boston: Bedford/St. Martin's, 2000.

Von Damm, Karen L. "Lost City Found." *Nature,* 12 July 2001, pp. 127–28.

Von Franz, Marie-Louise. *Creation Myths.* Dallas: Spring Publications, 1972.

Wallerstein, Immanuel. *The Modern World-System.* 3 vols. New York: Academic Press, 1974–89.

———. "World-System." In *A Dictionary of Marxist Thought,* edited by Tom Bottomore, pp. 590–91. 2nd ed. Oxford: Blackwell, 1991.

Walter, Malcolm. *The Search for Life on Mars.* Sydney: Allen and Unwin, 1999.

———, ed. *To Mars and Beyond: Search for the Origins of Life.* Canberra: National Museum of Australia, 2002.

Watson, Andrew M. *Agricultural Innovation in the Early Islamic World: The Diffusion of Crops and Farming Techniques, 700–1100.* Cambridge: Cambridge University Press, 1983.

Watson, James D. *The Double Helix: A Personal Account of the Discovery of the Structure of DNA.* 1968. Reprint, Harmondsworth: Penguin, 1970.

Watts, Sheldon. *Epidemics and History: Disease, Power, and Imperialism.* New Haven: Yale University Press, 1998.

Weber, Max. *The Protestant Ethic and the Spirit of Capitalism.* Translated by Talcott Parsons. 1930. Reprint, New York: Scribners, 1958.

Weinberg, Steven. *The First Three Minutes: A Modern View of the Origin of the Universe.* 2nd ed. London: Flamingo, 1993.

Wells, H. G. *The Outline of History: Being a Plain History of Life and Mankind.* 2 vols. London: George Newnes, 1920.

———. *A Short History of the World.* London: Cassell, 1922.

Wenke, Robert J. *Patterns in Prehistory: Humankind's First Three Million Years.* 3rd ed. New York: Oxford University Press, 1990.

White, J. Peter. "The Settlement of Ancient Australia." In *The Illustrated History of Humankind,* edited by Göran Burenhult. Vol. 1, *The First Humans: Human Origins and History to 10,000 BC,* pp. 147–51, 153–57. St. Lucia: University of Queensland Press, 1993.

White, J. Peter, and James F. O'Connell. *A Prehistory of Australia, New Guinea, and Sahul.* Sydney: Academic Press, 1982.

Whitmore, Thomas M., et al. "Long-Term Population Change." In *The Earth as Transformed by Human Action: Global and Regional Changes in the Biosphere over the Past 300 Years,* edited by B. L. Turner II et al., pp. 25–39. Cambridge: Cambridge University Press, 1990.

Wilkinson, David. "Central Civilization." In *Civilizations and World Systems: Studying World-Historical Change,* edited by Stephen K. Sanderson, pp. 46–74. Walnut Creek, Calif.: Altamira, 1995.

Wills, Christopher. *The Runaway Brain: The Evolution of Human Uniqueness.* New York: Basic Books, 1993.

Wilson, Edward O. *Biophilia.* Cambridge, Mass.: Harvard University Press, 1984.

———. *Consilience: The Unity of Knowledge.* London: Abacus, 1998.

———. *The Diversity of Life.* Harmondsworth: Penguin, 1992.

———. *The Future of Life.* New York: Alfred Knopf, 2002.

Wolf, Eric R. *Europe and the People without History.* Berkeley: University of California Press, 1982. [A superb, if sometimes difficult, history of the modern world by an anthropologist.]

———. *Peasants.* Englewood Cliffs, N.J.: Prentice-Hall, 1966.

Wolpoff, M. H., Wu Zinzhi, and A. Thorne. "Modern *Homo sapiens* Origins: General Theory of Hominid Evolution Involving the Fossil Evidence from East Asia." In *The Origins of Modern Humans: A World Survey of the Fossil Evidence,* edited by Fred H. Smith and Frank Spencer, pp. 411–83. New York: Alan Liss, 1984. [A definitive statement of their position regarding the single-species theory of hominine evolution.]

Wong, R. Bin. *China Transformed: Historical Change and the Limits of European Experience.* Ithaca, N.Y.: Cornell University Press, 1997.

World Commission on Environment and Development. *Our Common Future.* Oxford: Oxford University Press, 1987.

World Development Indicators. Washington, D.C.: World Bank, 2002.

World Resources, 2000–2001: People and Ecosystems: The Fraying Web of Life. Washington, D.C.: World Resources Institute, 2000.

Wright, Robert. *Nonzero: The Logic of Human Destiny.* New York: Random House, 2000.

Wrigley, E. A. *Continuity, Chance, and Change: The Character of the Industrial Revolution in England.* Cambridge: Cambridge University Press, 1988.

———. *People, Cities, and Wealth.* Oxford: Blackwell, 1987.

———. *Population and History.* London: Weidenfeld and Nicolson, 1969.

Wrigley, E. A., and R. S. Schofield. *The Population History of England, 1541– 1871: A Reconstruction.* Cambridge, Mass.: Harvard University Press, 1981.

Zvelebil, Marek. "Mesolithic Prelude and Neolithic Revolution." In *Hunters in Transition: Mesolithic Societies of Temperate Eurasia and Their Transition to Farming,* edited by Marek Zvelebil, pp. 5–15. Cambridge: Cambridge University Press, 1986.

———, ed. *Hunters in Transition: Mesolithic Societies of Temperate Eurasia and Their Transition to Farming.* Cambridge: Cambridge University Press, 1986.

INDEX

Abbasid Empire, 292, 299, 318, 323, 368, 431
Abbasid exchange, 370
Aborigines, Australian, 17, 365, 515n6; creation stories, 19–20; Dreamtime/Dreaming, 3, 20, 197, 515n10; Holocene, 212–15, 223, 227–28, 233, 284; languages, 184; term for Europeans, 364
absorption bands, 30–31, 48
Abu Hureyra, 236, 237
Abu-Lughod, Janet, 290, 369
acceleration, 183, 350–53, 440–41, 463, 471–72
accounting, 274–77, 321. *See also* writing
accretion, 60–62, 64
accumulation, 259, 306–9, 330–31, 366, 403–4; Afro-Eurasia, 360–61; modern Asia, 385; primitive, 400–401; states as sources of, 316–25. *See also* innovation
Achaemenid Empire, 298–99, 305, 307, 312, 317–19, 321, 327
acquired characteristics, 86
adaptation, 80, 134, 210; adaptive radiation, 125–27, 152–56, 163, 193–94; for agriculture, 216, 223, 225–30, 236, 237, 243; bacteria, 112; Baldwinian, 160–61, 166–67, 173;

chemical evolution, 94–95, 100–101; Charles Darwin and, 83–84, 86–88, 90, 104; defined, 83–84; to environment, 83–88, 94–95, 100–101, 104, 145, 160, 175, 178, 181–82, 190–91; eukaryotes, 114; to fire, 194–95; Holocene, 212, 215, 216; in human evolution, 131–32, 144–47, 150–57, 160–67, 171; modern humans *(Homo sapiens)*, 112, 144–47, 150–57, 160–67, 171, 180, 190–91, 193–94; for modernity, 359, 362–63, 391, 413; multicelled organisms, 121, 129; Paleolithic, 184, 190–91, 210, 215. *See also* collective learning; innovation
advertising, 447
Africa: agrarian era, 211–15, 218, 221, 238, 256, 284–85, 295, 298–99, 319, 338, 369–70, 378; Ethiopia, 149, 154–56, 158, 298, 300; geology, 69, 72–73, 109, 130, 155; human evolution, 127, 153–55, 159, 163–65, 170, 177–82, 202; modern era, 381–82, 384, 435, 451; Out of Africa hypothesis, 177; Paleolithic era, 152, 156, 190–91, 193, 197, 200, 202, 483; state formation, 294, 299–301, 337; Sudan, 214, 229, 294–95, 298–300. *See also* Afro-Eurasia; Egypt

factory, 418, 420–21, 437; division
of labor, 356, 420
Fagan, Brian, 228
family groups, 249; australopithecines,
158–59; early human, 182, 187–
89, 239–40; Neanderthals, 168; networks, 182, 184, 187; ranked lineages, 241, 260–61, 265–66, 272–73,
280; village, 239–41, 249; violence,
188–89. *See also* kinship
famine, 223, 265, 271, 311, 316, 352–
53, 435–36, 451
farmers, 284–87, 306, 335–38, 341,
374, 415–18; diet, 223, 224, 234,
303; small-scale, 336, 348, 483. *See
also* agriculture; early agrarian era;
horticulture; peasants; sedentism
feedback loops, 536n8; brain evolution, 162, 166–67, 173–75; Gaia
hypothesis, 128; negative, 44, 128,
310–11; positive, 162, 252–53
Fermi, Enrico, 486
Fertile Crescent, 220–22, 225, 230,
235–36, 259, 268–69. *See also*
Middle East
fertilizer, 255–56, 303, 376, 416; artificial, 433, 443; manure, 230, 239,
242, 255–56, 376, 416, 454
Feynman, Richard, 24, 26
finches, Galápagos, 87–88, 485
Finley, Moses, 322
fire, 194–97, 202, 526n39; cooking
over, 258; ecological innovation,
140; energy process, 110; extinction
and, 199, 529n42; fire stick farming,
194–95, 199, 202, 215, 254, 529n42,
531n59; Greek, 397; *Homo ergaster/
erectus*, 164, 194; oxygen and, 95,
113; signal, 307; swidden cultivation, 254
fish, 106, 132, 181; agrarian civilizations, 303, 340–41; cichlids, 288;
clothing from, 340–41; commercially raised, 388, 453, 460–61;
early forms, 67–68, 123–24, 494,
504; evolution, 123–24, 135;
exploitation, 461; shellfish, 181, 232,

233; threatened species, 142; trade,
341, 373. *See also* fishing
fishing, 179; agricultural societies,
217–18, 239, 259, 268; Easter Island,
473–74; eels, 226, 228; foragers, 228,
229, 340; trepang, 212; Upper Paleolithic, 228, 229
fitness, evolutionary, 87–88
Flanders, 369, 373, 395
flatworms, 122, 145
flax, 218–20, 399
flies, 91, 108, 120, 165
Flood, Josephine, 228
floods: adaptation and, 194–95; biblical,
68; forests preventing, 480; Kath,
368; mining and, 419; village walls
to control, 242
flour, 229
flowering plants, 106, 116, 122, 135,
222, 504
flu, 224
food: cooking, 164, 194, 258, 474;
famine, 223, 265, 271, 311, 316,
352–53, 435–36, 451; food chain,
110, 140, 218, 249; *Homo habilis*,
162–64; intensification and, 233,
253–54; milk, 125, 160, 216, 255;
modern era, 443, 452, 459–60;
Paleolithic, 185–87; and population
growth, 233, 353, 417–18, 476;
purchased, 348. *See also* agriculture; diet; fishing; foragers; fruit;
grain; hunting; legumes; meat;
tubers
foragers, 111, 221–42, 254, 279, 284–
85, 501, 524n19; affluent, 185–87,
196, 225–29, 231, 234, 501; agrarian
civilizations and, 284–85, 335–36,
338–41; agricultural revolution,
207–9, 223–24; Americas, 284, 483;
australopithecines, 158; carrying
capacity, 232; diet, 185–86, 223,
524n19; and domestication, 218,
221–25, 236, 237, 242–43; egalitarian, 240; elites, 249; environmental
impact, 242–43; exchange networks,
201, 222, 225, 233, 241, 336, 341,

Text:	10/13 Aldus
Display:	Interstate Light Condensed
Indexer:	Barbara Roos
Compositor:	Integrated Composition Systems
Printer:	Maple-Vail Manufacturing Group